Annals of Mathematics Studies
Number 209

The Structure of Groups with a Quasiconvex Hierarchy

Daniel T. Wise

PRINCETON UNIVERSITY PRESS

PRINCETON AND OXFORD

2021

Published by Princeton University Press
41 William Street, Princeton, New Jersey 08540
6 Oxford Street, Woodstock, Oxfordshire OX20 1TR

press.princeton.edu

All Rights Reserved

Library of Congress Cataloging-in-Publication Data

Names: Wise, Daniel T., 1971– author.
Title: The structure of groups with a quasiconvex hierarchy / Daniel T. Wise.
Description: Princeton : Princeton University Press, 2021. | Series: Annals of mathematics studies ; Number 209 | Includes bibliographical references and index.
Identifiers: LCCN 2020040293 (print) | LCCN 2020040294 (ebook) | ISBN 9780691170442 (hardback) | ISBN 9780691170459 (paperback) | ISBN 9780691213507 (ebook)
Subjects: LCSH: Hyperbolic groups. | Group theory.
Classification: LCC QA174.2 .W57 2021 (print) | LCC QA174.2 (ebook) | DDC 512/.2—dc23
LC record available at https://lccn.loc.gov/2020040293
LC ebook record available at https://lccn.loc.gov/2020040294

British Library Cataloging-in-Publication Data is available

Editorial: Susannah Shoemaker and Kristen Hop
Production Editorial: Nathan Carr
Production: Jacquie Poirier
Publicity: Matthew Taylor

This book has been composed in LaTeX

Printed on acid-free paper. ∞

Printed in the United States of America

10 9 8 7 6 5 4 3 2 1

to Yael

Contents

Acknowledgments

My research was supported by NSERC. The first version of this document was completed while on sabbatical at the Hebrew University in 2008-2009, and I am especially grateful to Zlil Sela and the Einstein Institute of Mathematics as well as the Lady Davis Foundation for its support. I am also grateful to Ian Agol, Macarena Arenas, Hadi Bigdely, Dave Futer, Mark Hagen, Frédéric Haglund, Jingyin Huang, Jason Manning, Eduardo Martinez-Pedroza, Denis Osin and his class, Piotr Przytycki, and Eric Swenson for many corrections. In addition, I wish to thank production editor Nathan Carr and illustrator Laurel Muller of Princeton University Press for their contributions.

This project would not have been possible without the continuing support of my darling wife Yael. I thank our children Tali, Steve, Ari, and Yoni for their patience. Finally, I thank my parents Batya and Mike for their continual confidence-building measures.

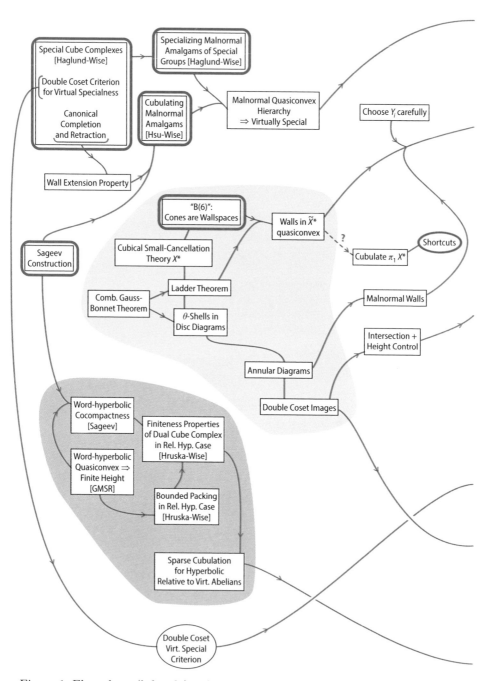

Figure 1. Flow chart (left side) indicating main points and some topical constellations.

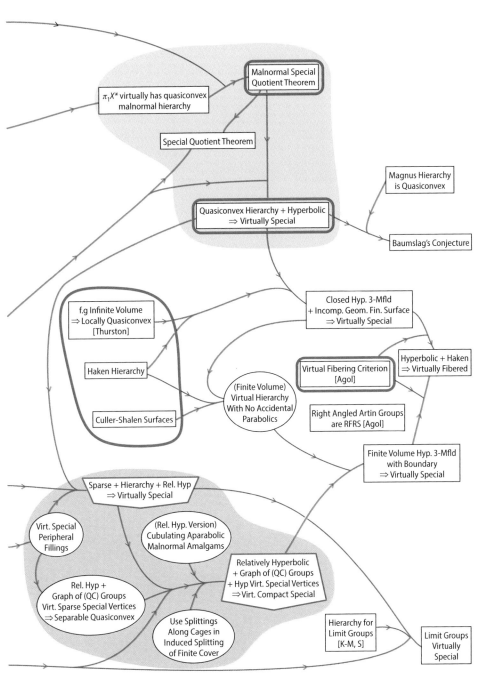

Figure 2. Flow chart (right side) indicating main points and some topical constellations.

The Structure of Groups with
a Quasiconvex Hierarchy

Chapter One

Introduction

This text has several parts:

In the first part of the text we develop a small-cancellation theory over cube complexes. When the cube complex is 1-dimensional, we obtain the classical small-cancellation theory, as well as the closely related Gromov graphical small-cancellation theory.

It is hard to say what the main result is in the first part, since it seems the definitions are more important than the theorems. For this and the second part, the reader might wish to scan the table of contents to get a feel for what is going on. We give the following sample result to give an idea of the scope here. In ordinary small-cancellation theory, when $W_1, \ldots, W_r$ represent distinct conjugacy classes, the presentation $\langle a, b, \ldots \mid W_1^{n_1}, \ldots, W_r^{n_r} \rangle$ is "small-cancellation" for sufficiently large n_i. In analogy with this we have the following:

C6-sample. *Let X be a nonpositively curved cube complex. Let $Y_i \to X$ be a local-isometry with Y_i compact for $1 \leq i \leq r$ such that each $\pi_1 Y_i$ is malnormal, and $\pi_1 Y_i, \pi_1 Y_j$ do not share any nontrivial conjugacy classes. Then $\langle X \mid \widehat{Y}_1, \ldots, \widehat{Y}_r \rangle$ is a "small-cancellation" cubical presentation for sufficiently large "girth" finite covers $\widehat{Y}_i \to Y_i$.*

Many other general small-cancellation theories have been propounded. For instance two such graded theories directed especially towards Burnside groups were produced by Olshanskii and McCammond. Stimulated by Gromov's ideas of small-cancellation over word-hyperbolic groups, there have been later important works of Olshanskii, followed by more recent theories "over relatively hyperbolic groups" by Osin [Osi06] and Groves-Manning [GM08]. The theory we propose is decidedly more geometric, and arguably favors explicitness over scope. However, although it may be more limited by presupposing a nonpositively curved cube complex as a starting point, it has the advantage of not presupposing (relative) hyperbolicity—yet some form of hyperbolicity must lurk inside for there to be any available small-cancellation.

In the second part of the text we impose additional conditions that lead to the existence of a wallspace structure on the resulting small-cancellation presentation. We can illustrate the nature of the results with the following sample:

B6-sample. *Let G be an infinite word-hyperbolic group acting properly and cocompactly on a CAT(0) cube complex. Let $H_1, \ldots, H_k$ be quasiconvex subgroups that are not commensurable with G. And suppose that each H_i has separable hyperplane stabilizers. There exist finite index subgroups $H_1', \ldots, H_k'$ such that the quotient $G/\langle\langle H_1', \ldots, H_k' \rangle\rangle$ has a codimension-1 subgroup.*

Here $\langle\langle A, B, \ldots \rangle\rangle$ denotes the normal closure of $\{A \cup B \cup \cdots\}$ in the group.

In the third part of the text, we probe further and seek a virtually special cubulation.

We then prove the following:

Theorem A (Special Quotient Theorem). *Let G be a word-hyperbolic group that is virtually the fundamental group of a compact special cube complex. Let $H_1, \ldots, H_r$ be quasiconvex subgroups of G. Then there are finite index subgroups $H_i' \subset H_i$ such that: $G/\langle\langle H_1', H_2', \ldots, H_r' \rangle\rangle$ is virtually special.*

We then prove the following:

Theorem B (Quasiconvex Hierarchy $\Rightarrow$ Virtually Special). *Let G be a word-hyperbolic group with a quasiconvex hierarchy, in the sense that it can be decomposed into trivial groups by finitely many HNN extensions and amalgamated free products along quasiconvex subgroups. Then G is virtually special.*

There are two important applications of the virtual specialness of groups with a quasiconvex hierarchy: It is applied to hyperbolic 3-manifolds with a geometrically finite incompressible surface to reveal their virtually special structure. This resolves the subgroup separability problem for fundamental groups of such manifolds. It also completes a proof that Haken hyperbolic 3-manifolds are virtually fibered. It is also applied to resolve Baumslag's conjecture on the residual finiteness of one-relator groups with torsion.

The fourth part of the text deals with groups that are hyperbolic relative to virtually abelian subgroups, and provides similar structural results for many such groups when they also have quasiconvex hierarchies.

Chapter Two

CAT(0) Cube Complexes

2.a   Basic Definitions

An *n-cube* is a copy of $[-\frac{1}{2}, \frac{1}{2}]^n$, and a 0-*cube* is a single point. We regard the boundary of an n-cube as consisting of the union of lower dimensional cubes. A *cube complex* is a cell complex formed from cubes, such that the attaching map of each cube is combinatorial in the sense that it sends cubes homeomorphically to cubes by a map modeled on a combinatorial isometry of n-cubes. The *link* of a 0-cube v is the complex whose 0-simplices correspond to ends of 1-cubes adjacent to v, and these 0-simplices are joined up by n-simplices for each corner of an $(n + 1)$-cube adjacent to v.

A *flag complex* is a simplicial complex with the property that each finite set of pairwise-adjacent vertices spans a simplex. A cube complex C is *nonpositively curved* if link(v) is a flag complex for each 0-cube $v \in C^0$.

Two-dimensional nonpositively curved complexes with one 0-cell, are a special case of the $C(4)$-$T(4)$ small-cancellation presentations that have old roots within combinatorial group theory. The nonpositively curved cube complexes were introduced to geometric group theory by Gromov in [Gro87] as a source of examples of high-dimensional metric spaces with nonpositive curvature. The supporting details of this theory were given by Moussong, Bridson, and Leary, in the locally finite, finite dimensional, and general cases. We refer to [Mou88, Lea13] but especially to [BH99] for a general account of CAT(0) geodesic metric spaces.

2.b   Right-Angled Artin Groups

Let Γ be a simplicial graph. The *right-angled Artin group* or *raag* or *graph group* $G(\Gamma)$ associated to Γ is presented by:

$$\langle\ v : v \in\ \text{vertices}(\Gamma)\ \mid\ [u, v] : (u, v) \in\ \text{edges}(\Gamma)\ \rangle.$$

For our purposes, the most important example of a nonpositively curved cube complex arises from a right-angled Artin group. This is the cube complex $C(\Gamma)$ containing a torus T^n for each copy of the complete graph $K(n)$ appearing in Γ [CD95, MV95]. The cube complex $C(\Gamma)$ is sometimes called a *Salvetti complex*.

Each added torus T^n is isomorphic to the usual product $(S^1)^n$ obtained by identifying opposite faces of an n-cube. Note that $\pi_1 C(\Gamma) \cong G(\Gamma)$ since the 2-skeleton of $C(\Gamma)$ is the standard 2-complex of the presentation above.

To see that $C(\Gamma)$ is nonpositively curved we must show that $\operatorname{link}(a)$ is a flag complex where a is the 0-cube of $C(\Gamma)$. Each vertex of $\operatorname{link}(a)$ corresponds to an element of $\Gamma^0 \times \{\pm 1\}$. A set of vertices form an n-simplex precisely if they correspond to a corner of an $(n+1)$-cube of c, which holds precisely if they correspond to $n+1$ distinct generators oriented arbitrarily, that is, an $(n+1)$-clique of Γ with a ± 1 associated to each vertex. It is then clear that $\operatorname{link}(a)$ is simplicial as the intersection of simplices is a simplex. Moreover, $\operatorname{link}(a)$ is a flag-complex, since the 2^n different ways of orienting the vertices of an n-clique correspond to the 2^n different corners of the associated n-cube of c, and hence each collection of pairwise-adjacent vertices spans a simplex of $\operatorname{link}(a)$.

2.c   Hyperplanes in CAT(0) Cube Complexes

Simply-connected nonpositively curved cube complexes are called *CAT(0) cube complexes* because they admit a CAT(0) metric where each n-cube is isometric to $[-\frac{1}{2}, \frac{1}{2}]^n \subset \mathbb{R}^n$; however we shall rarely use this metric.

The crucial characteristic properties of CAT(0) cube complexes are the separative qualities of their hyperplanes: A *midcube* is the codimension-1 subspace of the n-cube $[-\frac{1}{2}, \frac{1}{2}]^n$ obtained by restricting exactly one coordinate to 0. A *hyperplane* is a connected nonempty subspace of the CAT(0) cube complex C whose intersection with each cube is either empty or consists of one of its midcubes. The 1-cubes intersected by a hyperplane are *dual* to it. We will discuss immersed hyperplanes within a nonpositively curved cube complex in Section 6.a.

Remark 2.1. Hyperplanes in a CAT(0) cube complex C have several important properties [Sag95]:

(1) If D is a hyperplane of C then $C - D$ has exactly two components.
(2) Each midcube of a cube of C lies in a unique hyperplane.
(3) Regarding each midcube as a cube, a hyperplane is itself a CAT(0) cube complex.
(4) The union $N(D)$ of all cubes that D passes through is the *carrier* of D and is a convex subcomplex of C (see Section 2.d) that is isomorphic to $D \times I$.

   Here $I = [-\frac{1}{2}, +\frac{1}{2}]$ is a 1-cube

2.d   Geodesics and the Metric

Although we have defined the standard 1-cube to be a copy of $[-\frac{1}{2}, \frac{1}{2}]$, it will often be convenient to consider real intervals as 1-dimensional cube complexes whose vertices are the integer points. In particular, we let I_n denote the interval

$[0, n]$ subdivided so that all integers are vertices. A *length n path from x to y* in a cube complex X is a combinatorial map $I_n \to X$ where $0, n \mapsto x, y \in X^0$. A path is a *geodesic* if there is no shorter length path with the same endpoints. We emphasize that geodesics are almost never unique when $\dim(X) \geq 2$, indeed there are $n!$ distinct geodesics connecting vertices at opposite corners of an n-cube. We define the *distance* between 0-cubes in a connected nonpositively curved cube complex to be the length of a geodesic between them. As usual, this provides a genuine metric on the 0-cells of the 1-skeleton. Moreover we are then able to declare the distance $\mathsf{d}(A, B)$ between subcomplexes as the minimal distance $\mathsf{d}(a, b)$ where $a, b \in A^0, B^0$. We also define the *diameter* $\mathrm{diam}(Y)$ of a connected complex to be the supremum of the lengths of geodesics in Y.

The combinatorial viewpoint we have adopted does not use the CAT(0) comparison metric, and we refer to [BH99] for an extensive account of that viewpoint—for cube complexes and many other spaces.

2.e Properties of Minimal Area Cubical Disk Diagrams

This section was motivated by lectures of Andrew Casson from the University of Texas at Austin in the '80s (apparently on generalized $C(4)$-$T(4)$ presentations related to Heegaard decompositions). I am grateful to Yoav Moriah who shared his notes with me and to Michah Sageev who encouraged me to take a look at this. Part of this material was explained using the alternate viewpoint of "pictures" in [Sag95, Sec 4.1]. While the results are easy, I had not previously considered the relevance of disk diagrams to cubical complexes of dimension ≥ 3. The viewpoint here, and in particular Lemma 2.3, is due to Casson. We note that the properties listed in Remark 2.1 can be deduced from this viewpoint.

A *disk diagram* D is a compact contractible combinatorial 2-complex with a chosen planar embedding $D \subset \mathbb{R}^2$. Its *boundary path* or *boundary cycle* $\partial_{\mathsf{p}} D$ is the attaching map of the 2-cell containing the point at ∞ where we regard $\mathbb{R}^2 \cup \infty$ to be the 2-sphere. The disk diagram D is *trivial* if it consists of a single 0-cell. A *spur* of D is an open edge in ∂D that ends on a valence 1 vertex of ∂D. Note that there is a spur for each backtrack in $\partial_{\mathsf{p}} D$. A 1-cell of D is *isolated* if it does not lie on the boundary of any 2-cell. A 0-cell v of D is *singular* if $\mathrm{link}(v)$ is not isomorphic to a cycle, i.e., D does not look like $\mathbb{R}^2$ at v. The diagram D is *singular* if it has a singular 0-cell. Equivalently D is singular if it is not homeomorphic to a closed 2-ball, in which case D is either trivial, has a cut vertex, or consists of a single isolated edge.

We say D is a *square disk diagram* if it is a cube complex, i.e., all its 2-cells are squares. Many of the arguments below are by induction on $\mathsf{Area}(D)$ which equals the number of squares in D. (We note that there are instances where it is more natural to instead count the number of edges or even the number of cells in D.)

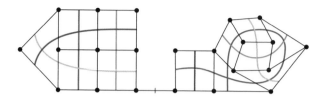

Figure 2.1. Dual curves in a square complex disk diagram.

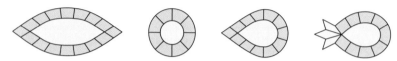

Figure 2.2. A bigon, nongon, monogon, and oscugon.

A *diagram in a complex* X is a combinatorial map $D \to X$ where D is a disk diagram. In this section, we study *cubical disk diagrams* which are disk diagrams in a nonpositively curved cube complex X. Of course, every cubical disk diagram is a square disk diagram.

We often use the following standard fact about the existence of disk diagrams (see [§2.2] [ECH$^+$92] or [LS77]):

Lemma 2.2 (van Kampen). *A closed combinatorial path $P \to X$ is nullhomotopic if and only if there exists a diagram $D \to X$ with $P \cong \partial_p D$ so that there is a commutative diagram:*

$$
\begin{array}{ccc}
\partial_p D & \to & D \\
\| & & \downarrow \\
P & \to & X
\end{array}
$$

Let D be a square disk diagram. The *dual curves* in D are (noncombinatorial) paths that are concatenations of midcubes of squares of D. In addition, the midcube of an isolated edge of D provides a dual curve that is a trivial path. Note that when $D \to \widetilde{X}$ is a disk diagram in a CAT(0) cube complex, each dual curve maps to a hyperplane of $\widetilde{X}$.

The 1-cells crossed by a dual curve are *dual* to it. Note that each midcube lies in a unique maximal dual curve (or cycle). One simply extends outwards uniquely across dual 1-cells. A *bigon* is a pair of dual curves that cross at their first and last midcubes. A *monogon* is a single dual curve that crosses itself at its first and last midcubes.

An *oscugon* is a single dual curve that starts and ends at distinct dual 1-cells that are adjacent but don't bound the corner of a square. A *nongon* is a single dual curve of length ≥ 1 that starts and ends on the same dual 1-cell, so it corresponds to an immersed cycle of midcubes. We refer the reader to Figure 2.2.

Figure 2.3. On the left is a smallest possible bigon. On the right is a monogon which must contain a smaller bigon.

Lemma 2.3. *Let $D \to X$ be a disk diagram in a nonpositively curved cube complex. If D contains a bigon, nongon, or oscugon, then there is a new diagram D' such that:*

(1) *D' and D have the same boundary path, so $\partial_p D' \to X$ equals $\partial_p D \to X$,*
(2) *$\mathsf{Area}(D') \leq \mathsf{Area}(D) - 2$, and*
(3) *pairs of edges on $\partial_p D'$ that lie on the same dual curve of D' are precisely the same as pairs of edges on $\partial_p D$ that lie on the same dual curve of D.*

Corollary 2.4. *No disk diagram contains a monogon.*
    If D has minimal area among all diagrams with boundary path $\partial_p D$, then D cannot contain a bigon, a nongon, or an oscugon.

Proof. The second statement follows immediately from Lemma 2.3. Consider a minimal area counterexample D to the first statement: So D contains an immersed rectangular strip $[-\frac{1}{2}, \frac{1}{2}] \times [0, n]$ of squares whose first and last square map to the same "cross-square," and this strip carries a dual curve σ at $\{0\} \times [0, n]$. We may assume $n > 2$ as if $n = 2$ then two adjacent edges at the corner of the cross-square are identified, and this violates the nonpositive curvature of the nonpositively curved cube complex X that D maps to, and if $n = 1$ then a square fails to embed in X. Choose m with $1 < m < n$. Then the 1-cube $\{\frac{1}{2}\} \times [m - 1, m]$ is dual to a dual curve λ which must cross σ a second time. We can therefore apply Lemma 2.3 to replace D by D' and obtain a smaller area diagram that is still a counterexample by Condition 2.3.(3). □

Proof of Lemma 2.3. **Reducing to the bigon case:** Consider a monogon, nongon, oscugon, or bigon within D that is *smallest* in the sense that the smallest subdiagram E containing it has minimal area. We first observe this smallest situation must arise from a bigon. Indeed, for a monogon, nongon, or oscugon, the associated dual curve α has length ≥ 1, for by the nonpositive curvature of the cube complex, squares locally embed, and so even for a monogon, the dual curve must pass through at least one more square besides its self-crossing square. Thus, as illustrated on the right in Figure 2.3, a second dual curve β crosses α and then must cross α a second time to exit. The pair α, β then provides a smaller situation. We next observe that α and β cannot contain a proper subpath that is a nongon or oscugon, for this would likewise lead to a smaller situation.

Figure 2.4. Some hexagon moves.

We are thus led to examine a *bigonal diagram* E which is a subdiagram with the property that each of its squares is either: one of the squares s_1, s_2 where α, β intersect; or a square with a midcube traversed by exactly one of α, β; or a square contained inside the bigon formed by α, β. Moreover, the bigonal diagram has the additional feature that the rectangles carrying α, β both embed.

Zipping a bigon: We now show that any bigonal subdiagram can be replaced by a disk diagram with the same boundary path but smaller area. Specifically, we will perform a slight modification to obtain a disk diagram with the same boundary but containing a smaller area bigonal diagram, and hence this smaller disk diagram itself can have its area reduced by 2.

The "base case" arises from two squares meeting along a corner as on the left in Figure 2.3. By nonpositive curvature, these two squares map to the same square in X, and hence we can remove this cancellable pair to decrease the area, by replacing the pair of squares by a pair of edges glued together at a point.

Observe that every dual curve in E other than α, β must pass through both α and β, since otherwise there would be a smaller bigon.

A *hexagon move* on a diagram D is the replacement of three squares forming a subdivided hexagon by an alternate three squares forming a subdivided hexagon. This corresponds to pushing a hexagon on one side of a 3-cube to obtain the hexagon on the other side.

The plan is to find a (certain type of) minimal triangle in the complement of the dual curves, and to then perform a hexagon move to obtain a new disk diagram with a smaller bigon as in Figure 2.4. The first type of minimal triangle has one side on α and one side on β and no dual curves passing through it. The second type has its base on α, and neither of its two other sides are subsegments of β.

If there is at least one crossing pair of dual curves as on the right of Figure 2.5, then we shall show below that the second type of triangle exists, and so we can perform a hexagon move of the second type. Hence by induction, the new diagram can have its area reduced by 2. If the bigon contains no crossing pair of dual curves as on the left in Figure 2.5, then the first type of triangle occurs, and so we can perform a hexagon move of the first type. We emphasize that a first type hexagon move can then be performed in either direction, i.e., at each corner (and this is the crucial point in obtaining Lemma 2.6 below).

Hexagon moves do not affect the boundary path (nor affect whether or not dual curves cross within the diagram); they simply adjust the route that dual curves take within a diagram, and hence Condition 2.3. (3) holds.

Figure 2.5. Dual curves of a bigonal diagram.

Figure 2.6. The directed graph Λ formed from dual curves cannot have a directed cycle.

A minimal triangle exists: The collection of dual curves within our bigon forms a graph Λ, and we make Λ into a directed graph by orienting all dual curves upwards from α to β, and thus orienting each edge of Λ (see the left of Figure 2.6). Observe that Λ has no directed cycle. Indeed, consider a directed cycle ξ. Suppose that ξ travels counterclockwise—as an analogous argument works in the clockwise case. Among the dual curves contributing edges to ξ, let σ denote the one having rightmost intersection with α. Let λ denote the next dual curve contributing an edge in the directed cycle ξ. Then λ would intersect α even further to the right which is impossible (see the middle of Figure 2.6). Here we use that each pair of dual curves intersect only once which follows from the minimality assumption on the bigon.

Each vertex of Λ (not on α, β) is the "top" of a triangle whose base is on α. Choose a vertex v that is minimal (excluding the leaf vertices on α) in the partial ordering induced by the directed graph with no directed cycles. Then the corresponding triangle Δ is our desired triangle of the second type. Indeed, if any other dual curve crosses either leg of Δ then there would be an even lower vertex u, which contradicts the minimality of v (as on the right of Figure 2.6). $\square$

We shall later use the term *shuffle* to refer to an adjustment of a disk diagram obtained through a finite sequence of hexagon moves.

Definition 2.5 (Cornsquare). Let $D \to X$ be a disk diagram. A *cornsquare* in D consists of a 2-cube c in D and dual curves α, β emanating from consecutive edges a, b of c that terminate on consecutive edges a', b' of $\partial_{\mathsf{p}} D$. The path $a'b'$ is the *outerpath of the cornsquare*.

We refer the reader to Figure 2.7. Note that we allow the possibility that there are squares in D between α, β. However, D can be shuffled so that there

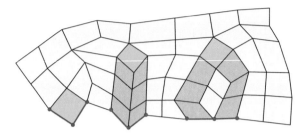

Figure 2.7. Three cornsquares in a disk diagram D. The outerpath of each is the corresponding red horizontal subpath of $\partial_{\mathsf{p}}D$. Note that D has other cornsquares besides these.

are no such squares, and moreover, following Lemma 2.6, D can be shuffled so that $a'b'$ actually forms the corner of a square.

The term "cornsquare" is short for "corner of generalized square" which captures the idea that there is a hidden square along $a'b'$, and although it might possibly be remote, it can be brought towards $\partial_{\mathsf{p}}D$ by shuffling. This notion arises again in Section 2.i and will play an important role in the more general context of Chapter 3.

The final part of the argument of Lemma 2.3 leads to the following useful point that we frequently employ.

Lemma 2.6 (Crossing pair has a square). *Let $D \to X$ be a diagram in a nonpositively curved cube complex. Suppose D contains a cornsquare whose outerpath is $a'b'$. Then there is another diagram $D' \to X$ with $\mathsf{Area}(D') \leq \mathsf{Area}(D)$ such that D' contains a square whose boundary path contains $a'b'$.*

In particular, let $D \to X$ be a diagram containing dual curves α, β that are dual to consecutive edges a', b' of a square, and also dual to edges a, b with a common endpoint. Then the images of a, b in X bound the corner of a square of X.

Proof. The "zipping bigon" part of the proof of Lemma 2.3 only used that there was a square on *one* corner of the bigonal diagram. The sequence of moves either push squares outwards through the top or bottom dual curves, or they push hyperplanes past a', b' resulting in a shorter bigon. The final stage of this sequence is a diagram consisting of a single square on a, b.   □

Remark 2.7. Let α, β be dual curves intersecting in a minimal area diagram $D \to X$. There is a cornsquare or spur in each of the four "quadrants" of D subtended by α, β. Indeed, let aPb be a subpath of $\partial_{\mathsf{p}}D$, with a dual to α and b dual to β, that does not contain a backtrack. Consider an innermost pair e_1, e_2 of edges in aPb whose dual curves σ_1, σ_2 are either equal or cross. Then e_1, e_2 must be consecutive and hence provide a cornsquare or spur. Indeed, an edge e_3

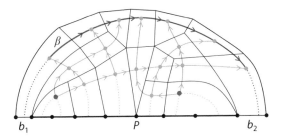

Figure 2.8. The digraph Λ consists of parts of the dual curves in D between $b_1 P b_2$ and β. The two vertices without ancestors correspond to squares with a corner on $b_1 P b_2$.

between e_1, e_2 would have a dual curve σ_3 that either returned to aPb between e_1, e_2 or that crosses σ_1 or σ_2, and this violates innermostness of e_1, e_2.

We now repeat the argument at the end of the proof of Lemma 2.3 to glean a bit more information:

Lemma 2.8 (Square or spur on each side). *Let $D \to X$ be a minimal area disk diagram in a nonpositively curved cube complex. Let β be a dual curve in D that starts and ends on edges b_1, b_2 where $b_1 P b_2$ is a subpath of $\partial_{\mathsf{p}} D$. Then $b_1 P b_2$ contains a length 2 subpath $e_1 e_2$ such that either $e_1 e_2$ is a backtrack at a spur of D, or $e_1 e_2$ bounds the corner of a 2-cube of D.*

Proof. It suffices to assume that β is innermost, in the sense that the curve dual to each edge of P must cross β. If P is trivial, then since D has no oscugon by Corollary 2.4, we see that b_1, b_2 traverse the same edge, which is a spur and we are done with $b_1, b_2 = e_1, e_2$. More broadly, innermostness allows us to assume that $b_1 P b_2 \to D$ embeds, for otherwise there would be a cutpoint in D, which would subtend a diagram containing a dual curve that violates the innermost assumption.

Consider the graph Λ whose vertices are centers of squares in the P-component of $D - \beta$, and whose edges are parts of dual curves joining centers of adjacent squares. We refer the reader to Figure 2.8. We direct Λ by directing β from b_1 to b_2, and directing all other dual curves from P to β. As in the end of the proof of Lemma 2.3, there is no directed cycle ξ in Λ. Indeed, ξ cannot have an edge on β, for then an edge of Λ is directed away from β which contradicts that dual curves are directed from P to β. Regard traveling up towards β and then in the direction of β as "clockwise." Suppose ξ were a cycle not containing an edge on β and assume it travels counterclockwise (a similar argument works in the clockwise case). Among the dual curves forming ξ, let σ denote the one which intersects P closest to b_2. As there are no bigons or monogons by Lemma 2.3, the next dual curve λ that ξ provides would have to intersect P closer to b_2, which is impossible.

Since Λ has no directed cycle, it has a vertex v with no ancestor. Then v is the center of a square s with consecutive edges $e_1 e_2$ forming a subpath of P. Indeed, each dual curve (other than β) travels from P to β, for otherwise we would either contradict that β is innermost, or there would be a bigon. Hence the two incoming dual curves at v arrive from edges of $b_1 P b_2$ that lie on s, as otherwise v would have an ancestor. $\qquad\square$

2.f   Convexity

Although the basic properties of CAT(0) cube complexes and their hyperplanes have been explained many times in the literature, the reviewers have asked me to sketch some of these properties from the diagrammatic viewpoint elaborated upon in Section 2.e.

A subcomplex $Y \subset X$ of the CAT(0) cube complex is *convex* if for each pair of vertices $a, b \in Y^0$, each (combinatorial) geodesic joining a, b lies in Y.

Lemma 2.9. *The intersection of two convex subcomplexes is a convex subcomplex.*

Proof. This follows immediately from the definitions. $\qquad\square$

Lemma 2.10 (Helly property). *Let X be a CAT(0) cube complex. Let $Y_1, \ldots, Y_n$ be finitely many convex subcomplexes. Suppose $Y_i \cap Y_j \neq \emptyset$ for each i, j. Then $\cap_{i=1}^n Y_i \neq \emptyset$.*

Proof. We first show this is true in the base case when $n = 3$. For $i \neq j$ let x_{ij} be a vertex in $Y_i \cap Y_j$. Let P_i be a geodesic in Y_i from x_{ki} to x_{ij}. Let D be a disk diagram for $P_1 P_2 P_3$. Finally, choose the above such that $\mathsf{Area}(D)$ is minimal.

Consider a square s in D. Since each P_i is a geodesic, no dual curve in D has both ends on the same P_i. Thus by the pigeon-hole principle, for some i, each dual curve through s has an end on P_i. We thus have a cornsquare on P_i and hence after shuffling we can reduce the area which is impossible. Thus D is a tripod, and its central point is an element of $Y_1 \cap Y_2 \cap Y_3$.

We now use the base case to help us prove the result by induction: For $1 \leq i < n$ let $Y_i' = Y_i \cap Y_n$. Since Y_i, Y_j, Y_n have pairwise nonempty intersection by hypothesis, the special case implies that $Y_i' \cap Y_j' = Y_i \cap Y_j \cap Y_n \neq \emptyset$. Thus $\cap_{i=1}^n Y_i = \cap_{i=1}^{n-1} Y_i' \neq \emptyset$ by induction. $\qquad\square$

An *immersion* is a local injection. A map $\phi : Y \to X$ between nonpositively curved cube complex is a *local-isometry* if it is an immersion and for each $y \in Y^0$, whenever u, v are ends of 1-cubes at y, if $\phi(u), \phi(v)$ form a corner of a 2-cube in X at $\phi(y)$, then u, v form a corner of a 2-cube in Y. A subcomplex that embeds by a local-isometry is *locally-convex*. A connected locally-convex subcomplex

$\widetilde{Y}$ of a CAT(0) cube complex $\widetilde{X}$ is called *convex*. Equivalently, a connected subcomplex $Y \subset X$ of a CAT(0) cube complex is convex if: for each cube c of X with $\dim(c) \geq 2$, if an entire corner of c lies in Y then all of c lies in Y. It can be deduced from the viewpoint in Section 2.e that for a CAT(0) cube complex $\widetilde{X}$, a subcomplex $\widetilde{Y} \subset \widetilde{X}$ is convex if and only if firstly: an n-cube lies in $\widetilde{Y}$ precisely when its $(n-1)$-skeleton lies in $\widetilde{Y}$, and secondly: P lies in $\widetilde{Y}$ whenever $P \to \widetilde{X}$ is a geodesic path whose endpoints lie in Y^0.

The combinatorial notion of convexity we employ here is consistent with the usual notion of convexity one encounters for geodesic metric spaces. Indeed, a subcomplex $\widetilde{Y} \subset \widetilde{X}$ is "combinatorially convex" as defined above precisely when it is "metrically convex" (in the CAT(0) metric) in the sense that $P \subset \widetilde{Y}$ whenever $P \to \widetilde{X}$ is a (not necessarily combinatorial) geodesic with endpoints in $\widetilde{Y}$.

Lemma 2.11 (Locally-convex $\Rightarrow$ convex). *Let X and Y be CAT(0) cube complexes. Let $Y \to X$ be a local-isometry. Then $Y \to X$ is an embedding, and its image is a convex subcomplex.*

In particular, a connected locally-convex subcomplex is convex.

Proof. Consider a geodesic $P \to X$ that is path-homotopic to a path $Q \to Y \to X$. Let D be a disk diagram between $P \to X$ and $Q \to X$. Assume $(\mathsf{Area}(D), |Q|)$ is minimal in the lexicographical order among all possible choices with P fixed. There is no cornsquare on Q for otherwise we could shuffle to obtain a smaller diagram D' between Q' and P. Thus each dual curve starting on Q ends on P and no two cross. Suppose D contains a square s. Then at most one end of one dual curve through s ends on Q, and so $|P| \geq |Q| + 2$, so P is not a geodesic. Thus D is a line and $P = Q$. Thus $Y \to X$ is an isometry, and in particular an embedding. Moreover, a geodesic P in X with endpoints in Y satisfies $P = Q \subset Y$. $\square$

Corollary 2.12 (Local-isometry π_1-injects). *If $Y \to X$ is a local-isometry of non-positively curved cube complexes, then $\pi_1 Y \to \pi_1 X$ is injective.*

Proof. If $\sigma \to Y$ represents a nontrivial element of $\pi_1 Y$ then its lift $\tilde{\sigma} \to \widetilde{Y}$ is not closed. Hence the image $\sigma \to X$ of σ represents a nontrivial element in $\pi_1 X$ since its lift $\tilde{\sigma} \to \widetilde{X}$ is also not closed, by Lemma 2.11. $\square$

2.g Hyperplanes and Their Carriers

Let M be the disjoint union of all midcubes of a cube complex X. Let $\bar{M}$ be the quotient of M obtained by identifying each midcube with the subcube of each larger midcube that it lies in. Note that the map $M \to X$ induces a continuous map $\bar{M} \to X$. An *(abstract) hyperplane* of X is a component U of $\bar{M}$.

The *(abstract) carrier* of U is defined to be $N(U) = U \times I$, and we define $N(U) \to X$ such that for each cube m of U, we have $m \times I$ maps isomorphically to the cube c containing m as a midcube.

Lemma 2.13 (Hyperplanes exist). *Let X be a CAT(0) cube complex. Every midcube of a cube of X lies in a hyperplane of X.*

Proof. Let U be a component of $\bar{M}$. Suppose that $U \to X$ is not injective. Then there is a pair of midcubes of U mapping to the same cube of X. We can assume these are 1-dimensional midcubes mapping to the same 2-cube s. Let a, b be the consecutive edges of the 2-cube that these midcubes end on. There is then a sequence of 1-midcubes joining them, and we thus obtain a corresponding rectangular strip R that starts with a and ends with b. Let P be a path along one side of this strip. By possibly extending R by adding a copy of s, we may assume that P is a closed path in X, and so P bounds a disk diagram $E \to X$. The union $D = E \cup_P R$ is another disk diagram that contains an oscugon associated to the dual curve carried by R. Applying Lemma 2.3, we obtain a new disk diagram D' with the same boundary path as D, but having no oscugons. Thus the initial and terminal edge of R are identified to a spur in D' under the composition $R \to D \to D'$. In particular they map to the same edge in X. This contradicts that a, b are not parallel in s.   □

Lemma 2.14 (CAT(0) hyperplane properties). *Let X be a CAT(0) cube complex.*

(1) *The map $N(U) \to X$ is an embedding for each hyperplane U.*
(2) *$N(U) \subset X$ is a convex subcomplex.*
(3) *Each hyperplane is simply-connected.*
(4) *Let U, V be hyperplanes of X that cross in the sense that they contain distinct midcubes in a cube. If U, V are dual to 1-cubes a, b that share a 0-cube, then a, b lie in a common 2-cube.*

Proof. We first show that the map $\phi : N(U) \to X$ is a local-isometry. Let a, b be edges at a 0-cube v of $N(U)$, and suppose $\phi(a), \phi(b)$ bound the corner of a square in X. If one of a, b is dual to U then a, b form the corner of a square in $N(U)$. If neither is dual to U, then they each form a square with the edge c dual to U at v, and thus by nonpositive curvature of X, there is a 3-cube bounded by the three squares with corners at $\phi(a), \phi(b), \phi(c)$. This 3-cube contains a midcube m that is part of U, and a, b bound the corner of a square parallel to m in $m \times I$.

As we have verified that $N(U) \to X$ is a local-isometry, the convexity of $N(U) \subset X$ holds by Lemma 2.11.

We now show that $N(U)$ and hence U is simply-connected. Consider an essential closed path $P \to N(U)$, such that $P \to X$ is nullhomotopic. Let $D \to X$ be a disk diagram with boundary path D. We moreover choose the above such that $(\mathsf{Area}(D), |P|)$ is minimal. If D has a spur then we can shorten P. Otherwise, an innermost pair of edges on P with crossing dual curves yields a cornsquare in D. By Lemma 2.6 we can shuffle to obtain a new disk diagram with the same area but with a genuine square having a corner on P. By local convexity, we can remove it to obtain a smaller area counterexample D', P'.

Property (4) holds by Lemma 2.6. Indeed, let a', b' denote the edges at the square s' where the hyperplanes U, V cross. Choose rectangular strips R_a, R_b that start at the square s' and end at the edges a, b. Let E be a disk diagram between paths P_a, P_b along the bases of R_a, R_b. Let $D \to X$ be the diagram formed by $R_a \cup_{P_a} D \cup_{P_b} R_b$. Then Lemma 2.6 provides a square at a, b.

Alternatively, we sketch an explanation depending on convexity (which in turn depended on Lemma 2.6): Note that $N(U) \cap N(V)$ is a convex subcomplex which contains a square s' as well as a vertex $a \cap b$. Consider a length n geodesic joining them, and then verify that it extends to a product $I \times I \times [0, n]$ where $I \times I \times \{0\}$ maps to s' and $I \times I \times \{n\}$ maps to a square at a, b. $\qquad \square$

Let U be a hyperplane in a CAT(0) cube complex X. Let $N^o(U)$ be the *open carrier* consisting of the open cubes intersecting U. We refer to each component of $N(U) - N^o(U)$ as a *frontier* of U. Each frontier is a subcomplex $U \times \{\pm \frac{1}{2}\}$ if we identify $N(U)$ with $U \times [-\frac{1}{2}, \frac{1}{2}]$.

The complement $X - U$ consists of two subspaces called *halfspaces*. Each halfspace is associated with two *combinatorial halfspaces*, namely, the smallest subcomplex containing it, and a largest subcomplex contained in it, and these are referred to as a *major halfspace* and *minor halfspace*. Note that the major and minor halfspaces meet along a frontier F, and are the closures of components of $X - F$.

Another convenient property of hyperplanes is that:

Corollary 2.15. *Let U be a hyperplane of the CAT(0) cube complex X. Each frontier of U is convex. Each major and minor halfspace is convex.*

Proof. Each subcomplex $U \times \{\pm \frac{1}{2}\}$ is a convex subcomplex of $N(U)$, and thus convex in X as it is a convex subcomplex of a convex subcomplex.

The closure of each component of $X - (U \times \{\pm \frac{1}{2}\})$ is convex since its boundary $U \times \{\pm \frac{1}{2}\}$ is convex. Indeed, it is sufficient to consider geodesics starting and ending on the boundary of a component, which is a frontier. $\qquad \square$

Corollary 2.16. *Let γ be a combinatorial path in a CAT(0) cube complex. Then γ is a geodesic if and only if the hyperplanes dual to the edges of γ are distinct. Thus the distance between 0-cells equals the number of hyperplanes separating them.*

Proof. Suppose γ is a geodesic, and suppose $e_1 \gamma' e_2$ is a subpath where e_1, e_2 are dual to the same hyperplane H. Then $e_1 \gamma' e_2$ has endpoints on one of the subcomplexes $H \times \{\pm \frac{1}{2}\} \subset N(H)$, which is convex by Corollary 2.15, but e_1 and e_2 do not lie in this subcomplex, and so its convexity is contradicted.

Conversely, suppose γ has the property that the hyperplanes dual to its edges are distinct. Let γ' be a shortest subpath of γ that is not a geodesic. Let σ be a geodesic with the same endpoints as γ'. Then $|\sigma| \geq |\gamma'|$ since each hyperplane dual to an edge of γ' must separate the endpoints of σ and is hence dual to an odd number of its edges. $\square$

The *convex hull* of a subset $S \subset X$ is the subcomplex $\mathrm{hull}(S) \subset X$ that is the smallest convex subcomplex of X containing S. Corollary 2.15 and Lemma 2.9 imply that $\mathrm{hull}(S)$ lies in the intersection of all minor halfspaces containing S. We will see from Lemma 2.19 that $\mathrm{hull}(S)$ equals this intersection.

For 0-cubes p, q in a CAT(0) cube complex X, the *interval* $\mathcal{I}(p,q)$ is defined by $\mathcal{I}(p,q) = \mathrm{hull}(\{p,q\})$.

Lemma 2.17. *For each 0-cube $k \in \mathcal{I}(p,q)$ we have $\mathsf{d}(p,k) + \mathsf{d}(k,q) = \mathsf{d}(p,q)$. Equivalently, each 0-cell $k \in \mathcal{I}(p,q)$ lies on a geodesic from p to q.*

Proof. By the triangle inequality, it suffices to verify that $\mathsf{d}(p,k) + \mathsf{d}(k,q) \leq \mathsf{d}(p,q)$. By Corollary 2.16, $\mathsf{d}(a,b)$ equals the number of hyperplanes separating a, b. First observe that each hyperplane separating exactly one of p, k and k, q, must also separate p, q. Secondly, we verify that no hyperplane H separates both p, k and k, q. Indeed, then $\{p, q\}$ lies in one halfspace of H but k lies in the other. Hence K lies in the minor halfspace of H containing $\{p, q\}$, but k does not, so $k \notin \mathcal{I}(p,q)$. $\square$

Lemma 2.18. *Let P, Q be convex subcomplexes of the CAT(0) cube complex X. Consider all paths that start at a vertex of P and end at a vertex of Q, and let γ have minimal length among all such paths. Then every edge of γ is dual to a hyperplane that separates P, Q.*

Proof. Let p, q be the endpoints of γ in P, Q. Let $\mathcal{I} = \mathcal{I}(p,q)$ be the interval consisting of the convex hull of $\{p, q\}$. Let H be a hyperplane that separates p, q but intersects P. Since $\mathcal{I}, N(H), P$ pairwise intersect, they must triply intersect by Lemma 2.10, so there exists a 0-cube $k \in \mathcal{I} \cap N(H) \cap P$. However $\mathsf{d}(k,q) < \mathsf{d}(p,q)$ by Lemma 2.17, and since $k \in P$ this contradicts the choice of p, q. $\square$

As mentioned earlier, minor halfspaces play the following useful role:

Lemma 2.19. *A subcomplex Y of a CAT(0) cube complex X is convex if and only if Y is the intersection of minor halfspaces.*

Proof. The intersection of minor halfspaces is convex by Corollary 2.15 and Lemma 2.9. We now show that if $Y \subset X$ is convex, then Y is the intersection of minor halfspaces. Let p be a 0-cube in $X - Y$. Let γ be a geodesic fom Y to p.

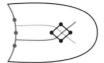 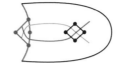 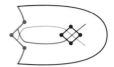

Figure 2.9. We first add a square along $Q \to Y$ to obtain a bigon, and are then able to reduce the area by 2.

By Lemma 2.18, each hyperplane U dual to an edge of γ separates Y, p. Thus a minor halfspace of U contains Y but not p.   □

2.h Splaying and Rectangles

We now describe several related properties concerning the dual curves in minimal area cubical disk diagrams. We emphasize that our treatment focuses on subcomplexes and exclusively considers paths that are combinatorial, as discussed in Section 2.d.

The following is implicit in the proof of Lemma 2.11.

Lemma 2.20 (Splayed). *Let $Y \subset X$ be a convex subcomplex of a CAT(0) cube complex. Let P be a path whose endpoints lie on Y, and let D be a disk diagram between P and Y, so there is a [geodesic] immersed path $Q \to Y$ with the same endpoints as P, and D is a diagram for PQ^{-1}. Suppose D has minimal area among all possible such choices fixing P and Y.*

Then there is no intersection in D between dual curves starting on distinct 1-cells of Q.

The statement of Lemma 2.20 holds with Q allowed to vary either among all such immersed paths, or among all such geodesics. Indeed, the argument by contradiction given below provides a lower area diagram D without increasing the length of Q.

Proof. We first show that when a and b are consecutive 1-cells in Q, then the dual curve in D starting at a is disjoint from the dual curve starting at b.

Suppose a, b are parallel in D to 1-cells a', b' that meet at the corner of a square c' in D. Since X is CAT(0), by Lemma 2.14.(4), the 1-cells a, b must also meet at a square c. Since Y is convex, we see that $c \subset Y$.

We can thus adjust the diagram D to obtain a new diagram D' formed by attaching c to Q along a, b. Now $\mathsf{Area}(D') = \mathsf{Area}(D) + 1$. However, D' contains a bigon, and therefore by Lemma 2.3, its area can be reduced by two, to obtain a new diagram D'' with $\mathsf{Area}(D'') < \mathsf{Area}(D)$. This would contradict the minimality of D. See Figure 2.9.

Suppose there is a dual curve in D that starts on a and ends on b. Since $Q \to Y$ is an immersed path, the edges a, b in D are distinct. The dual curve

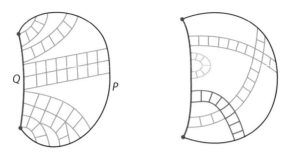

Figure 2.10. As on the left, dual curves emanating from edges of $Q \to D$ are splayed in the minimal area diagram D between P and the convex subcomplex $Y \subset X$. The configurations on the right cannot occur.

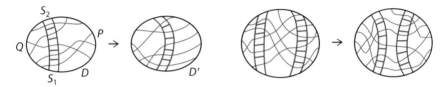

Figure 2.11.

thus provides an oscugon in D. By Lemma 2.3, there is a new diagram D' with the same boundary path, but $\mathsf{Area}(D') < \mathsf{Area}(D)$.

The general statement holds by considering an innermost pair of 1-cells whose dual curves are either equal or intersect. As proven above, these 1-cells cannot be adjacent. But any 1-cell on Q between them would give another dual curve, which either intersects one of these, or ends on another 1-cell of Q lying between them as in the right in Figure 2.10. This contradicts our innermost assumption.   □

Lemma 2.21 (Pushing beyond crossings)**.** *Let $D \to X$ be a minimal area disk diagram. Let S be a rectangular strip carrying a dual curve in D that starts and ends on 1-cells s_1, s_2 such that $\partial_{\mathsf{p}} D$ is of the form $s_1 P s_2 Q$. There exists a new diagram D' with $\partial_{\mathsf{p}} D = \partial_{\mathsf{p}} D'$ and $\mathsf{Area}(D') = \mathsf{Area}(D)$ such that s_1, s_2 are still connected by a strip S' but the dual curves emanating from S' to P are splayed: No two cross each other on the P side of S'.*

We refer to the left pair of diagrams in Figure 2.11 indicating the total transformation from D to D'.

Proof. This follows by repeatedly using hexagonal replacement moves. Consider an innermost pair of a, b of edges along ∂S whose dual curves cross on the side bounded by P. If they are not adjacent, then they are not innermost. Note that

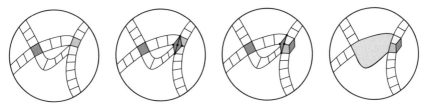

Figure 2.12.

the two dual curves cannot equal each other, or there would be a bigon with S, and thus the area can be reduced by Lemma 2.3.

Let c be the first square where the dual curves cross. Add a cancellable pair of copies c'', c' of c along a, b. This increases the area by 2, and increases the area between S and P by 2. Perform a hexagonal replacement along S and the contiguous copy c' of c to obtain S'; the area between P and S' is now one more than the area between P and S was. Finally the copy c'' of c has a bigon with c. We are able to reduce the area by 2. This area reduction is on the P side of S', and so the resulting diagram D' has the property that the area between P and S' has been reduced by one. See the sequence of pictures in Figure 2.12 for a single transformation. Performing this repeatedly yields a new diagram where S' has splayed strips on the P side, as claimed. □

Remark 2.22. We can apply Lemma 2.21 to understand the potential behavior between rectangular strips in disk diagrams. Let D be a diagram that has a pair of disjoint strips. Then we can replace it with a new diagram with the same boundary and at most as much area, such that the strips are moved inwards towards each other, but strips emanating from them are now splayed. See the transformation on the right in Figure 2.11.

This is particularly relevant when we consider a diagram between two convex subspaces Y_1, Y_2, and in particular, a diagram between a convex subspace and the carrier of a hyperplane. We are able to reach the conclusion of a "flat rectangle" between the rectangular strips.

A *grid* is a complex isomorphic to $I_m \times I_n$ for some m, n. We record the following easy observation. We revisit this later with pseudo-grids in Section 3.q.

Lemma 2.23. *A square disk diagram D is a grid if and only if:*

(1) $\partial_p D$ *is a concatenation* $\partial_p D = V_1 H_1 V_2 H_2$.
(2) *Each dual curve is either vertical and ends on* V_1, V_2 *or is horizontal and ends on* H_1, H_2.
(3) *Each horizontal dual curve crosses each vertical dual curve exactly once.*
(4) *Horizontal dual curves do not cross, and vertical dual curves do not cross.*

Corollary 2.24 (Grid with tails). *Let $D' \to X$ be a minimal area disk diagram in a nonpositively curved cube complex. Suppose $\partial_p D' = V_1' H_1' V_2' H_2'$, and D' has no spur or cornsquare with outerpath on V_1', H_1', V_2', or H_2'. Then D' is the union of a grid D together with a (possibly trivial) arc A_{ij} attached at the four corners of D, where $\partial_p D = V_1 H_1 V_2 H_2$, and the corners are at the four concatenation points.*

Proof. Suppose a dual curve σ starts on V_i' and ends on H_j'. We claim σ is trivial, as otherwise, σ crosses another dual curve σ', and since σ, σ' intersect at a single point, σ' has an endpoint on V_i' or H_j', so there is a cornsquare on V_i' or H_j'. Thus D' has four (possibly trivial) arcs $A_{ij} = V_i' \cap H_j'$, and removing these four arcs yields a diagram D that is a grid by Lemma 2.23. □

2.i Annuli

This section can be postponed until annuli arise in Chapter 14 and more importantly Section 5.0 and its sequels.

An *annular diagram* is a compact complex A with $\pi_1 A \cong \mathbb{Z}$ such that there is a chosen planar embedding $A \subset \mathbb{R}^2$. The annular diagram has two *boundary paths* or *boundary cycles* which correspond to the attaching maps of the two 2-cells that can be added to A to form a 2-sphere S^2.

An *annular diagram in a complex* X is a combinatorial map $A \to X$ where A is an annular diagram. It is natural to refer to A as an annular diagram *between* P_1 and P_2 as A indicates the homotopy between them as in the following standard analog of Lemma 2.2 (see [LS77]): Let $P_1 \to X$ and $P_2 \to X$ be closed paths in X. Then there is a homotopy between them if and only if there is an annular diagram $A \to X$, such that each $P_i \to X$ factors as $P_i \to A \to X$ where each $P_i \to A$ is a boundary path of A. Moreover, identifying the subdivided circles P_1 with $S^1 \times \{0\}$ and P_2 with $S^1 \times \{1\}$, the homotopy $S^1 \times [0,1] \to X$ factors as $S^1 \times [0,1] \to A \to X$.

We say A is *singular* if A is not homeomorphic to a cylinder, and we adopt the terminology used for disk diagrams: isolated 1-cell, singular 0-cell, spur, and so forth. We now turn to studying annular diagrams in a nonpositively curved cube complex X. The annular diagram is then a square complex, and we define its *dual curves* as we did for a disk diagram. We define a *cornsquare* in A as in Definition 2.5 except that we require that the dual curves emanating from the cornsquare do not enclose a boundary path of A—as for instance in the third annulus in Figure 2.13.

A *flat annulus* is an annular diagram A with the property that each dual curve is either closed or has an end on each boundary path of A, and that for each square, at least one of its dual curves is closed.

Lemma 2.25 (Flat annulus). *Let $A \to X$ be an annular diagram. Suppose there is no spur and no cornsquare with outerpath on a boundary path of A. Then A is a flat annulus.*

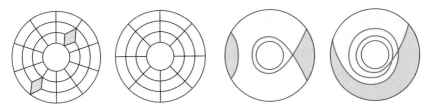

Figure 2.13. Two flat annuli are illustrated on the left. The second is a product. The first has only one closed dual curve, and it self-crosses several times. The third and fourth figure illustrate the situation where a dual curve has both ends on the same boundary path, in which case one can find a cornsquare.

We are especially interested in flat annuli that are minimal area, in which case they do not contain any bigons. We note however that a flat annulus A that is not a product will have a finite cover that contains bigons. The first annulus in Figure 2.13 illustrates a simple but typical example of the type of annulus examined in Lemma 2.25. This contrasts with the motivating case of a product, illustrated by the second annulus in Figure 2.13. The reader can imagine more elaborate examples.

Proof. Suppose d is a dual curve that starts and ends on the outer boundary path. We will show that there is a cornsquare. The analogous argument works when d starts and ends on the inner boundary path.

If d doesn't cross itself, then we choose the side of d not containing the inner boundary path. If d crosses itself, then we consider a minimal initial and terminal part of d that cross at some square, and choose the side of the diagram they subtend which does not contain the inner boundary path. The "chosen" sides are shaded in the third and fourth diagrams in Figure 2.13. In either case, our chosen side of the diagram contains an innermost pair of dual curves that cross each other, and within this lies the claimed cornsquare, as in Remark 2.7.

Now suppose there is a square s with two dual curves that end on the same boundary path of A. This forms a "triangle" whose top is in s and whose base is on ∂A. An innermost such triangle yields a cornsquare in A. □

Remark 2.26. The dual curves not ending on ∂A can travel around A, but we cannot always choose the annular diagram so that these dual curves do not self-cross. For instance, we refer the reader to the annular diagram at the very left of Figure 2.13. While minimal area of the diagram can help avoid some such self-crossing behavior, there is no way to avoid it in general, and we can only conclude that the "horizontal" dual curves travel "around" A, possibly multiple times. Of course, in the special case when immersed hyperplanes of X do not cross themselves, self-crossing of dual curves is impossible for any annular diagram mapping to X.

2.j   Annular Diagrams and Malnormality

This section can be postponed until its criterion is invoked in Chapter 10.

Definition 2.27 (Equivalent Annuli). Two annular diagrams $f_1 : A_1 \to X$ and $f_2 : A_2 \to X$ are *equivalent* if they have the same boundary cycles P, P' (in an orientation-preserving manner) so that the diagram below commutes, and there are lifts $\widetilde{f_i} : \widetilde{A}_i \to \widetilde{X}$ that restrict to the same lifts of $\widetilde{P}, \widetilde{P}'$.

$$
\begin{array}{ccccc}
 & & P \sqcup P' & & \\
 & \swarrow & \downarrow & \searrow & \\
A_1 & \to & X & \leftarrow & A_2
\end{array}
$$

Definition 2.28 (Geometric Malnormality). An inclusion $Z \to X$ of cell complexes is *malnormal* if there is no *essential* annular diagram $(A, \partial A) \to (X, Z)$ in the sense that any such map is equivalent to a map $(A', \partial A') \to (Z, Z)$. More generally, a map $Z \to X$ is *malnormal* if for each commutative diagram below on the left, there exists $A' \to Z$ such that the middle diagram commutes and such that the two annular diagrams on the right are equivalent. (Note that $\partial_{\mathsf p} A$ consists of two boundary cycles, and we identify $\partial_{\mathsf p} A' = \partial_{\mathsf p} A$.)

$$
\begin{array}{ccc}
\partial_{\mathsf p} A & \to & Z \\
\downarrow & & \downarrow \\
A & \to & X
\end{array}
\qquad
\begin{array}{ccc}
\partial_{\mathsf p} A & \to & Z \\
\downarrow & \nearrow & \\
A' & &
\end{array}
\qquad
\begin{array}{c}
A \to X \\
A' \to Z \to X
\end{array}
$$

A furtive use of Definition 2.28 shows that $Z \to X$ is π_1-injective on each component if it is malnormal. Indeed, a disk diagram in X for a path P in Z can be reconsidered as an annular diagram between P and the trivial path (which could be adjusted to a nontrivial path in Z to accommodate the annular diagram definition). The equivalent diagram $A' \to Z$ shows that P is already nullhomotopic in Z.

We now relate the geometric notion of malnormality with the notion from Definition 8.1 of a malnormal collection (of conjugacy classes) of subgroups:

Lemma 2.29 (Algebraic and geometric forms of malnormality). *Suppose $Z \to X$ is a map where X is connected, and Z is a disjoint union $Z = \sqcup_{i \in I} Z_i$. Then $\{\pi_1 Z_i\}$ is a malnormal collection in $\pi_1 X$ if and only if $Z \to X$ is malnormal.*

Proof. Suppose $b^{-1}ab = c$ for some nontrivial $a \in \pi_1 Z_i$ and $b \in \pi_1 Z_j$. Then there exists an annular diagram $A \to X$ with $P_i \to X$ representing a and factoring as $P_i \to Z_i \to X$, and with $P_j \to X$ representing b and factoring as $P_j \to Z_j \to X$. By the malnormality of $Z \to X$ there is an annular diagram $A' \to Z$ equivalent to A. Thus $i = j$ by connectivity, and A' provides a conjugacy between a and c in $\pi_1 Z_i$.

Conversely, if $A \to X$ is an annulus with $\partial_A \to \sqcup Z_i$, then by malnormality of $\{\pi_1 Z_i\}$ the two components of ∂A map to the same Z_i, and the conjugacy in Z_i is represented by $A' \to Z$ equivalent to $A \to Z$. $\qquad\square$

Lemma 2.30. *Let X be a nonpositively curved cube complex. Let $\{h_1, \ldots, h_k\}$ be a malnormal collection of immersed 2-sided hyperplanes in X. Let $\{w_1, \ldots, w_\ell\}$ be a collection of disjoint embedded 2-sided hyperplanes. Let $X' = X - N^o(\cup w_i)$. Let $\{h_{ij}\}$ be hyperplanes in X' mapping to $\{h_1, \ldots, h_k\}$.*

Then $\{h_{ij}\}$ is a malnormal collection of immersed hyperplanes in X'.

Diagrammatic proof. An essential annular diagram $A \to X'$ with boundary cycles mapping to $\cup N(h_{ij})$ projects to an annulus $A \to X' \to X$ with boundary cycles in $\cup N(h_i)$. By malnormality of $\cup N(h_i) \to X$, Lemma 2.29 implies that $A \to X$ is equivalent to an annulus $A' \to \cup N(h_i)$, and in particular, into a single $N(h_i)$. These equivalent annuli are illustrated on the left in Figure 2.14.

Suppose that A' is chosen so that it is cut by $\{w_1, \ldots, w_\ell\}$ in a minimal number of components. Note that all w-dual curves in A' are closed since $\partial A' \cap w = \emptyset$, and therefore each of these components is a dual w-circle that is either essential or nullhomotopic.

By considering innermost nullhomotopic w-circles first, each w-circle bounding a disc in A' can obviously be removed through a homotopy without introducing further w-circles. The essential w-circles come in "facing pairs" as in the annulus A' on the left of Figure 2.14 that are joined by a dual curve in a disk diagram C between the conjugators c, c' of A, A'. This diagram C is illustrated between the annuli A, A' in the left diagram.

The combinatorial path p along the outside of this dual curve is homotopic to the subpath b' of c' whose initial and terminal edges are dual to w_j for some j. Let E denote the subdiagram of C between p and b'. Considering lifts $\tilde{E}$ and $\tilde{A}' \subset N(\tilde{h}_i)$, we see that $\tilde{p}$ has the same endpoints as $\tilde{b}'$. We can choose a path $\tilde{b}''$ in $N(\tilde{w}_j) \cap N(\tilde{h}_i)$ that is path-homotopic to $\tilde{p}$ and hence $\tilde{b}'$. Let $\tilde{D}$ be the disk diagram in $N(\tilde{h}_i)$ between $\tilde{b}'$ and $\tilde{b}''$, and let D, b', b'' be their projections to X. We illustrate $\tilde{D}$ within the configuration of two crossing hyperplane carriers in $\tilde{X}$ in the middle of Figure 2.14. Since p doesn't cross any w we can assume that b'' has the same property. Moreover, we can choose D to avoid w-circles by removing them as above.

Finally we can drag A' along b' into D to obtain a new annular diagram $A'' \to N(h_i)$ as on the right of Figure 2.14. This turns the pair of essential circles into the single nonessential circle. We then remove this circle as we did earlier. $\qquad\square$

Upstairs proof. If there is an annulus $(A, \partial A) \to (X', \cup N(h_{ij}))$, then the malnormality hypothesis gives an equivalent annulus $(A, \partial A) \to (\cup N(h_i), \cup N(h_{ij}))$,

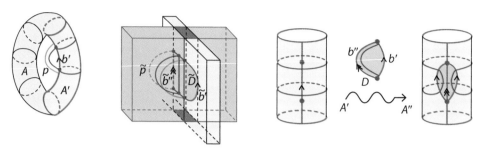

Figure 2.14. The diagrammatic proof of Lemma 2.30 is illustrated by the figures above. On the left, is the diagram C between conjugators c, c' of the annuli A, A', where an innermost pair of w-circles in A' yields a path p alongside a w-circle in C. The subpath $b' \subset c'$ with the same endpoints as p is path-homotopic through D to a path b'' in $N(h_i)$ that does not traverse any w-hyperplane. This allows us to produce a new diagram A'' with fewer essential w-circles, by pulling A' through D as on the right.

so these maps restrict to the same map on ∂A and have equal conjugators, or equivalently, they have lifts to $\widetilde{X}$ that restrict to the same lifts of ∂A. Our goal is to find another equivalent annulus of the form $(A, \partial A) \to (\cup N(h_{ij}), \cup N(h_{ij}))$.

We work in the universal cover $\widetilde{N}$ of the component $N = N(h_i)$ containing the image of A. Observe that $\widetilde{M} = \widetilde{N} - N^o(\cup \widetilde{w})$ is a convex subcomplex of $\widetilde{N}$. There is a $\pi_1 A$-equivariant retraction map $\widetilde{N} \to \widetilde{M}$, which takes a point $x \in \widetilde{N}$ to the unique point in $\widetilde{M}$ that is nearest to it in the CAT(0) metric.

The composition $\widetilde{A} \to \widetilde{N} \to \widetilde{M}$ yields the desired map $\widetilde{A} \to \widetilde{N}$.   □

2.k   Convex Cores

The following is essentially proven in [Hag08, SW15]. The exception is the second variant in the relatively hyperbolic case, which we explain in the proof. We refer to Definition 7.18 for quasiflats, and to Definition 7.21 for cosparseness.

Proposition 2.31. *Let X be a compact nonpositively curved cube complex with $\pi_1 X$ word-hyperbolic. Let $H \subset \pi_1 X$ be a quasiconvex subgroup, and let $C \subset \widetilde{X}$ be a compact subspace. Then there exists an H-cocompact subspace $\widetilde{Y} \subset \widetilde{X}$ with $C \subset \widetilde{Y}$.*

There are three variations on this where G is hyperbolic relative to virtually abelian subgroups.

Firstly, suppose G acts properly and cocompactly on $\widetilde{X}$. If H is full then for each compact subspace C, there is an H-cocompact convex subcomplex $\widetilde{Y}$ with $C \subset \widetilde{Y}$.

Secondly, generalizing fullness, suppose that for each maximal parabolic subgroup $P \subset G$, if $H \cap P$ is infinite, then $H \cap P$ cocompactly stabilizes a convex *subcomplex. Then for each compact subspace C, we can choose an H-cocompact convex subcomplex $\widetilde{Y}$ with $C \subset \widetilde{Y}$.*

Thirdly, relaxing cocompactness, suppose G acts properly and cosparsely on $\widetilde{X}$. If H is a quasi-isometrically embedded subgroup, then there exists $\widetilde{Y} \subset \widetilde{X}$ with $C \subset \widetilde{Y}$ and such that $\widetilde{Y}$ is H-cosparse.

We refer to the subcomplex $\widetilde{Y}$ of Proposition 2.31 as a *locally-convex core* containing C. Likewise, when C is a compact subspace of $H \backslash \widetilde{X}$, we term $H \backslash \widetilde{Y} \subset H \backslash \widetilde{X}$ a *locally-convex core* containing C.

We use the notation $\mathcal{N}_r(S)$ to denote the closed r-neighborhood of a subset S.

Remark 2.32. The structure of $\widetilde{Y}$ in the final statement in Proposition 2.31 is natural and follows from the argument used to prove [SW15, Thm 7.2] although a slightly weaker statement is expressed there.

Let $\{P_i\}$ be the subgroups of G with $|H \cap P_i| = \infty$, and let $\{\widetilde{F}_i\}$ be the corresponding quasiflats of G. When H is full, $\widetilde{Y} = \text{hull}(HC)$ has the property that:

$$(HC \cup_i H\widetilde{F}_i) \subset \widetilde{Y} \subset \mathcal{N}_r(HC \cup_i H\widetilde{F}_i). \tag{2.1}$$

When H has parabolics that are not finite index, in some cases the construction in [SW15] yields $\widetilde{Y}$ that does not contain entire quasiflats, because intersections $H \cap P^g$ might not be sufficiently diagonal to ensure the entire quasiflat lies in the intersection of halfspaces used to produce $\widetilde{Y}$. (We caution that the secondary claim in [SW15, Thm 7.2] that cosparseness holds with respect to a subspace containing each quasiflat having infinite coarse intersection with an H-orbit is incorrect, since there won't necessarily be finitely many H-orbits of hyperplanes.)

When H is full, it follows that for each $m \geq 0$ there is a uniform upperbound on $\text{diam}(\widetilde{Y} \cap \mathcal{N}_m(\widetilde{F}_k))$ unless $\widetilde{F}_k \subset \widetilde{Y}$. Indeed, if the latter statement does not hold, then $g_i \widetilde{F}_i \cap \mathcal{N}_m(\widetilde{F}_k)$ lies in a finite neighborhood of GK, and hence in a finite neighborhood of HC. (Here K is the compact subcomplex such that GK contains the intersection of distinct G-translates of the various $\widetilde{F}_k$.) Thus if $\widetilde{F}_k$ has infinite coarse intersection with $\widetilde{Y}$ then it has infinite coarse intersection with an orbit $H\widetilde{x}$, and so $|\text{Stab}_H(\widetilde{F}_k)| = \infty$ and hence $\widetilde{F}_k$ is one of the quasiflats included in $\widetilde{Y}$.

Proof of 2nd variant in Proposition 2.31. The proof of [SW15, Thm 7.2] provides the second inclusion of Equation (2.1) for a relatively quasiconvex subgroup H, without assuming fullness.

The hypothesis generalizing fullness is equivalent to the following assumption: if P is a maximal parabolic subgroup stabilizing a quasiflat $\widetilde{F}$, then $H \cap P$ cocompactly stabilizes a convex subcomplex of $\widetilde{F}$ (e.g., all of $\widetilde{F}$ when

$|H \cap P| < \infty$). This is because if $\widetilde{E}$ is convex and $(H \cap P)$-cocompact, then applying Proposition 2.33, so is $\widetilde{E}^{+r}$, and so for $r > \mathsf{d}(\widetilde{E}, \widetilde{F})$, we obtain a nonempty convex $(H \cap P)$-cocompact subcomplex $\widetilde{E}^{+r} \cap \widetilde{F}$.

The isolation property, that there exists K such that each $g\widetilde{F} \cap g'\widetilde{F}' \subset HK$, can be verified by using relatively thin triangles. Here is a sketch: Pick a basepoint $x \in \widetilde{X}$, and choose $m > 0$ such that for each quasiflat $\widetilde{F}$, if $\mathrm{Stab}_H(\widetilde{F})$ is infinite then $\widetilde{F} \cap \mathcal{N}_m(Hx) \neq \emptyset$. Let q be a point in $g\widetilde{F} \cap g'\widetilde{F}'$, and let p, p' be points in $gF \cap \mathcal{N}_m(Hx)$ and $g'F' \cap \mathcal{N}_m(Hx)$ that are closest to q. Consider the geodesic triangle $\Delta(qpp')$ using the CAT(0) metric. It is δ-thin relative to a point or a flat; see [HK05]. We claim that there is uniform $b > 0$ such that $\mathsf{d}(p, Hx) < b$, and hence the isolation property is satisfied by choosing K so that $\mathcal{N}_b(Hx) \subset HK$. Observe that qp and qp' have a uniformly bounded δ-overlap since otherwise $g\widetilde{F} = g'\widetilde{F}'$. Let s, s' be the last δ-close points along qp and qp'. If $\Delta(qpp')$ is thin relative to a point, then s, s' lie within δ of a point on pp', which itself lies in κ-neighborhood of Hx for some uniform $\kappa > 0$, as H is relatively quasiconvex. Suppose $\Delta(qpp')$ is thin relative to a flat $\widetilde{F}''$. If both sp and sp' fellow travel with $\widetilde{F}''$ for a long distance, then $\widetilde{F} = \widetilde{F}'' = \widetilde{F}'$, which is impossible. So at least one of these distances is uniformly small, say $|sp|$, and let t be the last point on sp that is δ-close to $\widetilde{F}''$. Hence t lies within δ of pp' and hence within $\delta + \kappa$ of Hx.

Finally, the existence of an H-cocompact convex subcomplex now follows from the convex subcomplex property of Lemma 7.37. $\qquad \square$

The following proposition describes properties of the *cubical thickening* Y^{+a} of Y. See [HW08, Lem 13.15] for a precise form of this construction.

Proposition 2.33 (Cubical thickening). *Let $\widetilde{Y} \subset \widetilde{X}$ be a convex subcomplex of a finite dimensional CAT(0) cube complex. For each $a \geq 0$ there exists $b \geq 0$ and a convex subcomplex $\widetilde{Y}^{+a}$ such that $\mathcal{N}_a(\widetilde{Y}) \subset \widetilde{Y}^{+a} \subset \mathcal{N}_b(\widetilde{Y})$. Moreover, assuming $\widetilde{X}$ has a proper group action, we can assume $\widetilde{Y}^{+a}$ is stabilized by $\mathrm{Stab}(\widetilde{Y})$.*

Note that when $X = S^1 \times S^1$ is the usual 2-dimensional torus, for most cyclic subgroups $H \subset \pi_1 X$, there does not exist a cocompact H-invariant convex subcomplex of $\widetilde{X}$. Indeed, the convex hull of a diagonal line in $\widetilde{X}$ is all of $\widetilde{X}$. The following example illustrates a more drastic failure:

Example 2.34. We now describe a finite dimensional example X where $F_2 \cong \pi_1 X$ but X does not admit a local-isometry $Y \to X$ where $\pi_1 Y$ is nontrivial and Y is compact.

Let $\bar{X}$ be the standard 2-complex of $\langle a_1, a_2, b_1, b_2 \mid [a_i, b_j] : i, j \in \{1, 2\} \rangle$, so $\bar{X}$ is isomorphic to the cartesian product of two bouquets of two circles and $\pi_1 \bar{X} \cong F_2 \times F_2$.

Let $X \to \bar{X}$ be the based covering space with $\pi_1 X = \langle a_1 b_1, a_2 b_2 \rangle$. Note that $\pi_1 X$ has trivial intersection with $\langle a_1, a_2 \rangle$ and with $\langle b_1, b_2 \rangle$. Consequently, for each nontrivial element $g \in \pi_1 X$, there is a unique flat plane F_g in $\widetilde{X}$ that is stabilized by g which acts by a diagonal translation. Moreover, F_g does not contain a nonempty proper convex subcomplex stabilized by g. It follows that for any local-isometry $Y \to X$ with Y compact, we have $\pi_1 Y = 1$.

2.1 Superconvexity

Definition 2.35. Let $\widetilde{X}$ be a metric space. A subset $\widetilde{Y} \subset \widetilde{X}$ is *superconvex* if it is convex and for any bi-infinite geodesic γ, if $\gamma \subset \mathcal{N}_r(\widetilde{Y})$ for some $r > 0$, then $\gamma \subset \widetilde{Y}$. Note that the intersection of superconvex subspaces is superconvex.

A map $Y \to X$ is *superconvex* if the induced map between universal covers $\widetilde{Y} \to \widetilde{X}$ is an embedding onto a superconvex subspace.

Lemma 2.36. *Let H be a quasiconvex subgroup of a word-hyperbolic group G. And suppose that G acts properly and cocompactly on a CAT(0) cube complex X. For each compact subcomplex $D \subset X$ there exists a superconvex H-cocompact subcomplex $K \subset X$ such that $D \subset K$.*

Proof. It follows from δ-hyperbolicity that any infinite geodesic lying in $\mathcal{N}_r(H\widetilde{x})$ actually lies in $\mathcal{N}_{2\delta}(H\widetilde{x})$. Now apply Proposition 2.31 to obtain a convex cocompact core Y containing $\mathcal{N}_r(H\widetilde{x})$. (Note that we use the combinatorial metric.)

To see that Y is superconvex, observe that any geodesic in a finite neighborhood of Y is actually contained in a 2δ neighborhood of $H\widetilde{x}$, and hence in Y. $\qquad\qquad\qquad\qquad\qquad\qquad\qquad\qquad\qquad\qquad\qquad\qquad\qquad\quad$ $\square$

We now consider superconvex quasiflats and generalize Lemma 2.36 to a relatively hyperbolic setting:

Lemma 2.37 (Superconvex quasiflat). *Let G be hyperbolic relative to virtually abelian subgroups $\{P_1, \ldots, P_r\}$. Suppose $G = \pi_1 X$ where X is a compact nonpositively curved cube complex. For each i, there exists a P_i-cocompact superconvex subcomplex $\widetilde{E}_i \subset \widetilde{X}$.*

More generally, let H be a full relatively quasiconvex subgroup of G. For each compact subcomplex $D \subset \widetilde{X}$ there exists a superconvex H-cocompact subcomplex $K \subset \widetilde{X}$ such that $D \subset \widetilde{K}$.

Lemma 2.37 can be deduced from [HW14, Prop 8.1] applied to the action of G on the wallspace $\widetilde{X}$ whose walls are associated to the hyperplanes. That also explains why quasiflats can be assumed to be superconvex in the cosparse cubulations that arise from the cubulation of a relatively hyperbolic group. See Definition 7.21 and Theorem 7.9.

We will use the following, which is proven in [WW17, Thm 3.6 and Cor 5.4]:

Lemma 2.38. *Let X be a compact nonpositively curved cube complex. Let $M \subset \pi_1 X$ be a f.g. virtually abelian subgroup, and suppose there does not exist an abelian subgroup $A \subset \pi_1 X$ with $[M : A \cap M] < \infty$ and $\mathrm{rank}(M) > \mathrm{rank}(A \cap M)$. Then there exists a local-isometry $Y \to R$ with Y compact and $\pi_1 Y$ mapping to M.*

Moreover, suppose X is a compact special cube complex. For every maximal abelian subgroup M of $\pi_1 X$, there exists a local-isometry $Y \to X$ with Y compact and $\pi_1 Y$ mapping to M.

Proof of Lemma 2.37. By Lemma 2.38, there is a compact nonpositively curved cube complex F_i and a local-isometry $F_i \to X$ such that $\pi_1 F_i$ maps to P_i. Since G is hyperbolic relative to virtually abelian subgroups, there exists r such that for any flat strip $[0, B] \times \mathbb{R} \subset \widetilde{X}$, if $\{0\} \times \mathbb{R} \subset F_i$ then $[0, B] \times \mathbb{R} \subset \mathcal{N}_r(F_i)$ (see, e.g., [HK05, Thm 1.2.3]). By Proposition 2.33, $\widetilde{E}_i = \widetilde{F}_i^{+s}$ has the desired property for sufficiently large s.

Note that $\widetilde{X}$ has the property that geodesic triangles are thin relative to flats, in the sense that there exists κ such that for any geodesic triangle pqr, the side pq lies in the union of qr, pr and some flat F. See [SW15, Prop 4.2] which is deduced from results in [DS05, Sec 8.1.3]. From this we deduce that there exists $s > 0$ such that for any geodesics $\gamma, \gamma' \subset \widetilde{X}$ if $\gamma \subset \mathcal{N}_r(\gamma')$ for some r, then either $\gamma \subset \mathcal{N}_s(\gamma')$ or γ and γ' both lie in $\mathcal{N}_s(F)$ for some maximal flat F of $\widetilde{X}$. This is proven by considering an arbitrary long parallelogram with vertices $\gamma(\pm n)$ and $\gamma'(\pm n)$ at distance $\leq s$ from each other. We subdivide the parallelogram into two long triangles, each of which must be either uniformly thin or uniformly thin relative to a flat. These flats must be the same for the two triangles (by isolated flats) and moreover the flats of arbitrarily large triangles must likewise agree. Hence if a flat is necessary at some point, it is the same flat for all parallelograms.

Apply Proposition 2.31 to obtain a convex cocompact core $\widetilde{Y}$ containing $\mathcal{N}_s(H\widetilde{x})$. To see that $\widetilde{Y}$ is superconvex, observe that for any geodesic $\gamma \subset \mathcal{N}_r(\widetilde{Y})$, there is a geodesic $\gamma' \subset \widetilde{Y}$ such that $\gamma \subset \mathcal{N}_r(\gamma')$ for some r. Hence either $\gamma \subset \mathcal{N}_s(\gamma') \subset \widetilde{Y}$ or $(\gamma \cup \gamma') \subset \mathcal{N}_s(F)$ for some flat F. But then $F \subset \widetilde{Y}$, and so $\gamma \subset \widetilde{Y}$. $\square$

Lemma 2.39. *Let $Y \subset X$ be a superconvex cocompact subcomplex of a locally finite CAT(0) cube complex with a proper group action. There is a nonnegative function $f : \mathbb{R} \to \mathbb{R}$ such that for any length $f(r)$ geodesic segment σ that lies in $\mathcal{N}_r(Y)$, the midpoint of σ lies in Y.*

Consequently, for any geodesic segment σ lying in $\mathcal{N}_r(Y)$, if σ' is obtained by removing the initial and terminal subsegments of length $\frac{f(r)}{2}$ then $\sigma' \subset Y$.

Proof. For simplicity (and sufficient utility) we give the proof in the setting where σ is a combinatorial geodesic and the combinatorial metric is being used.

Since Y is cocompact and X is locally finite, we see that $\mathcal{N}_r(Y)$ is cocompact for each $r \geq 0$. Suppose the statement of the lemma is false. Then there is a sequence $\{\sigma_i : i \in \mathbb{N}\}$ where each σ_i is a length $2i$ combinatorial geodesic segment in $\mathcal{N}_r(Y)$ whose midpoint m_i does not lie in Y. By cocompactness, and the pigeon-hole principle, we may translate and pass to a subsequence so that $\{m_i\}$ is a constant sequence. By König's infinity lemma, by passing to subgeodesics of a subsequence, we can assume that $\sigma_\infty = \cup \sigma_i$ is a bi-infinite geodesic. Hence $\sigma_\infty \subset Y$ by superconvexity, and thus $m_i \in Y$. $\qquad\square$

The following consequence of superconvexity plays a role in producing examples in Section 3.t by controlling wall-pieces.

Lemma 2.40. *Let Y be a superconvex cocompact subcomplex of the CAT(0) cube complex X. There exists $D \geq 0$ such that the following holds: Let $I_1 \times I_n \to X$ be a combinatorial strip. Suppose the base $\{0\} \times I_n$ of $I_1 \times I_n$ lies in Y, and suppose the distance between the endpoints of the base exceeds D, that is, $\mathsf{d}\big((0,0),(0,n)\big) \geq D$. Then $I_1 \times I_n$ lies in Y.*

Proof. Let $D = f(1)$ where $f : \mathbb{R} \to \mathbb{R}$ is the function in Lemma 2.39.

Let m be the midpoint of a geodesic in X joining the vertices of the rectangle not on the base hypothesized to lie in Y. That is, m is the midpoint of $(1,0) \in X$ and $(1,n) \in X$. We can assume m is a 0-cube by choosing one of the endpoints of the edges it lies in, if necessary. By Lemma 2.39, the point m lies in Y.

Note that m and $(1,0)$ and $(1,n)$ all lie on the same side of the hyperplane U dual to $I \times \{0\}$, whereas the base of the rectangle lies on the other side of that hyperplane.

Each hyperplane passing through the rectangle either cuts through the base of the rectangle and hence crosses Y, or is the hyperplane U which separates the top from the bottom, and hence separates m from a point in Y, and so crosses Y.

Since $Y \subset X$ is convex, Lemma 3.70 implies that any 0-cube not in Y is separated from Y by a hyperplane. It follows that the entire rectangle lies in Y. $\qquad\square$

Chapter Three

Cubical Small-Cancellation Theory

The goal of this chapter is to describe a "small-cancellation theory" for cubical presentations generalizing the usual small-cancellation theory for ordinary presentations.

3.a Introduction

To orient the reader towards our goals, we begin with parallel rough statements of the main theorem about classical small-cancellation diagrams and corresponding main theorem in cubical small-cancellation diagrams.

An *i-shell* in a disk diagram D is a 2-cell R with $\partial_p R = QS$, where the *outerpath* Q is a subpath of $\partial_p D$, and the *innerpath* S is the concatenation of exactly i maximal pieces in D. See Figure 3.1 for a diagram containing various i-shells. Any reduced disk diagram $D \to X$ satisfies the $C(6)$ condition provided that X is a $C(6)$ complex. In this context *reduced* means that there are no cancellable pairs of 2-cells, which holds in particular when $D \to X$ is minimal area.

Following the language in [MW02], the main theorem of the classical small-cancellation theory in the $C(6)$ case is summarized by:

Classical small-cancellation diagrams: Let D be a $C(6)$ disk diagram. Then either D is a single 0-cell or a single 2-cell, or D contains a total of at least 2π worth of the following types of positively curved cells along its boundary:

π for spurs, 0-shells, and 1-shells
$\frac{2\pi}{3}$ for 2-shells
$\frac{\pi}{3}$ for 3-shells

Moreover, if there are exactly two such features of positive curvature, then the diagram is a ladder. See Figure 3.1.

In our generalization, a disk diagram D is built from *cone-cells* together with 2-cubes, 1-cubes, and 0-cubes. The cone-cells play the role that 2-cells did in the classical case, but roughly speaking, the squares are subsumed in a thickened 1-skeleton. The diagram D is "reduced" if it is locally minimal area in a certain sense (no square bigons, no cornsquares on cone-cells), and it is "small-cancellation" if internal cone-cells are surrounded by many neighbors—generalizing the classical $C(6)$ condition that internal 2-cells have at least 6 sides.

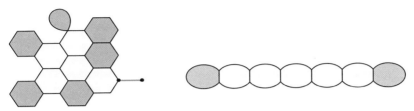

Figure 3.1. On the left in clockwise order we have a: 2-shell, 4-shell, spur, 3-shell, 1-shell, 1-shell, and 0-shell. On the right is a ladder.

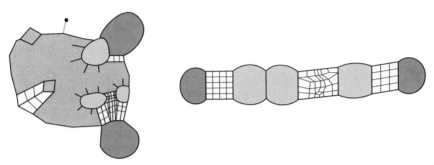

Figure 3.2. Two shells, two cornsquares, and a spur in a cubical small-cancellation diagram on the left. A cubical small-cancellation ladder is on the right.

The i-shells are replaced by positively curved cone-cells called *shells*, which we sometimes refer to with more precision as θ-*shells*, where θ denotes the curvature of the θ-shell (reversing the piece-focused terminology in[MW02]). Spurs continue to play the same role, but there is now another source of positive curvature: "cornsquares." Our main result is summarized by:

Cubical small-cancellation diagrams: Either D is a single 0-cell or cone-cell, or D contains at least 2π worth of positive curvature along its boundary of the form:

π for spurs
θ for θ-shells
$\frac{\pi}{2}$ for cornsquares

Moreover, if there are exactly two features of positive curvature, then D is a ladder. See Figure 3.2.

3.b Cubical Presentations

A *cubical presentation* $\langle X \mid \{Y_i\} \rangle$ has "generators" consisting of a connected nonpositively curved cube complex X, and "relators" consisting of a collection $\{Y_i \to X\}$ of local-isometries of connected nonpositively curved cube complexes.

The group of a cubical presentation is a quotient group $\pi_1 X / \langle\!\langle \{\pi_1 Y_i\} \rangle\!\rangle$ and is isomorphic to the fundamental group of a space X^* that we shall now discuss.

Associated to the cubical presentation $\langle X \mid \{\phi_i : Y_i \to X\} \rangle$ is a *coned-off space* X^* that is formed from the mapping cylinder $(X \cup \{Y_i \times [0,1]\})/\{(y_i, 1) \sim \phi_i(y_i)\}$ by identifying each $Y_i \times \{0\}$ to a single *conepoint* c_i. An application of the Seifert-Van Kampen lemma shows that the group of $\langle X \mid \{Y_i\} \rangle$ is isomorphic to $\pi_1 X^*$. The space X^* has a natural cell structure consisting of cubes in X together with "pyramids" consisting of cones of cubes. We refer to the image C_i of each $Y_i \times [0,1]$ as a *cone* of X^*, although more accurately, a cone is a map $C_i \to X^*$.

An ordinary presentation $\langle a_1, \ldots, a_s \mid R_1, \ldots, R_t \rangle$ whose relators are reduced and cyclically reduced contains the data for a cubical presentation: Its generators correspond to the nonpositively curved cube complex consisting of a bouquet of s circles, and its relators correspond to a collection of t immersed cycles. The group of an ordinary presentation is the quotient of the free group on the generators of the presentation modulo the normal subgroup generated by the relators, and this group is isomorphic to the fundamental group of the standard 2-complex of the presentation. This standard 2-complex would be isomorphic to the coned-off space if we subdivide the i-th 2-cell into $|R_i|$ distinct 2-simplices meeting around the conepoint which is a new 0-cell added at the center of the 2-cell.

Remark 3.1 (Language Abuse). We will use the pyramidal cells of the cones, or rather their 2-simplices, to initially discuss disk diagrams, but we will rarely use them later in the text. Instead we will especially focus on the "data" of the cubical presentations. While we shall often refer to each Y_i as a "cone," it really functions as the "attaching map" $Y_i \to X$ of the actual cones $C_i = \text{Cone}(Y_i)$, and perhaps it would have been more suitable to term each Y_i as a "relator."

When we refer to $\widetilde{X}^* = \widetilde{X^*}$, we sometimes have in mind only its underlying cube complex, which is the intermediate cover $\widetilde{X} \to \widetilde{X}^* \to X$ corresponding to the subgroup of $\pi_1 X$ equal to $\langle\!\langle \{\pi_1 Y_i\} \rangle\!\rangle$. However, sometimes we imagine $\widetilde{X}^*$ equipped with the various lifted cones $\{gY_i\}$. Formally $\langle \widetilde{X}^* \mid gY_i : \text{where } g, i \text{ vary} \rangle$ would give the data of another cubical presentation that covers X^*. We emphasize that the underlying cube complex of $\widetilde{X}^*$ plays the role of the 1-skeleton of the universal cover of a 2-complex, whereas the various gY_i play the role of the various 2-cells.

Definition 3.2 ($\text{Aut}_X(Y)$). Let $Y \to X$ be a local-isometry. Define $\text{Aut}_X(Y)$ to be the group of automorphisms $\phi : Y \to Y$ such that there is a commutative diagram:

$$
\begin{array}{ccc}
Y & \xrightarrow{\phi} & Y \\
& \searrow & \downarrow \\
& & X
\end{array}
$$

Note that $\mathrm{Aut}_X(\widetilde{Y})$ equals $\mathrm{Stab}_{\pi_1 X}(\widetilde{Y})$, and $\mathrm{Aut}_X(Y) \cong \left(\mathrm{Normalizer}_{\mathrm{Aut}_X(\widetilde{Y})}\right.$ $\pi_1 Y)/\pi_1 Y$. This is simplified in practice because of Convention 3.3.

The significance of $\mathrm{Aut}_X(Y)$ is that in the definition of "piece" we treat two lifts of $\widetilde{Y}$ as identical if they differ by an element of $\mathrm{Stab}_{\pi_1 X}(\widetilde{Y})$. This generalizes the way relators that are proper powers are treated in the classical case.

Convention 3.3. We shall always assume a cubical presentation $\langle X \mid \{Y_i\} \rangle$ has the property that $\pi_1 Y_i$ is normal in $\mathrm{Stab}_{\pi_1 X}(\widetilde{Y}_i)$ for each i. Equivalently, we shall assume that each element of $\mathrm{Stab}_{\pi_1 X}(\widetilde{Y}_i)$ projects to an element of $\mathrm{Aut}_X(Y_i)$.

3.c   Pieces

Let $\langle X \mid \{Y_i\} \rangle$ be a cubical presentation. An *abstract contiguous cone-piece* of Y_j in Y_i is an intersection $\mathcal{P} = \widetilde{Y}_j \cap \widetilde{Y}_i$ where either $i \neq j$ or where $i = j$ but $\widetilde{Y}_j \neq \widetilde{Y}_i$. So either we are considering an arbitrary lift of $\widetilde{Y}_j$ with $i \neq j$, or we are considering two distinct translates of the lift of the universal cover of Y_i.

An *abstract contiguous wall-piece* of Y_i is an intersection $N(H) \cap \widetilde{Y}_i$ where $N(H)$ is the carrier of a hyperplane H that is disjoint from $\widetilde{Y}_i$. Since we are not interested in empty pieces, we shall assume that H is dual to an edge with an endpoint on $\widetilde{Y}_i$.

Although we are primarily interested in contiguous cone-pieces and contiguous wall-pieces, it is natural to also consider noncontiguous pieces, since these arise naturally when we consider disk diagrams in cubical presentations. However, as we shall see, any abstract (noncontiguous) wall-piece or cone-piece is actually a subcomplex of an abstract contiguous wall-piece. To define noncontiguous pieces, outside of a disk diagram context, it will be convenient to utilize the following definition:

Definition 3.4 (Projections). Let U, V be subcomplexes of the CAT(0) cube complex $\widetilde{X}$. A cube c_v of V is the *projection of a cube of* U, if there is a cube c_u in U, and a combinatorial map $c \times I_m \to \widetilde{X}$ where c is a cube, such that $c \times \{0\} \mapsto c_u$ and $c \times \{m\} \mapsto c_v$ and no hyperplane intersecting V is dual to an edge $\{p\} \times I_m \to \widetilde{X}$ where p is a 0-cube of c.

The *projection* $\mathrm{Proj}_{\widetilde{X}}(U \to V)$ is the union of all cubes of V that are projections of cubes of U. The projection is *trivial* if it consists of a single 0-cube.

Two d-cubes c_1, c_2 are *parallel* in the CAT(0) cube complex $\widetilde{X}$ if they are dual to the same codimension-d hyperplane. That is, there is a combinatorial map $c \times I_m$ where c is a d-cube, and $c \times \{0\} \mapsto c_1$ and $c \times \{m\} \mapsto c_2$. All 0-cubes in $\widetilde{X}$ are parallel.

$\mathrm{Proj}_{\widetilde{X}}(U \to V)$ is the union of nearest cubes in V that are parallel to cubes of U.

Remark 3.5. In Definition 3.4, when U and V are convex, we can also assume that no hyperplane intersecting U is dual to an edge of $\{p\} \times I_m \to \widetilde{X}$. When c_v is a 0-cube, we can choose a shortest geodesic between c_v and U, and the hyperplanes dual to its edges are disjoint from U by Lemma 2.18. More generally, when c_v is a d-cube, there is a codimension-d hyperplane $H = \cap_{i=1}^d H_i$ whose carrier $N(H) = \cap_{i=1}^d N(H_i)$ has the property that $c_u \subset N(H) \cap U$ and $c_v \subset N(H) \cap V$. Note that $N(H) \cong c \times \widetilde{X}'$ where $\widetilde{X}'$ is itself a CAT(0) cube complex. We then choose $c \times I_m \to \widetilde{X}$ to be the composition $c \times I_m \to c \times \widetilde{X}' \subset \widetilde{X}$ where $I_m \to \widetilde{X}'$ is a geodesic between $U' = U \cap N(H)$ and $V' = V \cap N(H)$. No edge of I_m is dual to a hyperplane intersecting U' or V' by Lemma 2.18. Hence no hyperplane Z dual to $\{p\} \times I_m$ intersects U or V, by Lemma 2.10 applied to $Z, U, N(H)$ and $Z, V, N(H)$, after passing to the cubical subdivision of $\widetilde{X}$ so Z is a subcomplex.

Lemma 3.6. *Let U and V be convex subcomplexes of the CAT(0) cube complex $\widetilde{X}$. Then $\mathsf{Proj}_{\widetilde{X}}(U \to V)$ is convex, and each 0-cell $v \in \mathsf{Proj}_{\widetilde{X}}(U \to V)$ satisfies $\mathsf{d}(U, v) = \mathsf{d}(U, V)$.*

Proof. Let a, b be 0-cubes in $\mathsf{Proj}_{\widetilde{X}}(U \to V)$, and let γ_a and γ_b be paths showing a, b are projections of 0-cubes of V. By Remark 3.5, we may assume γ_a is a geodesic and each hyperplane dual to γ_a separates U, V. Likewise for γ_b. Thus each hyperplane dual to an edge of γ_a is dual to an edge of γ_b, and vice versa. Hence $|\gamma_a| = |\gamma_b|$.

Let P_U and P_V be paths in U, V so there is a disk diagram $D \to \widetilde{X}$ with $\partial_{\mathsf{p}} D = P_U \gamma_a P_V \gamma_b$, and choose P_U, P_V, D so that the area of D is minimized. Let α be a dual curve emanating from P_V, and observe that α must terminate on P_U, since no hyperplane crossing V can be dual to an edge of γ_a or γ_b. Let s be the rectangle carrying α. Every dual curve crossing α must have one end on γ_a and the other end on γ_b. Indeed, if it double crossed α then we could find a smaller area diagram by Lemma 2.3, and if it crossed P_U or P_V we would obtain a cornsquare, and there would be a lower area diagram by Lemma 2.6. It follows that $P_V \subset \mathsf{Proj}_{\widetilde{X}}(U \to V)$, since a side of s provides a path from U to V that isn't dual to any hyperplane intersecting V. We have shown that $\mathsf{Proj}_{\widetilde{X}}(U \to V)$ is connected, and moreover, that all 0-cubes of $\mathsf{Proj}_{\widetilde{X}}(U \to V)$ are at the same distance from U.

We now show that $\mathsf{Proj}_{\widetilde{X}}(U \to V)$ is locally-convex. Let $\{e_1, \ldots, e_d\}$ be 1-cubes of $\mathsf{Proj}_{\widetilde{X}}(U \to V)$ on a d-cube c_v of $\widetilde{X}$ meeting at a 0-cube $v \in c_v$. Let $\{H_i\}$ be hyperplanes dual to $\{e_i\}$, and for each i let $\dot{H}_i$ be the frontier of H_i containing v. Lemma 2.10 applied to $\{U, \dot{H}_1, \ldots, \dot{H}_d\}$ allows us to choose a 0-cube $u \in \dot{U} = U \cap \dot{H}_1 \cap \cdots \cap \dot{H}_d$. Let $\gamma \subset \cap_{i=1}^d \dot{H}_i$ be a shortest geodesic from u to v, and let $m = |\gamma|$. There is a map $c \times I_m \to \widetilde{X}$ with $c \times \{0\} \mapsto c_u$ and $c \times \{m\} \mapsto c_v$, and where $\{p\} \times I_m$ maps to γ. By Remark 3.5, let γ_v be a shortest geodesic from U to v such that no edge of γ_v is dual to a hyperplane of V. We claim that each hyperplane Z dual to an edge of γ is also dual to an edge of γ_v and hence does not cross V. Indeed, otherwise Z would intersect U, and letting $\dot{Z}$

denote the frontier of Z on the side of v, we see that $\dot{Z}$ intersects U and each $\dot{H}_i$, so Lemma 2.10 provides a 0-cube $z \in \dot{Z} \cap \dot{U} \cap \mathcal{I}(u, v)$. Since $z \neq u$, Lemma 2.17 gives $\mathsf{d}(v, z) < \mathsf{d}(v, u)$ which contradicts our choice of u.

Finally, convexity of $\mathsf{Proj}_{\widetilde{X}}(U \to V)$ follows from local convexity by applying Lemma 2.11 to the universal cover of the subcomplex $\mathsf{Proj}_{\widetilde{X}}(U \to V)$. $\qquad\square$

Let $\langle X \mid \{Y_i\} \rangle$ be a cubical presentation. An *abstract cone-piece* of Y_j in Y_i is a nontrivial component $\mathcal{P}$ of $\mathsf{Proj}_{\widetilde{X}}(\widetilde{Y}_j \to \widetilde{Y}_i)$, where we may regard $\widetilde{Y}_i \subset \widetilde{X}$ as a specific lift of the universal cover, and where $\widetilde{Y}_j \subset \widetilde{X}$ is an arbitrary lift when $i \neq j$, but where we require that $\widetilde{Y}_j \neq \widetilde{Y}_i$ when $i = j$. Each abstract cone-piece $\mathcal{P}$ comes equipped with a map $\mathcal{P} \to Y_i$, and is "tagged" by data indicating the relative positions $\widetilde{Y}_i, \widetilde{Y}_j$. One way to indicate the data is to choose a representative *connecting strip*. This is a rectangular strip $S \cong I \times I_n$ for some $n \geq 0$ with initial and terminal edges $I \times \{0, n\}$ mapping to 1-cells in the lifts of $\widetilde{Y}_i, \widetilde{Y}_j$ to $\widetilde{X}$ that are associated with the piece. There is leeway in the choice of strip. Assuming the basepoints are chosen appropriately, S corresponds to a double coset representative $\pi_1 Y_i g \pi_1 Y_j$.

Let $N(H)$ denote the carrier of a hyperplane H in $\widetilde{X}$ such that $H \cap \widetilde{Y}_i = \emptyset$. In parallel to the earlier definitions, an *abstract wall-piece* of H in Y_i is a nontrivial component $\mathcal{P}$ of $\mathsf{Proj}_{\widetilde{X}}(N(H) \to \widetilde{Y}_i)$. As above, there is some (possibly trivial) connecting strip S between $\widetilde{Y}_i$ and $N(H)$.

A *cone-piece* of Y_j in Y_i is a path $P \to \mathcal{P}$ in some abstract piece of Y_j in Y_i. Likewise a *wall-piece* of H in Y_i is a path $P \to \mathcal{P}$ in some abstract wall-piece. We will often refer to a *cone-piece in Y_i* to mean a path $P \to Y_i$ without any reference to the various choices involved, e.g., the lift of $\widetilde{Y}_j$ providing the abstract piece, and similarly we refer to *wall-pieces in Y_i*. We will almost always assume that our pieces are *nontrivial* in the sense that they have length ≥ 1. We will sometimes use the word *piece* to mean either a cone-piece or a wall-piece, and likewise for *abstract piece*. Finally, we use the term *contiguous wall-piece* and *contiguous cone-piece* as well as *contiguous piece* for paths in contiguous pieces.

Lemma 3.7 (Contiguous wall-pieces dominate noncontiguous pieces). *Suppose $\widetilde{Y}_j$ is disjoint from $\widetilde{Y}_i$. Then the abstract cone-piece of $\widetilde{Y}_j$ on $\widetilde{Y}_i$ is contained in a contiguous abstract wall-piece. Similarly, if $N(H)$ is the carrier of a hyperplane H that is not dual to an edge with an endpoint on $\widetilde{Y}_i$, then the abstract wall-piece of H on $\widetilde{Y}_i$ is contained in a contiguous abstract wall-piece.*

Proof. Let $\widetilde{A}, \widetilde{B}$ be convex subcomplexes of $\widetilde{X}$. If $\widetilde{A} \cap \widetilde{B} = \emptyset$ then we let γ be a shortest path from a vertex of $\widetilde{A}$ to a vertex of $\widetilde{B}$, and consider the hyperplane U dual to the first edge of γ. Since U separates $\widetilde{B}$ from $\widetilde{A}$ by Lemma 2.18, we have $\mathsf{Proj}_{\widetilde{X}}(\widetilde{B} \to \widetilde{A}) \subset \mathsf{Proj}_{\widetilde{X}}(N(U) \to \widetilde{A})$, and so it suffices to show that $\mathsf{Proj}_{\widetilde{X}}(N(U) \to \widetilde{A}) = N(U) \cap \widetilde{A}$.

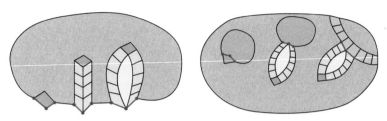

Figure 3.3. Cornsquares: The boundary of the diagram on the left has three illustrated cornsquares. The first is a genuine square with a corner on the boundary. The second is a remote square which could obviously be pushed to the boundary. The third is typical as the two rectangles from the square to the boundary bound a nontrivial square subdiagram. The diagram on the right illustrates cornsquares on cone-cells and on a rectangle.

We therefore focus on the case where $\widetilde{A} \cap \widetilde{B} \neq \emptyset$, in which case we show that $\mathsf{Proj}_{\widetilde{X}}(\widetilde{B} \to \widetilde{A}) = \widetilde{A} \cap \widetilde{B}$. Let σ be a geodesic from a 0-cube a in $\mathsf{Proj}_{\widetilde{X}}(\widetilde{B} \to \widetilde{A})$ to a 0-cube of $\widetilde{B}$, and suppose that $a \notin \widetilde{B}$, or equivalently, that $|\sigma| \geq 1$. Since $\widetilde{A} \cap \widetilde{B} \neq \emptyset$, we see that σ lies in $\widetilde{A}$. Then the first edge of σ is dual to a hyperplane crossing A, and hence a is not in $\mathsf{Proj}_{\widetilde{X}}(\widetilde{B} \to \widetilde{A})$, which is a contradiction. $\qquad\square$

Remark 3.8. Let us look ahead to see how pieces are related to diagrams (see Figure 3.7 and Section 3.e for the notion of "cone-cell").

As suggested by Figure 3.4, we will bound sizes of wall-pieces by bounding sizes of contiguous wall-pieces. Consider a wall-piece arising within a diagram as in the second diagram in Figure 3.4. Any cornsquare in the diagram with outerpath on the cone-cell could be pushed to be absorbed into the cone-cell (so wouldn't even exist in a reduced scenario). Any cornsquare with outerpath on one of the three bounding rectangles, could be pushed outside of the diagram so that we arrive at a grid (the third illustration) and one side of this grid lies on our cone-cell along the wall-piece of interest. We thus find that the remote wall-piece is also a wall-piece in a rectangle (in the fourth illustration) that is contiguous with the cone-cell.

This will enable us to produce examples by limiting the types of wall-pieces that must be examined to those arising from rectangles "based" on $\widetilde{A}$.

Moreover, by minimal area, we know that none of the squares in this contiguous rectangle can be absorbed into the cone-cell. This explains why we can ignore hyperplanes H that cross $\widetilde{A}$. Indeed, a fundamental property of CAT(0) cube complexes is that a hyperplane cannot *inter-osculate* with a convex subcomplex in the sense that it is dual to an edge of the subcomplex, and also dual to another edge with a single endpoint on the subcomplex (this is proven in Lemma 3.14). Thus, if the hyperplane H of the dual curve of a contiguous rectangle actually crossed $\widetilde{A}$, then each of the squares in this contiguous rectangle would be contained in $\widetilde{A}$, and hence there would be squares absorbable into the cone-cell and a violation of the minimality of area.

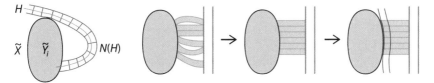

Figure 3.4. A noncontiguous wall-piece lies in a contiguous wall-piece.

3.d Some Small-Cancellation Conditions to Bear in Mind

The main hypothesis leading to a useful theory is an appropriate bound on lengths of pieces in each Y_i. There are various formulations: Let $\|Y_i\|$ denote the infimum of the lengths of essential closed paths in Y_i. Let $\operatorname{diam}(\mathcal{P})$ denote the diameter of an abstract piece $\mathcal{P}$. Let $\alpha > 0$ be a real number. The *absolute* $C'(\alpha)$ *condition* requires that $\operatorname{diam}(\mathcal{P}) < \alpha \|Y_i\|$ for each abstract piece $\mathcal{P}$ in each Y_i.

A more general (contextual) $C'(\alpha)$ *condition* requires that $|P|_{Y_i} < \alpha|R_i|$ whenever R_i is an essential cycle in Y_i that contains a piece P as a subpath. (Here we use $|P|_{Y_i}$ to denote the distance in $\widetilde{Y}_i$ between the endpoints of $\widetilde{P}$; thus the above inequality is implied by $\operatorname{diam}(\mathcal{P}) < \alpha|R_i|$ when $P \to Y_i$ factors through $\mathcal{P}$.) We will employ the absolute condition merely to verify the contextual condition. This contextual condition differs from the absolute condition both by focusing on pieces (which are paths) instead of abstract pieces, and by measuring pieces against cycles they occur in, instead of measuring them against arbitrary cycles. There is a related condition, $C^{\natural}(\alpha)$, which requires that $|P|_{Y_i} \leq \alpha \|Y_i^{\circledast}\|$. Its hypothesis is often the one that is functioning in our arguments in the later sections, but we have avoided structuring the exposition around it.

The *combinatorial* $C(n)$ *condition* requires that no essential closed path in Y_i is the concatenation of fewer than n pieces.

Remark 3.9 (Graded Theory). We will develop a graded small-cancellation theory in Sections 3.u and 3.y.

3.e Disk Diagrams and Reduced Disk Diagrams

Let $P \to X$ be a closed path that is nullhomotopic in X^*. We can consider various choices $\psi : (D, P) \to (X^*, X)$ of disk diagrams in X^* with boundary path P. Note that the 2-cells of D are either squares of X, or are triangles in some cone C_i of X^*. Moreover, since $P \to X$ avoids conepoints, each 0-cell a_i mapping to a conepoint c_i is internal to D, and hence the triangles adjacent to a_i (which must map to C_i) are grouped in cyclic collections meeting around a_i to form a subspace A_i that is a cone on its bounding cycle. We refer to each a_i as a

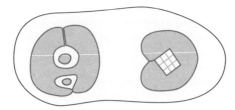

Figure 3.5. Without small-cancellation hypotheses, cone-cells might not have embedded boundary cycles.

Figure 3.6. Removing a cancellable pair of triangles around a conepoint.

conepoint of D. Note that $\partial A_i \to X$ factors through $Y_i \to X$. We refer to A_i as a *cone-cell* of D.

Remark 3.10. We caution the reader that ∂A_i might not embed, and moreover, A_i might not be an actual (sub)disk diagram. See Figure 3.5. However, by performing a very simple reduction, we can assume that the boundary path of A_i has no internal backtracks (see Figure 3.6). In the presence of adequate small-cancellation conditions, and minimal complexity properties of D, each A_i will be homeomorphic to a disk, as we will explain later in Section 4.a.

The *complexity* of a disk diagram $D \to X^*$ is the ordered pair

$$\big(\#\text{Cone-cells}(D), \#\text{Squares}(D)\big).$$

We use the *lexicographical order* where $(a_1, b_1) < (a_2, b_2)$ if either $a_1 < a_2$ or if $a_1 = a_2$ and $b_1 < b_2$. We say $D \to X^*$ has *minimal complexity* if there does not exist a lower complexity diagram $D' \to X^*$ with $\partial_{\mathsf{p}} D' = \partial_{\mathsf{p}} D$.

Definition 3.11 (Reduced). A disk diagram $D \to X^*$ in a cubical presentation is *reduced* if all of the following conditions hold. See Figure 3.8.

(1) There is no bigon in a square subdiagram of D.
(2) There is no cornsquare whose outerpath lies on a cone-cell of D.
(3) There does not exist a cancellable pair of squares.
(4) There is no square s in D with an edge on a cone-cell A mapping to the cone Y, such that $(A \cup s) \to X$ factors as $(A \cup s) \to Y \to X$.

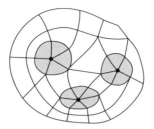

Figure 3.7. Cone-cells in $D \to X^*$.

(5) For each internal cone-cell A of D mapping to a cone Y, the path $\partial_p A$ is essential in Y.
(6) There does not exist a pair of combinable cone-cells in D.

A cone-cell A of D is *internal* if $\partial_p A$ does not traverse any edge of ∂D, otherwise it is *external*.

Note that $D \to X^*$ is reduced if it has minimal complexity. Indeed, a violation of any of the conditions for being reduced implies that there exists a new disk diagram with lower complexity but the same boundary path. It follows that reduced diagrams exist since diagrams with a given nullhomotopic boundary path exist. However, we caution that we shall sometimes consider reduced diagrams that are not of minimal complexity, namely, diagrams that have an external cone-cell which is replaceable. Finally, we note that Condition (3) is very rarely needed, and almost all statements function without it.

When we later discuss "graded small-cancellation theory," Condition (5) will instead state that $\partial_p A$ does not bound a disk diagram all of whose cone-cells have grade less than the grade of Y. For metric small-cancellation (see Definition 3.55) the condition is simplified to requiring that $\partial_p A \to Y^{\circledast}$ is essential.

We have discussed in Lemma 2.3 how a bigon in a square subdiagram allows one to reduce the number of squares. We refer to the square s of Condition (4) as an *absorbable square* and the act of replacing A by the cone-cell whose boundary path surrounds $A \cup s$ as *absorbing a square*. A cornsquare whose outerpath lies on a cone-cell A is defined analogously to a cornsquare whose outerpath lies on $\partial_p D$ and we refer to Definition 2.5 and the cornsquare in the middle of the right diagram in Figure 3.3. If there is a cornsquare whose outerpath lies on a cone-cell, then after shuffling, we obtain a new diagram with the same complexity but having an actual square with two consecutive edges on A, and thus by the local convexity of the cone Y that A maps to, we see that this square is absorbable. A cone-cell A is *replaceable* if $\partial_p A$ is nullhomotopic in its associated cone Y, in which case we can *replace* the cone-cell by a square diagram and decrease the complexity. A *combinable pair* of cone-cells in D is a pair of cone-cells A_1, A_2 mapping to the same cone Y of X^* such that $\partial_p A_1$ and $\partial_p A_2$ both pass through a vertex p of D, and regarding p as their basepoint, they map to closed paths at the same point of Y. We may thus combine these two cone-cells to a single

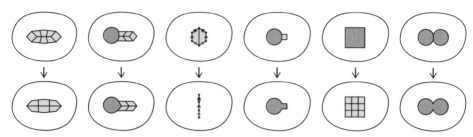

Figure 3.8. The six replacements above correspond to the way the complexity can be decreased if $D \to X^*$ is not reduced.

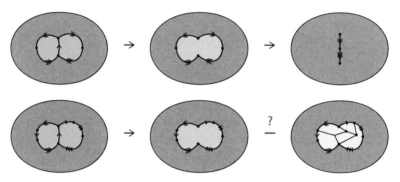

Figure 3.9. A pair of cone-cells can sometimes be combined and then possibly replaced by a square diagram.

cone-cell mapping to Y whose boundary path is the concatenation $\partial_\mathsf{p} A_1 \partial_\mathsf{p} A_2$. (We discuss a generalization of this in Lemma 3.16.) We sometimes use the term *cancellable pair* to cover two squares that can be cancelled, or a square that can be absorbed, or a pair of combinable cone-cells (which might not be replaceable afterwards).

Remark 3.12 (Cancelling versus Combining). A "classical" cancellable pair in a disk diagram consists of a pair of 2-cells meeting along a 1-cell in a "mirror image" fashion. The two 2-cells are removed together with an open arc where they meet, and this is "zipped up" so only a "surgical scar" is left behind. In our more general situation, this is broken up more explicitly into two steps: Combining the two cone-cells to a single cone-cell, and then possibly replacing this cone-cell by a square disk diagram. For a cancellable pair, this disk diagram has zero area. (See Figure 3.9.) In general, the new cone-cell might not be replaceable by a square diagram since its boundary path might be essential in the cone. Moreover, even if it is replaceable, it will usually require some squares. The boundary path of the new cone-cell arising in the classical case is a nullhomotopic path in a circle. However in our generalization, even in the case where cones are circles, we might combine two cone-cells wrapping around

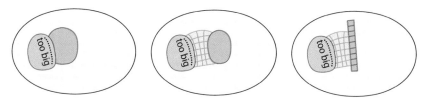

Figure 3.10. A path in the diagram whose endpoints are too distant to be a genuine piece will lead to a reduction in the complexity of the diagram. The left case is similar to the classical small-cancellation theory. It is a degenerate version of the middle case, which is itself covered by the rightmost case.

the same circle p and q times, so the boundary of the new cone-cell might be essential as it wraps around $(p - q)$ times.

While Lemma 3.16 can be used to reduce the complexity of a diagram whenever two cone-cells share an impossibly big piece, we must use a somewhat different argument to reduce the complexity when a cone-cell has an impossibly big wall-piece.

In this case, we first rearrange the square part of the diagram so that our original rectangle is pushed past any cornsquares that lie between it and the cone-cell (this uses Lemma 2.21). We then find that the piece of our original rectangle actually lies in a piece of some other rectangle whose dual curve is dual to a 1-cell adjacent to our cone-cell. It is thus on such pieces that we place small-cancellation hypotheses. This discussion is described more formally in the following:

Lemma 3.13 (Contiguous wall-pieces dominate). *Let D be a diagram, and let A be a cone-cell in D, and let S be a rectangular strip of D, let $Q = q_1 q_2 \ldots q_k$ be a path on ∂A, and suppose each rectangular strip R_i starting at the edge q_i of Q ends at a square of S, so that R_1 ends at the first square, and R_k ends at the last square. Suppose that the squares of R_i, R_j are distinct for $i \neq j$. Let E be the subdiagram bounded by R_1, Q, R_k, and S, and suppose E is a square diagram.*

There exists a new diagram E' with the same boundary path as E, such that E' contains a rectangular strip T' whose first square lies on q_1, and such that each rectangular strip Q'_i emerging from q_i passes through T'.

Proof. See Figure 3.11. We push the strip across any crossing dual curves of E that cross S and the dual curve of either Q_1 or Q_k. This is done in Lemma 2.21 where the path P (there) corresponds to a path $U_1 Q U_k$ where U_1, U_k are external paths of R_1, R_k.

This gives us a new diagram E' and a new strip S'. The strip T' emerging from the square adjacent to q_1 has the desired property. ☐

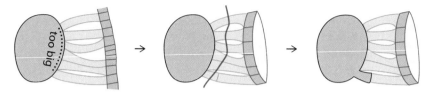

Figure 3.11. A big wall-piece yields an adjacent hyperplane with a big piece, which must then cross $\widetilde{Y}_i$ and hence yields an absorbable square as in Lemma 3.15.

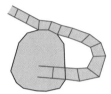

Figure 3.12. A subcomplex inter-osculating with a hyperplane.

We thus see that any hyperplane is behind a hyperplane whose convex hull touches the cone-cell. Consequently, if non-crossing hyperplanes have bounded projections, we get a bound on sizes of pieces unless there is a square absorption.

The following result rules out the behavior in Figure 3.12. It holds by applying Lemma 2.18 to the hyperplane separating $\widetilde{Y}$ from a 0-cell at distance 1 from it.

Lemma 3.14 (No inter-osculating hyperplanes). *Let $\widetilde{Y} \subset \widetilde{X}$ be a convex subcomplex. Let U be a hyperplane that osculates with $\widetilde{Y}$ in the sense that it has a dual 1-cell with exactly one 0-cell in $\widetilde{Y}$. Then U cannot also be dual to a 1-cell that is contained in $\widetilde{Y}$.*

Consequently:

Lemma 3.15. *Let A be a cone-cell of D that maps to the cone Y. Suppose that D contains a square s with an edge e on ∂A, such that in a lift to $\widetilde{X}$ of s with $\widetilde{e}$ on ∂A, an adjacent 1-cell $\widetilde{e}'$ of $\partial \widetilde{s}$ is dual to a hyperplane U that is also dual to a 1-cell that lies in $\widetilde{Y}$. Then $\widetilde{s}$ lies in $\widetilde{Y}$.*

Consequently, s can be absorbed into A to reduce the complexity of D.

The following generalization that *remotely combines* cone-cells is not part of the main exposition, as we shall only need to combine cone-cells when their boundary paths pass through a common vertex (and actually have a nontrivial common subpath). It is illustrated in Figure 3.14:

Lemma 3.16 (Remotely combining cone-cells). *Suppose that A_i and A'_i map to the same cone Y_i, and suppose there is an embedded path $P \to D$ whose endpoints*

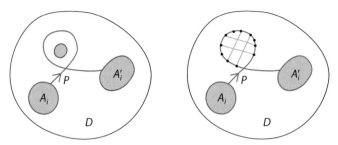

Figure 3.13. The path P wraps around a cone-cell in the diagram on the left, so we cannot combine A and A'. There is no problem on the right, as P wraps around a square diagram.

are the basepoints of ∂A_i and $\partial A'_i$, such that $P \to X$ is homotopic in X to a path $P' \to X$ that factors through Y_i. Then there exists a lower complexity disk diagram with the same boundary path.

Remark 3.17. While Lemma 3.16 still holds if P self-intersects around a square subdiagram as on the right of Figure 3.13, it may fail if P self-intersects around a subdiagram containing other cone-cells, as in the diagram on the left. This second type of self-intersection cannot exist in a minimal complexity disk diagram under the small-cancellation condition we will examine below.

Proof. Let $K \to X$ be a square disk diagram for the path-homotopy between P and P'. (A path-homotopy in X implies a path-homotopy in Y_i by π_1-injectivity.) Cut D along P, and insert two copies of K doubled along P'. Consider the subdiagram $B = A_i \cup P_i \cup A'_i$. Observe that we can cut along ∂B, and substitute a cone from a single conepoint associated with Y_i. We have thus reduced complexity. $\square$

Having not pursued the idea of remotely combining cones, the following idea is not used in the sequel, but the idea of ranking the cone-cells will be pursued and developed later in the context of graded small-cancellation theory in Section 3.u.

Remark 3.18 (Generalized remote combination). When the path P of Lemma 3.16 is only homotopic into Y within X^*, then the replacement provides new cone-cells in the new diagram as in Figure 3.15. We have drawn a picture of a *drum* which is a thickened disc that has a rectangle around the outside, and a disk diagram with cones on each membrane. The new diagram is obtained from the old by placing the drum along the connecting square diagram, and then pushing upwards through the drum. See Figure 3.15.

Assign a linear *ranking* to the cones $Y_1, Y_2, \ldots$ and let X^*_r denote $\langle X \mid Y_1, \ldots, Y_r \rangle$. Suppose any (hyperplane) path $P \to \widetilde{X}^*$ that starts and ends on the

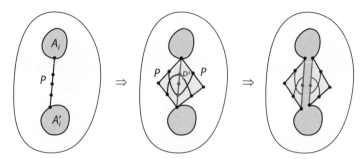

Figure 3.14. Remotely combining cone-cells: If D contains two cone-cells A_i, A_i' that are joined by a path P that is path-homotopic in X to a path P' mapping to the same cone as A_i, A_i', then we can replace A_i, A_i', P by a single cone-cell at the expense of possibly adding some squares.

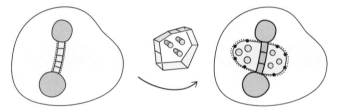

Figure 3.15. Pushing across a drum to obtain a graded replacement.

same lift of some Y_i is actually path-homotopic with a path $P' \to Y_i \subset \widetilde{X}^*$ where the path-homotopy actually occurs in X_{i-1}^*. Under this hypothesis, we can use a *ranked complexity* on disk diagrams that: assigns a rank of 0 to each square; counts cone-cells according to the ranks of their cones; and uses the ordering where higher rank cells have the greatest value. In this case, the combination above actually decreases complexity.

3.f Rectified Disk Diagrams

Consider the disk diagram $(D, P) \to (X^*, X)$. An *open cone-cell* A of D associated to some conepoint a is the union of open cells whose closures intersect a. Accordingly, A is the union of a together with a sequence of open 1-simplices and 2-simplices cyclically arranged about a. The *external boundary* of $I_n \times [-\frac{1}{2}, \frac{1}{2}]$ is $I_n \times \{-\frac{1}{2}, +\frac{1}{2}\}$, and its *initial* and *terminal* 1-cells are $\{0\} \times [-\frac{1}{2}, \frac{1}{2}]$ and $\{n\} \times [-\frac{1}{2}, \frac{1}{2}]$. A *rectangle* of D is a combinatorial map $I_n \times [-\frac{1}{2}, \frac{1}{2}] \to D$ that is injective except perhaps at the external boundary. We note that a rectangle contains (part of) a *dual curve* of D at $I_n \times \{0\}$, and that the above injectivity requirement implies that this dual curve embeds. Unless specifically indicated, we shall always assume that a rectangle is *nondegenerate* in the sense that $n \geq 1$.

We shall now assume that D is *nontrivial* in the sense that it doesn't consist of a single 0-cell. Thinking of D as embedded in S^2, we will regard its complementary 2-cell A_∞ as the *cone-cell at infinity*.

We now assume there is a *linear ordering* on the cone-cells of D, and we shall assume that A_∞ is last, so for instance, we can label the cone-cells $A_1, \ldots, A_m$, A_∞. Any choice will be adequate for our purposes, though distinct choices may lead to somewhat different rectified cell structures for D. We will also choose a linear ordering on the 1-cells in the attaching map of each cone-cell. Again, different choices will lead to slightly different results, but for instance, a first 1-cell together with a counterclockwise ordering is adequate.

Combining these choices and using the lexicographical ordering, we obtain a linear ordering on the set $\mathcal{S}$ of 1-cells in the attaching maps of all the cone-cells. (This is essentially an ordering on a subset of the 1-cells of D except where both sides of a 1-cell lie on a cone-cell, in which case it will not play an important role since it will only yield a degenerate rectangle below.) We use these orderings to determine which rectangles of D are *admitted*: Beginning with the first 1-cell in our ordering, we apply the following procedure to each 1-cell e as we proceed through the sequence. Traveling outwards from e away from its cone-cell, there is a maximal rectangle consisting of a sequence of distinct squares. This rectangle will either terminate at another 1-cell on the boundary of some cone-cell (possibly the same cone-cell that contained its initial 1-cell e), or it will terminate on the external boundary of some rectangle that was previously admitted, or it will terminate on its own external boundary, specifically, on the boundary of one of its own squares that had appeared earlier in the sequence. Following our ordering of 1-cells, we proceed in this way until each 1-cell in $\mathcal{S}$ lies in a (possibly degenerate) rectangle. We note that a degenerate rectangle may arise when either our 1-cell appears in two ways among the attaching maps (so it has a cone-cell on each side) or when it has a cone-cell on one side, and a square belonging to a previously admitted rectangle on the other side. In this case, we do not add any rectangle, but we will continue to refer to such a 1-cell as a degenerate rectangle. Each admitted rectangle has a linear orientation directed from its initial 1-cell to its terminal 1-cell.

We now describe a *rectified disk diagram* $\bar{D}$ that we will use to study D. Let $I_n \times (-\frac{1}{2}, \frac{1}{2})$ denote the *internal part* of the (possibly degenerate) rectangle $I_n \times [-\frac{1}{2}, \frac{1}{2}]$. Let E denote the subspace of D consisting of the union of each open cone-cell and the internal part of each admitted rectangle. Note that the internal part of a degenerate rectangle is an open 1-cell.

We refer to Figure 3.17.

For each trivial component (meaning a single 0-cell) of $D - E$ we have a 0-cell in our rectified disk diagram $\bar{D}$. For each nontrivial component F of $D - E$ we have a cycle in our rectified disk diagram that is an embedded copy of $\partial_p F$. In a certain sense the "0-skeleton" of $\bar{D}$ is the disjoint union of boundary cycles of components of $D - E$. More precisely, the 0-skeleton of $\bar{D}$ consists of the 0-cells in the disjoint union of boundary paths. And we have also added the 1-cells within the nontrivial boundary cycles.

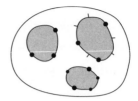
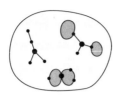

Figure 3.16. Nontrivial singular diagrams of $D - E$ "inflate" to the shard 2-cells of $\bar{D}$.

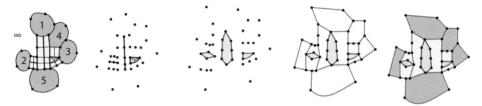

Figure 3.17. A diagram D, the complement $D - E$, the "0-skeleton" of $\bar{D}$ together with the shards, its "1-skeleton," and its "2-skeleton."

For each admitted rectangle, its initial and terminal 1-cells are also open 1-cells of $\bar{D}$. These are identical for a degenerate rectangle.

We have formed the entire 1-skeleton of $\bar{D}$. Its 0-skeleton consists of two types of 0-cells: those from trivial and nontrivial boundary cycles of components F of $D - E$. Its 1-skeleton consists of two types of 1-cells: the 1-cells of each $\partial_p F$ and the intial/terminal 1-cells of admitted rectangles.

We shall now describe the three types of 2-cells of $\bar{D}$. For each nontrivial component F of $D - E$ we add an open 2-cell $\bar{F}$ called a *shard*.

For purposes of exposition, we focus on a situation where each shard is simply-connected. According to Remark 3.19 this is guaranteed in the setting of a minimal complexity diagram under small-cancellation hypotheses.

The boundary path of F is a combinatorial path in D, that we arranged to be an embedded cycle in the 1-skeleton of $\bar{D}$, and we attach an open 2-cell $\bar{F}$ along this cycle. We emphasize that the boundary path of each nontrivial shard is the concatenation of external subpaths of admitted rectangles. The process of inflating shards is illustrated in Figure 3.16.

Thus, an open 1-cell e of D (that is not an initial or terminal 1-cell of an admitted rectangle) will contribute zero, one, or two open 1-cells to $\bar{D}$ according to the number of ways it lies along the external boundary of an admitted rectangle. A 0-cell v of D will contribute d distinct 0-cells of $\bar{D}$ provided that v lies in d ways on the external boundaries of admitted rectangles.

The other two types of 2-cells are more straightforward: We add the admitted nondegenerate rectangles as 2-cells and then the cone-cells.

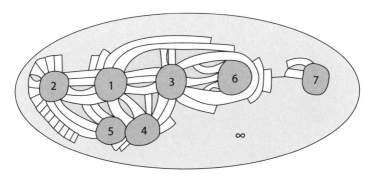

Figure 3.18. The rectified diagram: cone-cells in orange, rectangles in yellow, shards in blue. Shards corresponding to components containing no squares are not indicated.

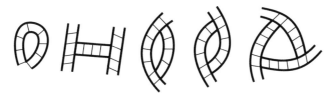

Figure 3.19. Some impossible configurations of admitted rectangles. The first diagram is impossible under the assumption that there are no cone-cells inside and this square diagram has minimal area. The other diagrams are impossible because of the ordered logic of admitted rectangles.

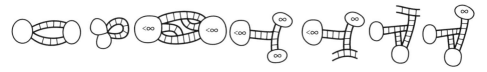

Figure 3.20. More impossible configurations of admitted rectangles and cone-cells. The first diagram is missing some admitted rectangles. The second diagram is impossible assuming that the bounded region is a shard and has minimal area.

There is a map $\bar{D} \to D$ that is combinatorial on the 1-skeleton, on the rectangles and on the cone-cells (after subdivision), but is not combinatorial on the shards (since the image of a shard can be a singular subdiagram of D).

We close this section by describing some configurations that cannot arise in a rectified diagram $\bar{D}$. Some impossible configurations are illustrated in Figure 3.19. The leftmost such configurations excludes many other cases from consideration. The rightmost configurations are impossible because of the ordered way in which we admitted rectangles. Additional impossible configurations are illustrated in Figure 3.20.

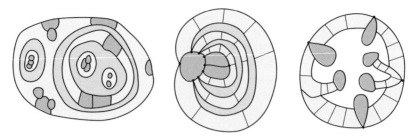

Figure 3.21. Two innermost non-simply-connected complementary components on the left, a simple possibility for D' in the middle, and a complicated possibility for D' on the right (partially illustrated).

Remark 3.19 (Back to the Future). Under sufficiently strong small-cancellation conditions which we will later examine, we will be able to see that the components of $D - E$ are square disk diagrams. Specifically, we must rule out the possibility that some component is not simply-connected. If there is a non-simply-connected component then there is an *innermost* one, in the sense that it doesn't separate the boundary of the diagram from an even deeper component that is not simply-connected. This innermost component would bound a rectified disk diagram D' (containing at least one conepoint) with E' as above, such that $D' - E'$ has simply-connected components. The boundary of D' has a particularly restrictive nature. Assuming that D is reduced (which ensures the same for D'), together with the small-cancellation conditions we will impose, will imply by Theorem 3.46 that D' does not exist.

The subdiagram D' has the property that it contains at least one conepoint, and since it is innermost, $D' - E'$ has simply-connected components, and so the construction (that we are in the midst of) can proceed. However, the boundary of D' consists entirely of subpaths of external boundaries of rectangles. See Figure 3.21 for some possibilities. The conclusion of Theorem 3.46 proscribes this, since $\bar{D}'$ has neither a shell, nor an explicit cornsquare (see below).

Cornsquares of D can be "hidden" by the rectification $\bar{D}$, as when they are exposed, at least one of the rectangles originates from the cone-cell at infinity. We are led to the following definition:

An *explicit cornsquare* in $\bar{D}$ arises from a pair of rectangles R_1, R_2 that bound a shard, such that R_2 has a terminal 1-cell on the external boundary of R_1, and $\partial_{\mathsf{p}} D$ has a subpath $e_1 e_2$, where e_1 is the initial or terminal 1-cell of R_1, and e_2 is the initial 1-cell of R_2. In this case, the cornsquare is "at" the square of R_1 where R_2 terminates, and $e_1 e_2$ is its outerpath.

Definition 3.20 (Standard and Semi-standard). A rectification $\bar{D}$ of a disk diagram D is *semi-standard* if it arises from an ordering of 1-cells on the boundaries of cone-cells that is induced from a linear ordering of the cone-cells with the infinite cone-cell last, followed by a cyclic linear ordering of each cone-cell.

Figure 3.22. In a semi-standard rectification the scenario on the left implies one of the four subsequent scenarios. The three scenarios on the right are impossible.

The rectification is *standard* if it is semi-standard and has the following additional property: For each cone-cell C, if some admitted rectangle with one end on an edge of $\partial_p C$ has its other end on an edge of a cone-cell, then the admitted rectangle at the first edge of $\partial_p C$ has its other end on an edge of a cone-cell.

While a semi-standard ordering is induced by several variable parameters, a standard ordering of the 1-cells of the various $\partial_p C$ must be chosen with a bit more care: One admits rectangles of the i-th cone-cell after choosing its first 1-cell (and orientation of its ∂_p). If the first rectangle terminates on the external boundary of an admitted rectangle, but the rectangle at some other edge e of $\partial_p C$ has its other end on a boundary edge of a cone-cell, then we rechoose the first edge to be e.

Lemma 3.21. *Suppose $\bar{D}$ is a semi-standard rectification of the disk diagram D. Let e_1, e_3 be adjacent 1-cells on the boundary of a cone-cell C, and suppose e_1, e_3 are initial edges of admitted rectangles R_1, R_3 that are oriented outwards from C. Suppose R_3 ends on a rectangle R_2 which implicitly crosses R_1. Then e_1 is the first 1-cell of ∂C and e_3 is the last. Consequently, there can be at most one such configuration for each cone-cell C.*

We refer the reader to Figure 3.22 which indicates the hypothesis situation on the left diagram, and the possible outcomes in the 2nd, 3rd, 4th, and 5th diagrams. The last three diagrams indicate configurations that are consequently impossible.

Proof. Let e_2 denote the initial 1-cell of R_2, and observe that $e_1 \leq e_2 < e_3$. Consequently, by our hypothesis on the structure of the ordering of 1-cells on boundaries of cone-cells, we see that e_2 lies on ∂C. Moreover, the cyclical orientation is then determined, and hence since e_1, e_3 are adjacent with $e_1 < e_3$ and the ordering increasing in the other direction from e_3, we see that e_1 is first as illustrated. □

Remark 3.22 (Rectification of Annular Diagrams). For an annular diagram $D \to X^*$, the rectification process follows the same procedure, and there are few differences from the disk diagram case. The primary difference is that it is possible (see Figure 3.23) to have a single annular shard separating the boundary

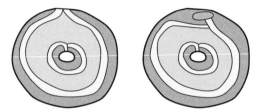

Figure 3.23. It is possible for there to be an annular shard in the rectification of an annular diagram. Such a shard would have two internal angles that are each either 0 or $\frac{\pi}{2}$. Hence such a shard would have curvature $\leq -\pi$.

cycles. However, the curvature of such a shard is $\leq -\pi$, and hence if such an annular shard existed, then D is forced to have features of positive curvature by Theorem 3.23. We note, however, that Remark 3.19 continues to apply to show that any shard not separating the boundary cycles of an annular diagram is simply-connected when X^* is a small-cancellation complex.

3.g   Gauss-Bonnet Theorem

Let E be a combinatorial complex embedded in the sphere. Suppose an *angle* consisting of a real number $\sphericalangle(c)$ is assigned to each corner of each 2-cell of E.

The *curvature* of a 0-cell v of E is defined to be

$$\kappa(v) = 2\pi - \sum_{c \in Corners(v)} \sphericalangle(c) - \pi\chi(\mathrm{link}(v)). \qquad (3.1)$$

Note that when E is locally a surface without boundary at v then $\mathrm{link}(v)$ is a circle and so the final correction term vanishes yielding $\kappa(v) = 2\pi - \sum_{c \in Corners(v)} \sphericalangle(c)$. Similarly, when E looks like a surface with boundary at v, we have $\kappa(v) = \pi - \sum_{c \in Corners(v)} \sphericalangle(c)$. The *curvature* of a 2-cell f of E with $|f|$ sides is the sum of the angles of the corners of f minus the angle sum for a Euclidean polygon with the same number of corners:

$$\kappa(f) = \sum_{c \in Corners(f)} \sphericalangle(c) - (|f| - 2)\pi$$

Alternatively, letting $\mathrm{defect}(c) = \mathrm{defect}(\sphericalangle(c)) = \pi - \sphericalangle(c)$ we have:

$$\kappa(f) = 2\pi - \sum_{c \in Corners(f)} \mathrm{defect}(c). \qquad (3.2)$$

A simple computation verifies the following well-known fact lying at the heart of small-cancellation theory (see for instance [MW02]).

Theorem 3.23 (Combinatorial Gauss-Bonnet). *Let E be a finite 2-complex embedded in the sphere, with an angle assigned at each corner of each 2-cell. Then:*

$$2\pi\chi(E) = \sum_{v \in Vertices(E)} \kappa(v) + \sum_{f \in 2\text{-}cells(E)} \kappa(f).$$

3.h Assigning the Angles

We shall now assign angles to the corners of the 2-cells of $\bar{D}$ aiming to obtain nonpositive curvature at internal 0-cells, at rectangles, and at shards, and also at the cone-cells provided that certain small-cancellation conditions are met. There are actually two main angle assignments we will discuss here: The first is the *split-angling*, where the angle assigned to a corner of a cone-cell will depend upon the neighboring cells at that corner—its personality changes to suit its surroundings. The second angle assignment is the *type-angling*, where cone-cells are treated a bit more like regular Euclidean polygons. In both cases, small-cancellation conditions we will examine later will provide nonpositive curvature of cone-cells. However, whereas the shards are automatically nonpositively curved in the split-angling, we will have to hypothesize this (later) for the type-angling.

As we have just indicated, in the type-angling the angle assignments depend upon a *typing* of the cone-cells, but the reader should focus especially on the case where each cone-cell has type 6 except for the infinite cone-cell A_∞ whose type is ∞. The typing is a map from the set of cone-cells to $\mathbb{N} \cup \{\infty\}$, and in practice, this will be induced by a map from the set of relators $\{Y_i\}$ to $\mathbb{N}$.

The rectangles of $\bar{D}$ have the usual Euclidean angles—we assign an angle of $\frac{\pi}{2}$ to each of their constituent squares, and so the four corners at the initial and terminal 1-cells have angle $\frac{\pi}{2}$ and all other corners have angle π.

The reader should think of the ∞ cone-cell A_∞ as having an angle of π at each of its corners.

Split-angling assignment to cone-cell corners: We now describe the split-angling. All internal corners of cone-cells are assigned an angle of $\frac{\pi}{2}$ with several exceptions that we list and illustrate in detail below. Though there might appear to be a dizzying array of cases, they are actually simple degenerations and variations of several very natural choices modeled on familiar Euclidean scenarios, and we refer the reader to Figure 3.24.

Consider a pair of adjacent 1-cells on the boundary of a cone-cell. We say the pair of associated admitted rectangles *end in parallel on an admitted rectangle*, if the (possibly degenerate) subdiagram bounded by these rectangles is a shard: In particular, there is no cone-cell inside it. (It is possible that the terminal 1-cells of the rectangles are not adjacent but they will be in the following case.) The pair of admitted rectangles *end in parallel on a cone-cell* if there is a shard bounded by these two rectangles, and the edges they end on are adjacent to

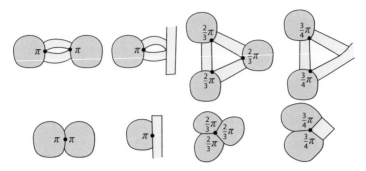

Figure 3.24. Internal cone-angles for the split-angling are modeled by the four cases above: All variants of the four cases above are provided in Figures 3.25, 3.26, 3.27, and 3.28.

Figure 3.25. Parallel rectangles ending on the same cone-cell give π angles at cone-cell corner.

each other. This includes the degenerate case where one or both rectangles are degenerate.

The qualifier *implicitly* indicates a related situation where one of the rectangles travels through a square alongside which the target rectangle terminates. (See the bottom layers of Figures 3.26 and 3.28.)

An angle of $\frac{\pi}{2}$ is assigned to cone-cell corners except for the following cases:

(1) π when the associated admitted rectangles end in parallel on the same cone-cell. (See Figure 3.25.)
(2) π when the associated admitted rectangles (implicitly) end in parallel on the same admitted rectangle. (See Figure 3.26.)
(3) $\frac{2}{3}\pi$ when they end on a pair of finite cone-cells that are also adjacent along an admitted (possibly degenerate) rectangle, such that these three bound a (possibly degenerate) triangular shard. (See Figure 3.27.)
(4) $\frac{3}{4}\pi$ when they (implicitly) end on a rectangle-cone-cell combination. (See Figure 3.28.)
(5) We emphasize that $\frac{\pi}{2}$ is assigned in Cases (3) and (4) when either of the other cone-cells is infinite. (See Figure 3.29.)
(6) 0 is assigned at a corner when the rectangles terminate at a singular vertex on C_∞. (See Figure 3.30.)

We note that the choice of a 0 angle at a singular vertex is not very critical— we could have allowed $\frac{\pi}{2}$ here without much effect on the theory. In particular

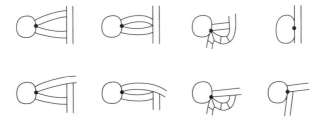

Figure 3.26. Rectangles (implicitly) parallel to the same admitted rectangle have π angle at the cone-cell corner.

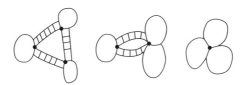

Figure 3.27. Three adjacent cone-cells.

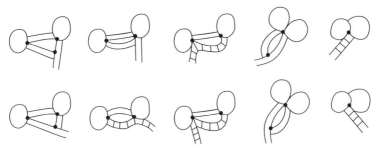

Figure 3.28. $\frac{3}{4}\pi$ is assigned to the internal angle between edges whose emerging rectangles end in parallel on a cone-cell/rectangle pair.

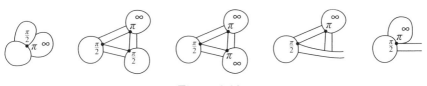

Figure 3.29.

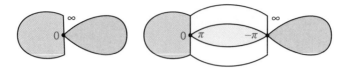

Figure 3.30. An angle of 0 is assigned to a corner facing a singular 0-cell.

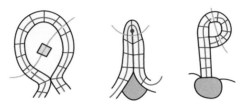

Figure 3.31. Monogonal shards do not exist if D is reduced.

Theorem 3.31 would hold equally well with a $\frac{\pi}{2}$ choice here. But the 0 angle permits more general coverage in the hypothesis of Theorem 3.43.

Curiously, the bigonal shard in this case takes angles of $-\pi, \pi$.

Definition 3.24 (Type-Angling). In the *type-angling*, for a type g cone-cell, we place internal angles of $\frac{g-2}{g}\pi$ in place of the angles $\frac{2}{3}\pi$ and $\frac{3}{4}\pi$ listed above in the split-angling. Note that this gives an angle of π when $g = \infty$.

In order to guarantee nonpositive curvature at vertices and at shards, we shall later be forced to make certain hypotheses on the possible collections of typed cone-cells and squares surrounding a shard.

3.i   Nonpositive Curvature of Shards

In this section we will use minimality properties of the diagram D to conclude that shards have nonpositive curvature. In particular we will assume that there are no bigons in the square subdiagrams, and we will assume that there are no cornsquares on cone-cells, which holds under the blanket assumption that D is reduced.

The corners of a shard are of two types: *dull* corners which lie along a pair of edges on the external part of a single admitted rectangle, and otherwise *sharp* corners. The dull corners are not very interesting and we simply assign to them an angle of π.

The sharp corners of a shard will have an angle of $\frac{\pi}{2}$ except in a few special cases that we discuss: Observe that there is no *monogonal shard* consisting of a nontrivial shard bounded by a single admitted rectangle (see Figure 3.31), for then either the shard contains no squares and then there would be a backtrack along its boundary, or there would be some square inside the shard and hence a

Figure 3.32. Some simple bigonal shards. The second diagram can occur with an infinite cone-cell. The third and fourth diagrams are impossible.

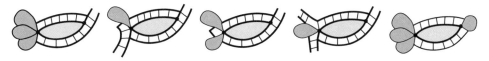

Figure 3.33. Three additional squares/cone-cells at one corner, and one at the other.

bigon of dual curves in the nonrectified D, and this would permit a square area reduction by Lemma 2.3.

Consider a *bigonal shard* having exactly two sharp corners and hence exactly two admitted rectangles with external edges along its boundary. The 0-cells at the sharp corners where these rectangles meet already have at least two $\frac{\pi}{2}$ corners (from squares in these admitted rectangles, which have an edge along the sharp corners). If each of these 0-cells has at least two additional "admitted squares" (i.e., from within admitted rectangles) and/or cone-cells, then there is already a total angle sum of at least 2π around each and so we can assign an angle of 0 to each sharp corner of our shard.

It is impossible for one of these corners to have no further cone-cell or admitted square alongside it, for then the two admitted rectangles would be the same (they would continue around that corner), and we would have a monogonal shard—which was excluded earlier.

Suppose each of these sharp corners has exactly one additional square/cone-cell around it. There are essentially four scenarios illustrated in Figure 3.32, and the third and fourth are impossible. In the first case, we can assign an angle of 0 to each sharp corner, since the corresponding internal corners of the cone-cells are both π. The second case is the most interesting: For a finite cone, it is impossible since it provides a cornsquare on a cone-cell. In the case where it is an infinite cone A_∞, we assign angles of $+\frac{\pi}{2}$ at the corner opposite the square, and assign an angle of $-\frac{\pi}{2}$ at the corner opposite the cone-cell at infinity. The positive curvature at the latter corner will be important later on, and the situation will be referred to as a *cornsquare* on ∂D.

Let us now consider the possibility where one sharp corner has exactly one further cone-cell/square, and the other sharp corner has three or more. In this case, we put an angle of $+\frac{\pi}{2}$ on the corner with one additional cone-cell/square, and we put an angle of $-\frac{\pi}{2}$ on the corner with three or more. A non-exhaustive selection of the scenarios is illustrated in Figure 3.33.

We now consider the possibility of one additional cone-cell/square at one sharp corner, and two additional cone-cells/squares at the other. First let us

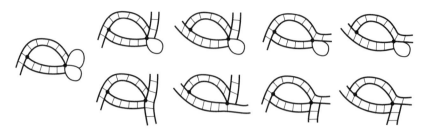

Figure 3.34. Square at one sharp corner, two cells at other.

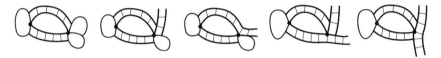

Figure 3.35. Cone-cell at one sharp corner, two additional cells at other.

suppose there is a single additional square at one sharp corner. The various conceivable possibilities are illustrated in Figure 3.34, but some of these cannot arise. In the (possible) cases we assign an angle of $+\frac{\pi}{2}$ at the left corner of the bigon and $-\frac{\pi}{2}$ at the right corner of the bigon. The bottom four cases are impossible: The leftmost three do not exist since an admitted rectangle must have an initial 1-cell on a cone-cell. The rightmost bottom case does not exist because there is no consistent way of ordering the initial 1-cells of these three rectangles. In the top four cases, the internal angle of the corner of the cone-cell at our bigon is π since this is the case of two rectangles (implicitly) terminating on the same rectangle. We note however that the second case above (from the left) cannot exist because it has a rectangle without an initial 1-cell on a cone-cell.

We now suppose there is a single additional cone-cell at one sharp corner, and two additional cone-cells and/or squares at the other. The five possibilities that can arise are illustrated in Figure 3.35. In the fourth and fifth cases an angle of 0 is assigned to each sharp corner, as these two cases correspond to a π angle for the corner of the cone-cell since its rectangles (implicitly) terminate on another admitted rectangle. For the split-angling, we assign $\frac{2}{3}\pi$ to each cone-cell corner in the first case and hence use $\pm\frac{\pi}{3}$ for the sharp bigon corners, and in the second and third cases we assign $\frac{3}{4}\pi$ to each cone-cell corner and hence use $\pm\frac{\pi}{4}$ for the sharp bigon corners.

We emphasize that except for the cases discussed above, the corners of a shard receive a $\frac{\pi}{2}$ angle.

We now consider a *triangular shard* which is bounded by three admitted rectangles. As in the bigonal shard case, if one of its sharp corners has two additional cells, then we can assign an angle of 0 to that sharp corner, and an angle of $\frac{\pi}{2}$ to the remaining two corners. (Note that there must be at least one additional cell at each corner.) So, let us assume that each corner has only one

Figure 3.36. Locally possible configurations of triangular shards with a single additional cell at each corner.

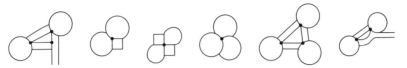

Figure 3.37. In most cases, when there is an infinite cone around a shard, then the shard (or one of its boundary vertices) will have negative curvature depending on whether we use the split-angling or the type-angling.

additional cell. There are four cases according to whether there are zero, one, two, or three cone-cells.

Case zero, where each of these additional cells are squares is easily seen to be impossible: up to symmetry the locally possible situations are indicated in the first and second diagrams (counting from the left) in Figure 3.36. In the first there is no possible consistent way of ordering the initial 1-cells of the bounding admitted rectangles, and in the second there is a rectangle with no initial 1-cell on a cone-cell.

The possibilities for the case where there is exactly one cone-cell are illustrated in the third, fourth, and fifth diagrams (counting from the left) in Figure 3.36. The fifth diagram is impossible. The third and fourth diagrams are cases where the rectangles emerging from adjacent edges of a cone-cell (implicitly) end on the same rectangle. Thus the internal angle of this cone-cell is π, and we can assign an angle of zero to the sharp corner of the shard at the cone-cell, and angles of $\frac{\pi}{2}$ to each of the other two sharp corners.

The possibilities for the cases where there are two or three cone-cells are illustrated in the sixth and seventh diagrams (from the left) in Figure 3.36. In the sixth case, in the split-angling, we use $\frac{3}{4}\pi$ for the internal cone-cell corners and $\frac{\pi}{4}, \frac{\pi}{4}, \frac{\pi}{2}$ for the sharp triangular shard corners. In the seventh case, we use $\frac{2}{3}\pi$ for each internal cone-cell corner, and $\frac{\pi}{3}$ for each shard corner.

Remark 3.25. There are a variety of shards with infinite cone-cells at the corners such that the natural angle assignments lead to negative curvature of the shards and/or the vertices. Some of these are described in Figure 3.37.

Remark 3.26 (Nonpositive Shard for Type-Angling). For the type-angling, we assign an angle of $\frac{g_i-2}{g_i}\pi$ to each type g_i cone-cell corner, and then assign complementary angles of $\frac{2}{g_i}\pi$ to the opposing sharp corners so that the 0-cells have zero

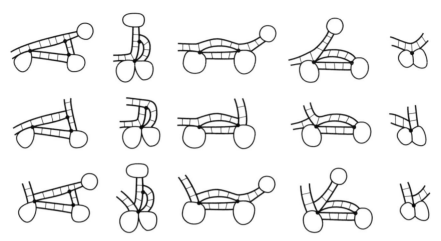

Figure 3.38. Some cases requiring a $-\frac{\pi}{2}$ angle. These figures should only be viewed by readers who are 18 years or older.

curvature. Hence there is a condition to check on the types of the cells around a shard which would imply that its curvature is nonpositive. For instance, in the first case of Figure 3.35, the condition is $\frac{2}{g_1}\pi + \frac{2}{g_2}\pi + \frac{2}{g_3}\pi \geq 2\pi$, but in the second and third cases, the condition is $\frac{2}{g_1}\pi + \frac{2}{g_2}\pi + \frac{2}{4}\pi \geq 2\pi$ (as a square has been substituted for a cone-cell). Conditions ensuring nonpositive curvature of other shards in the type-angling case are similar.

3.j   Tables of Small Shards

Roughly twenty cases arise in the table in Figure 3.39. These are organized as they are obtained from the four figures on the left by contracting one or three parts of the rectangles around the triangle. Note that contracting two of these three sides results in a rectangle around a monogonal shard, and this is not possible when D is reduced, since it has no square bigons.

One of the two ways that a 0-cell on the boundary of the cone-cell at infinity can have positive curvature is when it is the corner of a square. The other way is more interesting as it involves an opposing bigon with angles $\pm\frac{\pi}{2}$ (see Figure 3.40) and leads to a cornsquare. In a certain sense, one case is a degenerate version of the other; i.e., the actual square case is a degenerate version of the general cornsquare case.

3.k   Nonpositive Curvature of Cone-Cells via Small-Cancellation

We now discuss a metric small-cancellation condition on cone-cells in $\bar{D}$ that implies the nonpositive curvature of each of the cone-cells. There are similar but more complex combinatorial conditions that are a bit more general. The

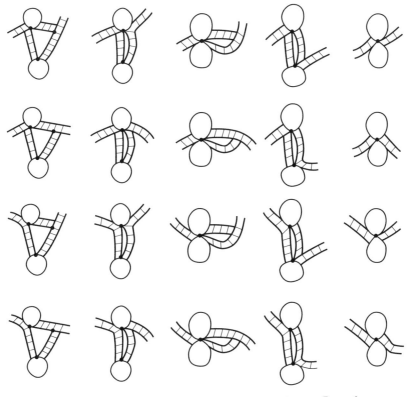

Figure 3.39. Some more cases requiring a 0 or $-\frac{\pi}{2}$ angle

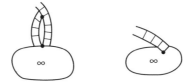

Figure 3.40. Cornsquares: The nontrivial shard on the left requires angles of $\pm\frac{\pi}{2}$ in the bigon. Both cases have curvature $\frac{\pi}{2}$ at the vertex on the boundary of A_∞.

conditions are couched in terms of the "projections" (within a disk diagram) of cone-cells, rectangles and combinations of these onto a given cone-cell in $\bar{D}$. We refer to these as *cone-pieces* and *wall-pieces* in the boundary path of a cone-cell. Each such piece consists of a subpath of the boundary path of the cone-cell, that is a concatenation of edges, all of whose rectangles end in parallel on the same other cone-cell (respectively, other rectangle). That they end in *parallel* means that they bound a shard (possibly together with the rectangle where they end).

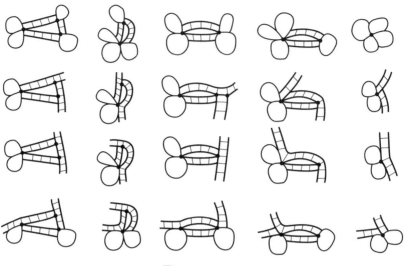

Figure 3.41.

The curvature of a cone-cell with p sides is $\sum_{i=1}^{p} \lessdot_i - (p-2)\pi$ or equivalently, it is $2\pi - \sum (\pi - \lessdot_i) = 2\pi - \sum \text{defect}(\lessdot_i)$.

The internal angle $\lessdot$ between 1-cells of a cone-cell is related to the destination of the corresponding rectangles. If they both (implicitly) end on the same rectangle, or on the same cone-cell, then the internal angle is π so its defect is 0. We obtain positive deficiencies when there is a *transition* between the ends of these rectangles in the sense that they don't end in parallel on the same cone and don't end (implicitly) on the same rectangle. Note that in the split-angling, the internal cone-cell angles that are not equal to π are $\frac{2}{3}\pi, \frac{3}{4}\pi, \frac{\pi}{2}$. In the type-angling, they are $\frac{g-2}{g}\pi$.

Definition 3.27 (Destination Notation). Consider the sequence of cone-cells and rectangles around a cone-cell. We combine together emerging rectangles with similar destination, and use the following notation $CWCCWWWW$ for a sequence of cone-cell and rectangle destinations around our cone-cell. We will abuse the notation in the following way: Each C or W will also denote a path in the boundary of our cone-cell that is the concatenation of the initial or terminal edges of rectangles that start or end on the cone-cell C or rectangle W. We refine this further by using notation $\hat{C}$ and $\check{C}$ to mean that the rectangles are oriented from our cone-cell upwards, or from the destination cone-cell towards our cone-cell. The notation $\bar{C}$ means that the rectangles are degenerate, and $\tilde{C}$ denotes the cone-cell at infinity. We also use $C\overrightarrow{W}$ to mean that $\overrightarrow{W}$ emerges from C, and likewise $\overleftarrow{W}C$. Finally, we put numbers between terms: $C_{\frac{\pi}{3}}C$ and $C_{\frac{\pi}{2}}W$ and $W_{\frac{\pi}{2}}W$ and so forth, to indicate the defect of the angle in our cone-cell at the corresponding internal corner where there is a destination transition. We refer to Figure 3.42 for an example of this notation.

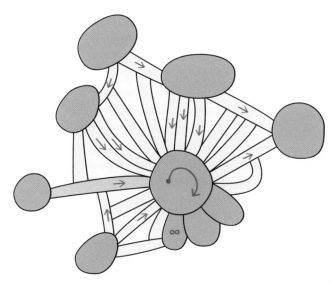

Figure 3.42. Clockwise from the yellow rectangle edge, expressing defects associated to the split-angling we have:

$$\check{C}_{\frac{\pi}{4}}\check{C}_{\frac{\pi}{4}}\overleftarrow{W}_{\frac{\pi}{2}}\overrightarrow{W}_{\frac{\pi}{4}}\check{C}_{\frac{\pi}{4}}\overrightarrow{W}_{\frac{\pi}{4}}\hat{C}_{\frac{\pi}{4}}\bar{C}_{\frac{\pi}{3}}\bar{C}_{\frac{\pi}{2}}\tilde{C}_{\frac{\pi}{2}}\overleftarrow{W}_{\frac{\pi}{4}}\check{C}_{\frac{\pi}{4}}\overrightarrow{W}_{0}$$

Lemma 3.28 (Consecutive 0 defect transitions). *Let $\bar{D}$ be a semi-standard rectification of a reduced diagram D. A cone-cell has at most one sequence of four consecutive pieces with defect 0 at each transition.*

Furthermore, if $\bar{D}$ is a standard rectification, then there do not exist three consecutive defect 0 piece transitions, and all three consecutive piece transitions with defect 0 are of the form $(W_0 C_0 W)$.

Proof. There is at most one $W_0 W$ by Lemma 3.21, and hence we can exclude the cases $W_0 W_0 W_0 C$ and $C_0 W_0 W_0 W$.

Observe that $C_0 W_0 C$ is impossible since it means that an admitted rectangle has both ends on admitted rectangles. With more precision, observe that $C_0 W$ can only arise in the form $C_0 \overleftarrow{W}$ in the sense that the admitted rectangle associated to W has its terminal edge on the admitted rectangle between our cone-cell and the cone-cell appearing in the notation $C_0 W$. (This reexplains why $C_0 W_0 C$ cannot arise, as the W would be directed both to the left and to the right.) Similarly, $C_0 W_0 W$ would be of the form $C_0 \overleftarrow{W}_0 \overleftarrow{W}$ and $W_0 W_0 C$ would be of the form $\overrightarrow{W}_0 \overrightarrow{W}_0 C$. We are thus able to exclude the scenario $C_0 W_0 W_0 C$. The remaining scenarios are of the form $W_0 W_0 C_0 W$ and $W_0 C_0 W_0 W$, and at most one of these can occur since as noted above there is at most one $W_0 W$.

Finally, neither of these are possible with the additional assumption on the order of admitted rectangles, since if $W_0 W_0 C_0 W$ occurs, then it means there exists a rectangle (which might have already been admitted) between our cone-cell and another, and the first edge in the ordering of our cone-cell must either have a rectangle terminating on it or terminating from it. However, Lemma 3.21 applied to the W_0W subsequence insists that the first edge has an admitted rectangle terminating on a rectangle, and we obtain a contradiction. $\qquad\square$

We now establish a condition that leads to a quick conclusion:

Theorem 3.29. *Let $D \to X^*$ be a reduced disk diagram. Consider a standard rectification $\bar{D}$. Using the split-angling, an internal cone-cell C has nonpositive [negative] curvature unless $\partial_{\mathsf{p}}C$ is the concatenation of ≤ 15 pieces arising in $\bar{D}$.*

The idea of the proof of Theorem 3.29 is to pick up some defect for each transition between distinct pieces. Since the smallest nonzero defect is $\frac{\pi}{4}$, it looks like nonpositive curvature holds unless $\partial_{\mathsf{p}}C$ is the concatenation of at most 7 pieces. However, the flaw with this argument is that there can be a 0 defect at certain transitions, for instance, the transition between a cone-piece and a wall-piece where rectangles implicitly end on the same rectangle. We refer the reader to Theorem 3.40 and Problem 3.41. We remedy this flaw by "clustering" pieces together in a certain way to reach the slightly weaker conclusion.

Definition 3.30 (Cluster). Consider an internal cone-cell C in a rectified reduced diagram $\bar{D}$. Its boundary cycle $\partial_{\mathsf{p}}C$ is the concatenation of wall-pieces and cone-pieces, denoted as in Definition 3.27. (We caution that the C's in the notation are associated to pieces with other cone-cells in the diagram.) A *cluster* is a maximal sequence of such pieces where the defect between consecutive terms is 0. We decompose $\partial_{\mathsf{p}}C$ into clusters which we indicate with parentheses, e.g.,

$$(W_0C_0W)_{\frac{\pi}{2}}(C)_{\frac{\pi}{3}}(C)_{\frac{\pi}{4}}(W_0W)_{\frac{\pi}{2}}(W_0C)_{\frac{\pi}{3}}(C)_{\frac{\pi}{4}}(C_0W)_{\frac{\pi}{2}}(W_0C_0W)_{\frac{\pi}{2}}.$$

Proof of Theorem 3.29. Consider the decomposition of $\partial_{\mathsf{p}}C$ into clusters. As $\bar{D}$ is a standard rectification, Lemma 3.28 ensures that each cluster has length at most 3.

Suppose $\partial_{\mathsf{p}}C$ is not the concatenation of fewer than 24 pieces in $\bar{D}$.

Then there are at least 8 clusters, and traveling clockwise, we associate defects with the subsequent cluster, so there is a defect of at least $\frac{\pi}{4}$ for each cluster, giving a total defect of $\geq 2\pi$.

If there are more than 8 clusters, then the defect sum would exceed 2π and hence the curvature would be negative. The only way to have exactly 8 clusters is if each cluster is of the form $(\overrightarrow{W}_0C_0\overleftarrow{W})$. However, the transition between two consecutive clusters is of the form $(\overrightarrow{W_0C_0W})_{\frac{\pi}{2}}(W_0C_0W)$ so there is actually $\frac{\pi}{2}$ contribution to the defect instead of just $\frac{\pi}{4}$. Hence the curvature of an internal cone-cell is negative.

The above proof functions under the assumption that $\partial_p C$ is not the concatenation of fewer than 16 [or 17] pieces with very little change. The key is to observe that certain clusters actually lie behind fewer pieces than immediately apparent. Specifically: $(W_0 C_0 W)$ is actually the concatenation of at most two wall-pieces, and $(W_0 C)$ and $(C_0 W)$ lie behind a single wall-piece. (Note that $(\overrightarrow{W}_0 \overleftarrow{W})$ still requires two pieces.) We thus find that each cluster is associated to the concatenation of at most two pieces but contributes a defect of at least $\frac{\pi}{4}$. Consequently, when $\partial_p C$ requires at least 16 pieces, there would be at least 8 clusters, and hence a total defect of $\geq 2\pi$. Similarly, the total defect would exceed 2π if $\partial_p C$ requires at least 17 pieces.   □

The following is more difficult and depends on results in Section 3.1.

Theorem 3.31. *Let $D \to X^*$ be a reduced disk diagram. Consider a standard rectification $\bar{D}$. Using the split-angling, an internal cone-cell C has nonpositive [negative] curvature unless $\partial_p C$ is the concatenation of at most 11 [12] pieces arising in D.*

While Theorem 3.29 employed pieces that explicitly arise in the rectified diagram $\bar{D}$, the hypothesis in Theorem 3.31 is instead concerned with arbitrary pieces in $\partial_p C$. Such pieces might not arise from the same rectification of $\bar{D}$.

Proof. There is an orientation on each wall-piece determined by the orientation of the rectangle. By Lemma 3.34, it suffices to prove the result for internal cone-cells that do not self-collide. Consequently, by Lemma 3.35, we can assume the angle defect between successive wall-pieces is $\frac{\pi}{2}$.

   The angle defect between an outgoing wall-piece and a cone-piece is $\geq \frac{\pi}{4}$. If the angle defect between an incoming wall-piece and a cone-piece is 0, then the cone-piece must actually be contained in the wall-piece, otherwise the angle defect is $\geq \frac{\pi}{4}$ there as well. The angle defect between consecutive cone-pieces is $\geq \frac{\pi}{3}$.

   Accordingly, we associate $\frac{\pi}{6}$ to the outgoing vertex of each wall-piece. For a cone-piece, either it contains angle defects of one of $\frac{\pi}{4}, \frac{\pi}{2}, \frac{\pi}{3}$ on each side and we associate $\frac{\pi}{12}$ to it from each side (the most interesting case here is $\frac{\pi}{4} = \frac{\pi}{6} + \frac{\pi}{12}$), or it has an angle of 0 on one side, in which case it lies in an extension of the incoming wall-piece on that side and we include it as part of that (extended wall-piece). If this happens on each side of the cone-piece, then we include it on just one of these sides. We illustrate this extension in Figure 3.43.

   Thus $\partial_p C$ is the concatenation of wall-pieces with associated angle defects of $\frac{\pi}{6}$, and cone-pieces with associated angle defects of $\geq \frac{\pi}{6}$.

   If $\partial_p C$ is not the concatenation of ≤ 11 such pieces, then there are at least 12 such pieces, and hence a total defect of $\geq 2\pi$. The argument for negative curvature is similar, as there are at least 13 such pieces, and hence a total defect of $\geq \frac{13}{6}\pi$.   □

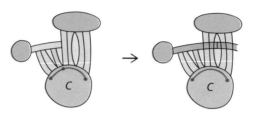

Figure 3.43. An explicit wall-piece extends to include a cone-piece.

Figure 3.44. Diagram $\bar{D}_C$ obtained from a self-colliding cone-cell C. (Ignore the extra rectangle portion in the first and third cases.)

Looking ahead, we obtain the following conclusion about small-cancellation presentations for which we refer to Definition 3.47. [See Definition 3.55 for the graded $C'(\frac{1}{12})$ condition.]

Theorem 3.32. *If X^* is $C(12)$ [respectively, $C(13)$], then X^* is a [negatively curved] small-cancellation presentation. In particular, the former holds when $|P| \leq \|Y\|$ whenever P is a piece in a cone Y, and the latter holds when X^* is a [graded] $C'(\frac{1}{12})$ cubical presentation.*

Proof. When X^* is [graded] $C'(\frac{1}{12})$, reduced diagrams $D \to X^*$ have the property that internal cone-cells are not the concatenation of fewer than 13 pieces. Hence the internal cone-cells have nonpositive curvature by Theorem 3.31. The nonpositive curvature of the shards in the rectification of a reduced disk diagram is explained in Section 3.i. □

3.1 Internal Cone-Cells That Do Not Self-Collide

Definition 3.33 (Self-Collision). Let C be a cone-cell in a rectified diagram $\bar{D}$. We say C *self-collides* if there is either an admitted rectangle whose initial and terminal 1-cell lies on C, or there is an admitted rectangle whose initial 1-cell lies on C, but whose terminal 1-cell lies on the external boundary of another admitted rectangle with initial or terminal 1-cell on C. (These two admitted rectangles could be the same.) That C does not self-collide generalizes the idea of 2-cells in a disk diagram having boundary paths that do not traverse the same edge twice.

We refer the reader to Figure 3.44 for various self-collisions. We also include the case of a rectangle starting and ending on C, and a rectangle of C that collides with itself. We omit the case of a self-collision associated to a backtrack in $\partial_{\mathsf{p}}C$, since we work under the assumption that all attaching maps of cone-cells are immersions.

The following gives a criterion for avoiding self-colliding cone-cells:

Lemma 3.34 (Avoiding self-collisions). *Consider a rectified disk diagram $\bar{D}$ arising from a diagram D. Suppose angles have been assigned to the corners of $\bar{D}$ so that: all rectangle corners have angle $\frac{\pi}{2}$, nontrivial shards and internal 0-cells have nonpositive curvature, and each cone-cell angle is $\leq \pi$.*

If nonpositive curvature holds for all internal cone-cells of $\bar{D}$ that do not self-collide, then no cone-cell of $\bar{D}$ self-collides. Hence all cone-cells have nonpositive curvature.

Note that the closure of an internal cone-cell could contain 0-cells of ∂D.

Proof. An internal cone-cell C of $\bar{D}$ that self-collides determines a diagram $\bar{D}_C$ obtained from the immersed subdiagram subtended by C together with the one or two admitted rectangles where the collision takes place. The angle assignment on $\bar{D}_C$ is induced from the assignment on $\bar{D}$.

Consider a minimal counterexample, that is, a self-colliding cone-cell C such that $\bar{D}_C$ contains a minimal number of cone-cells. Minimality implies that, with the possible exception of C itself, no cone-cell of $\bar{D}_C$ self-collides. Indeed, if B is a cone-cell in $\bar{D}_C$ that self-collides then this provides a smaller diagram $\bar{D}_B$, since C is not contained in $\bar{D}_B$, but colliding rectangles of B are unable to encircle C since either C is an external cone-cell of $\bar{D}_C$ or the colliding rectangle(s) of C surround C, but cannot be crossed by the colliding rectangle(s) of B.

Suppose C is an external cone-cell of $\bar{D}_C$. To simplify the computation, we reassign angles of C at the interior of $\partial C \cap \partial D_C$ to be π, however, if we had retained the original angles we would obtain the same result after cancellation with the positive curvatures at the associated external 0-cells. Observe that $\kappa(C) \leq \pi$, and at most one other 0-cell on ∂D has nonzero curvature (of $\pm \frac{\pi}{2}$), and all other cells have nonpositive curvature, and so Theorem 3.23 is contradicted.

Suppose C meets $\bar{D}_C$ at a single 0-cell v. Let $\triangleleft$ be the internal angle of C at v. Then $\kappa(v) = -\triangleleft$, and $\kappa(C) = 2\pi - \sum \text{defect} \leq 2\pi - (\text{defect}(\triangleleft)) \leq \pi + \triangleleft$. All other curvatures are nonpositive except possibly for one that is $\pm \frac{\pi}{2}$. Hence, again, Theorem 3.23 is contradicted.

Finally, suppose C is internal. Then $\kappa(C) \leq 2\pi$, but there is a 0-cell u with $\kappa(u) = -\frac{\pi}{2}$. Hence, again, Theorem 3.23 is contradicted. $\square$

Lemma 3.35 (Cone-cell ordered rectified diagrams). *Let $\bar{D}$ be a rectified diagram induced from an ordering on the bounding 1-cells of cone-cells of D that arises from an ordering of the cone-cells with C_∞ last. And suppose there are no self-colliding cone-cells (e.g., because Lemma 3.34 applies).*

If the 1st subdiagram of Figure 3.22 is contained in $\bar{D}$, then it can only arise from the 2nd, 3rd, 4th or 5th subdiagrams of Figure 3.22.

Figure 3.45. Using the argument of Lemma 3.21, if the middle scenario arises, then the order of admitted rectangles ensures that the top rectangle must also emerge from the cone as on the right, and hence there is a self-collision. Thus if there are no self-collisions, as guaranteed by Lemma 3.34, then only the left scenario can arise.

Note that the initial hypothesis of Lemma 3.35 is a bit milder than requiring that $\bar{D}$ be a standard rectification.

Proof. This follows from the logic of the rectified diagram construction of Section 3.f. Indeed, let e_2 be the initial edge of the horizontally positioned admitted rectangle, let e_1 be the initial edge of the rectangle it terminates on, and let e_3 be the initial edge of the rectangle terminating on the horizontally positioned rectangle. Then $e_1 \leq e_2 \leq e_3$. Thus e_2 must lie on $\partial_{\mathsf{p}}C$, and is either equal to e_1 or intermediate between e_1 and e_3. $\square$

Corollary 3.36. *Suppose the cone-cell C does not self-collide in the rectification $\bar{D}$ of D. There does not exist a sequence $R_1, R_2, \ldots, R_k$ of outgoing admitted rectangles whose initial edges form a subpath $e_1 e_2 \cdots e_k$ of the boundary path $\partial_{\mathsf{p}}C$, such that $R_1, R_2, \ldots, R_{k-1}$ end in parallel on an admitted rectangle R that itself terminates on the external boundary of R_k. See Figure 3.45.*

We use the notation introduced in Definition 3.27 for the following:

Remark 3.37. Suppose there are no self-colliding cone-cells. Then $\overleftarrow{W}_0\overleftarrow{W}$ and $\overrightarrow{W}_0\overleftarrow{W}$, and $\overrightarrow{W}_0\overrightarrow{W}$, and $\overrightarrow{W}_0\hat{C}$, and $\hat{C}_0\overleftarrow{W}_0$ cannot arise because of Corollary 3.36.
$\overleftarrow{W}_0\overrightarrow{W}$ cannot arise since rectangles initiate on a cone-cell, not on another rectangle. Likewise $\overleftarrow{W}_0\hat{C}$ and $\overleftarrow{W}_0\check{C}$ and of course $\overleftarrow{W}_0\bar{C}$ cannot arise, nor the same in opposite order. $\overleftarrow{W}_{\frac{\pi}{4}}\hat{C}$ and $\hat{C}_{\frac{\pi}{4}}\overrightarrow{W}$ cannot arise following the proof of Corollary 3.36.

Continuing with the constraints indicated in Remark 3.37, we now provide a table with minimal defects:

Table 3.38.

(1) $\tilde{C}_{\frac{\pi}{2}}C$

(2) $\overleftarrow{W}_{\frac{\pi}{2}}\hat{C}$ and $\hat{C}_{\frac{\pi}{2}}\overrightarrow{W}$

(3) $\overleftarrow{W}_{\frac{\pi}{4}}\bar{C}$ and $\bar{C}_{\frac{\pi}{4}}\overrightarrow{W}$

(4) $W_{\frac{\pi}{2}}W$

(5) $C_{\frac{\pi}{4}}C$

(6) $\bar{C}_{\frac{\pi}{3}}\bar{C}$

(7) $C_{\frac{\pi}{4}}\overleftarrow{W}$ and $\overrightarrow{W}_{\frac{\pi}{4}}C$

(8) $\check{C}_0\overleftarrow{W}$ and $\overrightarrow{W}_0\check{C}$

Now consider the type-angling: Observe that if a type $g \neq \infty$ cone-cell has at least g transitions, then the defect sum is $\geq 2\pi$ and it has nonpositive curvature.

Proposition 3.39. *Consider the type-angling on $\bar{D}$. Suppose that for each g_1, g_2, g_3 type cells meeting around a vertex, bigonal shard, or triangular shard, we have $\frac{1}{g_1} + \frac{1}{g_2} + \frac{1}{g_3} \leq 1$ (where $g_3 = 4$ for a square). Then each 0-cell has nonpositive curvature.*

Suppose that $|W| \leq \frac{1}{g}|C|$ for each wall-piece on the cone-cell C and $|C'| \leq \frac{1}{g}|C|$ for each cone-cell piece on the cone-cell C. Then each cone-cell has nonpositive curvature.

Note that the first hypothesis in Proposition 3.39 holds when each type is ≥ 8.

3.m More General Small-Cancellation Conditions and Involved Justification

The goal of this section is to give a generalization of Theorem 3.31. This generalization, stated in Theorem 3.40, will not be used in the sequel.

Theorem 3.40. *Assign the split-angling to $\bar{D}$ and suppose the following conditions hold for each cone-cell. Then each internal finite cone-cell in $\bar{D}$ has nonpositive curvature, and if the inequalities are strict then each has negative curvature.*

(1) *Each (contiguous) cone-piece is $\leq \frac{1}{6}$ and each wall-piece is $\leq \frac{1}{24}$.*

(2) *Each (contiguous) cone-piece is $\leq \frac{1}{8}$ and each wall-piece is $\leq \frac{1}{16}$.*

(3) *Each (contiguous) cone-piece is $\leq \frac{1}{12}$ and each wall-piece is $\leq \frac{1}{12}$.*

Moreover, if $\dim(X) = 2$, then one obtains nonpositive curvature (respectively, negative if strict) if either:

(1) *Each contiguous cone-piece is $\leq \frac{1}{6}$ and each wall-piece is $\leq \frac{1}{12}$.*

(2) *Each contiguous cone-piece is $\leq \frac{1}{8}$ and each wall-piece is $\leq \frac{1}{8}$.*

(3) *Each contiguous cone-piece is $\leq \frac{1}{12}$ and each wall-piece is $\leq \frac{1}{6}$.*

Proof. We prove the following statement: Let $\frac{1}{4} = \frac{1}{c} + \frac{1}{r}$ with $c \geq 6$. If the following both hold then each internal finite cone-cell in $\bar{D}$ has nonpositive curvature, and if the inequalities are strict then each has negative curvature.

(1) Each contiguous cone-piece is $\leq \frac{1}{c}$ of the systole
(2) and each wall-piece is $\leq \frac{1}{2r}$ of the systole.

In fact, a slightly more general combinatorial condition ensuring that cone-cells are negatively curved is proven by the argument below. It requires the more general hypothesis that any circumnavigation of a cone-cell by α cone-pieces and β wall-pieces satisfies $\frac{\alpha}{c} + \frac{\beta}{2r} > 1$ (respectively, ≥ 1 for nonpositive curvature).

We will focus on the case where $\dim(X)$ is arbitrary. The strengthening that occurs when $\dim(X) = 2$ will be apparent at the end of the proof where we instead use that wall-pieces are $\leq \frac{1}{r}$ of the systole.

Following Definition 3.30, we will combine the pieces into clusters and distribute the defect at the transitions so that they lie within the two parenthesized clusters. We emphasize that no parentheses are assigned when there is a 0 defect at the transition—as the adjacent pieces will lie in the same cluster in this case. When the transitional defect is nonzero, we always separate the adjacent pieces and distribute the defect: When the defect is $\frac{\pi}{4}$ and there is a C on one side, then the defect is distributed into two separate clusters using $\frac{\pi}{4} = \frac{\pi}{c} + \frac{\pi}{r}$ with the $\frac{\pi}{c}$ share on the C side and the $\frac{\pi}{r}$ share on the W side. In all other cases, the transitional defect is divided equally between the resulting neighboring clusters.

We illustrate the clusters and defect distribution by applying it to the following example:

$$\hat{C}_{\frac{\pi}{3}} \check{C}_{\frac{\pi}{4}} \overrightarrow{W}_0 \check{C}_0 \overleftarrow{W}_{\frac{\pi}{4}} \bar{C}_{\frac{\pi}{3}} \bar{C}_{\frac{\pi}{2}} \bar{C}_{\frac{\pi}{3}} \bar{C}_{\frac{\pi}{2}} \overrightarrow{W}_0 \check{C}_{\frac{\pi}{2}} \overrightarrow{W}_{\frac{\pi}{4}} \bar{C}$$

$$\hat{C}_{\frac{\pi}{6}})(_{\frac{\pi}{6}} \check{C}_{\frac{\pi}{c}})(_{\frac{\pi}{r}} \overrightarrow{W}_0 \check{C}_0 \overleftarrow{W}_{\frac{\pi}{r}})(_{\frac{\pi}{c}} \bar{C}_{\frac{\pi}{6}})(_{\frac{\pi}{6}} \bar{C}_{\frac{\pi}{4}})(_{\frac{\pi}{4}} \bar{C}_{\frac{\pi}{6}})(_{\frac{\pi}{6}} \bar{C}_{\frac{\pi}{4}})(_{\frac{\pi}{4}} \overrightarrow{W}_0 \check{C}_{\frac{\pi}{4}})(_{\frac{\pi}{4}} \overrightarrow{W}_{\frac{\pi}{r}})(_{\frac{\pi}{c}} \bar{C}$$

To obtain the nonpositive curvature of the cone-cells, we check that the total defect is at least 2π. To see this, we verify that a defect of $\geq \frac{2\pi}{p}$ is associated to any cluster contributing a $\frac{1}{p}$ portion of the cone-cell systole.

This is immediate for elementary clusters: For a wall-piece cluster the minimum associated defect occurs at $(_{\frac{\pi}{r}} W_{\frac{\pi}{r}})$, and thus the minimum defect is $\geq \frac{2\pi}{r}$ which is double the required $\frac{2\pi}{2r}$. Likewise, for a contiguous or noncontiguous cone-piece cluster, since $c \geq 6$, the minimum associated defect occurs at $(_{\frac{\pi}{c}} C_{\frac{\pi}{c}})$ which gives a defect of $\frac{2\pi}{c}$ for a $\frac{1}{c}$ contribution to the circumference.

For compound clusters containing one or two 0 defects we rely on the following two observations (the second of these observations is illustrated on the left in Figure 3.46):

(1) $(\check{C}_0 \overleftarrow{W})$ and $(\overrightarrow{W}_0 \check{C})$ each lie behind the extension of the single wall-piece $\overleftarrow{W}$ and $\overrightarrow{W}$.
(2) $(\overrightarrow{W}_0 \check{C}_0 \overleftarrow{W})$ lies behind the configuration $\overrightarrow{W}^{\times} \overleftarrow{W}$ of two crossing wall-pieces indicated in Figure 3.46.

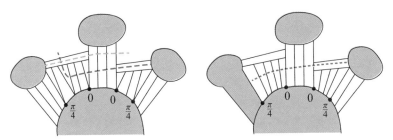

Figure 3.46. The concatenation of three pieces within $C_{\frac{\pi}{4}}W_0C_0W_{\frac{\pi}{4}}C$ is dominated by the concatenation of two wall-pieces. If $\dim(X)=2$, it is dominated by a single wall-piece.

Consequently, each double cluster has size bounded by one wall-piece which is $\leq \frac{1}{2r}$ of the circumference and the associated defect is minimized at $(\frac{\pi}{r}\check{C}_0\overleftarrow{W}\frac{\pi}{r})$ (and analogous) giving a defect of $\frac{2\pi}{r}$ which is double the needed $\frac{2\pi}{r}$. Similarly, the triple cluster has size bounded by two wall-pieces which is $\leq \frac{2}{2r}$ of the circumference and the minimal associated defect occurs at $(\frac{\pi}{r}\overrightarrow{W}_0\check{C}_0\overleftarrow{W}\frac{\pi}{r})$ which gives a defect of $\frac{2\pi}{r}$ as needed.

When X is 2-dimensional, the triple cluster has size bounded by a single wall-piece, and thus allows us to improve things to assume that wall-pieces have size $\leq \frac{1}{r}$, as we previously had double the excess that we needed in cases besides the triple clusters. □

Problem 3.41. Does nonpositive curvature (negative if $<$) hold in the following two cases?

(1) Each contiguous cone-piece is $\leq \frac{1}{8}$ and each wall-piece is $\leq \frac{1}{8}$.
(2) Each contiguous cone-piece is $\leq \frac{1}{6}$ and each wall-piece is $\leq \frac{1}{12}$.

We refer the reader to Figure 3.47 for a diagram limiting possible generalizations of Theorem 3.40. We note that the $4,8,8$ Euclidean tiling, and the tiling obtained by subdividing its square's edges, suggest that each statement in Problem 3.41 is sharp if true.

3.n   Informal Discussion of the Limits of the Theory

While the $C(6)$ condition leads to small-cancellation for classical presentations, for the "graphical presentations" where $\dim(X)=1$, we cannot expect $C(6)$ to suffice in general. Indeed, consider the snub octahedron: it is a tiling of the sphere where each vertex is surrounded by a square and two hexagons. In any reasonable sense, each hexagon is not the concatenation of fewer than 6 pieces. Let D be the disk diagram obtained from the complement of a square in the

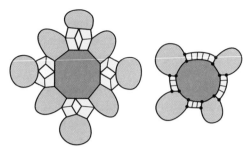

Figure 3.47. The diagram suggests that Theorem 3.40 is sharp when $\dim(X) = 2$.

above tiling of the 2-sphere. There does not seem to be a natural way to interpret things so that D has a positively curved shell.

Another example to bear in mind is the snub icosahedron, which has hexagons surrounding pentagons, and is otherwise similar.

The snub-$(4,4)$ tiling of the plane, consisting of squares surrounded by octagons, has nonpositive curvature and seems like a reasonable limit here. An obvious type-angling works fine for it, and (assuming the squares are already in X) the split-angling appears to work no matter how the admitted rectangles are oriented.

3.o Nonpositively Curved Angling Rules

An *angling rule* for a cubical presentation $\langle X \mid \{Y_i\} \rangle$ is a rule for assigning angles to corners of cone-cells in rectified diagrams $\bar{D}$ arising from diagrams $D \to X^*$. Two examples of angling rules are the split-angling and the type-angling which were described in Section 3.h.

There is one subtle additional requirement that is also satisfied by the split-angling and type-angling. Suppose D' is a subdiagram of D, and its rectification $\bar{D}'$ is a subdiagram of $\bar{D}$. We require that each angle of each internal corner of a cone-cell C of $\bar{D}'$ be equal to the corresponding angle of C in $\bar{D}$ whenever its outgoing rectangles have the same targets in D as they do in $\bar{D}$. This requirement is rarely needed, but supports the proof of Lemma 5.10. In that case, the above requirement concerns the situation where D is obtained from D' by adding a cone-cell or rectangle to D along a maximal external subpath of $\partial_{\mathsf{p}} C$.

An angling rule on X^* is *nonpositively curved* if the following holds for each rectified diagram $\bar{D}$ arising from a reduced diagram $D \to X^*$ and an ordering of the cone-cells where the cone-cell at infinity is last:

(1) Each internal cone-cell, shard, and internal vertex has nonpositive curvature.
(2) The rectangles have the usual angles: $\frac{\pi}{2}$ at the four corners, and π elsewhere.
(3) $0 \le \sphericalangle \le \pi$ for each $\sphericalangle$ in a cone-cell.

Figure 3.48. Singular doubly-external, doubly-external, singly-external, and nil-external corners.

(4) For consecutive edges of the boundary path of a cone-cell, if their associated admitted rectangles have distinct (implicit) destinations then the internal angle at that corner of the cone-cell has nonzero defect.

(5) The angle at a corner of a cone-cell at a vertex on ∂D equals 0 when it is singular doubly-external, equals π when it is nonsingular doubly-external, equals $\frac{\pi}{2}$ when it is singly-external, and is > 0 when it is nil-external.

The angling rule is *negatively curved* if we strengthen the Condition (1) above to require that internal cone-cells have negative curvature. A cone-cell is *external* or *internal* according to whether or not its boundary contains a 1-cell in ∂D.

The four classes of external corners of a cone-cell are defined via Figure 3.48. The corner corresponds to two consecutive edges in the boundary path of the cone-cell, and the four classes are according to whether these two edges are non-consecutive external edges of $\partial_p D$, consecutive external edges of $\partial_p D$, exactly one is external, or neither is external but their common vertex is on ∂D.

3.p   Positive Curvature along Boundary

The possible positively curved cells are:

(1) A single isolated 0-cell has curvature 2π.
(2) A single isolated cone-cell has curvature 2π.
(3) The 0-cell at the end of a spur has curvature π.
(4) The 0-cell at the center of the outerpath of a cornsquare has curvature $\frac{\pi}{2}$.
(5) A *shell* C is a positively curved external cone-cell. Its boundary path is a concatenation QS where the *outerpath* Q is a subpath of the boundary path of the diagram, and the *innerpath* S has all its open 1-cells in the interior of the diagram. The curvature of C equals π minus the sum of the defects of the angles along interior(S).

We will occasionally use the more precise term θ-*shell* to describe a cone-cell C with an outerpath Q that is a subpath of $\partial_p D$, with an innerpath S whose open 1-cells are internal, such that θ is π minus the sum of the defects of internal angles along interior(S). Accordingly, the θ-shell is *negatively curved* if $\theta < 0$ and it is *nonnegatively curved* if $\theta \geq 0$, and it is *positively curved* if $\theta > 0$ in which case we simply use the term shell as above.

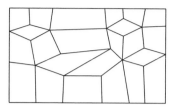

Figure 3.49. A pseudo-grid.

There is a degenerate case of a shell whose innerpath S is trivial. This arises when C is attached to the rest of the diagram at a cut vertex. Perhaps somewhat artificially, it also arises when the cone-cell C is the entire diagram, in which case we can regard C as a shell whose innerpath S, is the trivial path at some vertex on ∂C, and whose outerpath is $\partial_{\mathsf{p}} C$ starting and ending at that vertex.

3.q   Ladder Theorem

A *pseudo-grid* between ν and μ is a square disk diagram E where $\partial_{\mathsf{p}} E = \nu\rho\mu^{-1}\varrho^{-1}$ such that each dual curve starting on ν ends on μ, and vice versa, and where no dual curves starting on ν cross each other. Moreover, we assume that E has no bigons. We permit E to contain cornsquares, but these can only have outerpath on ρ or ϱ. When there are no cornsquares on ρ or ϱ, the pseudo-grid is an actual grid by Lemma 2.23. The pseudo-grid is *trivial* if it consists of a single vertex. It is *degenerate* if ρ and ϱ are trivial, in which case E is an arc. See Figure 3.49.

Definition 3.42 (Ladder). A *ladder* is a disk diagram D with the property that there is a sequence of $n \geq 2$ closed cone-cells and/or vertices $C_1, C_2, \ldots, C_n$ that are ordered so that the closure of C_j separates C_i from C_k when $i < j < k$. The diagram D is an alternating union of cone-cells (or vertices) and pseudo-grids E_i (possibly trivial or degenerate) in the following sense:

(1) $\partial_{\mathsf{p}} D$ is a concatenation $P_1 P_2^{-1}$ where the initial and terminal points of P_1 lie on C_1 and C_n, respectively.
(2) $P_1 = \alpha_1 \rho_1 \alpha_2 \rho_2 \cdots \alpha_n$ and $P_2 = \beta_1 \varrho_1 \beta_2 \varrho_2 \cdots \beta_n$.
(3) $\partial_{\mathsf{p}} C_i = \mu_i \alpha_i \nu_i^{-1} \beta_i^{-1}$ for each i where μ_1 and ν_n are trivial paths.
(4) $\partial_{\mathsf{p}} E_i = \nu_i \rho_i \mu_{i+1}^{-1} \varrho_i^{-1}$ for $1 \leq i < n$ and E_i is a pseudo-grid from ν_i to μ_{i+1}.

There are two degenerate cases for E_i that should be noted: Firstly, it is possible that E_i is a vertical arc, and that $\nu_i = \mu_{i+1}$. Secondly, it is possible that E_i is a horizontal arc, in which case $\rho_i = \varrho_i$ and ν_i, μ_{i+1} are trivial.

We refer to Figure 3.50 for help with the notation, and to Figure 3.51 for pictures of various ladders. A ladder is *nonsingular* if it has no cut vertex, spur,

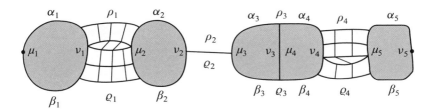

Figure 3.50. Ladder definition notation.

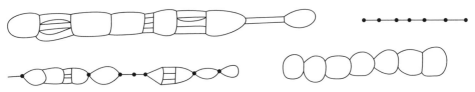

Figure 3.51. Some ladders.

or isolated 1-cell. We regard a disk diagram consisting of a single 0-cell or cone-cell as a *trivial ladder*.

Theorem 3.43 (Ladder Theorem). *Let $D \to X^*$ be a reduced diagram. Suppose the rectified diagram $\bar{D}$ was created using an ordering of 1-cells induced by an order of cone-cells (with C_∞ last). And suppose that X^* has a nonpositively curved angling rule. If $\bar{D}$ has exactly two positively curved cells then D is a ladder.*

In practice, it is often natural to assume that D has no cornsquares on its boundary. In this case, we deduce that the ladder D of Theorem 3.43 has the additional property that its pseudo-grids are genuine grids.

Proof. As there are exactly two positively curved cells, each has curvature exactly π. Indeed, referring to the curvature formulas in Section 3.g, following Equation (3.1), the $0 \leq \sphericalangle$ hypothesis imply that the curvature of a boundary 0-cell v is $2\pi - \pi\chi(\mathrm{link}(v)) - \sum \sphericalangle \leq \pi - \sum \sphericalangle \leq \pi$. Applying Equation (3.2), the hypotheses on external corner angles together with the $\sphericalangle \leq \pi$ hypothesis imply that the curvature of an external cone-cell is $\leq \pi$. Here each nontrivial boundary arc provides a defect of π from the two $\frac{\pi}{2}$ singly-external corners, and all other angles have nonpositive defect since $\sphericalangle \leq \pi$.

The only π curvature 0-cell arises from a spur. The π-curvature cone-cells arise when the external cone-cell has a single external boundary path with a defect of $\frac{\pi}{2}$ on each end and a nontrivial innerpath whose internal corners have angle π, or when the external boundary path is the entire boundary path and the innerpath is trivial, in which case there is a defect of π at the singular doubly-external corner where the boundary path starts and ends. Note that the

Figure 3.52. It must be a ladder.

outerpath cannot be trivial, for we have hypothesized that all internal cone-cells have nonpositive curvature.

Since each other cell has nonpositive curvature, Theorem 3.23 implies that each other cell has zero curvature.

We focus on the case where D is nonsingular, as an inductive argument would allow us to string together individual ladders to obtain a new ladder. Note that these individual ladders might consist of cone-cells or isolated edges.

Consider a positively curved cell C_1, which must be a shell with a nontrivial innerpath since we are dealing with the nonsingular case. Let e_1, e_2 denote the initial and terminal 1-cells of the innerpath of C_1. Let R_1, R_2 denote the rectangles at e_1, e_2. (It is possible that $e_1 = e_2$ and hence $R_1 = R_2$.) Traveling along $R_1 \cap \partial \bar{D}$ (see Figure 3.52 for various scenarios) we see that $\partial \bar{D}$ proceeds along the entire (possibly degenerate) external path on one side of R_1 until it ends on a rectangle or cone-cell. More precisely, the final vertex on ∂D of the external path of R_1 is incident with either two squares or with a square and a cone-cell. Indeed, another incoming (possibly degenerate and thus including another cone-cell) rectangle would give us a 0-cell on $\partial \bar{D}$ with negative curvature. The same reasoning holds for R_2.

Note that R_1 cannot terminate on the external boundary of R_2 (or vice versa), and R_1 cannot terminate on C, for then there would either be a corn-square on C_1 which contradicts that D is reduced, or R_1, R_2, C would not bound a square subdiagram, and hence the rectangles emerging from C_1 would not all end in parallel on the same cone-cell or rectangle, in which case, by hypothesis, the innerpath of C_1 would have a positive angle at some corner, and so $\kappa(C) < \pi$. We thus find that the sequence of rectangles emerging from the innerpath of C_1 have the same (implicit) destination which is either a cone-cell C_2 or a rectangle R.

We now show that they cannot all end (implicitly) on the same rectangle R. Note that R cannot have both ends on $\partial_{\mathsf{p}} D$ because of the order in which rectangles are admitted, and likewise it cannot have one end on $\partial_{\mathsf{p}} D$ and the other end terminating on R_1 or R_2. Thus R extends to provide an additional square beyond the square where some R_i terminates on it, or it ends there at a cone-cell. Consider the vertex v on $\partial \bar{D}$ where R_i ends on R. As we just explained, there must be an additional square or a cone-cell along the other boundary edge at v, and all nil-external cone-cell angles are ≥ 0. Moreover, no shard could have a sharp corner at v, since its rectangles on either side would terminate at a pair of consecutive edges at v, and this is impossible in our case, for one of these boundary edges is an external edge of R_i. Thus there are no negative angles coming from sharp corners of shards at v, and so the angle sum at v is $\geq \frac{3\pi}{2}$, so $\kappa(v) \leq -\frac{\pi}{2}$, which is impossible.

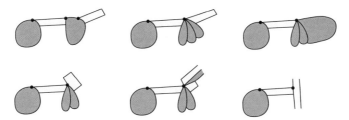

Figure 3.53. Woof Woof: R_1 can only terminate as in the top left figure. The remaining scenarios are impossible as they yield a negatively curved vertex of $\partial \bar{D}$.

We have deduced that all the rectangles leaving the innerpath of C_1 end parallel to each other on the cone-cell C_2.

Each $R_i \cap \partial D$ consists of a complete external path of R_i, since if a cone-cell or rectangle is attached along the interior of this path, then there would be a vertex with negative curvature. (If $R_1 = R_2$, then $R_i \cap \partial D$ consists of two complete external paths.) We conclude that the subdiagram consisting of the rectangles between C_1 and C_2, and the shards between them, forms a pseudo-grid from a path $\nu_1 \to C_1$ to a path $\mu_2 \to C_2$.

If $\kappa(C_2) = \pi$ then we have arrived at a length 2 ladder, as the same reasoning applies in the reverse direction.

Otherwise $\kappa(C_2) = 0$, and we will show that $\partial C_2 \cap \partial \bar{D}$ consists of exactly two nontrivial components. There cannot be one component, for then, as D is nonsingular, we would have $\kappa(C_2) = \pi$, which resulted in a length 2 ladder above. A component starting at an endpoint of μ_2, i.e., intersecting R_1 and R_2, cannot be trivial, for since nil-external angles are > 0 and since there must be a further rectangle or singly-external corner of a 2-cell at such a vertex, we would have an angle sum $> \pi$ and hence a negatively curved vertex. Finally, if we remove the pseudo-grid E_1 and the cone-cell C_1, we obtain a rectified diagram with exactly two positively curved features, one of which is C_2. The result is again a ladder by induction. And since our pseudo-grid E_1 is attached along a subpath of the outerpath of C_2 that is disjoint from its innerpath, the diagram D is a ladder as well.   □

Remark 3.44. The hypothesis of $\vartriangleleft > 0$ at nil-external corners of cone-cells forces cone-cells in a ladder to intersect the boundary in nontrivial subpaths.

With a weaker $\vartriangleleft \geq 0$ hypothesis, it is possible to have strings of cone-cells that look like bigons and triangles having 0 angles along nil-external corners of the boundary as in Figure 3.54.

Remark 3.45. There are similar results for annular diagrams obtained in Section 5.p. Note that an annular diagram is treated so that it has two cone-cells at infinity, and these both occur at the end in the ordering that is used to choose

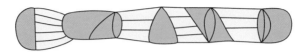

Figure 3.54. If nil-external corners are merely required to have nonnegative angles, then we would reach a similiar conclusion where cone-cells in ladders may intersect the boundary along trivial arcs.

rectangles in the rectification. We assume that the angling rule functions the same way for an annular diagram as for a disk diagram (this is the case for the split-angling).

The main difference is that in order to obtain an annulus that is as thin as a ladder, we must assume that internal cone-cells have negative curvature, and that (harder to obtain) external cone-cells cannot have nonpositive curvature unless they touch both the inner and outer infinite cones. Otherwise we obtain a thickness 2 situation that is harder to control.

3.r   Trichotomy for Reduced Diagrams

Theorem 3.46 (Diagram Trichotomy). *Suppose $\bar{D}$ is a rectified disk diagram arising from a reduced diagram $D \to X^*$ and nonpositively curved angling rule. Then one of the following hold:*

(1) $\bar{D}$ *consists of a single 0-cell or a single closed cone-cell.*
(2) $\bar{D}$ *is a ladder.*
(3) $\bar{D}$ *has at least three shells and/or spurs and/or cornsquares along $\partial\bar{D}$. Moreover, if there is no shell or spur, then $\bar{D}$ must have at least four cornsquares.*

Proof. This follows from Theorem 3.23 since curvatures are $\le \pi$ except for the two cases in (3.46), and curvatures of internal cells and shards are nonpositive by hypothesis. The case where there are only two positively curved cells was treated in Theorem 3.43. The possibility of exactly three cornsquares (and no other features of positive curvature) is excluded since each would contribute $\frac{\pi}{2}$.   □

Definition 3.47 (Small-Cancellation Presentation). We will refer to $\langle X \mid \{Y_i\}\rangle$ as a *cubical small-cancellation presentation* if it is equipped with a nonpositively curved angling rule. We use the term *negatively curved small-cancellation presentation* to mean that X^* has a negatively curved angling rule.

3.s   Examples

(1) Ordinary $C'(\frac{1}{6})$ small-cancellation theory where the cube complex is a graph and the relators are circles.

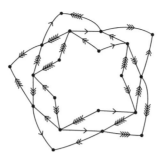

Figure 3.55. The cubical presentation of Example 3.48 is obtained from a 5 generator raag by adding the relator Y above.

(2) Graphical small-cancellation theory. This was first observed in [RS87] but was rediscovered subsequently by Gromov [Gro03, Oll06].
(3) A $C'(\frac{1}{12})$ cubical presentation. See Theorem 3.32.
(4) Small-cancellation theory over right-angled Artin groups with ordinary relators. (This gives a variety of simple examples that don't utilize relative hyperbolicity of $\pi_1 X$. But one must be careful about the cores.)
    Specifically, a small-cancellation set of words alternating substantially around the generators. Pieces correspond to subwords that equal each other (after shuffling).
(5) Many Artin groups have useful cubical presentations and we refer to Section 5.r. For instance, this is the case for every Artin group associated to a Coxeter group with no exponent of degree 2, and more generally, this appears to work if there is no triangle of type: $(2, 2, n)$, $(2, 3, 3)$, $(2, 3, 4)$, or $(2, 3, 5)$. Moreover, it seems we can allow degree 2 exponents if we exclude degree 3, but this requires more complicated generalized relators that are centralizers.

Example 3.48. An interesting example arises from the 4-string braid group B_4. The kernel of its homomorphism to the underlying Coxeter group S_4 is called the 4-string pure braid group P_4. It is shown in [DLS91] that $P_4 \cong G \times \mathbb{Z}$ where G is the following quotient of a right-angled Artin group:

$$\langle a_1, \ldots, a_5 \mid [a_i, a_{i+1}] : \text{coefficients modulo } 5 \rangle \ / \ \langle\langle a_1 a_2^{-1} a_3 a_4^{-1} a_5 a_1^{-1} a_2 a_3^{-1} a_4 a_5^{-1} \rangle\rangle.$$

This yields an interesting cubical presentation, after we replace the length 10 cycle with a local-isometry $Y \to X$ where X is the nonpositively curved square complex $\langle a_1, \ldots, a_5 \mid [a_i, a_{i+1}] : \text{coefficients modulo } 5 \rangle$ and where Y is obtained from the length 10 cycle above by adding ten squares where their corners appear (see Figure 3.55).

Its nontrivial contiguous cone-pieces are squares and edges. Its contiguous wall-pieces are paths of length at most 2. There is a concatenation of five wall-pieces or five cone-pieces that equals the generator of $\pi_1 Y$. It comes close to being a small-cancellation presentation but fails to satisfy the $B(6)$ property studied in Chapter 5. Higher degree (e.g., $n > 5$) generalizations of this example work better.

Example 3.49. Consider the presentation $\langle a, b, c \mid (ab)^2, (bc)^2, (ca)^2, (aaabbbccc)^2 \rangle$ which differs slightly from a surprising presentation for an index 3 subgroup of $PSL(2, \mathbb{Z}[\frac{1}{2}])$ discovered by Cameron Gordon. (His presentation has $a^3 b^3 c^3$ not raised to a power.)

There is an obvious homomorphism to $\mathbb{Z}_2^3$ in which the obvious torsion elements survive, and we can then collapse pairs of 2-cells corresponding to the relators. The result is a cubical presentation $\langle X \mid Y_1, Y_2, Y_3, Y_4 \rangle$ where X is an orientable genus 3 surface built from squares with 6 meeting around each 0-cell, and each Y_i is built from a lift of $(a^3 b^3 c^3)^2$ by adding six squares at the corners corresponding to the transitions ab, bc, ca, ab, bc, ca. All maximal cone-pieces are either of the form aa, bb, cc or are one of the added squares. Besides the wall-pieces consisting of single 1-cells, there are wall-pieces of the form cb^{-1}, ba^{-1}, and ac^{-1}.

Thus the $C(6)$ condition is satisfied with the split-angling. There is an obvious (antipodal) wallspace structure on each Y_i, and it appears that together with it the $B(6)$ condition holds as well. We study the $B(6)$ condition later in Definition 5.1.

3.t Examples Arising from Special Cube Complexes

The following results were originally motivated by the observation made in [Wis03] that for any finite immersed graph $\Lambda \to \Gamma$, there is a finite cover $\hat{\Lambda}$ such that $\langle \Gamma \mid \hat{\Lambda} \rangle$ satisfies Gromov's graphical $\frac{1}{6}$ small-cancellation theory (see [RS87]).

Theorem 3.50. *Let X be a compact nonpositively curved cube complex. Let $H_1, \ldots, H_k$ be residually finite subgroups of $\pi_1 X$, and for each i, let $Y_i \to X$ be a local-isometry with $\pi_1 Y_i \cong H_i$ and Y_i compact. Suppose each $\widetilde{Y}_i \subset \widetilde{X}$ is superconvex. Suppose that there is an upperbound on the diameters of intersections between distinct translates of $\widetilde{Y}_i, \widetilde{Y}_j$ in $\widetilde{X}$ (we allow $i = j$ here).*

Then for each $\alpha > 0$ there are finite covers $\widehat{Y}_i$ such that $\langle X \mid \widehat{Y}_1, \ldots, \widehat{Y}_k \rangle$ is $C'(\alpha)$.

Proof. By Lemma 2.40, let D be a uniform upperbound on the diameter of a flat strip whose base lies on $\widetilde{Y}_i$, but whose dual hyperplane does not cross $\widetilde{Y}_i$. Note that D also bounds lengths of noncontiguous pieces between disjoint translates of $\widetilde{Y}_i, \widetilde{Y}_j$. Let E be a uniform upperbound on the diameters of intersections of distinct translates of $\widetilde{Y}_i, \widetilde{Y}_j$. By possibly passing to a finite index supergroup of each H_i, we can assume that each $H_i = \mathrm{Stab}(\widetilde{Y}_i)$.

By residual finiteness, for each i, we can choose a finite regular cover $\widehat{Y}_i \to Y_i$ such that $\|\widehat{Y}_i\| > \frac{1}{\alpha}\max(D, E)$. □

We refer to Definition 8.1 for malnormal collections of subgroups, and we refer to Chapter 6 for the notion of a special cube complex.

Corollary 3.51. *Let X be a virtually special compact cube complex. Suppose $\pi_1 X$ is word-hyperbolic. Let $H_1, \ldots, H_k$ be quasiconvex subgroups of $\pi_1 X$ that form a malnormal collection. Then for each $\alpha > 0$ there are finite index subgroups $H'_i \subset H_i$ that are represented by local-isometries $\widehat{Y}_i \to X$ with $\widehat{Y}_i$ compact such that $\langle X \mid \widehat{Y}_1, \ldots, \widehat{Y}_k \rangle$ satisfies $C'(\alpha)$.*

Proof. By Lemma 2.36, there is an H_i-cocompact superconvex subcomplex $\widetilde{Y}_i \subset \widetilde{X}$ for each i. There is an upperbound on the overlap between translates of $\widetilde{Y}_i$ and $\widetilde{Y}_j$ by the malnormality assumption. Each H_i is residually finite since $\pi_1 X$ is virtually special. We can thus apply Theorem 3.50. □

Lemma 3.52 (Malnormal controls overlap). *Let X be a compact nonpositively curved cube complex (with $\pi_1 X$ word-hyperbolic). For $1 \leq i \leq r$, let $Y_i \to X$ be a local-isometry with Y_i compact, and assume the collection $\{\pi_1 Y_1, \ldots, \pi_1 Y_r\}$ is malnormal. Then there is a uniform upperbound D on the diameters of intersections $g\widetilde{Y}_i \cap h\widetilde{Y}_j$ between distinct $\pi_1 X$-translates of their universal covers in $\widetilde{X}$.*

Lemma 3.52 can be interpreted as saying that there is an upperbound on diameters of contiguous cone-pieces in $\langle X \mid Y_1, \ldots, Y_r \rangle$. In practice, one applies Lemma 3.52 under the assumption that X is compact and $\pi_1 X$ is word-hyperbolic, for this enables the existence of compact cores Y_i for quasiconvex subgroups, by Proposition 2.31. A deeper investigation of Lemma 3.52 in the quasiconvex malnormal and hyperbolic situation, shows that there is a uniform upperbound on the sizes of all cone-pieces—not just the contiguous ones. This relationship between malnormality and pieces is concealed by the noncontiguous cone-pieces which can sometimes be ignored in the small-cancellation theory since they are hidden behind and thus controlled by hyperplanes in a superconvex situation.

One way to prove Lemma 3.52 is to consider the nondiagonal components of $(\sqcup Y_i) \otimes_X (\sqcup Y_i)$. Each of these is contractible by malnormality, and of finite diameter since the finitely many Y_i are compact. We refer the reader to Section 8.b.

Proof. If $h\widetilde{Y}_i \cap g\widetilde{Y}_j$ has infinite diameter then applying the pigeon-hole principle, there is an infinite periodic path lying in both $h\widetilde{Y}_i$ and $g\widetilde{Y}_j$. This yields an infinite order element in both $(\pi_1 Y_i)^h$ and $(\pi_1 Y_j)^g$, thus violating malnormality unless $i = j$ and $hg^{-1} \in \pi_1 Y_i$. □

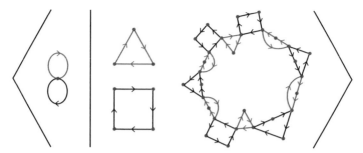

Figure 3.56. There are two grade-1 cones, and one grade-2 cone. All non-absorbable pieces are between the grade-2 cone and itself. The example doesn't satisfy $C(p)$ for very large p, but suggests a plethora of further examples along similar lines. For a presentation of the form $\langle b, c \mid b^m, c^n, W \rangle$, one can attach copies of b^m and c^n to a copy of W along its syllables. When $m = 2$ and $n = 3$, one needs $|W|$ quite large before typical words give graded small-cancellation presentations.

3.u Graded Small-Cancellation

Thus far we have emphasized the version of cubical small-cancellation theory where cone-pieces between cones are small unless the associated cones in $\widetilde{X}$ are equal. We will now turn to a generalization which requires that cone-pieces be small unless one cone lies within the other.

This form of cubical small-cancellation theory is used explicitly in Section 12.c. Simple examples are illustrated in Figures 3.56, 3.57, 3.58, and 3.59. The small-cancellation theory of disk diagrams generalizes by requiring that if a cone-cell boundary path is the concatenation of few pieces, then the boundary path is already nullhomotopic using the lower grade cone-cells. We can thus replace this higher grade cone-cell by a diagram with lower grade cone-cells and squares. It is thus natural to use minimal *graded complexity* which is the sequence counting the number of cone-cells of decreasing grades, terminated by the number of squares. A lexicographical ordering is used here so $(\ldots, 0, 0, 3, 6, 5) > (\ldots, 0, 0, 3, 5, 17)$ etc.

For a graded cubical presentation, a reduced diagram $D \to X^*$ is defined as in Definition 3.11 with one slight adjustment. For a cone-cell A of D that is associated to a cone Y of X^*, we say A is *replaceable* if $\partial_p A$ bounds a disk diagram consisting of lower grade cone-cells and squares, and hence obtain a lower complexity disk diagram with the same boundary path as D. In particular, A is replaceable if $\partial_p A$ is nullhomotopic in $Y^\circledast$, the induced subpresentation, described in Definition 3.53.

We say X^* is a *graded small-cancellation presentation* if there is a nonpositively curved angling rule for the rectified reduced diagrams. We describe some concrete examples in Section 3.v, but our main source of examples of graded small-cancellation cubical presentations arise from the metric small-cancellation conditions discussed in Definition 3.55.

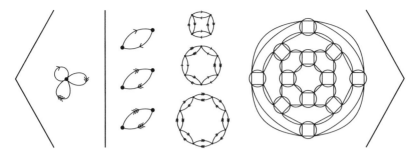

Figure 3.57. Cubical presentation for the 2-3-4 triangle group.

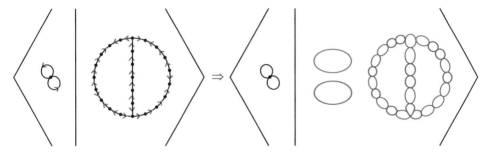

Figure 3.58. A cubical presentation obtained by substituting.

Our main tools—the trichotomy of Theorem 3.46 as well as Theorem 3.43 which characterizes ladders—are essentially about diagrams, so they continue to hold in the graded context.

3.v   Some Graded Examples

We describe several concrete examples of graded cubical presentations. The reader can keep these examples in mind when considering the definitions in the next sections.

Coxeter groups: A Coxeter group yields a graded presentation. See Figure 3.57. Its generators are the usual generators. Its grade i relators are Cayley graphs of the i-generator Coxeter subgroups. For a relator Y of grade ≥ 3, the subpresentation $Y^{\circledast}$ is already simply-connected so $\|Y^{\circledast}\| = \infty$.

Substituting: We refer to Figure 3.58. Let $\langle B \,|\, C \rangle$ denote a graphical presentation where B is a bouquet of circles u, v, and $C \to B$ is an immersed theta-graph. Let X be a new bouquet of circles x, y, and let U, V denote immersed circles, and "substitute" copies of U, V for edges labelled by u, v in C and then fold to obtain a graph Y. Even preserving "orientation," there are many ways of doing this, and one could choose basepoints, or just do it randomly. When $\langle B \,|\, C \rangle$ and $\langle X \,|\, U, V \rangle$ are sufficiently small-cancellation, then so is the cubical presentation $\langle X \,|\, U, V, Y \rangle$ where U, V have grade 1 and Y has grade 2.

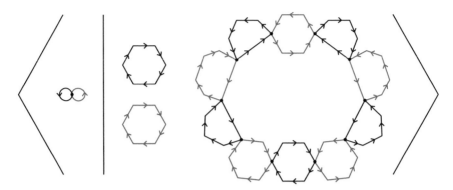

Figure 3.59. Haglund's group acting properly and cocompactly on a $(6,10)$ Gromov polyhedron.

Gromov polyhedra: Haglund's theory of "Gromov polyhedra" fits nicely into this category. These are CAT(0) 2-complexes whose 2-cells are p-gons and whose vertex links are complete graphs K_r. Haglund shows how to produce many such examples with a proper cocompact group action. For instance, Haglund begins with a free product $A * B$ of finite groups, with generators consisting of the full set of nontrivial elements $\{A-1\}, \{B-1\}$. He then adds relators $R_1, \ldots, R_k$ with the property that each generator $a \in \{A-1\}$ or $b \in \{B-1\}$ appears exactly once, with the extra symmetry condition that if u and u^{-1} both appear in R_i then there is an automorphism of R_i sending one occurrence to the other. If $r = |A| = |B|$ then each link will be isomorphic to K_r, and if p is the syllabic length of each relator, then each 2-cell will have p sides. For instance, Haglund provides the following example where u, u^{-1} never recur: $\langle a, b \mid a^{17},\ b^{17},\ ab^4a^2b^5a^{12}b^6a^9b^7,\ a^3b^{16}a^4b^2a^6b^{14}a^{10}b^9 \rangle$.

Haglund's examples naturally lead to cubical presentations that are graded, where X is a wedge on a bouquet of circles for the generators (optionally wedged along a new edge) and the grade 1 relators are the Cayley graphs $\Gamma(A), \Gamma(B)$ of A, B, and the grade 2 relators are copies of the circles corresponding to the R_i, but with copies of $\Gamma(A), \Gamma(B)$ attached along each generator. Haglund's symmetry condition implies that the maximal pieces only occur between these grade 1 relators, and are merely copies of $\Gamma(A), \Gamma(B)$.

Figure 3.59 illustrates one of Haglund's examples: $\langle a, b \mid a^6, b^6, b^4a^5b^5a^2b^3a^4 bab^2a^3 \rangle$. We used a single generator for each $A, B \cong \mathbb{Z}_6$, but the reader can redraw with a wedge of two bouquets of 5 circles, grade 1 relators isomorphic to 6-simplices, and a grade 2 relator that looks like a string of 6-simplices glued around a decagon.

3.w   Graded Metric Small-Cancellation

This section describes a metric small-cancellation condition for graded cubical presentations. We will show how to produce examples in Lemma 12.4.

Definition 3.53 (Graded Presentations and Subpresentations). A *graded cubical presentation* $\langle X \mid \{Y_i\} \rangle$ is equipped with a *grading* of its cones, which is a map from the set of cones to $\mathbb{N}$. We shall sometimes use the notation $\langle X \mid \{Y_{ij}\} \rangle$ to indicate that Y_{ij} has grade i.

For a cone Y of X^*, we define the *subpresentation* $Y^{\circledast}$ to be the cubical presentation induced by $Y \to X$ and $X^*_{\mathrm{grade}(Y)-1}$ where $X^*_{\mathrm{grade}(Y)-1}$ is the sub-presentation of X^* that includes all cones of grade lower than the grade of Y. In many cases arising in our applications, $Y^{\circledast}$ is the cubical presentation whose base is Y and whose cones are either contractible, or consist of lower grade cones of X^* that properly factor through Y.

Remark 3.54 (Reduced in Graded case). Reduced diagrams in the graded setting are defined as before except that we adjust Definition 3.11.(5) to state that: For each internal cone-cell A of D mapping to a cone Y, the path $\partial_{\mathsf{p}} A$ is essential in $Y^{\circledast}$.

For a path $S \to Y$ in a cone of a graded cubical presentation X^*, we let $|S|_Y$ denote the distance in $\widetilde{Y}^{\circledast}$ between the endpoints of a lift $\widetilde{S}$ to $\widetilde{Y}^{\circledast}$. We emphasize that the cones of $Y^{\circledast}$ play no role in $|S|_Y$ except to determine the covering space $\widetilde{Y}^{\circledast}$. The combinatorial distance is measured within the underlying cube complex of $\widetilde{Y}^{\circledast}$, so no shortcuts are allowed through the conepoints. We will occasionally also use the notation $|P|_Y$ to denote the diameter within $\widetilde{Y}^{\circledast}$ of the image of P, where $P \to Y$ admits a lift $P \to \widetilde{Y}^{\circledast}$.

Definition 3.55 (Graded Metric Small-Cancellation). The graded cubical presentation $\langle X \mid \{Y_i\} \rangle$ satisfies $C'(\alpha)$ if for each piece P between $g_i \widetilde{Y}_i$ and $g_j \widetilde{Y}_j$ either $|P|_{Y_i} < \alpha \|Y_i^{\circledast}\|$ and $|P|_{Y_j} < \alpha \|Y_j^{\circledast}\|$, or $\mathrm{grade}(Y_i) \leq \mathrm{grade}(Y_j)$ and $g_i \widetilde{Y}_i \subset g_j \widetilde{Y}_j$ and there is a map $Y_i \to Y_j$ such that the commutative diagram below holds, or vice versa, i.e., $\mathrm{grade}(Y_j) \leq \mathrm{grade}(Y_i)$ and $g_j \widetilde{Y}_j \subset g_i \widetilde{Y}_i$ etc.

$$
\begin{array}{ccccc}
g_i \widetilde{Y}_i & \to & Y_i & & \\
\downarrow & & \downarrow & \searrow & \\
g_j \widetilde{Y}_j & \to & Y_j & \to & X
\end{array}
$$

We also require a similar bound on wall-pieces, and as usual it suffices to describe it for contiguous wall-pieces. Suppose $P \to Y_i$ is a contiguous wall-piece that lies in the intersection $N(H) \cap \widetilde{Y}_i$ where H is a hyperplane that does not intersect $\widetilde{Y}_i$. Then we require that $|P|_{Y_i} < \alpha \|Y_i^{\circledast}\|$.

Definition 3.56. A cubical presentation X^* has *well-embedded cones* if the following properties are satisfied:

(1) Let Y_1 be a cone of X^*. Then $Y_1 \to X$ is injective.
(2) Let Y_1, Y_2 be cones in X^*. Then $Y_1 \cap Y_2$ is connected.
(3) Let Y_1, Y_2, Y_3 be cones in X^*. If $Y_i \cap Y_j \neq \emptyset$ for each i, j then $Y_1 \cap Y_2 \cap Y_3 \neq \emptyset$.

Example 3.57. Let X be a graph that is a 3-cycle with edges a, b, c so that abc is a path in X. Properties 3.56. (1), (2), and (3) fail for the following cubical presentations which satisfy $C'(\alpha)$ for each $\alpha > 0$:

(1) $\langle X \mid abc \rangle$.
(2) $\langle X \mid ab, c \rangle$.
(3) $\langle X \mid a, b, c \rangle$.

Moreover, let $Z^* = \langle X \mid X, a, b, c \rangle$. Then Z^* is $C'(\alpha)$ but $\widetilde{Z}^* = Z^*$ fails to have well-embedded cones. Note that we then have $X^\circledast = \langle X \mid a, b, c \rangle$.

A graded cubical presentation X^* has *small subcones* if for each cone Y of X^*, and each cone Z of $Y^\circledast$, we have $\mathrm{diam}(Z) < \frac{1}{3} \|Y^\circledast\|$. Note that when X^* is $C'(\frac{1}{3})$ the small subcone property is equivalent to requiring that $\mathrm{diam}(Z) < \frac{1}{3} \|Y^\circledast\|$ whenever Z and Y are cones of X^* with $Z \to X$ factoring as $Z \to Y \to X$.

Lemma 3.58 (Well-embedded cones). *Let X^* be a $C'(\frac{1}{12})$ graded metric cubical presentation with finitely many grades and with small subcones. Then $\widetilde{X}^*$ has well-embedded cones.*

Remark 3.59. A *maximal cone* of X^* is a cone Y such that there does not exist a cone Y' with $Y \to X$ properly factoring as $Y \to Y' \to X$. Lemma 3.58 generalizes Lemma 4.1 in the sense that, as can be seen from the proof, Properties 3.56.(1)-(3) hold for maximal cones without any hypotheses.

Proof of Lemma 3.58. In each part of the proof we will use that in a disk diagram in a $C'(\frac{1}{12})$ cubical presentation, the outerpath of a non-replaceable shell is not the concatenation of fewer than 5 pieces, since by Lemma 3.70, the innerpath cannot consist of fewer than 8 pieces,

Suppose Property 3.56.(1) for $\widetilde{X}^*$ is false, so some $Y_1 \to \widetilde{X}^*$ is not embedded. Let $P \to Y_1$ be a path that is not closed, but whose lift to $\widetilde{X}^*$ is closed. Let D be a disk diagram with $\partial_p D = P$, and assume that $(\mathsf{Comp}(D), |P|)$ is minimal among all possible such examples. A spur or cornsquare would lead to a lower complexity example. By Theorem 3.46, either D is a single cone-cell which we shall regard as a shell with trivial innerpath, or D contains at least one shell R with outerpath Q and innerpath S. Let Y be the cone supporting R. Since the outerpath of a shell cannot consist of at most two cone-pieces, either $Y_1 \to X$ factors through $Y \to X$ or $Y \to X$ factors through $Y_1 \to X$. In the latter case, we can absorb R into Y_1 to obtain a lower complexity disk diagram with R removed and S substituted for Q in P. In the former case, Y_1 factors through Y. If $P \to Y_1 \to Y$ is not closed then we have a lower complexity counterexample, obtained by removing R and substituting S for Q in P. So we suppose that $P \to Y_1 \to Y$ is closed. Since $\mathrm{diam}(Y_1) < \frac{1}{3} \|Y\|$ we see that $P \to Y$ is nullhomotopic and bounds a disk diagram $D' \to Y^\circledast$. Hence there is a lower complexity counterexample as $\mathsf{Comp}(D') < \mathsf{Comp}(D)$ since $\mathrm{grade}(Y) < \mathrm{grade}(Y_1)$.

We now prove Property 3.56.(2) for $\widetilde{X}^*$. Let Y_1, Y_2 be a counterexample to Property 3.56.(2) for $\widetilde{X}^*$, where $P_1 \to Y_1$ and $P_2 \to Y_2$ are paths joining distinct components of $Y_1 \cap Y_2$, and so that there is a disk diagram $D \to \widetilde{X}^*$ with $\partial_\mathsf{p} D = P_1 P_2$, and such that the complexity $(\mathsf{Comp}(D), |\partial_\mathsf{p} D|)$ is minimal among all such counterexamples. Note that D cannot have a spur at either end since we could then shorten both P_1 and P_2, and D cannot have a spur in the interior of either P_i. Likewise, D cannot have a cornsquare whose outerpath lies in the interior of either P_i, or following Lemma 2.6 we could produce a lower complexity counterexample. By Theorem 3.46, D is either a single cone-cell which we subsequently regard as a shell with trivial innerpath at an endpoint of P_1, P_2, or D has a shell R with innerpath S and outerpath Q, such that Q either lies in one of P_1, P_2 or in their concatenation $P_1 P_2^{-1}$. But the outerpath Q cannot be the complement of at most three pieces. Hence the cone-cell Y supporting R either factors through some Y_i, or some Y_i factors through Y. Without loss of generality, suppose $i = 1$. In the former case, we obtain a lower complexity diagram by substituting S for Q and omitting R. In the latter case $Y_2' = Y \cap Y_2$ is connected since otherwise there is a lower complexity counterexample whose diagram is obtained by substituting S for Q and omitting R. Since Y_2' is connected, we may assume that we chose P_2 to be a path in Y_2', i.e., if necessary replace it by a path in Y_2' with the same endpoints. We will show below that $P_1 P_2 \to Y^\circledast$ is nullhomotopic. and so there is a disk diagram $D' \to Y^\circledast$ with $\partial_\mathsf{p} D' = P_1 P_2$. Since D has a Y cone, but all cones of $D' \to Y^\circledast$ have lower grade, we see that $\mathsf{Comp}(D') < \mathsf{Comp}(D)$, which contradicts our minimal choice of D.

Consider the lift $\widetilde{P}_1$ of $\widetilde{P}_1$ to $\widetilde{Y}^\circledast$ and let $\widetilde{P}_1' \to \widetilde{Y}^\circledast$ be a geodesic with the same endpoints. Note that $P_1 \to Y^\circledast$ is path-homotopic to $P_1' \to Y^\circledast$ where P_1' is the projection of $\widetilde{P}_1'$. Note that $|P_1'| = |P_1|_{Y^\circledast} < \frac{1}{3}\|Y^\circledast\|$. Similarly, let $P_2' \to Y^\circledast$ be the projection of a geodesic $\widetilde{P}_2' \to \widetilde{Y}^\circledast$ that is path-homotopic to $P_2 \to Y_2' \to Y^\circledast$. Note that either Y_2' is a cone of $Y^\circledast$ in which case by hypothesis we have $|P_2'| < \frac{1}{3}\|Y^\circledast\|$, or Y_2' is a piece in which case $|P_2'| < \frac{1}{12}\|Y^\circledast\|$. Thus the closed path $P_1 P_2$ is homotopic in $Y^\circledast$ to the closed path $P_1' P_2'$. However, $P_1' P_2' \to Y^\circledast$ is nullhomotopic because of the following inequality:

$$|P_1' P_2'| = |P_1'| + |P_2'| < \frac{1}{3}\|Y^\circledast\| + \frac{1}{3}\|Y^\circledast\| < \|Y^\circledast\|.$$

Verification of Property 3.56.(3) for $\widetilde{X}^*$ is similar: Consider a counterexample Y_1, Y_2, Y_3 with pairwise nonempty intersection, and a disk diagram D with $\partial_\mathsf{p} D = P_1 P_2 P_3$, and suppose $(\mathsf{Comp}(D), |\partial_\mathsf{p} D|)$ is minimal among all such counterexamples. We can assume D has no spur and at most three cornsquares. Consequently, there must be a shell R. Its outerpath Q cannot be the concatenation of four or fewer subpaths of the $\{P_i\}$. Hence the cone-cell Y supporting R either factors through some Y_i, or some Y_i factors through Y. The former case is impossible since we would obtain a lower complexity diagram with the same properties. We may thus assume that some Y_i factors through Y. Without loss of generality assume $i = 1$. Since $Y_1 \hookrightarrow Y$ the triple Y, Y_2, Y_3 have nonempty pairwise intersection. Hence they triply intersect, for otherwise there is a counterexample with lower complexity obtained from D by substituting S for Q and omitting R. By Property 3.56.(2) for $\widetilde{X}^*$, we have $Y_2' = Y_2 \cap Y$ and $Y_3' = Y_3 \cap Y$ are connected.

We will show below that $P_1 P_2 P_3 \to Y^{\circledast}$ is nullhomotopic, and hence bounds a disk diagram $D' \to Y^{\circledast}$, which contradicts that D is minimal complexity since $\mathrm{grade}(Y) < \mathrm{grade}(Y_1)$. For notational ease, let $Y_1' = Y_1$. As in the proof of Property 3.56.(2), for each i we choose a geodesic $\widetilde{P}_i' \to \widetilde{Y}^{\circledast}$ with the same endpoints as a lift $\widetilde{P}_i$ of $P_i \to Y$, and we let $P_i' \to Y$ be its projection. Moreover $|P_i'| < \frac{1}{3} \|Y^{\circledast}\|$ for each i. Indeed, if Y_i' is a cone of Y then $|P_i'| = |\widetilde{P}_i'| = |P_i|_{Y^{\circledast}} < \frac{1}{3} \|Y^{\circledast}\|$. And if Y_i' is a piece (so $i = 2$ or $i = 3$) then $|P_i'| = |\widetilde{P}_i'| = |P_i|_{Y^{\circledast}} < \frac{1}{12} \|Y^{\circledast}\|$.

As P_i' and P_i are path-homotopic in $Y^{\circledast}$ for each i, we see that $P_1 P_2 P_3 \to Y^{\circledast}$ is homotopic to the closed path $P_1' P_2' P_3' \to Y^{\circledast}$. We complete the proof by showing that $P_1' P_2' P_3' \to Y^{\circledast}$ is nullhomotopic. This follows from the following inequality:

$$|P_1' P_2' P_3'| = |P_1'| + |P_2'| + |P_3'| < \frac{1}{3} \|Y^{\circledast}\| + \frac{1}{3} \|Y^{\circledast}\| + \frac{1}{3} \|Y^{\circledast}\| < \|Y^{\circledast}\|. \qquad \square$$

3.x   Missing Shells and Injectivity

Definition 3.60 (Map of Cubical Presentations). Let $\langle A \mid \{B_j\} \rangle$ and $\langle X \mid \{Y_i\} \rangle$ be cubical presentations. A *map between cubical presentations* $f : A^* \to X^*$ is a local-isometry $f : A \to X$ such that for each j there is an induced map $f : B_j \to Y_{f(j)}$ so that there is a commutative diagram:

$$
\begin{array}{ccc}
B_j & \to & Y_{f(j)} \\
\downarrow & & \downarrow \\
A & \to & X
\end{array}
$$

This data corresponds to a combinatorial map $A^* \to X^*$ sending the base to the base by a local-isometry, sending conepoints to conepoints, and sending vertical 1-cells to vertical 1-cells etc. When A^* and X^* are graded, then we require that the map between cones is grade-preserving.

Finally, when X^* has an angling rule, we assume the angling rule for A^* is induced by the angling rule for X^*.

Definition 3.61 (No Missing Shells). Let X^* be a cubical small-cancellation presentation. A map of cubical presentations $f : A^* \to X^*$ has *no missing shells* if the following holds:

For any non-replaceable shell R in a reduced diagram $D \to X^*$, if the outer-path Q of R lifts to A, then its innerpath S is path-homotopic in Y_i to a path S' so that the lift of Q extends to a lift of $QS' = \partial_{\mathsf{p}} R'$ to $\partial_{\mathsf{p}} R' \to B_j$ for some j with $i = f(j)$, and we have the following commutative diagram:

$$
\begin{array}{ccc}
Q & \to & A & \leftarrow & B_j \\
& \searrow & \nearrow & \downarrow \\
& & \partial_{\mathsf{p}} R' & \to & Y_i
\end{array}
$$

Remark 3.62 (Graded Generalization). Definition 3.61 generalizes as follows to the graded case:

(1) The innerpath $S \to Y_i$ is path-homotopic to $S' \to X$ via a disk diagram $D_S \to X^*_{\mathrm{grade}(Y_i)-1}$.
(2) The outerpath Q is path-homotopic in A^* to a path $Q' \to A$ via a disk diagram $D_Q \to A^*$ that projects to a disk diagram $D_Q \to X^*_{\mathrm{grade}(Y_i)-1}$.
(3) $Q'S' = \partial_{\mathsf{p}} R'$ lifts to a cone of A^*.

Typically S and Q are actually path-homotopic to S' and Q' in $Y_i^{\circledast}$.

The intuition here is that R can be replaced by $D_S \cup_S R \cup_Q D_Q$ to form a diagram whose complexity differs from D only at the count of cone-cells whose grade is less than $\mathrm{grade}(Y_i)$ in the motivating case where we actually have $D_Q \to Y_i^{\circledast}$.

Definition 3.63 (Induced Presentation). Let $\langle X \mid \{Y_i\} \rangle$ denote a cubical presentation. Let $A \to X$ be a local-isometry. The *induced presentation* is $\langle A \mid \{A \otimes_X Y_i\} \rangle$ which we shall denote by A^*. There is a natural map $A^* \to X^*$, which is our most common example of a map between cubical presentations.

When X^* is graded we shall grade A^* by assigning each component of $A \otimes_X Y_i$ the grade of Y_i.

Note that it is possible for a cone-cell of A^* to arise in more than one way from $A \otimes_X Y_i$, and even from Y_i and Y_j of different grades. One can either omit the higher grade copies, or leave these harmless redundant cone-cells. They won't appear internally in reduced diagrams.

In a graded setting we are often interested in the presentation induced by X_n^* which is the cubical presentation having only cones of grade $\leq n$. For instance, this naturally arises when A is one of the cones of X, and $n = \mathrm{grade}(A) - 1$.

Remark 3.64. Suppose $A^* \to X^*$ has no missing shells and $C \to A$ is a local-isometry, and let $C^* \to X^*$ be the map of the presentation induced by the composition $C \to A \to X$. It is possible for $C^* \to X^*$ to have missing shells. Indeed, consider the case where $C \to A$ is the outerpath of a shell of some cone-cell of A.

Definition 3.65 (Liftable Shells). Within the setting of Definition 3.61 (or rather, Remark 3.62), we say $A^* \to X^*$ has *liftable shells* if $QS \to Y_i$ lifts to $QS \to B_j$

for some $B_j \to Y_i$ and cone B_j of A^*, whenever Q is the outerpath of a non-replaceable shell. This stronger property does not allow a homotopy of either Q or S.

No missing shells and liftable shells coincide when $\dim(X) = 1$. One natural consequence of liftable shells is the following result that follows from the definitions and the universal property of the fiber-product (see Definition 8.8) applied to the liftable shell:

Lemma 3.66. *Let $A \to X$ and $B \to X$ be based local-isometries, and let C be the based component of $A \otimes_X B$. Suppose $A^* \to X^*$ and $B^* \to X^*$ have liftable shells. Then $C^* \to X^*$ has liftable shells.*

The following provides a criterion for liftable shells, and hence for the no missing shell property:

Lemma 3.67. *Let $\langle X \mid \{Y_i\} \rangle$ be a small-cancellation cubical presentation with short innerpaths (see Definition 3.69). Let $A \to X$ be a local-isometry and let A^* be the associated induced presentation. Suppose that for each i, each component of $A \otimes_X Y_i$ is either a copy of Y_i or is a contractible complex K with $\operatorname{diam}(K) \leq \frac{1}{2}\|Y_i\|$. Then the natural map $A^* \to X^*$ has liftable shells.*

In the graded case we have the following more general formulation: Suppose that for each component K of $A \otimes Y_i$, either K maps isomorphically to Y_i or $\operatorname{diam}(K) \leq \frac{1}{2}\|Y_i^\circledast\|$ and $\pi_1 K^* = 1$ where K^* is induced by $K \to Y_i^\circledast$.

Note that the graded generalization is arranged so that K lifts to $\widetilde{Y}_i^\circledast$. A natural scenario of the generalization is the special case when each component K of $A \otimes Y_i$ is either a copy of Y_i or satisfies $\operatorname{diam}(K) \leq \frac{1}{2}\|Y_i^\circledast\|$ with K either a contractible cube complex or a copy of a cone Y_j with $\operatorname{grade}(Y_j) < \operatorname{grade}(Y_i)$.

The analogous statement holds for medium innerpaths when each $\operatorname{diam}(K) < \frac{1}{2}\|Y\|$.

Proof. Consider a shell R with outerpath Q and innerpath S. If the two maps $A \leftarrow Q \to Y$ determine a map $Q \to A \otimes_X Y$ whose image lies in a component K that is a copy of Y, then the shell lifts. The other possibility will lead to a contradiction: Let $Q' \to K$ denote a geodesic that is path-homotopic to $Q \to K$ in K^*. So $|Q'| \leq \operatorname{diam}(K) \leq \frac{1}{2}\|Y^\circledast\|$ where the first inequality holds since $\pi_1 K^* = 1$ and the second holds by hypothesis. Let $S' \to \widetilde{Y}^\circledast$ be a geodesic with the same endpoints as $S \to \widetilde{Y}^\circledast$, so S' is path-homotopic to S in $Y^\circledast$, and note that $|S'| = |S|_{Y^\circledast} < |Q'|$ by short innerpaths. Both Q', Q and S', S are path-homotopic in $Y^\circledast$, and so $Q'S' \to Y^\circledast$ is essential since $QS \to Y^\circledast$ is essential. But then $Q'S' \to Y^\circledast$ is nullhomotopic since $|Q'S'| = |Q'| + |S'| < |Q'| + |Q'| \leq \frac{1}{2}\|Y^\circledast\| + \frac{1}{2}\|Y^\circledast\|$. $\qquad\square$

Theorem 3.68. *Let $f : A^* \to X^*$ be a map of cubical presentations, and suppose X^* is small-cancellation. If f has no missing shells then f is π_1-injective. Moreover the map $\widetilde{A}^* \to \widetilde{X}^*$ is injective on the cubical part.*

Proof. A path is *essential* if its lift to the universal cover is not closed. Suppose some essential path in A^* projects to a path in X^* that is not essential. Let (D, P) be such an example of minimal complexity. So $P \to A^*$ is essential, but projects to a path $P \to X^*$ bounding a disk diagram $D \to X^*$ that has minimal complexity among all such (D, P), and moreover having $|P|$ minimal among all those with D having minimal complexity.

By Theorem 3.46, D must have a cell with positive curvature. We can obviously exclude the case where D is trivial. If D has a cornsquare on P, then applying Lemma 2.6 to the subdiagram bounded by this cornsquare, we see that there is a new diagram with at most the same complexity as D but such that P contains the outerpath of this square. Since $A \to X$ is a local-isometry, this square lies in A, and so we can push P across the square to produce a smaller counterexample (D', P'). If D has a spur then we can remove two edges from P to obtain a smaller counterexample, and similarly, we can assume that there are no backtracks in $\partial_p D$ along the outerpath of some shell, for then we could fold to expose and remove a spur. Note that either a spur or a cornsquare must arise if D is a square diagram, when by the local-isometry hypothesis, the entire diagram D would lift to A.

Now suppose that D has some cone-cell. If D consists entirely of this cone-cell, say associated with $Y_i \to X$, then since there are no missing shells, P is path-homotopic to P' in A^* and P' lifts to a closed path in some B_j, and hence $P \to A^*$ is not essential.

If D has a shell R associated with $Y_i \to X$ and with outerpath Q and innerpath S, then since $f : A^* \to X^*$ has no missing shells, we see that Q is path-homotopic to Q' in A^*, and S is path-homotopic to S' in X^* through a diagram D_S with $\text{complexity}(D_S) < \text{grade}(R)$, and the path $Q' \to A$ factors through a map $Q' \to B_j$ that extends to a map $Q'S' = \partial_p R' \to B_j$ such that $f(j) = i$. This contradicts the minimality of the complexity of D. Indeed, when we modify P by Q' by S', we obtain an essential path $P' \to A$ that is null-homotopic in X^*, but bounding the lower complexity diagram obtained from D by removing R and adding D_S. $\qquad\square$

3.y   Short Innerpaths and Quasiconvexity

This section is used prominently in Chapter 14. Corollary 3.72 is used in the proof of Theorem 12.1 and Theorem 13.1. The material could probably be used to give a simplified treatment of parts of Chapter 5.

A useful property of the classical $C'(\frac{1}{6})$ theory is generalized in the form of the following hypothesis that requires that innerpaths of shells be short.

Definition 3.69 (Short Innerpaths). A small-cancellation presentation X^* has *short innerpaths* if for each positively curved shell Y with essential closed path $SQ \to Y^\circledast$, we have $|S|_Y < |Q|$.

Equivalently, for each essential closed path $SQ \to Y^\circledast$ with $\{\{S\}\}_Y < \pi$, we have $|S|_Y < |Q|$. (See Section 5.b and Figure 5.1 for $\{\{S\}\}_Y$.)

We are occasionally interested in related conditions:

X^* has *medium innerpaths* if $|S|_Y \le |Q|$ when $\{\{S\}\}_Y < \pi$ as above.

X^* has *tight innerpaths* if $|S|_Y < |Q|$ when $\{\{S\}\}_Y \le \pi$ as above.

X^* has *tiny innerpaths* if $|S'|_Y < |Q'|$ when $\{\{S\}\}_Y < \pi$, and where $Q = S_1 Q' S_2$ and $S' = S_2 S S_1$ and S_1, S_2 are wall-pieces (that are possibly trivial).

Note that if X^* is a small-cancellation presentation with short innerpaths, then Dehn's algorithm shows that the number of cone-cells in a reduced diagram $D \to X^*$ is bounded by $|\partial_\mathsf{p} D|$.

Lemma 3.70 (Pieces and $\{\{S\}\}$). *There exists α such that if $\langle X \mid \{Y_i\} \rangle$ is $C'(\alpha)$ then it has short innerpaths under the split-angling. In particular, this holds using the standard rectification when $\alpha = \frac{1}{16}$.*

Moreover, if X^* is $C'(\frac{1}{12})$, then we have the following more detailed statement: For any path $S \to R$ where R is a cone-cell in a reduced diagram $D \to X^*$ with ∂_R mapping essentially to the cone Y_i, if $\{\{S\}\}_{Y_i} \le \pi$ then S is the concatenation of at most 8 pieces, and if $\{\{S\}\}_{Y_i} < \pi$ then S is the concatenation of at most 7 pieces.

Consequently, short innerpaths holds when X^* is $C'(\frac{1}{14})$, tight innerpaths holds when X^* is $C'(\frac{1}{16})$, and tiny innerpaths holds when X^* is $C'(\frac{1}{18})$.

We first give a quick proof of the initial statement under the $C'(\frac{1}{16})$ hypothesis but under the additional assumption that we are using a standard rectification. We then give a more complicated proof of the second statement in the $C'(\frac{1}{14})$ case that depends on the details of the proof of Theorem 3.40.

Proof in $C'(\frac{1}{16})$ case. We assume that $\bar{D}$ is a standard rectification, and follow the argument of the proof of Theorem 3.31. By Lemma 3.28, all clusters of consecutive pieces with 0 defect at the transitions are of the form (C), (W), $(C_0 \overleftarrow{W})$, $(\overrightarrow{W}_0 C)$, $(W_0 W)$ and $(\overleftarrow{W}_0 C_0 \overrightarrow{W})$. And the smallest nonzero defect contribution in the split-angling is $\frac{\pi}{4}$. We deduce that a path $S \to Y_i$ with $\{\{S\}\}_{Y_i} \le \pi$ is the concatenation of at most 10 pieces, since it would have at least 5 clusters. (With a bit more care, we see that 5 clusters would imply a $W)_{\frac{\pi}{2}}(W$ and hence a $\frac{\pi}{2}$ defect, and so there are actually at most 9 pieces—and this can be improved further.) To see that $\{\{S\}\}_{Y_i} < \pi$ implies at most 8 pieces, observe that if there were 9 pieces, then S would decompose into at least 5 clusters, since each consists of at most two pieces as above. This yields four doses of at least $\frac{\pi}{4}$ of defect. As above, either there are at least 5 clusters, or there are two consecutive clusters that consist of two pieces and hence a $W)_{\frac{\pi}{2}}(W$. Thus there are actually at most

8 pieces. We conclude, that in the standard rectification, short innerpaths holds for $C'(\frac{1}{16})$ cubical presentations. □

Proof in the $C'(\frac{1}{12})$ case. We now show that when X^* satisfies $C'(\frac{1}{12})$, the innerpath S of a positively curved shell is the concatenation of at most 7 pieces. Thus X^* has short innerpaths if it satisfies $C'(\frac{1}{14})$.

The proof presumes that the reader is conversant with the cluster method used in the proof of Theorem 3.40, as we shall apply it in the case $c = 12, r = 6$.

The pieces of the path S are broken into clusters as in the proof of Theorem 3.40. Since we are only counting angle defects at the internal vertices of S, the initial and terminal clusters contribute a defect of at least $\frac{\pi}{6}$ for at most two pieces, or $\frac{\pi}{12}$ for at most one piece: Each other cluster contributes at least $\frac{\pi}{3}$ if it is covered by two pieces, or $\frac{\pi}{6}$ if it is covered by one piece. Thus the maximal number n of pieces corresponding to an innerpath S with $\{\{S\}\} \leq \pi$ satisfies: $4\frac{\pi}{12} + (n-4)\frac{\pi}{6} \leq \pi$. Hence $n \leq 8$. Similarly if $\{\{S\}\} < \pi$ then the strict inequality gives that S is the concatenation of at most $n \leq 7$ pieces. □

Theorem 3.71. *Let $\langle X \mid \{Y_i\} \rangle$ be a small-cancellation presentation with short innerpaths. Let $\langle A \mid \{B_j\} \rangle$ be another cubical presentation, and suppose the map $A^* \to X^*$ has no missing shells.*

Let $p, q \in \widetilde{A}^*$ (not conepoints) and let $\bar{p}, \bar{q}$ be their images in $\widetilde{X}^*$. Let γ' be an arbitrary geodesic joining $\bar{p}, \bar{q}$. Then there exists a geodesic $\gamma \to \widetilde{X}^*$ that is path-homotopic to γ' in X. And there is a path $\sigma \to \widetilde{A}^*$ between p, q such that the image $\bar{\sigma}$ of σ in $\widetilde{X}^*$ has the following property: $\bar{\sigma}\gamma^{-1}$ is the boundary path of a ladder $L \to \widetilde{X}^*$. If the ladder doesn't consist of a single cone-cell or vertex, then p and q lie in the interior of the outerpath of the first and last cone-cells and/or spurs of L on each end.

Proof. Let D be a minimal complexity diagram between paths σ and γ, where σ varies among paths in $\widetilde{A}^*$ with endpoints p, q, and where γ varies among all geodesics in $\widetilde{X}$ that are path-homotopic to γ' (so we aren't varying among geodesics in $\widetilde{X}^*$).

Observe that D has no cornsquares along γ or along σ. Likewise D has no outerpath of a shell along either σ or γ. It has none along σ for such a shell would map to a cone of $\widetilde{A}^*$ and thus σ could be passed through it. It has none along γ for such a shell would violate that γ and hence γ' is a geodesic by the short innerpaths hypothesis.

There are thus only two positively curved cells: A spur or cone-cell at $\bar{p}$ and a spur or cone-cell at $\bar{q}$. The result follows from Theorem 3.43. □

Corollary 3.72. *Let A^* and X^* be compact, and let X^* be a small-cancellation presentation with short innerpaths. Suppose $A^* \to X^*$ has no missing shells. Then $\widetilde{A}^* \to \widetilde{X}^*$ is a quasi-isometry. Moreover, if $\widetilde{X}$ is δ-hyperbolic, and the maximal diameter of a cone of X^* is κ, then $\widetilde{A}^* \to \widetilde{X}^*$ is a $(\delta + \kappa)$-quasiconvex.*

Proof. The map $\widetilde{A}^* \to \widetilde{X}^*$ is an embedding (on the underlying cubical complexes) by Theorem 3.68.

Consider a ladder as in Theorem 3.71, such that $(\mathsf{Comp}(D), |\sigma|)$ is minimal among all choices with the same endpoints as $\sigma \to \widetilde{A}^*$.

For the first statement, we show that the lengths of γ and σ are proportional since for some uniform μ, each cone-cell has at most μ edges on σ and at least one edge on γ, and note that pseudo-grids have the same length at their top and bottom. Let μ be an upperbound on the total number of vertices in a cone. Thus if a subpath σ' of σ lying in some cone Y_i has length exceeding μ then we see that it has a subpath mapping to a closed path of Y_i. If this closed path is essential, then it lifts to a closed path in a cone of A^*, and if it is not essential then it bounds a square diagram in $\widetilde{A}$. Either way, we violate the minimality of our choice.

For the second statement, we follow the terminology of Theorem 3.71, and examine the ladder it provides. Note that all points in each cone-cell lie within κ of each other within the cone it maps to, and hence within $\widetilde{X}^*$. Moreover, our minimality assumption ensures that each pseudo-grid is a grid (as cornsquares can be pushed out).

Each grid has the property that its left side has endpoints within κ of each other. Each 0-cube at the top of the grid is separated from its corresponding 0-cube at the bottom by the exact same dual curves that separate the endpoints of the left path. Consequently, by Corollary 2.16, the distance between corresponding points at the top and bottom are all the same as the distance between the endpoints of the left side. We have thus established the κ-fellow traveling of γ' and σ in $\widetilde{X}^*$, and we are done since the δ-fellow traveling of γ and γ' in $\widetilde{X}$ and hence in $\widetilde{X}^*$ holds by δ-hyperbolicity of $\widetilde{X}$. $\qquad\square$

Remark 3.73. The condition in Theorem 3.71 can be replaced by medium innerpaths: $|S|_Y \le |Q|$, but we must then allow γ to vary in $\widetilde{X}^*$ and not just in $\widetilde{X}$. So we conclude that γ exists, and still have a quasi-isometric embedding conclusion.

Lemma 3.74 (Convexity). *Let X^* be a small-cancellation presentation with short innerpaths. Suppose $A^* \to X^*$ has no missing shells.*

Let $2\alpha + \beta \le \frac{1}{2}$ where $\alpha, \beta > 0$. Suppose $|P|_{Y_i} < \alpha \|Y_i\|$ whenever P is a cone-piece of a translate of $\widetilde{Y}_j$ in a translate of $\widetilde{Y}_i$ (so one is not contained in the other).

Suppose that for any path P in the intersection of translates $\widetilde{Y}_i, \widetilde{A}$ in $\widetilde{X}$, either $|P|_{Y_i} < \beta \|Y_i\|$ or $\widetilde{Y}_i \subset \widetilde{A}$.

Then $\widetilde{A}^* \to \widetilde{X}^*$ embeds as a convex subcomplex.

Proof. Let γ be a geodesic in $\widetilde{X}^*$ whose endpoints lie on $\widetilde{A}^*$. Consider a minimal complexity diagram D between γ and a variable path σ in $\widetilde{A}^*$. As there is at most one cornsquare at each end of D, we see that if D has no cone-cells then D must be an arc so $\gamma = \sigma$.

There is a geodesic path γ' in D with the same endpoints as γ such that γ', γ together bound a maximal square subdiagram of D. Let E denote the subdiagram of D bounded by σ, γ'. Observe that E cannot have any cornsquares along σ or γ'. The former is impossible since σ is allowed to vary in $\widetilde{A}^*$, so a cornsquare can be pushed through it. The latter is impossible since we could again push past the square to make E smaller.

There are no outerpaths of shells along either γ' or σ. The latter is excluded since $A^* \to X^*$ has no missing shells by hypothesis. The former is excluded by our hypothesis that X^* has short innerpaths, so if such a shell existed it would contradict our hypothesis that γ' and hence γ is a geodesic. We conclude that there are at most two features of positive curvature in E, so E is either a single cone-cell or 0-cell or ladder by Theorem 3.43.

In fact we seek to conclude that $\gamma' = \sigma$ so E is an arc. And we now apply our strengthened α, β hypothesis to reach this conclusion.

Consider a cone-cell C in E that maps to the cone Y. The path $\partial_{\mathsf{p}} C$ is the concatenation $wxyz$ where w and y are cone-pieces (but w is trivial when C is an initial cone-cell, and y is trivial when C is a terminal cone-cell of the ladder), and x is a path in $\widetilde{C} \cap \widetilde{A}$, and z is a subpath of γ'. By our hypotheses, $|wxy|_Y \leq |w|_Y + |x|_Y + |y|_Y < \alpha\|Y\| + \beta\|Y\| + \alpha\|Y\| \leq \frac{1}{2}\|Y\|$. The diagram D and hence E is reduced, so $wxyz$ is essential and hence $|w|_Y + |x|_Y + |y|_Y + |z|_Y \geq \|Y\|$. Thus since $|wxy|_Y < \frac{1}{2}\|Y\|$ we see that $|z|_Y > \frac{1}{2}\|Y\|$ which contradicts that γ' is a geodesic in $\widetilde{X}^*$. $\square$

The following strengthens Theorem 4.1 in the metric small-cancellation setting:

Corollary 3.75. *Let $\langle X \mid \{Y_i\} \rangle$ be a $C'(\frac{1}{14})$ cubical presentation. Then for each cone Y_j, the lift $Y_j \subset \widetilde{X}^*$ embeds as a convex subcomplex.*

The same statement holds in the graded case when Y_j has maximal grade in the sense that Y_j is not a proper subcone of another cone.

Proof. Apply Lemma 3.74 in the case $\alpha, \beta = \frac{1}{14}$ and $A = Y_j$. $\square$

Chapter Four

Torsion and Hyperbolicity

4.a Cones Embed

The following theorem is lurking in the preliminaries to the proof of Theorem 3.46, but we deduce it from its conclusion:

Theorem 4.1. *Suppose $\langle X \mid \{Y_i\} \rangle$ is a small-cancellation presentation. Then each Y_i embeds in $\widetilde{X}^*$.*

Proof. Consider a disk diagram E whose boundary path P is a nonclosed immersed path in Y_i, and suppose the complexity of E is minimal among all diagrams whose boundary path lifts to a path with the same endpoints as the lift $\widetilde{P} \to \widetilde{Y}_i$. Let $C \to Y_i$ be a cone-cell such that P is a subpath of $\partial_p C$ which is immersed except perhaps at the endpoints of P. (We could use $\partial_p C = PP^{-1}$, and we would actually have to use something similar if Y_i is simply-connected.) Let $D = E \cup_P C$, as in Figure 4.1. Observe that D is reduced since a cornsquare on C would allow us to find a lower complexity diagram than E, and if some cone-cell C' of E formed a combinable pair with C then we could find a lower complexity diagram than E (roughly speaking, use $E - C'$).

By Theorem 3.46, either firstly: $\bar{D}$ consists of a single 0-cell or cone-cell, or secondly: $\bar{D}$ contains two or more spurs and/or cornsquares or shells with outerpath on $\partial_p D$. The first possibility contradicts that $\widetilde{P}$ has distinct endpoints in $\widetilde{Y}_i$. The second possibility is impossible since $\bar{D}$ has exactly one cone-cell along its boundary. $\qquad\square$

4.b Torsion

Theorem 4.2 (Torsion in $\pi_1 X^*$). *Let $\langle X \mid \{Y_i\} \rangle$ be a small-cancellation presentation where the shells arising in Definition 3.47 satisfy the following strengthened condition: For any shell R' associated to a cone Y and with outerpath Q' and innerpath S', we have $|Q'|_Y - |S'|_Y$ exceeds twice the diameter of a maximal cone-piece of Y plus four times the diameter of a maximal wall-piece of Y. Then if $g^n = 1$ for some $g \in \pi_1 X^*$, then g is conjugate to an element of $\mathrm{Aut}_X(Y_i)$ for*

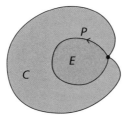

Figure 4.1. The diagram $D = E \cup_P C$.

some i. Thus g is represented by a closed path $\gamma \to X$ such that γ^n is a closed path in some Y_i.

As is clear from the proof, a more accurate statement of Theorem 4.2 need only compare the systole associated to the homotopy class $\partial_{\mathsf{p}} R$ with the cone-pieces and wall-pieces arising from overlaps with the convex hull of $\partial_{\mathsf{p}} R$ in $\widetilde{X}$.

Remark 4.3. The hypothesis of Theorem 4.2 holds when X^* is $C'(\frac{1}{20})$. Indeed, in that case, following Lemma 3.70, for an innerpath S' of a positively curved shell R' with outerpath Q', the innerpath S' is the concatenation of at most 7 pieces. We then observe that $|Q'|_Y - |S'|_Y > (\frac{13}{20} - \frac{7}{20})\|Y\| = \frac{6}{20}\|Y\|$.

Proof. Let γ be a closed combinatorial path in X that represents an element mapping to the conjugacy class of g, and moreover assume γ is a shortest such closed path among such choices. Thus γ^n is nullhomotopic in X^*. By possibly subdividing we can assume that $\pi_1 X$ acts without inversions on the hyperplanes of $\widetilde{X}$, and in particular, the bi-infinite geodesic γ^∞ is a geodesic [Hag07].

Consider a minimal complexity disk diagram D for γ^n. Note that D must contain a cone-cell, for otherwise γ would represent a torsion element in $\pi_1 X$ which is thus trivial and so $g = 1$. Let A be an annular diagram obtained as the local convex hull of $\partial_{\mathsf{p}} D$ in D, and let D' be the subdiagram of D, such that $D = D' \cup A$. Moreover, we assume that D, D', A is chosen such that the complexity of D' is minimized. Hence D' has no cornsquare, for any such square could be absorbed into A. By Theorem 3.46, the diagram D' has at least one shell R' with outerpath Q' and innerpath S' (we regard the case where D' is a single cone-cell as a shell with a trivial innerpath).

We now explain how to use the hypothesized strengthened nature of R'. Let $R = R'$, but regard it as a shell for D with outerpath Q and innerpath S, such that $S = US'V$ where U, V are (possibly trivial) wall-pieces. Here we use that every dual curve of A has one end on each boundary cycle of A.

We refer to the diagram on the right of Figure 4.2. Letting Y denote the cone-cell that R maps to, our extra hypothesis ensures that $|Q|_Y - |S|_Y$ exceeds

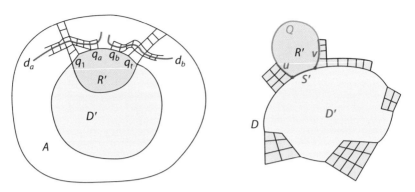

Figure 4.2. Finding a shell whose innerpath starts and ends with a wall-piece.

four times the maximal diameter of a wall-piece of Y plus twice the maximal diameter of a cone-piece of Y.

We refer to the diagram on the left of Figure 4.2 for the notation involved in the following argument. Consider the first and last edges q_1, q_t of $Q' = q_1 q_2 \cdots q_t$. For each i let s_i denote the dual curve emanating from q_i. By the construction of A, each of its squares forms a cornsquare on $\partial_p D$. Thus each s_i ends on $\partial_p D$. Moreover, s_i, s_j cannot cross within A, for if they crossed on one side of A then we would obtain a cornsquare on Q and violate minimal complexity, and if they crossed on the other side of A then: in case D' had another shell then its outerpath would be the concatenation of at most two wall-pieces, and in case D' were a single cone-cell then there would again be a cornsquare on R, and this would violate minimal complexity. For each i, if there is a square c_i at q_i then let d_i be the dual curve crossing s_i in c_i. Note that at most one end of d_i can be on $\partial_p D$ since γ^∞ is a geodesic. If d_i crosses both s_1, s_2 then Q' is a single wall-piece which is impossible. And d_i cannot end on Q' for then there is a cornsquare on Q' which violates minimal complexity. Thus one end of each d_i crosses one of s_1, s_t and the other ends on the part of $\partial_p D$ subtended by s_1, s_t. Finally, let a be the largest index such that d_a crosses s_1, and let b be the smallest index such that d_b crosses s_t. Observe that $a < b$ for otherwise, Q' is the concatenation of at most two wall-pieces. Thus there is no square at q_c for $a < c < b$, and hence q_c is on $\partial_p D$. We conclude that there is a single wall-piece $U = q_1 \cdots q_a$ associated to d_a, and a single wall-piece $V = q_b \cdots q_t$ associated to d_b, and $Q' = UQV$ where $Q = q_{a+1} \cdots q_{b-1}$ is a subpath of $\partial_p D$.

Consider the lift $D \to \widetilde{X}^*$ of $D \to X^*$, and let $\widetilde{Y}_j$ be the cone that R maps to. If $g \in \mathrm{Stab}(\widetilde{Y}_j)$ then we reach one of our desired conclusions, so assume $g \notin \mathrm{Stab}(\widetilde{Y}_j)$.

We now describe in turn the three possible contradictions illustrated in Figure 4.3.

If $|Q| \le |\gamma|$ then our choice of γ is contradicted, since $|S| = |US'V|$ and $|U| + |S'| + |V| < |Q|$ by hypothesis.

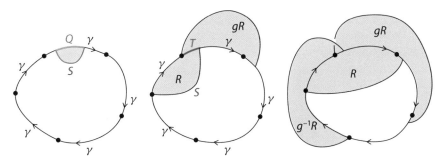

Figure 4.3. Three contradictory scenarios arising from the shell.

If $|\gamma| < |Q| < 2|\gamma|$ then considering $gQ \subset g\widetilde{Y}_j$, we find that $Q = \gamma'T$ for a cyclic permutation γ' of γ and a cone-piece T. This contradicts our choice of γ as γ' can be replaced by the path $ST^{-1} = US'VT^{-1}$ which is shorter since $|Q| - |T| > |US'VT^{-1}|$. Indeed, $|Q'| > |S'| + 2|U| + 2|V| + 2|T|$ by hypothesis, and so $|Q| - |T| = |Q|' - |U| - |V| - |T| > |S'| + |U| + |V| + |T|$.

If $2|\gamma| \leq |Q|$ then Q is the concatenation of two cone-pieces which contradicts our assumption that $|Q'|$ exceeds the sum of diameters of two cone-pieces and two wall-pieces. □

Corollary 4.4. *Let* $\langle X \,|\, \{Y_i\} \rangle$ *satisfy the hypothesis of Theorem 4.2 (e.g.,* X^* *is* $C'(\frac{1}{20})$). *Suppose* $\widehat{X} \to X$ *is a finite regular cover such that each* $Y_i \to X$ *lifts to an embedding in* $\widehat{X}$. *Then* $\pi_1 X^*$ *is virtually torsion-free.*

Proof. Let $\widehat{X}^*$ be the cover of X^* induced by $\widehat{X}$. By Theorem 4.2, a nontrivial torsion element of $\pi_1 X^*$ is conjugate to a closed path γ such that γ^n is a closed path in some cone Y_i. Note that γ does not lift to a closed path in Y_i. Thus since $Y_i \to X$ lifts to an embedding $Y_i \to \widehat{X}$, we see that γ does not lift to a closed path in $\widehat{X}$, and hence it is not closed in $\widehat{X}^*$. Thus γ does not represent an element of $\pi_1 \widehat{X}^*$. □

Perhaps the weakness of the proof of Theorem 4.2 is that it relies on minimal length instead of minimal area.

Problem 4.5. Does Theorem 4.2 hold for an arbitrary small-cancellation presentation? In particular, does it hold for a $C'(\frac{1}{12})$ cubical presentation?

4.c Hyperbolicity

The goal of this section is to obtain hyperbolicity of cubical small-cancellation quotients of word-hyperbolic groups. The criterion for hyperbolicity we use is the following result of Papasoglu proven in [Pap95]:

Proposition 4.6. *Let Γ be a graph. Suppose each geodesic bigon in Γ is ϵ-thin, in the sense that each side lies in the ϵ-neighborhood of the other side. Then each geodesic triangle is δ-thin, and so Γ is δ-hyperbolic.*

The following is generalized further in Theorem 14.2:

Theorem 4.7. *Let X^* be a small-cancellation presentation with short inner-paths. Suppose $\pi_1 X$ is word-hyperbolic and X^* is compact. Then $\pi_1 X^*$ is word-hyperbolic.*

Proof. We will show that there is a constant ϵ such that for any pair of geodesics γ_1, γ_2 with the same endpoints in $\widetilde{X}^*$, the bigon γ_1, γ_2 is ϵ-thin. Consequently, $\widetilde{X}^*$ is δ-hyperbolic by Proposition 4.6 and hence $\pi_1 X^*$ is word-hyperbolic since it acts properly and cocompactly on the cubical part of $\widetilde{X}^*$.

Let E be a minimal complexity diagram between γ_1 and γ_2. We refer to Figure 14.2. For each i, let D_i be a square subdiagram between γ_i and a geodesic λ_i in E having the same endpoints as γ_i. Let D be the subdiagram bounded by λ_1, λ_2. And finally, choose D_1, D_2 so that they have maximal area without λ_1 and λ_2 crossing each other. Since $\widetilde{X}$ is δ-hyperbolic, the geodesics γ_i, λ_i form an α-thin bigon for some constant α depending only on δ. Since $\widetilde{X}^*$ is small-cancellation, the diagram D must be a ladder by Theorem 3.43. Here we use that X^* has short innerpaths to ensure there is no shell with outerpath on λ_1 or λ_2. Finally, there exists β such that the bigon λ_1, λ_2 is β-thin. Indeed, the cones of X^* are compact so the parts of λ_1, λ_2 that travel on opposite sides of a cone-cell of D must lie within a uniformly bounded distance of each other. And for each grid in D, we see that its top and bottom are within ω of each other, where ω is an upperbound on the diameter of a wall-piece in X^*. In conclusion, γ_1, γ_2 forms an ϵ-thin bigon where $\epsilon = (\alpha + \beta + \omega)$. $\square$

Chapter Five

New Walls and the $B(6)$ Condition

5.a Introduction

In this section, we impose further hypotheses on the cones Y_i in a cubical presentation $X^* = \langle X \mid \{Y_i\} \rangle$. The main hypotheses are that each Y_i is a wallspace whose walls are collections of hyperplanes, and furthermore, the walls in Y_i have certain "convexity" properties—in the sense that any path in Y_i that starts and ends on the same wall, is either homotopic into that wall, or is "long" from a piece-count viewpoint. The wallspace cones and their properties allow us to define walls in $\widetilde{X}^*$, and these walls are the central focus of the section.

We define a notion of "length" of a path in a cone in Section 5.b and then describe the $B(6)$ wallspace structure on cones in Section 5.c. The construction of walls in $\widetilde{X}^*$ and quasiconvexity properties of these walls in $\widetilde{X}^*$ are examined in Sections 5.e, 5.f and 5.h. Conditions that imply that the set of walls is sufficiently rich to "fill" $\widetilde{X}^*$ are examined in Sections 5.k and 5.l. Malnormality properties of the wall stabilizers are treated in Sections 5.n, 5.o, 5.p, and 5.q.

5.b Total Defects of Paths in Cones

In classical $C(6)$ small-cancellation theory, a quick measure of the extent to which a path $P \to \partial R$ travels around R is the infimal number n where $P = P_1 \cdots P_n$ is an expression of P as the concatenation of pieces between other relators and our relator R. This can be thought of concretely in terms of a reduced diagram as on the left of Figure 5.1. If we assign $\frac{2\pi}{3}$ angles at the internal corners of R along P, then we obtain a total defect of $(n-1)\frac{\pi}{3}$. We shall now generalize the "piece-length" of P in R by focusing on the "total defect."

Let $\langle X \mid \{Y_i\} \rangle$ be a cubical presentation with an angling rule. Consider a path $P \to Y_i$. The *defect of P in Y_i* which we denote by $\{\!\{P\}\!\}_{Y_i}$ is the infimum of the sum of defects of angles along internal vertices of the path P in a cone-cell C_i mapping to Y_i within angled rectified diagrams $\bar{D}$ that have P as a path on the boundary of C_i and where each open edge of P is internal in D, and where $D \to X^*$ is reduced. See Figure 5.1.

It may be that some edge e of $P \to Y_i$ cannot arise within the interior of a reduced diagram D. Indeed, this happens precisely when e does not lie in any

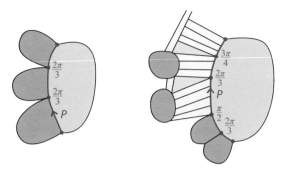

Figure 5.1. The total defect $\{\{P\}\}_{Y_i}$ of the path $P \to Y_i$ on the right is $\leq \frac{\pi}{3} + \frac{\pi}{2} + \frac{\pi}{3} + \frac{\pi}{4}$.

piece. In this case, the infimum of defects along P as it arises in such diagrams is the infimum of the empty set. Thus $\{\{P\}\}_{Y_i} = +\infty$.

For 1-cells e, e' in Y_i, we define $\{\{e, e'\}\}_{Y_i}$ to be $\inf(\{\{P\}\}_{Y_i})$ where $P \to Y_i$ is a path whose first and last 1-cells are e, e'. We define $\{\{v, v'\}\}_{Y_i}$ similarly for 0-cells v, v' in Y_i. For hyperplanes E, E' of Y_i, we define $\{\{E, E'\}\}_{Y_i}$ to equal $\inf(\{\{e, e'\}\}_{Y_i})$ where e, e' vary among 1-cells dual to E, E'.

A relationship between piece-length and $\{\{S\}\}$ is described in Lemma 3.70. It is shown there that if X^* is $C'(\frac{1}{12})$ then $\{\{S\}\} \leq \pi$ implies that S is the concatenation of at most 8 pieces, and similarly if $\{\{S\}\} < \pi$ then S is the concatenation of at most 7 pieces.

5.c   Generalization of the $B(6)$ Condition

We will now add more elaborate structure to generalize other aspects of the classical $C'(\frac{1}{6})$-$T(3)$ and $C'(\frac{1}{4})$-$T(4)$ metric small-cancellation theories, to higher dimensions along the lines of the $B(6)$-$T(3)$ and $B(4)$-$T(4)$ theories considered in [Wis04].

Definition 5.1 (Generalized $B(6)$). A cubical presentation $\langle X \mid \{Y_i\} \rangle$ satisfies the $B(6)$ *condition* if it satisfies the following conditions:

(1) (Small-Cancellation) $\langle X \mid \{Y_i\} \rangle$ is a small-cancellation presentation.
(2) (Wallspace Cones) Each Y_i is a wallspace (see Definition 7.1) where each *wall* in Y_i is the union $\sqcup H_j$ of a collection of disjoint embedded 2-sided hyperplanes, and there is an embedding $\sqcup N(H_j) \hookrightarrow Y_i$ of the disjoint union of their carriers into Y_i. Each such collection separates Y_i. Each hyperplane in Y_i lies in a unique wall.
(3) (Hyperplane Convexity) If $P \to Y_i$ is a path that starts and ends on vertices on 1-cells dual to a hyperplane H of Y_i and $\{\{P\}\}_{Y_i} < \pi$ then P is path-homotopic in Y_i to a path $P' \to N(H) \to Y_i$.

Figure 5.2. Hyperplane convexity vs. wall convexity in cones.

(4) (Wall Convexity) Let S be a path in Y that starts and ends with 1-cells dual to the same wall of Y. If $\{\{S\}\}_Y < \pi$ then S is path-homotopic into the carrier of a hyperplane of that wall.
(5) (Equivariance) The wallspace structure on each cone Y is preserved by $\mathrm{Aut}_X(Y)$.

For brevity, the term $B(6)$ will always refer to the "generalized $B(6)$ condition" of Definition 5.1, and not to the $B(6)$ condition for 2-complexes studied in [Wis04].

Remark 5.2 (Graded $B(6)$). When X^* is a graded cubical presentation, we require that wallspaces on cones be *consistent* in the following sense: If Y_i and Y_j are cones of X^*, and the map $Y_i \to X$ factors as $Y_i \to Y_j \to X$, then we require that the walls of Y_i map to walls of Y_j, and that their corresponding halfspaces map to halfspaces. Consistency is essentially a generalization of Condition 5.1.(5).

Note that Condition 5.1.(5) ensures that the wall equivalence relation (defined below) on hyperplanes in $\widetilde{X}^*$ agree locally with the wall structure on each Y.

Note that Condition (3) is not implied by Condition (4) since the latter requires that the path start and end with dual 1-cells, and not merely start and end at endpoints of dual 1-cells (see Figure 5.2).

Definition 5.3 (Embedding Properties of Hyperplanes in Cones). The following conditions hold for B(6) cubical presentations. In total, we find that for each hyperplane H of a cone Y, the carrier $N(H) \cong H \times I$ and there is an embedding $N(H) \hookrightarrow Y$.

(1) (2-sided) Each hyperplane of each Y_i is 2-sided, in the sense that its dual 1-cubes can be oriented so that dual 1-cubes that are opposite sides of a 2-cube are oriented the same way.
(2) (No self-intersection) No hyperplane H in Y_i is dual to all 1-cubes on the boundary of a 2-cube in Y_i.
(3) (No self-osculation) No hyperplane H in Y_i is dual to two distinct 1-cubes that share a 0-cube.

Remark 5.4. In view of Lemma 3.70, when X^* is $C'(\frac{1}{12})$, Conditions (3) and (4) follow if the following holds for each path $P \to Y$ that is the concatenation of at most 7 pieces: If P starts and ends at endpoints of 1-cells dual to hyperplanes in the same wall of a cone Y then these hyperplanes are the same and P is path-homotopic into its carrier.

Lemma 5.5. *If X^* is $B(6)$ then X^* has medium innerpaths.*

Moreover, if X^* is $B(6)$ and has the following additional property then X^* has short innerpaths: For each closed essential path P in a cone Y of X^*, there exists a wall W' of Y containing at least four hyperplanes dual to edges of P.

Proof. Let C be a shell in a reduced diagram $D \to X$, let Q and S denote its outerpath and innerpath, and let Y be the cone supporting C. We will show that $|S|_Y$ is equal to the number of walls of Y that separate the endpoints of S. As each such wall separates the endpoints of Q and is thus dual to some edge of Q, we see that $|Q|$ is greater than or equal to the number of such walls.

Consider a wall W of Y that separates the endpoints of S. We now show that in the lift $\widetilde{S} \to \widetilde{Y}$, all lifts of edges associated to W are dual to the same hyperplane of $\widetilde{Y}$. Let S' be a subpath of S that starts and ends with edges dual to W. Note that $\{\{S'\}\}_Y \leq \{\{S\}\}_Y < \pi$. By Definition 5.1.(4), the path S' is path-homotopic to a path in the carrier of a single hyperplane of W. The path-homotopy lifts to a path-homotopy of $\widetilde{S'}$, and so the edges in $\widetilde{S'} \to \widetilde{Y}$ are also dual to the same hyperplane.

We now deduce the second claim from the previous argument utilizing the hypothesized wall W' which contributes at most one to $|S|_Y$ but at least three to $|Q|$, and hence $|Q| > |S|_Y$. Indeed, S can traverse edges dual to at most one hyperplane of W'. But $|S|_Y$ equals the number of walls of Y separating the endpoints of S, and hence W' is dual to at least three distinct edges of Q. $\square$

5.d   Cyclic Quotients and the $B(6)$ Condition

Cyclic quotients are especially accessible to the $B(6)$ small-cancellation theory because a wallspace structure on the cones is then often readily achievable. We refer the reader to Corollary 5.48.

One subdivides the cube complex X, so that the local-isometry $Y \to X$ (representing a cone associated to the cyclic subgroup to quotient) has twice the number of separating hyperplanes. Now hyperplanes (that don't already separate) can be paired in a manner respecting almost all small-cancellation conditions provided a girth condition is satisfied.

An interesting example is a twisted product $Y = B \rtimes S$ where B is a CAT(0) ball and S is a subdivided circle (this is just a B-bundle over S). Presumably $B \rtimes S$ is a product when X is special, but in general, it is possible to have some hyperplanes that wrap multiply around the circle base space.

For Dehn fillings, it is natural to use the entire infinite cylinder covering a torus as the relator; see Theorem 15.6. So, this is a case where the compactness hypothesis on the cones is not appropriate.

Relative hyperbolicity gives an upperbound on diameters of pieces between planes. This gives an upperbound on diameters between (distinct) cylinders. A Dehn filling corresponding to an element that is long enough relative to these pieces and the walls gives us a small-cancellation quotient, and should be virtually special because of the subdivision and paired splicing.

These ideas can be generalized to relators with large abelian automorphism groups. See also Theorem 15.6.

5.e Embedding Properties of the Cones and Hyperplane Carriers

The aim of this section is to show that certain very short circuits of cones and hyperplane carriers do not exist in $\widetilde{X}^*$. We have in mind the situations offered by the split-angling, though similar statements hold in most cases for the type-angling.

Recall that Theorem 4.1 states that in a small-cancellation cubical presentation $\langle X \mid \{Y_i\} \rangle$, each Y_i embeds in $\widetilde{X}^*$. We will now augment this with several related embedding and intersection results.

Lemma 5.6. *Let $\langle X \mid \{Y_i\} \rangle$ be a small-cancellation presentation. Then the intersection of two cones in $\widetilde{X}^*$ is connected.*

If we strengthen the hypothesis as stated in Theorem 3.46.(3) to instead state that there be at least four features of positive curvature, one can deduce using an analogous argument that three pairwise intersecting cones must triply intersect.

Proof of Lemma 5.6. Suppose that $Y_1 \cap Y_2$ is not connected. Let D be a minimal complexity disk diagram between immersed paths $\gamma_1 \to Y_1$ and $\gamma_2 \to Y_2$ that start and end on points p, q in distinct components of $Y_1 \cap Y_2$. In particular, D is of minimal complexity among all such possibilities where the paths γ_1, γ_2 are allowed to vary among immersed paths in Y_1, Y_2 starting and ending on p, q. We claim that D is a degenerate ladder with no 2-cells, and more specifically, $\gamma_1 = \gamma_2$.

Let $C_1 \to Y_1$ be a cone-cell whose boundary path contains γ_1 (e.g., $\partial_{\mathsf{p}} C_1 = \gamma_1 \gamma_1^{-1}$) and define C_2 analogously. Consider the diagram $E = C_1 \cup_{\gamma_1} D \cup_{\gamma_2} C_2$. See Figure 5.3.

The minimal complexity of D ensures that E is reduced. However, if D contains a square or cone-cell then E violates Theorem 3.43. $\square$

Lemma 5.7 (Intersections of cones). *Let $\langle X \mid \{Y_i\} \rangle$ be a small-cancellation presentation. The intersection between distinct cones in $\widetilde{X}^*$ is CAT(0).*

Proof. Note that each cone-cell embeds by Theorem 4.1. If either Y_i or Y_j is contractible, then the intersection is a locally-convex subcomplex of a CAT(0) cube complex, and hence itself a CAT(0) cube complex by Lemma 2.11. So, suppose that neither Y_i nor Y_j are contractible. Consider a closed essential path

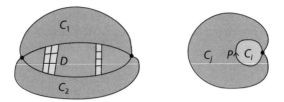

Figure 5.3. On the left is $E = C_1 \cup D \cup C_2$. On the right is $D = C_i \cup_P C_j$.

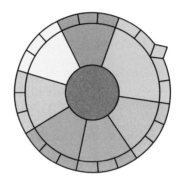

Figure 5.4. A self-crossing hyperplane: The simply-connected cubical presentation above has one cone-cell consisting of a 7-cycle in the center, and it is surrounded by seven further cone-cells that contain some 1-cubes and some 2-cubes.

P in their intersection. We regard $P \to Y_i$ as the boundary path of a cone-cell C_i mapping to Y_i. Likewise regard $P^2 \to Y_j$ as $\partial_p C_j$ where C_j is a cone-cell mapping to Y_j. Consider the disk diagram $D = C_i \cup_P C_j$ obtained by gluing these cone-cells together. See Figure 5.3. The diagram $D \to X^*$ is reduced but shows that X^* is not a small-cancellation presentation since it violates the conclusion of Theorem 3.43. □

Example 5.8 (A self-crossing hyperplane). In Figure 5.4 we illustrate a 2-dimensional cubical presentation $\langle X \mid Y_1, \dots, Y_8 \rangle$. Observe that all cones are embedded and have well-behaved hyperplanes, and $\widetilde{X}^* = X^*$, but there is a hyperplane in $\widetilde{X}^*$ that self-crosses. By using the n-fold generalization of this 7-fold picture, and by subdividing along the outside, one can produce analogous examples such that $C'(\alpha)$ is satisfied for arbitrary $\alpha > 0$.

Definition 5.9 (No Acute Corners). An angling rule for X^* has *no acute corners* if each cone-cell angle is $\geq \frac{\pi}{2}$ (except for singular doubly-external corners). We note that the split-angling has no acute corners, as does the type-angling provided that each type is ≥ 4.

No acute corners ensures that in a small-cancellation presentation, the outer-path of a non-replaceable shell cannot consist of a piece because of the following:

Lemma 5.10 (No acute consequences). *Let X^* be a small-cancellation presentation with no acute corners. Let $D \to X^*$ be a reduced diagram containing a non-replaceable cone-cell C. Suppose each boundary component of $\partial_{\mathsf{p}} C$ is a wall-piece or cone-piece. Then there is a reduced diagram E containing a copy of C as an internal cone-cell, and $\kappa_E(C) \geq \kappa_D(C)$.*

Proof. We form E from D by attaching corresponding rectangles and cone-cells along the parts of $\partial_{\mathsf{p}} C$ that are wall-pieces and cone-pieces. (A cone-piece Q extends to a cone-cell whose boundary path is QQ^{-1}.) Note that E is reduced since C is not replaceable.

Now $\partial_{\mathsf{p}} C = Q_1 S_1 \cdots Q_k S_k$ where the external boundary paths are the Q_i. Let c_i and c_i' denote the corners at $Q_i S_i$ and $S_i Q_{i+1}$ (coefficients mod k).

As there are no acute corners, assuming the corners at c_i, c_i' are nonsingular, we have $\sphericalangle(c_i), \sphericalangle(c_i') \geq \frac{\pi}{2}$, and so: $\kappa_D(C) = 2\pi - \sum_{i=1}^{k} \left((\pi - \sphericalangle_D(c_i)) + (\pi - \sphericalangle_D(c_i')) \right) - \sum_{c \in S_i} (\pi - \sphericalangle(c)) \leq 2\pi - \sum_{i=1}^{k} (\frac{\pi}{2} + \frac{\pi}{2}) - \sum_{c \in S_i} (\pi - \sphericalangle(c)) \leq 2\pi - \sum_{i=1}^{k} \left((\pi - \sphericalangle_E(c_i) + (\pi - \sphericalangle_E(c_i')) \right) + \sum_{c \in S_i} (\pi - \sphericalangle_E(c)) = \kappa_E(C)$. A similar statement holds when there are singular corners. $\square$

Reformulated appropriately, the following is a special case of Theorem 3.68.

Lemma 5.11 (Embedding hyperplane carriers). *Let $\langle X \mid \{Y_i\} \rangle$ be a small-cancellation presentation with no acute corners.*

Suppose Condition 5.1.(3) holds. Let H be an immersed hyperplane in the nonpositively curved cube complex of $\widetilde{X}^*$.

(1) *If Condition 5.3.(1) holds then H is 2-sided.*
(2) *If Condition 5.3.(2) holds then H does not self-intersect.*
(3) *If Condition 5.3.(3) holds then H does not self-osculate.*

In conclusion, if all three conditions hold, then $N(H) \to \widetilde{X}^*$ is an embedding and $N(H) \cong H \times I$.

The *carrier* of an immersed hyperplane $H \to Z$ is the union of copies of cubes of Z whose midcubes are the cubes of H. These copies of cubes are glued together along subcubes in $N(H)$ as they are in Z. The result, $N(H)$, is a (possibly twisted) I-bundle over H. Note that $N(H)$ is nonpositively curved, and that $N(H) \to Z$ is a local-isometry.

Example 5.12. When the immersed hyperplanes of X are not 2-sided, in general, H might not be 2-sided so $N(H) \not\cong H \times I$. For instance, let X be a Moebius strip obtained by identifying opposite sides of a square with a twist, and let Y_n denote a connected n-fold cover of X. Then $\langle X \mid Y_n \rangle$ is a small-cancellation presentation for each n, as it satisfies $C'(\alpha)$ for each $\alpha > 0$ since there are no pieces. However, the cube complex of $\widetilde{X}^*$ is a Moebius strip for odd n.

Proof of Lemma 5.11. We will concentrate on establishing the main conclusion where all three conditions are satisfied. Although we give a self-contained argument, we are essentially verifying that $N(H) \to X^*$ has no missing shells, so the conclusion holds by Theorem 3.68.

Consider two points $p, q \in N(H)$ that map to the same point of $\widetilde{X}^*$. Let $D \to X^*$ be a disk diagram of minimal complexity among all diagrams whose boundary path $P \to N(H)$ starts on p and ends on q. If D has a cornsquare, then since $N(H) \to X$ is a local-isometry, we see that this square could be absorbed into $N(H)$ and a lower complexity diagram could be produced.

Suppose that D has a shell C with boundary path QS, with innerpath S and outerpath Q and such that C maps to the cone Y. First consider the case where $Q \to N(H)$ does not pass through any dual 1-cell of $H \to N(H)$. Then either Q is a wall-piece in the cone Y or the rectangle of Y containing Q along its external boundary maps into Y. Since C is non-replaceable, the former case is impossible by Lemma 5.10. In the latter case, Condition 5.1.(3) implies that $S \to Y$ is path-homotopic in Y to a path $Q' \to N(H)$ so C can be replaced by a lower complexity diagram between S and Q', and Q can be replaced by Q', and we have a lower complexity counterexample.

We next consider the case where Q passes through a dual 1-cell of H. Then there exists $L \to N(H)$ where $L \cong I_n \times I$ is a rectangle that is dual to H and $Q \to N(H)$ factors as $Q \to L \to N(H)$, and by the local convexity of the cone Y, we have $L \to \widetilde{X}^*$ actually factors as $L \to Y \to \widetilde{X}^*$. Therefore Condition 5.1.(3) again implies that C can be replaced by a square diagram, to reduce the complexity as above.

We emphasize that in the special case of the above situation where S is trivial and Q is a wall-piece in Y, and C is replaced by a square diagram, we see that the endpoints p', q' of Q in $N(H)$ are actually equal to each other, and moreover, following Lemma 2.3, the 1-cubes dual to H at $p' = q'$ must be equal to each other as well. In the case where Q is not a wall-piece (this also includes the situation where Q passes through a dual 1-cell) we see that Q lies on a rectangle L dual to H such that $L \to C$. It is in this case that we must employ Condition 5.3.(3) to see that the 1-cubes dual to H at p', q' are actually equal in C (as L maps to a cylinder in C) and so these 1-cubes are equal in $\widetilde{X}^*$ and hence in H itself.

In particular, we are able to remove C, Q from D, P to obtain a lower complexity diagram D', P' such that P' has the same endpoints as P in $N(H)$.

We now focus on the 2-sidedness of H to see that $N(H) \cong H \times [-\frac{1}{2}, \frac{1}{2}]$. Consider a minimal complexity diagram D whose boundary path P is a path on $N(H)$ that passes through an odd number of dual 1-cubes of H. The additional presence of Condition 5.3.(1), implies that for any hyperplane H_i of Y_i, any

Figure 5.5.

closed path in $N(H_i)$ passes through dual 1-cubes of H_i an even number of times. Therefore, when QS is the boundary path of a cone-cell C in Y_i that lies in $N(H_i)$, we see that Q and S pass through dual cubes of H the same number of times (where H_i is a component of $H \cap Y_i$). In particular, when S is the trivial path, then the smaller complexity diagram D', P' that we obtained above has the property that P and P' have the same parity.

When D is a square diagram, it is clear that P passes through an even number of dual 1-cubes of H, since the dual curves in D provide a pairing of such dual 1-cubes. $\square$

The following result will play a fundamental role in understanding the walls of $\widetilde{X}^*$ by revealing combinable pairs of cone-cells in certain disk diagrams.

Lemma 5.13. *Let X^* be a small-cancellation cubical presentation that satisfies Hypothesis 5.1.(3). Suppose X^* has no acute corners. Then each hyperplane H in $\widetilde{X}^*$ has connected intersection with each cone $Y \subset \widetilde{X}^*$.*

Moreover, if $H \cap Y \neq \emptyset$ then $N(H) \cap Y$ is equal to $N(H \cap Y)$ which denotes the carrier in Y of the hyperplane $H \cap Y$ of Y.

Remark 5.14. When each Y_i is a pseudograph (see Definition 9.8) in the sense that its hyperplanes are CAT(0), one can reach the following stronger result: Any path $P_h \to N(H)$ whose endpoints lie on Y, is path-homotopic through a square diagram in $\widetilde{X}$ to a path $P_y \to Y$. This can be used to simplify some of the proofs below in the pseudograph setting. However such a square diagram might not exist in general. For instance, the rectangle at the top of Figure 5.5 cannot be pushed towards the cone at the bottom without passing through three essential cone-cells. The reader can think of the light intermediate diagram as being made of squares.

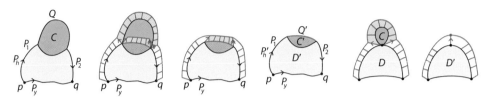

Figure 5.6. The four diagrams on the left illustrate how a hyperplane can be pushed across a cone-cell whose outerpath is not a real wall-piece. The right two figures illustrate the situation when the innerpath is trivial.

Proof. Let p, q be points in $N(H) \cap Y$ and let $D \to X^*$ be a diagram between paths $P_h \to N(H)$ and $P_y \to Y$ from p to q, and suppose D has minimal complexity among all such diagrams and paths joining p, q. We can moreover assume that P_h has no backtracks.

Consider the diagram $E = D \cup_{P_y} C_y$ where $C_y \to Y$ is a cone-cell with $\partial_p C_y = P_y P_y^{-1}$. The diagram E is reduced since an absorbable square or cornsquare on C_y or a cone-cell that is combinable with C_y would allow us to produce a diagram as above with lower complexity than D. By Theorem 3.46, E is either a cone-cell, a ladder, or has at least three positively curved features. If E is a cone-cell then D is an arc, so $P_y = P_h$ which is our desired conclusion. Otherwise E has at least two shells and/or squares and/or spurs, so at least one of these is along P_h.

We already excluded a spur on P_h, and if there were a cornsquare along P_h, then by local convexity of $N(H)$, after shuffling we could absorb a square to produce a lower complexity diagram.

We are left to consider the case of a shell C_i mapping to a cone Y_i, with $\partial_p C_i = QS$ where $Q \subset P_h$ is the outerpath and S is the innerpath and $\{\{S\}\}_{Y_i} < \pi$.

If H and Y_i are disjoint in $\widetilde{X}^*$ then Q is the outerpath of the non-replaceable shell C_i which violates Lemma 5.10.

Thus we can assume that the ladder in $N(H)$ containing Q lies in Y_i. Condition 5.1.(3) implies that S is path-homotopic in Y_i to a path $Q' \to N(H \cap Y_i)$. So we can let $C' \to Y_i$ be a diagram whose boundary path is SQ'. Letting $P_h = P_1 Q P_2$, we observe that $P'_h = P_1 Q' P_2$ is still a path in $N(H)$. This uses that hyperplanes of Y_i do not self-osculate to see that the 1-cells dual to Y_i at the initial points of Q and Q' must be the same, and likewise for the 1-cells at the terminal points. By replacing the path Q with the path Q', and replacing the cone-cell C_i with a diagram $C'_i \to Y_i$, we obtain a lower complexity diagram $D' \to X^*$ between P_y and P'_h. We refer the reader to Figure 5.6.

We now show that $N(H \cap Y) \to (N(H) \cap Y)$ is locally surjective on 1-skeleta. Consider an edge e in $N(H) \cap Y$ with a vertex v in $N(H \cap Y)$. If e is dual to H then $e \subset N(H \cap Y)$. Suppose e is not dual to H but e is instead on a square $s \subset N(H)$ containing another edge e' at v that is dual to $H \cap Y$. Since e and e' are both in Y, local convexity implies that s is in Y. Thus H contains the midcube of s parallel to e, and so s is in $N(H \cap Y)$.

Figure 5.7. A self-crossing wall in $\widetilde{X}^*$.

We now verify that $N(H) \cap Y = N(H \cap Y)$. Obviously $N(H \cap Y) \subset (N(H) \cap Y)$. To see that $(N(H) \cap Y) \subset N(H \cap Y)$, observe that since $N(H) \cap Y$ is path-connected as above, hence $N(H \cap Y)$ and $N(H) \cap Y$ have the same 1-skeleton by local-surjectivity, and hence they have the same cubes since they are locally-convex. □

5.f Defining Immersed Walls in X^*

Condition 5.1.(2) allows us to define an equivalence relation on hyperplanes in $\widetilde{X}^*$ that is generated by $A \sim B$ provided that for some translate of some cone Y_i in $\widetilde{X}^*$, we have $A \cap Y_i$ and $B \cap Y_i$ lie in the same wall of Y_i.

The *walls* of $\widetilde{X}^*$ are defined to be collections of hyperplanes of $\widetilde{X}^*$ corresponding to equivalence classes. Our main goal in this section is to show that these walls embed and separate the cube complex of $\widetilde{X}^*$.

Example 5.15. The simply-connected complex in Figure 5.7 whose five 2-cells are wallspaces, indicates that walls can self-cross without the small-cancellation hypothesis of Definition 5.1.

Definition 5.16. Let X^* be a $B(6)$ cubical presentation. Let W be a wall in $\widetilde{X}^*$. The *structure graph* Γ_W of a wall W is a bipartite graph whose 0-cells Γ_W^0 are in two classes Γ_W^h and Γ_W^c where Γ_W^h consists of *hyperplane vertices* corresponding precisely to the hyperplanes in W, and Γ_W^c consists of *cone vertices* corresponding precisely to the cones that are intersected by some hyperplane of W. Two vertices of Γ_W are connected by an edge precisely when the corresponding spaces have a nonempty intersection.

A combination of various results proven below will show that:

Theorem 5.17. *Let Γ_W be the structure graph of a wall W in $\widetilde{X}^*$.*

(1) Γ_W *is a tree.*
(2) *If $u \neq v \in \Gamma_W^h$ then the corresponding spaces are disjoint.*

(3) If $u \neq v \in \Gamma_W^c$ then the corresponding spaces U, V are disjoint unless u, v are both adjacent to some $h \in \Gamma_W^h$ corresponding to a hyperplane H and $U \cap V$ contains a 1-cell dual to H.

(4) If $u \in \Gamma_W^c$ and $h \in \Gamma_W^h$ then the corresponding spaces U, H are disjoint unless u, h are adjacent, and U contains a 1-cell dual to H.

The first statement of Theorem 5.17 is proven in Theorem 5.20 and the final three statements in Theorem 5.17 follow from the following:

Theorem 5.18. *Let X^* satisfy the $B(6)$ condition.*

Distinct midcubes of a square (and hence of a cube) cannot lie in hyperplanes of the same wall W.

If H_1, H_2 are hyperplanes in a wall W, and Y is a cone, then $H_1 \cap Y$ and $H_2 \cap Y$ lie in the same wall of Y.

Note that Theorem 5.18 provides a stronger conclusion than Lemma 5.11 under a similar hypothesis.

Proof. Suppose that there is an alternating sequence $H_0, Y_1, H_1, Y_2, H_2, \ldots,$ Y_r, H_r of hyperplanes and cones such that H_{i-1}, H_i belong to the same wall of Y_i for $1 \leq i \leq r$ but H_0, H_r are distinct hyperplanes that pass through the same square or cone Z of $\widetilde{X}^*$. We permit backtracking here, so it is possible that $H_i = H_{i+1}$ for various i. This facilitates the proof, which hinges upon minimal area instead of minimal length.

Let P be a closed path that travels along the corresponding hyperplane carriers and cones, and starts and ends on the square or cone Z. Let $D \to X^*$ be a diagram for P, and assume that D, P is minimal in the sense that D has minimal complexity among all such alternating sequences, paths P, and diagrams for P.

Our argument has two stages: In the first stage we show that if D has minimal complexity, then D can be augmented to form a "collared diagram" E which is obtained by wrapping an annular ladder A around D such that A contains a self-intersecting path of W. In the second stage we show that absorbing and combining preserves this collar structure, and so by passing to a minimal complexity collared diagram of this type, we see that it cannot exist.

Observe that P is a concatenation of (possibly trivial) paths $P_0^h P_1^y P_1^h P_2^y \cdots$ $P_r^y P_r^h P^z$ where each $P_i^y \to Y_i$ is a path in a cone, and each $P_i^h \to N(H_i)$ is a path in the carrier of a hyperplane, and $P^z \to Z$ is a path in the square or cone Z that W crosses in two locally inequivalent ways (see Figure 5.11). When Z is a square containing midcubes belonging to H_0 and H_r, it actually lies in $N(H_0) \cap N(H_r)$, and we can therefore assume that P^z is trivial in this case by possibly absorbing P^z into P_0^h (or into P_r^h).

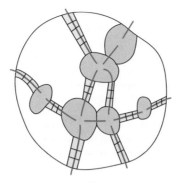

Figure 5.8. The "dual graph" in $D \to \widetilde{X}^*$ associated to the preimage of a wall W of $\widetilde{X}^*$.

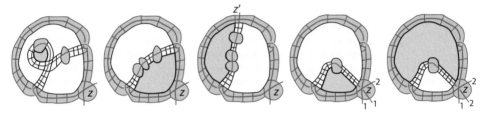

Figure 5.9. W-ladders in D yield lower complexity counterexamples.

We will show that P cannot contain a 1-cell e dual to some hyperplane in W in the sense that P_i^h does not pass through any 1-cell dual to H_i, and P_i^y does not traverse any 1-cell dual to the wall of W containing the hyperplanes $Y_i \cap H_{i-1}$ and $Y_i \cap H_i$. Indeed, if P traversed such an edge e, then we would be able to produce a lower complexity counterexample. The 1-cell e would be dual in D to a "dual curve" w of the wall W. Indeed, there is an immersed ladder $L \to D$ that is the concatenation of squares and cone-cells, such that L jumps across opposite 1-cells of a square, and L jumps across opposite 1-cells of a cone-cell in the sense that they are dual to the same wall of the ambient cone. We refer to such a ladder $L \to D$ associated to a wall W as a W-$ladder$. Since each cone-cell and each square contains an even number of 1-cells dual to W, there is actually a "dual graph" (instead of a dual curve) whose vertices internal to D have even valence, and so there must be an even number of vertices on the boundary, and hence some dual curve starts and ends on $\partial_p D$. See Figure 5.8. There are several cases to consider, each leading to a lower complexity counterexample. We refer the reader to Figure 5.9.

Suppose w crosses itself within D by passing through the same square in two ways or by passing through the same cone-cell in two 1-cells that map to 1-cells dual to hyperplanes in distinct walls of the ambient cone. Then a minimal such self-crossing dual curve w is contained in a subdiagram of D, and hence

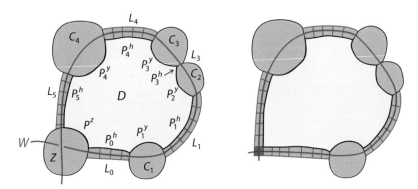

Figure 5.10. The collared diagram $E = D \cup_P A$.

provides a lower complexity counterexample: It bounds a path P' which bounds a subdiagram D' of D.

Otherwise, we can choose w to simply start and end on boundary 1-cells of ∂D, and we obtain a lower complexity counterexample in one of the following ways:

If w ends on a 1-cell on P_i^y that is dual to the wall of Y_i containing $H_{i-1} \cap Y_i$ and $H_i \cap Y_i$ then there is a lower complexity counterexample on the side of $D - w$ containing P^z. Similarly, if w ends on a 1-cell on P_i^h that is dual to the hyperplane H_i in $N(H_i)$, then there is a lower complexity counterexample on the side of $D - w$ containing P^z. In the above two cases, we have the same self-intersection, and are merely taking a shortcut through the diagram.

If w ends on a 1-cell on P_i^y that is not dual to the wall of Y_i containing $H_{i-1} \cap Y_i$ and $H_i \cap Y_i$, then $w \cup P$ bounds a lower complexity counterexample on the side of $D - w$ not containing P^z. In this case the self-intersection is new and is at the cone Y_i. Similarly, if w ends on a 1-cell on P_i^h that is not dual to the hyperplane H_i in $N(H_i)$, then $w \cup P$ bounds a lower complexity counterexample on the side of $D - w$ not containing P^z. In this case the self-intersection is new and is at the square in $N(H_i)$ corresponding to the 1-cell dual to the end of w and a 1-cube dual to H_i.

The final possibility is that w ends on a 1-cell f of P^z, and so Z is a cone and not a square by our earlier assumption. In this case, f is dual to a hyperplane that is distinct from at least one of the walls at Z corresponding to $H_0 \cap Z$ and $H_r \cap Z$. If it is distinct from the wall containing $H_0 \cap Z$, then there is a lower complexity counterexample on the side of $D - w$ containing the initial point of P. If it is distinct from the wall containing $H_r \cap Z$, then the side of $D - w$ containing the terminal part of P bounds a smaller complexity counterexample.

We have produced D such that $P = \partial_p D$ does not pass through a dual 1-cell of W in the sense above. The next step of the proof is to "augment" D by adding an annular ladder A along P to obtain a new diagram $E = D \cup_P A$. We refer the reader to Figure 5.10.

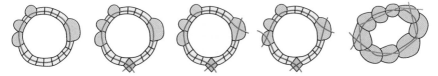

Figure 5.11. On the left are collared diagrams with zero, one, two, and three corners. The fifth diagram is a 1-cornered collared diagram in two different ways.

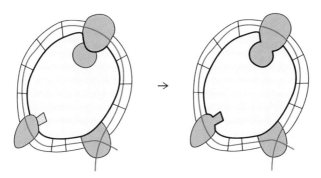

Figure 5.12. Absorbing and combining with external cone-cells.

Construction 5.19 (Collaring). The annulus A is a concatenation of rectangles $L_i \to N(H_i)$ and cone-cells $C_i \to Y_i$ and a cone-cell or square $z \to Z$. Each L_i is the unique (possibly degenerate) ladder in $N(H_i)$ that contains P_i^h as a (possibly trivial) external arc. Each C_i is chosen so that $\partial_p C_i$ extends the path P_i^y so that it starts and ends on the terminal and initial 1-cells of L_{i-1} and L_i that are dual to $H_{i-1} \cap Y_i$ and $H_i \cap Y_i$ respectively. Note that the orientations of these 1-cells are consistent since P_i^y doesn't pass through other 1-cells dual to W.

We refer to E as a *collared diagram* in the sense that it has a single dual curve of a wall passing all the way around within its external 2-cells, except for one *corner* 2-cell where there is a transition, as the two hyperplanes do not belong to the same wall in that cone. More generally, we will later consider *collared diagrams with k-corners* in the sense that they have exactly k such corners. The diagram E is a 1-cornered collared diagram. These ideas were treated in a simpler setup in [OW11]. See Figure 5.11.

When Z is a square, we let z be a copy of Z, and when Z is a cone we let z be a cone-cell such that $\partial_p z$ contains the path $e_0 P^z e_r$ where e_0 is the initial 1-cell of L_0 and e_r is the terminal 1-cell of L_r, oriented consistently.

Reductions preserve collar structure: Having obtained the 1-corner collared diagram E, we reach the next stage of the argument which is to obtain a reduced 1-corner collared diagram. The idea is to show that if E is not reduced, then we can produce a lower complexity 1-cornered collared diagram E'. The essential thing to verify is that the collar structure is preserved. Observe that reductions not involving an external 2-cell have no effect on the collared structure. We refer the reader to Figure 5.12.

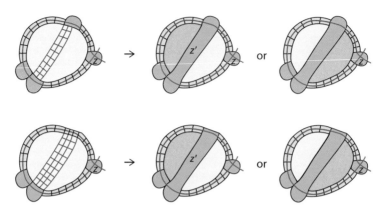

Figure 5.13. If there is a combinable pair of cone-cells as on the first diagram, then there is a smaller complexity counterexample as shaded in the second or third. The second diagram is the case where the two walls in the cancellable pair are distinct, and the third where the two walls are the same. The fourth diagram indicates a fake wall-piece, which is absorbed analogously in the fifth and sixth diagrams.

We now list the relevant cases to consider:

Absorption of an internal square into an external cone-cell
Absorption of an external square into an external cone-cell that is adjacent to it in the collar
Combination of internal cone-cell with an external cone-cell
Combination of external cone-cells that are adjacent in the collar
Absorption of internal square into an external cone-cell
Absorption of external square into an internal cone-cell
Combination of external cone-cells that are not adjacent in the collar
Absorption of an external square in an external cone-cell when they are not adjacent in the collar

Absorption into a cone-cell has little effect, since it does not affect the diagram much, and likewise, combining adjacent cone-cells in the collar or a cone-cell in the collar with an internal cone-cell has little effect, since it essentially redraws the cone-cell boundaries.

Absorption between external cells that shortcut the boundary is another interesting case that can lead to a substantially different diagram. We refer the reader to Figure 5.13.

An interesting situation that is not precluded by reduced is the replacement of an external cone-cell by a square diagram. In this case, we might have to shave off one side of the rectangle containing the dual curve corresponding to the wall on the collar. See Figure 5.14. Another interesting situation violating minimal complexity but not precluded by reduced, consists of a pair of squares folding to the same square of X, such that one is external and the other is not,

Figure 5.14. Replacement of external cone-cells: at a non-corner and at a corner. It is actually unnecessary to consider this case since reduced does not preclude replaceable external cone-cells.

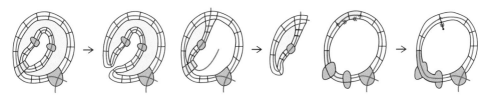

Figure 5.15. Some other available complexity reductions.

or such that both are external (see Figure 5.15). The latter case is analogous to the shortcut smaller diagram discussed above, and in the former case one actually passes to a substantially different diagram according to the following cases: If the W-ladder dual to the 1-cell of the internal square self-crosses (not illustrated) then there is an obvious lower complexity example. Otherwise, the W-ladder either closes with itself, or ends on the boundary. Consideration of the various possibilities leads to a lower complexity diagram (after some gluing).

Curvature along ∂E: Having produced a minimal complexity 1-cornered collared diagram E whose collar is a W-ladder, we shall now reach a contradiction.

Moreover, the dual curve crosses itself at the corner of the collar, in the form of two distinct midcubes in a square or nonequal walls in a cone-cell. We refer the reader to Figure 5.16.

By construction, E has no spur, and E has no cornsquare (except possibly a square corner) since pushing it out of E would provide a lower complexity example. In fact, if the collar contains at least one cone-cell, then there can be no explicit cornsquare anywhere, as all squares in the collar lie on rectangles admitted from an external cone-cell or an internal cone-cell. A positively curved shell C (that is not a cone-cell corner) has innerpath S path-homotopic to the carrier of a hyperplane in the cone Y of C by Condition 5.1.(4). Therefore, letting $\partial_p C = QS$ we can replace C by a square diagram with boundary path $e_1 S e_2 Q'$ such that Q' lies along the boundary of the hyperplane carrier in Y, and thus obtain a lower complexity collared diagram. We refer to the sequence of third, fourth, and fifth diagrams in Figure 5.16.

We conclude that only a corner can provide positive curvature, and there is at most one. However, by Theorem 3.46, we see that there are at least three corners, unless the diagram is a ladder or a single cone-cell. However our 1-cornered

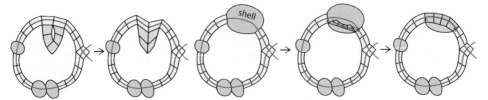

Figure 5.16. A cornsquare or shell in E would yield a lower complexity counterexample.

collared diagram cannot be a ladder since it only has one positively curved cell, and it cannot be a single cone-cell since that contradicts Definition 5.1.(2). □

We now prove the first statement listed in Theorem 5.17.

Theorem 5.20. *Let W be a wall, then Γ_W is a tree.*

Proof. This is similar to the proof of Theorem 5.18 but also uses its conclusion.

Let P be a closed path that is the concatenation of an alternating sequence of paths $P_1^y P_1^h P_2^y \cdots P_r^h$, where each $P_i^y \to Y_i$ is a cone path and each $P_i^h \to N(H_i)$ is a path in a hyperplane carrier, each hyperplane H_i belongs to W, and H_i intersects Y_i and Y_{i+1} in dual 1-cells (subscripts modulo r).

The path P induces a closed path $\bar P \to \Gamma_W$ whose vertices correspond to the above subpaths of P and whose edges are transitions between these subpaths. Conversely, given a closed path $\bar P \to \Gamma_W$, we can choose a path P as above that induces $\bar P$.

We must show that any closed $\bar P \to \Gamma_W$ is null-homotopic. We argue by contradiction to see that no such path exists, so we consider a minimal complexity diagram D whose boundary path P induces an essential path $\bar P \to \Gamma_W$.

Observe that P cannot traverse a 1-cell e that is dual to W in the sense that e either lies in P_i^y and is dual to a hyperplane in a wall of $H_{i-1} \cap Y_i$ or $H_i \cap Y_i$, or e lies in P_i^h and e is dual to $H_i \subset N(H_i)$. Indeed if such a 1-cell existed, then as in the proof of Theorem 5.18, we consider its dual graph in D, and find that we can choose a generalized dual curve w carried by a W-ladder in D, such that w does not cross itself. Note that while it is possible for the dual graph to have high even valence where it might bifurcate at cone-cells, Theorem 5.18 shows that it cannot genuinely cross itself, and in particular, if some path in the dual graph comes back to the same cone-cell then it returns along a hyperplane in the same wall of that cone. We are therefore able to choose a simple curve w in this dual graph that starts on e and ends on a 1-cell e' on P. Moreover, this curve is carried by a W-ladder within D. If e' corresponds to a different wall in Y_i or corresponds to a different hyperplane crossing H_i in $N(H_i)$, then we obtain a self-crossing wall, which violates Theorem 5.18. Thus e' is also dual to W, and after cyclically permuting, the path P is homotopic to the concatenation $P_1 P_2$ of two paths, and consequently $\bar P$ is homotopic to the concatenation $\bar P_1 \bar P_2$ of

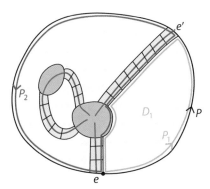

Figure 5.17. If P traversed an edge e dual to a hyperplane of W then there would be an essential path $\bar{P}_i \to \Gamma_W$ induced by a path P_i bounding a lower complexity diagram.

two closed paths. Indeed, $\bar{P}$ is obtained from $\bar{P}_1 \bar{P}_2$ by removing the sequence of backtracks in Γ_W corresponding to the sequence of rectangles and cone-cells on the W-ladder carrying w that the common path of P_1, P_2 travels through. At least one of $\bar{P}_1, \bar{P}_2$ must be essential in Γ_W since $\bar{P}$ is essential. However, the diagrams D_1, D_2 for P_1, P_2 are lower complexity than D, so this is impossible. We refer the reader to Figure 5.17.

Since P does not traverse a 1-cell dual to W as above, we see that the diagram D can be augmented as in Construction 5.19 to form a collared diagram E with no zero corners as in the first diagram in Figure 5.11.

In analogy with how we proceeded in the proof of Theorem 5.18, we can then choose a 0-cornered collared diagram $E = D \cup_P A$ where the collar A is an annular diagram containing a closed immersed W-ladder. (More precisely, A is a W-annuladder as in Definition 5.59.) Note that $\partial_\mathsf{p} E$ is parallel to P within A, and has a parallel decomposition into paths on rectangles and cone-cells, and thus $\partial_\mathsf{p} E$ induces the same path $\bar{P}$.

We will now follow the steps to reduce this diagram and see how the induced path changes. The relevant steps are when an external square is absorbed into a cone-cell (that is either internal or external but not adjacent within A), and when two external cone-cells are combined (whether they are adjacent in A or not). Note that we can ignore the reductions that don't affect the structure of the W-annuladder collar, for instance, reductions primarily involving internal cells or the removal of a square bigon with one dual curve on the collar. When an internal cone-cell absorbs an external square, the W-annuladder increases by the addition of a cone-cell and a rectangle, but the induced path only changes by the addition of a backtrack (from a hyperplane vertex to a cone-vertex and back). When an external cone-cell absorbs an external square, the induced path is again modified by a backtrack, but we can then break the resulting diagram into two smaller 0-cornered W-collared diagrams whose induced paths concatenate to form the original induced path. Thus at least one of them must be essential if the original path was essential, and we can pass to that smaller diagram. When

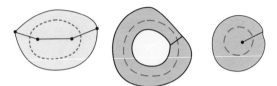

Figure 5.18. Base cases.

an external cone-cell is combined with an external cone-cell that is adjacent
to it within the collar, then the induced path is modified by the removal of a
backtrack (which ventured into and then out of the hyperplane vertex associated to the degenerate rectangle between these consecutive cone-cells). When
an external cone-cell is combined with a non-adjacent external cone-cell, the
diagram can be broken into two smaller W-collared diagrams as above, and the
original induced path is the concatenation of their induced paths, and so at least
one of these smaller diagrams must have an essential induced path. When two
consecutive external squares forming a bigon are cancelled there is no effect on
the induced path.

We draw attention to the "base cases." These are where the diagram is a
square bigon consisting of two squares that meet along three edges, or where
the diagram consists of a single cone-cell and it is adjacent to itself in the collar structure. See Figure 5.18. In the square case, the original induced path is
already trivial. In the cone-cell case, the original induced path is a backtrack,
and hence null-homotopic. Note that in this latter case, by Theorem 4.1, the
part of the diagram bounded by the cone-cell is already null-homotopic in the
cone associated to the cone-cell. And although we cannot combine this cone-cell
with itself, the internal part of the diagram must have a reduction within itself,
or with this cone-cell, unless it is trivial, and so the minimal scenario is the
degenerate situation depicted on the right of Figure 5.18.

Finally, Theorem 3.43 applied to a reduced spurless diagram that is not a
single cone-cell or 0-cell will provide a spur, a cornsquare, or a shell.

Spurs cannot exist since the diagram is collared, and cornsquares and shells
are dealt with as follows: Expelling a cornsquare with outpath on the diagram
has no effect on the induced path. Consider a shell C associated to a cone Y.
By Definition 5.1.(4), its innerpath is path-homotopic in Y to a hyperplane in
Y associated to the wall of Y crossed by W. We may thus replace C by a square
diagram with a rectangle along its external boundary, and this corresponds to
removing a backtrack in the induced path. $\square$

Definition 5.21 (Carrier). The *carrier* of the wall W is as follows: Let B denote
the disjoint union of the carriers $N(H_i)$ of its constituent hyperplanes together
with all cones intersected by these hyperplanes. We then form a quotient of B
by identifying pairs of subspaces $N(H)$ and Y along their intersection $N(H) \cap Y$
in $\widetilde{X}^*$, and identifying Y_i and Y_j along their intersection in $\widetilde{X}^*$ provided that

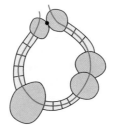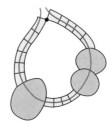

Figure 5.19. The diagram E arising in the proof of Theorem 5.22 has one collar, but two adjacent external cells that could be positively curved.

Y_i and Y_j share a 1-cell dual to some hyperplane H in W. These intersections were proven to be connected in Lemma 5.6 and Lemma 5.13. Let $N(W)$ denote the resulting quotient, and note that there is an induced map $N(W) \to \widetilde{X}^*$.

Theorem 5.22 (Carriers Embed). *Let X^* satisfy the $B(6)$ condition. For each wall W, the map $N(W) \to \widetilde{X}^*$ is an embedding.*

Proof. This is again a variant of the proof of Theorem 5.18.

Suppose there is a nonclosed path $P \to N(W)$ that maps to a closed path in $\widetilde{X}^*$. We choose such a path whose diagram $D \to X^*$ has minimal complexity.

We verify as above that P cannot pass through a dual 1-cell of one of the hyperplanes of W, for then there would either be a self-intersection which is precluded by Theorem 5.18, or we could cleave off part of D to obtain a new path P' with the same endpoints in $N(W)$ such that P' had a smaller complexity diagram D'.

We then augment D by Construction 5.19 to obtain a collared diagram E with one collar but two possible positively curved cells as depicted in Figure 5.19.

By passing to a minimal complexity such diagram, we can assume it is reduced and has no other features of positive curvature. The key point to verify here is that the collar structure is preserved when cancellable pairs are removed or absorbed. When there is a fake wall-piece or cone-piece from a rectangle or cone-cell in the collar on another side of the diagram (as in Figure 5.13) the absorbed wall from within the collar is the same as the wall in the absorbing cone, for otherwise we would contradict Theorem 5.18. Therefore we are able to cleave off a closed dual curve in the wall, and this provides a new path P' with the same endpoints as P, so the structure (and the original distinct pair of points in the carrier) are preserved.

Other cases are treated similarly.

Note that the first diagram in Figure 5.15 provides a new path $P' \to N(W)$ with the same endpoints, but with a lower complexity diagram. However, the second diagram in Figure 5.15 cannot occur, since it would yield a self-crossing which was ruled out in Theorem 5.18. Some variants of this are indicated in Figure 5.20. The first diagram explains how a cancellable pair of squares yields

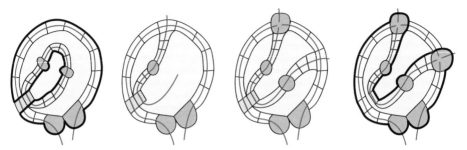

Figure 5.20. Some scenarios where P traverses a second edge dual to W.

a new path P' with a lower complexity diagram, the second and third yield self-crossing contradictions, and the fourth diagram indicates that no inner edge of the collar is dual to our wall, or there would have been a new path P' with a lower complexity diagram.

We conclude that E must be a ladder by Theorem 3.43, and so the two corners must be adjacent cone-cells corresponding to cones that share a 1-cell dual to W. Indeed, they are already adjacent in the diagram. We emphasize that, as concluded in Theorem 3.43, both of these positively curved cells must be cone-cells, and so the first and third figures in Figure 5.19 are excluded.

Thus E is a ladder consisting of a pair of intersecting cone-cells (or a single cone-cell in the special case where $Y_1 = Y_r$) whose intersection also intersects a hyperplane of W. But $Y_1 \cap Y_r$ is connected by Lemma 5.6, so $\bar{P}$ is null-homotopic in Γ_W. Here we use that distinct hyperplanes of $W \cap Y_i$ cannot lie in the same contiguous cone-piece by Condition 5.1(4). $\qquad\square$

Remark 5.23 (2-Sidedness). Since hyperplanes are 2-sided in their carriers by Lemma 5.11, carriers of walls in cones are 2-sided by hypothesis, and each carrier is a tree union of cones and hyperplane carriers by Theorem 5.20, we see that each wall W is 2-sided in $N(W)$ and hence separates $N(W)$.

Since $N(W)$ embeds in $\widetilde{X}^*$ by Theorem 5.22, we see that W is locally 2-sided in $\widetilde{X}^*$ and actually locally separates $\widetilde{X}^*$ since $N(W) \to \widetilde{X}^*$ is a local-isomorphism along W by construction.

Now the 2-sidedness of the walls in $\widetilde{X}^*$ follows from their embeddedness, local 2-sidedness in $N(W)$ which embeds, and the simple-connectivity and hence 1-acyclicity of $\widetilde{X}^*$.

5.g   No Inversions

Even when each wall W of $\widetilde{X}^*$ is 2-sided, it might be that $\mathrm{Stab}(W)$ does not also stabilize each of the halfspaces. We show how to avoid this with the following definition and lemma:

Definition 5.24 (No Inversions). The cubical presentation X^* has *no inversions* if the following conditions are satisfied:

(1) The hyperplanes in X are 2-sided.
(2) There is a choice of positive and negative side of each hyperplane, and a choice of positive and negative side of each wall in each Y, and these two notions are globally consistent, in the sense that the positive side of each hyperplane H equals the positive side of each wall represented by $H \cap Y$.
(3) $\mathrm{Aut}_X(Y)$ acts on Y without interchanging the sides of any wall of Y.

Lemma 5.25 (No inversions). *Suppose that $\langle X \mid \{Y_i\}\rangle$ satisfies the $B(6)$ conditions and has no inversions. Then for each wall W, the group $\mathrm{Stab}(W)$ acts without inversions in the sense that it stabilizes each component of $N(W) - W$.*

Proof. Consider the quotient $N(\bar{W}) = \mathrm{Stab}(W)\backslash N(W)$ where $\bar{W} = \mathrm{Stab}(W)\backslash W$. It is formed by gluing the various $N(\bar{H}_k)$ and $\bar{Y}_{ij}$ together, where $N(\bar{H}_k) = \mathrm{Stab}(H_k)\backslash N(H_k)$ and $\bar{Y}_{ij} = \mathrm{Aut}_{ij}\backslash(Y_i)$, where Aut_{ij} is the stabilizer of the j-th wall W_{ij} of the cone Y_i that is crossed by W. Note that Y_i could appear in W in several different ways corresponding to various walls in translates of Y_i where W slices through it. By hypothesis, $N(\bar{H}) = \bar{H} \times I$, and $\bar{W}_{ij}$ separates $\bar{Y}_{ij}$. Moreover, by hypothesis the gluing maps of the various $N(\bar{H}_k)$ and $\bar{Y}_{ij}$ respect the positive and negative sides. Thus $\bar{W}$ separates $N(\bar{W})$ and hence $\mathrm{Stab}(W)$ preserves the two sides of $N(W)$. $\qquad\square$

5.h   Carriers and Quasiconvexity

We now discuss some further terminology and notions related to walls. A wall W *crosses* a cone C if $W \cap C$ is nonempty. Hence, following Definition 5.21 and Theorem 5.18, the carrier of a wall W of $\tilde{X}^*$ is the subspace that is the union of all hyperplane carriers containing hyperplanes of W, together with all cones that W crosses.

The most important cones in a carrier are those cones C crossed *essentially* by W in the sense that $W \cap C$ consists of two or more hyperplanes of C. Many of the properties of walls and carriers are explainable for the *essential carrier* which only includes essentially crossed cones, but we have decided to retain the point of view including all crossed cones instead of just the essential ones. The essential carrier is a bit less "bumpy" then the carrier. In the "classical case" of $B(6)$ complexes, all crossed cones are essential.

It is a consequence of Theorem 5.18 that the carrier deformation retracts to the wall that is obtained by adding geometric walls that cone-off each wall in each cone via $C_w(X^*)$ as we shall explore in Construction 10.4. Furthermore, the carrier is simply-connected and is separated by this resulting geometric wall.

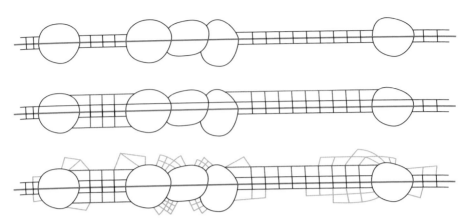

Figure 5.21. A carrier $N(W)$, a thickened carrier $T(W)$, and an extended carrier $E(W)$.

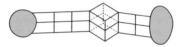

Figure 5.22. It seems harmless to include squares in T whose entire 1-skeleton lies in T.

Definition 5.26 (Thickened and Extended Carrier). The *thickened carrier* $T(W)$ of W is the union of the carrier $N(W)$ together with each grid $R = I_n \times I_m$ whose left and right boundary paths $\{0, n\} \times I_m$ lie on cones Y_1, Y_2 of $N(W)$, and such that for some $0 \le k < m$ the rectangle $I_n \times [k, k+1]$ is dual to a hyperplane in W.

The *extended carrier* $E(W)$ is the cubical local convex thickening of the thickened carrier $T(W)$. It is obtained by repeatedly adding any cube with an entire corner already present.

We refer to Figure 5.21 for heuristic illustrations of $N(W)$, $T(W)$, and $E(W)$ when X is 2-dimensional.

Definition 5.27. A *W-ladder* in a diagram is a ladder containing a dual curve mapping to a wall W of $\widetilde{X}^*$; it generalizes a rectangle containing a dual curve. Since a cone-cell can offer multiple continuations of a W-ladder, we will often choose continuations that yield a simple dual curve, and more specifically, a *simple W-ladder* is an embedding on its interior. On the other hand, we will sometimes consider an entire dual graph instead of a dual curve. We sometimes use the term *wall-ladder* to avoid specifying the associated wall W.

In Definition 5.59 we will consider W-*annuladders*, which are W-ladders that form an annulus and carry a closed dual curve.

Definition 5.28 (Eject). A diagram D is *semi-collared* by a W-ladder L, if $D = E \cup_P L$ where E is a diagram and P is a subpath of both $\partial_{\mathsf{p}} E$ and $\partial_{\mathsf{p}} L$. Note that if D is a collared diagram with n-corners then it is semi-collared by n wall-ladders.

Let D be semi-collared by the W-ladder L. A cone-cell C of L is *ejectable* if the following holds: Firstly, $\partial_{\mathsf{p}} C = QS$ where Q is a subpath of $\partial_{\mathsf{p}} D$ and S starts and ends with edges a, b dual to W. Secondly, there is a square diagram C' such that $\partial_{\mathsf{p}} C' = Q'S$, and C' contains a rectangle R and $\partial_{\mathsf{p}} R \cap \partial C'$ contains the subpath $aQ'b$. To *eject* C is to replace C by C', so L is replaced by a lower complexity W-annuladder L' with R replacing C, and a modified cone-cell with boundary path $(Q')^{-1}Q$ is then outside of L. See Figure 5.27.

By Definition 5.1.(4), in the $B(6)$ case we can eject a positively curved shell C lying on a partial collar W-ladder L when the innerpath S of C traverses edges dual to W near its beginning and end. Likewise, by Definition 5.55(2), in the $B(8)$ case we can eject C when it is a nonnegatively curved shell.

Theorem 5.29 (Thickened Carrier Isometric Embedding). *Assume X^* satisfies the $B(6)$ condition. The thickened carrier of a wall isometrically embeds in $\widetilde{X}^*$.*

Proof. We can assume without loss of generality that γ is disjoint from T and hence from N except at its endpoints. We refer to Figure 5.23 for a guide to the notation used in the proof.

Step 1: Setting up the diagram   We first show that there exists a disk diagram F containing (generalized) ladders $L_N \subset L_T$ at the top and our geodesic γ at the bottom. This is something we have done routinely when γ starts and ends on N. Note that F is semi-collared by a W-ladder in the sense of Definition 5.28.

We begin by extending γ to $\gamma^+ = \gamma_0 \gamma \gamma_t$ where γ_0 is a minimal length path ending on the initial vertex of γ and starting with a 1-cell dual to W, and likewise γ_t is a minimal length path starting on the terminal vertex of γ and ending with a 1-cell dual to W. We emphasize that γ_0 and γ_t can be chosen to travel through sequences of 1-cells in T that are dual to hyperplanes joining (the same) pair of cones in $\widetilde{X}^*$. When the endpoint of γ lies on a cone of $N(W)$ then we choose γ_i to be a path in this cone to a dual 1-cell of W.

For $i = 0$ and $i = t$, there is a grid R_i containing γ_i on one side and bounded by a rectangle mapping to a hyperplane of W on one side and ending on a path in a cone-cell of $N(W)$ on a third side. When γ_i lies on a cone of W (because the corresponding endpoint of γ lies on that cone), then R_i is degenerate, and is simply a copy of γ_i. Likewise, when γ_i is a single 1-cell (since γ ended on the carrier of a hyperplane of N) then we just let R_i be a copy of γ_i.

There is a sequence of cones and rectangles in N that start and end at the initial and terminal 1-cells of γ^+. We choose the first and last such rectangles to be those already lying in the grids R_0 and R_t. We consider the concatenation $V_0 U_0 V_1 U_1 \cdots V_t$ of paths where V_i is an external boundary arc of R_i, and where

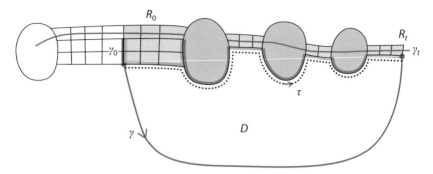

Figure 5.23. Notation used in the proof of Theorem 5.29.

$U_i \to Y_i$ is a path from one dual 1-cell to the other that travels on the same side of W as γ. We then consider a disk diagram D between γ and $V_0 U_0 V_1 U_1 \cdots V_t$. Adding on the above sequence of grids, cone-cells, and rectangles supporting the path $V_0 U_0 V_1 U_1 \cdots V_t$, we can augment D to obtain a disk diagram F that is "collared by a ladder on one side."

Step 2: Estimating the distance The ladder $L_T \to F$ contains the sequence of grids, cone-cells, and rectangles, and along the top of L_T is the ladder L_N mapping to $N(W)$. Let τ be the path on D that "opposes" γ, so that τ is also a path along the "bottom" of L_T. See Figure 5.23.

Having shown above that one such diagram exists, we now assume that we have chosen (D, E) to minimize the following lexicographical complexity:

$$\mathsf{Comp}(D, F) = \big(\#\text{Cone-cells}(D), \#\text{Cone-cells}(F), |\tau|\big)$$

We emphasize that $\#\text{Cone-cells}(D)$ equals the number of cone-cells in F that are *not* also in L_N (or equivalently in L_T). Note that we are not assuming that F is reduced or has minimal complexity in the usual sense. However, the minimality of Comp guarantees that no cone-cell in F can be replaced by a square diagram, and adjacent cone-cells cannot be combined. We are also not assuming that τ is the shortest possible path in some F, D, but rather, that it is shortest possible among all such paths arising from F, D with minimal numbers of cone-cells.

We will show below that for each edge e in τ, the generalized dual graph to e in D is a graph that ends on e in τ, and whose remaining endpoints all lie on γ. Since there are an odd number of such remaining endpoints, we see that $|\gamma| \geq |\tau|$, and since τ lies in T, our claim is proven.

Step 3: Choosing a minimal counterexample Suppose that the dual graph ends at another edge e' in τ for some E, D with Comp minimal as above. Let K be the wall-ladder in D carrying a dual curve that starts and ends on the edges e, e' of τ. Let τ' be the subpath of τ whose initial and terminal edges are e, e'. Let κ be the path on K with the same endpoints as τ' but such that κ doesn't pass through the dual curve of K, and in particular doesn't pass through

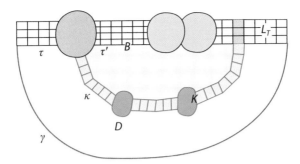

Figure 5.24. If a generalized dual curve starts and ends on τ then we obtain a 2-cornered collared diagram B.

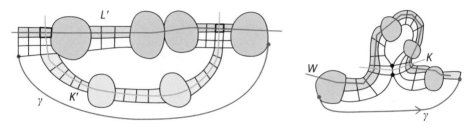

Figure 5.25.

e, e'. We extend the wall-ladder K in D carrying a dual curve that starts and ends on τ, to a wall-ladder K' that starts and ends on squares and/or cones of $L_N \subset L_T$. Let L' denote the subladder of L_N that begins and ends at the intersections with K'. See the diagram on the left of Figure 5.25. Let B be the subdiagram that is *bicollared* by K', L', in the sense that it is a 2-cornered collared diagram where K' and L' are the ladders in B that start and end at the corners. See Figure 5.24.

Among all minimal Comp examples for a given γ (or even a geodesic with the same endpoints as γ) exhibiting the contradictory feature represented by K, let us choose a contradiction with the additional property that B has minimal usual complexity $\big(\text{Cone-cells}(B), \text{Squares}(B)\big)$.

We will show by contradiction that such a minimal counterexample cannot exist. Note that in the degenerate case where $e = e'$ illustrated on the right of Figure 5.25, there is automatically a smaller choice of B corresponding to an innermost backtracking wall-ladder.

Step 4: Minimality of B implies square diagram We now focus on the subdiagram B that is bicollared by K', L'—we emphasize that $K', L' \subset B$. We shall show that any cone-cell in B actually lies in L'.

Note that we cannot immediately apply Theorem 3.43 as we don't know that B is reduced along L' and has no shells along L'.

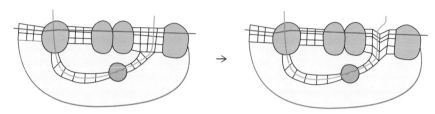

Figure 5.26.

The minimality of B implies that there are no cornsquares on cone-cells within B, for they can be absorbed. There is also no cornsquare within B whose outerpath lies on ∂B except perhaps at the two corners of B where L', K' meet. Indeed, if B has a cornsquare on a rectangle in K', then we can push K' upwards across this square while decreasing the area in B. Similarly, if there is a cornsquare in B with outerpath along the top of L', then we can push the square upwards through L' (and hence L_T) until it is outside of F, and reduce the area of B. Neither of these moves increases Comp, $|\tau|$, and the second move decreases Comp. See Figure 5.26 for a trickier-than-usual pushing of a cornsquare.

Suppose that B contains a cone-cell C that does not lie in L'. Let us examine what happens to it as we perform a sequence of reductions to B. It is impossible for any cone-cell C to be combined with another cone-cell by minimality of Comp, and minimality of $\#\text{Cone-cells}(D)$ makes it impossible for C to eventually work its way into L' as we proceed through the following procedures. Suppose after some sequence of reductions $B \mapsto \dot{B}$, a cone-cell C becomes a shell $\dot{C}$ with outerpath Q along $\dot{\kappa} = \kappa$ and innerpath S. Let Y be the supporting cone, so $\{\!\{S\}\!\}_Y < \pi$. And note that $S = S_1 f S' f' S_2$ has a subpath $f S' f'$ where f, f' are 1-cells on $\partial \dot{C}$ that are dual to the dual curve of K, and also lie on the previous and next cells of K respectively. It is important to note that f, f' must be oriented in the same way with respect to this hyperplane, otherwise the path S' would contain another edge dual to this hyperplane, and following the associated wall-ladder in D, we would be able to construct a backtracking wall-ladder K_0 bounding a subdiagram B_0 of B. By Condition 5.1.(4), the path $f S' f'$ is path-homotopic in Y to a path in the carrier of the hyperplane of Y dual to f, f', and hence S' is path-homotopic to a local geodesic J in this carrier, and let $V \to Y$ denote the diagram between S' and J. Let J' denote the corresponding local geodesic on the other side of J, so J, J' together form the side of a rectangle M in the carrier with initial and final 1-cells f, f'. We refer the reader to Figure 5.27.

The path $S_2 Q S_1 J'$ bounds a cone-cell C', and we replace $\dot{C}$ by the union $C' \cup_{J'} M \cup_J V$. This ejection (see Definition 5.28) has the effect of substituting the rectangle M for $\dot{C}$ in $\dot{K}$, and hence decreases the number of cone-cells in the resulting bicollared diagram replacing $\dot{B}$. This violates our minimality assumption on B.

Consequently, since after reducing there are no shells along $\dot{\kappa}$, and likewise, none along the part of $\dot{B}$ collared by $\dot{L}'$ (since the same argument applies), we

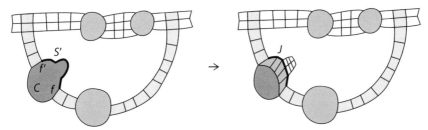

Figure 5.27. Cone-cell ejection: $fS'f' \subset \partial_\mathsf{p}$ at left, and the adjustment associated to the path-homotopy from S' to J at right.

see that there are at most two positively curved cells at the corners of $\dot{B}$, and so $\dot{B}$ is a ladder by Theorem 3.43. In particular, $\dot{B}$ has no internal cone-cells either, so all cone-cells of B have been accounted for as arising in L'. We conclude that there are only squares in B outside of L', and in particular K is a rectangle.

Step 5: Minimality implies that B is very thin   Let A be the part of B between κ and τ', so $K \subset A$. We will show that $A = K$.

Since D is reduced, there are no square bigons in $A \subset D$.

Each dual curve in A that starts on $\tau' - \{e, e'\}$ ends on a 1-cell on κ. Indeed, if such a dual curve ended on another edge of τ', then its corresponding wall-ladder could be used instead of K, and would provide a lower complexity counterexample violating the minimality of B. Moreover, each dual curve in A starting on κ must end on τ' for otherwise there would be a bigon, and this was excluded earlier. In summary, each dual curve in A travels between τ' and κ, except for the dual curve within K itself. Hence $|\kappa| = |\tau'| - 2$.

Suppose that A contains a square s that doesn't lie in K. Then each dual curve of s has one end on κ and the other end on τ'. Thus some corner of s has a pair of dual curves on κ. This contradicts Lemma 2.20.

Step 6: Reducing $|\tau'|$ and hence $|\tau|$ We now refer to Figure 5.28 on the left, as well as the resulting absorptions on the right.

If L' contains only squares then τ' can be replaced by the shorter path κ and K' can be pushed through L' (using bigon removals). This reduces the length of τ' and hence violates the minimality of τ.

If L' contains cone-cells, then we can assume it is already reduced without affecting any of our minimized quantities. Let R' denote the part of L_T that is subtended by K, so R' is the union of L' and various additional rectangles in L_T. We then observe that the ladder K' wraps around L'.

Suppose the first square of K has one edge on a square of L_T and one edge on a cone-cell C of L_T. If the rectangle of K alongside C cannot be absorbed into C, then the innerpath S of C is the concatenation of two pieces: a wall-piece associated with an initial part of K' followed by a cone-piece. (If C is the unique cone-cell, then S is a single wall-piece.) Letting Y denote the cone supporting C we have $\{\!\{S\}\!\}_Y < \pi$. This leads to a contradiction, because Condition (4) would then imply that S is homotopic into the carrier of a hyperplane associated to

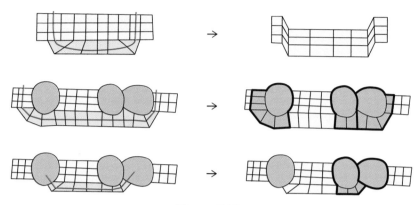

Figure 5.28.

the wall of $L_T \cap Y$, which would imply that the cone-cell C could be replaced by a square diagram, violating the minimality of #Cone-cells(F).

Thus the contiguous rectangle is absorbable into C, and this includes the entire initial rectangle of K'.

We can thus assume that K starts and ends on cone-cells. But then Theorem 3.43 implies that B becomes a ladder after reducing, and hence each rectangle in K contiguous with a cone-cell of L_T is absorbed into that cone-cell upon reducing. The path τ' can thus be replaced by κ, yielding a shorter counterexample. Note that in the degenerate case where L_T consists of a single cone-cell C it is immediate that all of K absorbs into C since there are cornsquares in K on C (one at each side). □

Corollary 5.30 (Walls quasi-isometrically embed). *Let X^* be $B(6)$. Suppose that pieces in cones have uniformly bounded diameter. Then for each wall W, the map $N(W) \to \widetilde{X}^*$ is a quasi-isometric embedding.*

Proof. By Theorem 5.29, it suffices to show that $N(W) \to T(W)$ is a quasi-isometric embedding. Let $M \in \mathbb{N}$ be one more than the maximal diameter of a wall-piece. The hypothesis implies that $T(W)$ lies in an M-neighborhood of $N(W)$. Since $N(W) \subset T(W)$ it is immediate that the combinatorial length metric satisfies $\mathsf{d}_N(p,q) \leq \mathsf{d}_T(p,q)$. Any $\gamma \to T(W)$ can be expressed as an alternating concatenation of paths on cone-cells and paths on grids. Thus to see that $\mathsf{d}_T(p,q) \leq M\mathsf{d}_N(p,q)$ it suffices to show that the above bound holds when p,q are endpoints of a grid subpath σ. This obviously holds when σ is trivial for then $p = q$. Otherwise, a path $\sigma \to R \subset T$ can be replaced by a path $\sigma' \to N$ as follows: If σ has an endpoint on $N(W)$ then $|\sigma| = |\sigma'|$ where σ' is the path that travels vertically in the grid, and then travels along the bottom of the grid (which is $N(W)$). Otherwise, we let σ' be the path that travels along the two sides and bottom, so $|\sigma'| \leq |\sigma| + 2(M-1)$, and hence $|\sigma'| \leq 2M|\sigma|$. □

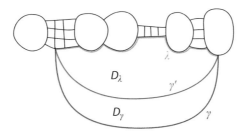

Figure 5.29. To show the geodesic γ is square-homotopic to a geodesic $\lambda \to T(X)$, we break up the diagram D between them into a largest possible square part D_γ between γ and γ', and a remaining part D_λ between γ' and λ. We then show that D_λ is an arc.

Theorem 5.31 (Extended Carrier Convexity). *Suppose X^* is $B(6)$ and has short innerpaths. Let W be a wall in $\widetilde{X}^*$. If $\gamma \to \widetilde{X}^*$ is a geodesic whose endpoints lie in $T(W)$ then γ is square-homotopic to a geodesic $\lambda \to T(W)$. Thus $\gamma \subset E(W)$.*

Consequently, each cone of $\widetilde{X}^*$ is convex.

I originally hoped to prove that $E(W)$ is convex, but there was a gap in the original version of this proof, and only the above weaker statement was covered.

Proof. Let γ be a geodesic in $\widetilde{X}^*$ that starts and ends on the thickened carrier T of W. As provided in the proof of Theorem 5.29, let λ be a path on T with the same endpoints such that the disk diagram $D \to \widetilde{X}^*$ between them has minimal Comp among all such possibilities for D, λ. Note that λ can be chosen to be on a minimal (#cone-cells, #squares) length ladder in T containing a ladder in N.

Let γ' denote a geodesic in D with the same endpoints as γ such that γ' and γ together bound a square subdiagram of D. Thus D is the union of two diagrams D_λ and D_γ that meet along γ' (see Figure 5.29). We will assume that D_λ has minimal area among all such possible choices with γ fixed.

There is no cornsquare in D_λ with outerpath on γ', for then we could pass a square across γ' to increase $\mathsf{Area}(D_\gamma)$ and decrease $\mathsf{Area}(D_\lambda)$. Since X^* has short innerpaths, there cannot be a shell whose outerpath is on the geodesic γ'.

Consider the diagram F bounded by γ' on one side, and bounded by and including the ladder L_T in $T(W)$ consisting of the sequence of cone-cells and grids between them (this isn't quite a ladder since we allow for an initial and/or terminal grid). We will show that γ' lies on L_T by showing that D_λ is an arc. This shows that $T(W)$ is isometrically embedded since $|\gamma'| = |\gamma|$. Moreover, since γ and γ' lie in the local (square) convex hull of each other, we see that γ lies in $E(W)$ as well, thus proving the theorem.

Without loss of generality, we can prove the statement for a subpath γ'_0 of γ' obtained by ignoring the initial and terminal subpaths of γ' that already lie on L_T. We let F_0 denote the resulting diagram, and observe that cropping γ

in this way would necessitate a corresponding cropping of initial and terminal rectangles of L_T.

As above, minimality of complexity implies that there are no outerpaths of cornsquares at the "top" boundary path along L_T. Condition 5.1.(4) implies there is no shell C within L_T, i.e., one whose outerpath is at the "top," since then the innerpath S and outerpath Q of C satisfy $\{\{S\}\}_C < \pi$ and so S is path-homotopic in the ambient cone to the carrier of a hyperplane and thus a rectangle U of length at most $|S| - 2$. So the C with $\partial_p C = QS$ can be replaced by a square diagram with boundary path SQ' where Q' is along one side of the ladder U. This shortens the ladder L_N in N at the top of L_T in the sense that U replaces a cone-cell, and hence violates our earlier minimality assumption.

The lack of positively curved shells claimed above holds with the sole exception of a cone-cell and/or square at either end of $L_N \subset L_T$. Note that cropping removed the possibility of a second cornsquare. Moreover, the second dual curve of a square at the end of L_N must terminate on γ' (and not on a cone in L_T). Indeed, as explained in the proof of Theorem 5.29, all wall-ladders emerging from λ terminate on γ (after passing through γ').

As there are at most two positively curved cells, Theorem 3.43 implies that F_0 is a single cone-cell, or a (genuine) ladder. Thus γ' lies on this (cone-cell) or ladder and $\gamma' = \lambda$ as claimed.

The secondary conclusion that cones are convex holds because if γ has endpoints on a cone Y, then the primary conclusion implies that there is a square diagram between γ and a geodesic λ with the same endpoints, and moreover, λ lies on a ladder such that (#cone-cells, #square) is minimized. Hence λ lies in Y, and so γ lies in Y by local convexity. $\square$

Corollary 5.32. *If X^* is $B(6)$ and has short innerpaths and X is 1-dimensional then carriers are convex.*

Proof. This holds by Theorem 5.31 as $N(W) = E(W)$ when $\dim(X) = 1$. $\square$

5.i Bigons

Definition 5.33. Let Y be a cone intersected by a hyperplane H. Let y and h be the associated vertices in Γ_W. The *wallray* $[Y, H)$ based at Y in the direction of H consists of the part of W corresponding to the largest subtree $\Gamma_{[Y,H)} \subset \Gamma_W$ such that $\Gamma_{[Y,H)}$ contains the edge (y, h) and such that y is a leaf of $\Gamma_{[Y,H)}$. The wallray is *carried* by the corresponding subspace of $N(W)$, which we denote by $N[Y, H)$.

Theorem 5.34 (Intersection of Wallrays). *Suppose X^* is $B(6)$. Let Y be a cone and let H_1, H_2 be distinct hyperplanes in $\widetilde{X}^*$ with dual 1-cells in Y. Let $h_1 = Y \cap H_1$, and let $h_2 = Y \cap H_2$. Suppose $\{\{h_1, h_2\}\}_Y > 0$, and hence h_1, h_2 are not dual to 1-cells e_1, e_2 that lie in the same cone-piece or wall-piece in Y.*

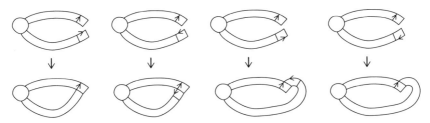

Figure 5.30. There are four cases that allow us to avoid producing a Moebius strip.

Then the wallrays $[Y, H_1)$ and $[Y, H_2)$ do not have crossing hyperplanes besides possibly H_1, H_2, and they do not both intersect any cone or square outside of Y.

The following corollary is easier to prove than Theorem 5.34.

Corollary 5.35. *Let W_1, W_2 be walls intersecting a cone Y. Suppose that $\{\{W_1 \cap Y, W_2 \cap Y\}\}_Y > 0$. Then W_1 and W_2 do not have hyperplanes that intersect in any cone or square outside of Y.*

Proof of Theorem 5.34. If the statement were false, then there would be a pair of paths $\bar{P}_i \to \Gamma_{[Y, H_i)}$ whose initial points are the leaf vertex y associated to Y, and whose terminal points are vertices v_1, v_2 where either $v_1 = v_2$ is a cone-vertex associated to a cone K that is not Y, or v_1, v_2 are hyperplane vertices corresponding to hyperplanes that cross in a square K. Accordingly, for each i, we can choose a W_i-ladder L_i consisting of a sequence of cone-cells and squares carrying hyperplanes of W_i (technically, L_i might not be a true ladder as it can end with a square) whose first cone-cell is a cone-cell A that maps to Y and whose last cone-cell or square, B, maps to K. Note that by concatenating the boundary paths of first cone-cells for L_1, L_2, we can assume they have the same initial cone-cell A.

Moreover, we can do this so that $L_1 \cup L_2$ forms an annular diagram. Note that we must avoid forming a Moebius strip. In the cone-cell case we let e_i be the first edge dual to W_i in the final cone-cell of L_i, and we let $P \to K$ be a path whose first edge is $e_1^{\pm 1}$ and whose last edge is $e_2^{\pm 1}$ and then we use the cone-cell whose boundary path is PP^{-1}, and note that the initial part of each L_i can be attached at a copy of e_i to obtain an annular diagram. In the square case we can choose to identify along an edge instead of along an entire square (see Figure 5.30).

Let D be a diagram where $\partial_p D$ is equal to a boundary path of $L_1 \cup L_2$. Let $E = D \cup_{\partial_p D} (L_1 \cup L_2)$ be formed by identifying along these boundary paths. We thus obtain a bicollared diagram, where the two ladders forming the bicollar meet at the corners A and B associated to Y and K.

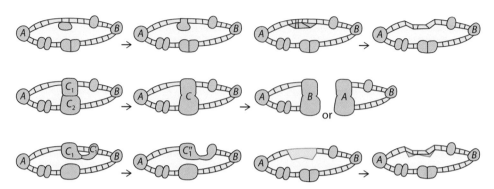

Figure 5.31. Various complexity reductions for a bicollared diagram: Absorbing an external square into an internal cone-cell; removing a square bigon with one side on the collar; combining two cone-cells (or a cone-cell and an external square), and choosing an appropriate bicollared subdiagram; shortcutting by combining a cone-cell and another cone-cell (or absorbing a square) that is not adjacent but within the same collar; replacing a shell by a path-homotopy to a rectangle.

We will show that there exists a reduced bicollared diagram with the same properties, and then apply Theorem 3.43 to a minimal complexity such diagram to obtain a contradiction.

This is very similar to the proof of Theorem 5.22: Absorption of a square in one collar into a cone-cell in the other will lead to a smaller counterexample by cutting along this cone-cell after absorbing (see Figure 5.31). Indeed, if this cone-cell is associated to Y, then we instead use the subdiagram starting at it and ending at B. And if this cone-cell is not associated to Y then we retain the subdiagram ending at it and starting at A. The same argument works for the combination of two cone-cells in L_1 and L_2. Combination or absorption of a cone-cell with a nonconsecutive cone-cell or square in L_i leads to a shortcut, and we can lop off the remainder. (The collar structure persists.) Removal of a square bigon where one side is a rectangle in a collar yields a rectangle again. Note that it is impossible for the two rectangles carrying this bigon to be our two collars since both their dual curves travel through the initial cone-cell. Similarly, combination or absorption between the initial cone-cell and terminal cone-cell or square is impossible (since we assumed that one is not contained in the other when we argued by contradiction). The last thing to observe is that at each stage the collars have the desired property, that they represent ladders in $N[Y, H_1)$ and $N[Y, H_2)$ that do not venture into other parts of $N(W)$. However, this is clear when we pass to a subdiagram obtained by a shortcut, and otherwise the only reduction move above that changes the route of these ladders in $\Gamma_{[Y,H_i)}$ is when an internal cone-cell absorbs a square in a collar, and this just adds a backtrack in $\Gamma_{[Y,H_i)}$.

The case where the terminal cell is a square that is combined with the initial cone-cell is impossible, by hypothesis. Moreover, the terminal corner cannot be

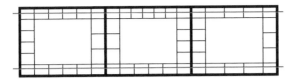

Figure 5.32. There is a ladder containing three cone-cells in the complex above such that the first and last cone-cells are connected by a pair of hyperplanes without a square diagram between them.

a cornsquare whose outerpath is on the initial cone-cell, for this would imply that $\{\{H_1, H_2\}\}_Y = 0$ (since as mentioned above this square cannot be absorbed into the initial cone-cell). Similarly, the extreme case where a terminal cone-cell is combined with the initial cone-cell is impossible.

We now consider a reduced bicollared diagram F that starts at a cone-cell associated to Y and ends at a square not mapping to Y or at a cone-cell in a cone that is different from Y, and such that the collars are associated to H_1 and H_2 emerging from the cone-cell at Y. Let us assume that F has minimal complexity among all possible such choices. It follows that except for the first and last cone-cell, no external cone-cell can be a shell C'. Indeed, if there were a shell C' associated to a cone Y' and having outerpath Q' and innerpath S', then we could pass to a lower complexity diagram with our property by replacing C' by a diagram $R \cup D'$ where $D' \to Y'$ is a path-homotopy from S to a curve S' that is an external boundary path of a rectangle R carrying a dual curve of a hyperplane in the wallray $N[Y, H_i)$, and Q is replaced by the opposite external boundary path Q' of R. This preserves the desired bicollared structure since it corresponds to removal of a backtrack in the corresponding path to $\Gamma_{[Y, H_i)}$. An explicit cornsquare cannot exist in the rectification since both collars contain the initial cone-cell, and hence each square in a collar lies in some rectangle emerging from a cone-cell of the diagram.

Since there are at most two features of positive curvature but at least two 2-cells, Theorem 3.43 asserts that the reduced diagram F must be a ladder. However, then $\{\{H_1, H_2\}\}_Y = 0$ which is impossible. $\qquad\square$

Remark 5.36. In general, it is possible for two hyperplanes H_1, H_2 to cross two different cones Y_1, Y_2 such that Y_1, Y_2 are not joined by a grid, but rather have a sequence of cones between them. This can be avoided by extra hypotheses on the hyperplanes within a cone. We refer the reader to Figure 5.32.

5.j   Square Cones

The goal of this section is the following consequence of Theorem 5.34 that explains how walls in $\widetilde{X}^*$ can interact. Its proof relies on Lemma 5.38 which is a technical trick showing that small-cancellation is preserved when adding a certain simply-connected cone.

Corollary 5.37 (Connected intersection). *Suppose X^* is $B(6)$. Let W_a, W_b be distinct walls of $\widetilde{X}^*$ that do not intersect the same cone. Either $W_a \cap W_b = \emptyset$ or $W_a \cap W_b = H_a \cap H_b$ where H_a is a hyperplane of W_a and H_b is a hyperplane of W_b.*

Proof. Suppose H_a, H_b are hyperplanes of W_a, W_b that intersect. Let s be a square whose midcubes are contained in H_a and H_b. Applying Lemma 5.38, the cubical presentation X_s^* is also $B(6)$ and $\widetilde{X}_s^*$ has the same walls as $\widetilde{X}^*$. Moreover, $W_a = [Y_s, H_a)$ and $W_b = [Y_s, H_b)$. Since W_a, W_b do not intersect the same cone in $\widetilde{X}^*$, the same holds for H_a, H_b, and hence $\{\{H_a, H_b\}\}_{Y_s} > 0$. Hence Theorem 5.34 asserts that $[Y_s, H_a) \cap [Y_s, H_b)$ equals $H_a \cap H_b$. $\qquad\square$

Lemma 5.38. *Let s be a square of $\widetilde{X}$. Let Y_s be the carrier of the codimension-2 hyperplane dual to s, so $Y_s = N(H_a) \cap N(H_b)$ where H_a, H_b are the hyperplanes of $\widetilde{X}$ containing midcubes of s.*

Let $X^* = \langle X \mid \{Y_i\} \rangle$ be a cubical presentation, and suppose that $H_a \cap H_b$ does not cross any lift of any Y_i. Consider $X_s^* = \langle X \mid Y_s, \{Y_i\} \rangle$. If X^* is small-cancellation then so is X_s^*. Moreover if X^* has no acute corners then X_s^* has no acute corners. If X^* is $B(6)$ [or $B(8)$] then so is X_s^*.

Proof. We will explain how an angling rule for X^* induces an angling rule for X_s^*. Consider a reduced diagram $D_s \to X_s^*$. Let $B_1, \ldots, B_m$ denote the cone-cells of $D_s \to \widetilde{X}_s^*$ that are associated with lifts of Y_s. Each B_i is external since if B_i were internal then D_s would not be reduced since B_i is replaceable, as Y_s is simply-connected. For each i, let E_i be a maximal subpath of $\partial_{\mathsf{p}} B_i$ that is a subpath of $\partial_{\mathsf{p}} D_s$.

Let $D \to X^*$ be the diagram obtained from D_s by collapsing each B_i along a free face consisting of a maximal boundary arc. Specifically, we let $D_0 = D_s$, and for each $i \geq 1$, we obtain D_i from D_{i-1} by collapsing B_i along E_i, and finally $D = D_m$. (Since some B_i might be multiply external, there are multiple choices for E_i and hence possibly multiple results for D.) In the extraordinary case where $E_i = \partial_{\mathsf{p}} B_i$ in which case B_i is a cone-cell attached at a cut vertex where B_i has a doubly-external corner, we deformation retract B_i to a spur consisting of one of the edges at this corner.

We form a diagram $D_r \to X^*$ from D as follows: For each nontrivial maximal contiguous cone-piece P_{ij} between B_i and a cone-cell C of D, we attach a rectangle R_{ij} along P_{ij}. Note that we can choose R_{ij} to be a rectangle carrying a dual curve of one of the two hyperplanes associated to the lift of Y_s that B_i maps to. And since at most one of these hyperplanes crosses the cone of C, we can choose R_{ij} to be dual to a hyperplane not crossing that cone, and so no square of R_{ij} is absorbable into C. We refer to Figure 5.33.

Note that D_r is reduced since D_s is. In particular, D_r contains no square bigons, since a dual curve bounding a square diagram that starts and ends on

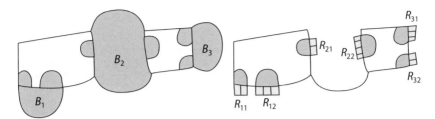

Figure 5.33. On the left is D_s and on the right is D_r.

some R_{ij} would imply the same configuration in D_s starting and ending on B_i and there would thus be a cornsquare on a cone-cell.

For purpose of the rectification, we order the cone-cells B_i consistently with the cone-cell at infinity of D_s and similarly order the 1-cells of the cone-cell at infinity whose boundary is ∂D_r.

Let $\bar{D}_r$ be the rectification of the reduced diagram $D_r \to X^*$, and assign angles to the corners in $\bar{D}_r$ according to the angling rule of X^*. This determines an angle assignment for $\bar{D}_s$ except for the corners of the $\{B_i\}$. These are assigned angles as follows: $\frac{\pi}{2}$ at any nil-external or singly-external corner, and 0 at a singular doubly-external corner, and π elsewhere.

Note that a corner of a cone-cell of D is singular doubly-external in D_r if and only if it is singular doubly-external in D_s.

An internal cone-cell of D_s has nonpositive curvature since it has nonpositive curvature in D_r. By construction of D_r, a corner of a cone-cell of D is singly-external or doubly-external.

An internal 0-cell of D_s that is disjoint from $\cup B_i$ must have nonpositive curvature in D_s since it does in D, unless it is the endpoint of a spur or the central vertex of the outerpath of a cornsquare.

A 0-cell of D_s that is disjoint from D must have nonpositive curvature since it lies in the interior of a boundary arc of some B_i, and hence has a single π corner.

Let v be a 0-cell of D_s that intersects both ∂D and $\cup B_i$. If v has nonpositive curvature in D, then v has nonpositive curvature in D_s. Indeed, if $\text{link}_D(v)$ is disconnected, then $\kappa(v) \leq 0$ since all angles are ≥ 0. We may thus assume that all angles at v are nonpositive since no corner at v is singular doubly-external. Hence each component of $\text{link}_{D_s}(v) - \text{link}_D(v)$ is either an open edge with an angle of π or is the union of a vertex and open edge with an angle of $\frac{\pi}{2}$. If v has positive curvature in D then either v is the end of a spur or the central vertex of the outerpath of a cornsquare. In the spur case, we can assume that v lies along at least two cone-cells, B_i, B_j, since we may assume each $\partial_p B_i$ is immersed in D_s. Thus $v \in \partial D_s$ and there are at least two singly-external corners. In the cornsquare case, it is impossible for its outerpath to lie on $\partial_p B_i$ for any B_i, as then D_s would not be reduced, hence there is at least one singly-external corner at v, so the angle sum is $\geq \pi$, and so $\kappa(v) \leq 0$ whether or not $v \in \partial D_s$.

For a cone-cell $Y \neq Y_s$, the value of $\{\{P\}\}_Y$ is the same with respect to X_s^* as it is with respect to X^*, since by construction, the corners of a cone-cell C associated to Y have the same angles with respect to a reduced diagram $D_s \to X_s^*$ as they do with respect to another reduced diagram $D_r \to X^*$.

The wallspace structure on Y_s has exactly one hyperplane per wall. Conditions 5.1.(3) and 5.1.(4) are immediate since Y_s is CAT(0) and every wall is a hyperplane. The $B(6)$ structure on $\langle X \mid Y_s, \{Y_i\} \rangle$ has precisely the same walls as the $B(6)$ structure on $\langle X \mid \{Y_i\} \rangle$.

The argument for negatively curved small-cancellation and $B(8)$ is similar. $\square$

5.k   1-Dimensional Linear Separation

The wallspace $\widetilde{X}^*$ satisfies the *linear separation property* if $\#(p, q) \geq L\mathsf{d}(p, q) - M$ for some constants $L, M \geq 1$. Here $\#(p, q)$ denotes the number of walls separating the 0-cells p, q, and $\mathsf{d}(p, q)$ is the distance between p, q in the 1-skeleton of $\widetilde{X}$.

Definition 5.39. Let X^* be a cubical presentation with X 1-dimensional. We say $p, q \in Y^0$ are *strongly separated* by the wall w of Y if no 1-cell e dual to a hyperplane of w lies in the same piece as p or q.

The cone Y in X^* has the $\frac{\pi}{2}$-*strong separation property* if p, q are strongly separated by some wall whenever $\{\{p, q\}\}_Y \geq \frac{\pi}{2}$.

Lemma 5.40. *Let $\langle x_1, \ldots \mid R_1, \ldots \rangle$ be an ordinary presentation, satisfying the classical $B(6)$ small-cancellation condition: so more than half a relator cannot be the concatenation of fewer than four pieces, and we assume (by subdividing) that the relators have even length, and each wall of R_i is an antipodal pair of edges. Then using the split-angling, each relator has the $\frac{\pi}{2}$-strong separation property.*

Note that the split-angling is the same as the type-angling here with each 2-cell having type 6.

Proof. Suppose p, q are points on the boundary of a 2-cell (i.e., a cone) that do not lie in the concatenation of at most two pieces. Let P be a *piece-neighborhood* of p in the sense that P contains every closed edge that lies in a piece with p. Similarly, let Q be a piece-neighborhood of q. By hypothesis, $P \cap Q = \emptyset$. By possibly exchanging the notation, assume that $|Q| \geq |P|$. Let w_1, w_2 be the walls dual to the edges immediately before and after Q. If w_1 and w_2 are both dual to edges of P then $|P| > |Q|$ which is impossible. If neither w_1 nor w_2 is dual to an edge of P, then P is separated from Q by one or both of w_1, w_2. If one of w_1, w_2 is dual to an edge of P then the other separates Q from P. We refer the reader to Figure 5.34. $\square$

Figure 5.34. The diagram on the left indicates the notation, and it is then obvious that if Q passes through both walls then $|Q| \geq |P| + 2$. The diagrams on the right indicate some of the possible combinatorial fashions that one of the walls w_1, w_2 might separate.

Theorem 5.41. *Suppose X is 1-dimensional, and $\langle X \mid \{Y_i\} \rangle$ is a cubical presentation satisfying $B(6)$, where the angling rule is either the split-angling or the type-angling. Suppose X^* has short innerpaths, and that the $\frac{\pi}{2}$-strong separation property holds for each cone. Suppose there is a uniform upperbound on the diameters of cones. Then $\widetilde{X}^*$ satisfies the linear separation property.*

Proof. Let γ be a geodesic in $\widetilde{X}^*$. Let e_1 be a 1-cell on γ. We will show that either the wall W_1 in $\widetilde{X}^*$ that is dual to e_1 crosses γ in no other 1-cell, or else there is a 1-cell e_2 within a uniformly bounded distance of e_1, so that the wall W_2 dual to e_2 crosses γ in no other 1-cell. Either way, this shows that each edge of γ is within a uniform distance of an edge dual to a wall crossing γ exactly once. This easily implies the linear separation property.

Let e_1' be a second 1-cell of γ that is dual to W_1, and let γ_1 be the subpath of γ that starts with e_1 and ends with e_1'. By Corollary 5.32, the geodesic γ_1 lies in $N(W_1)$. Let $L_1 \to N(W_1)$ be a ladder that contains γ_1, so L_1 starts on the dual 1-cell e_1 and ends on the second dual 1-cell e_1'. See the left diagram in Figure 5.36. We assume L_1 has a minimal number of cone-cells (so consecutive 1-cells dual to W_1 are distinct). Let C_1 be its first cone-cell. Let p, q denote outermost points in $\gamma \cap Y$ where Y is the cone supporting C_1. We will show that $\{\!\{p, q\}\!\}_Y \geq \frac{\pi}{2}$.

Let γ' denote the subpath of γ joining p, q. Suppose there is a path $\sigma \to Y$ from p to q with $\{\!\{\sigma\}\!\}_Y < \pi$. Since γ is a geodesic in $\widetilde{X}^*$, we must have $|\gamma'| \leq |\sigma|$. Since $\{\!\{\sigma\}\!\}_Y < \pi$ we see that σ and γ' are path-homotopic in Y by the short innerpath hypothesis. Consequently, if $\{\!\{\sigma\}\!\}_Y < \frac{\pi}{2}$ then $\{\!\{\gamma'\}\!\}_Y \leq \{\!\{\sigma\}\!\}_Y < \frac{\pi}{2}$ by the following observation: $\{\!\{\alpha\}\!\}_Y \geq \{\!\{\beta\}\!\}_Y$ whenever X is 1-dimensional and $\alpha \to Y$ and $\beta \to Y$ are path-homotopic, and β is an immersion. Indeed, as depicted in Figure 5.35, the number of internal corners along α does not increase as backtracks are folded outwards. Thus our claim is clear for the type-angling. For the split-angling, two defects of $\frac{\pi}{3}$ and/or $\frac{\pi}{2}$ can become a single defect of $\frac{\pi}{3}$ or $\frac{\pi}{2}$. Therefore the total defect does not increase when folding.

We now show that $\{\!\{\gamma'\}\!\}_Y > \frac{\pi}{2}$. Concatenate the part of γ' ending at the second cone-cell C_2 of L_1, with a path γ'' consisting of part of the piece between

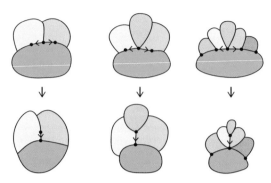

Figure 5.35. If $\alpha \to Y$ folds to $\beta \to Y$ then $\{\{\alpha\}\}_Y \geq \{\{\beta\}\}_Y$.

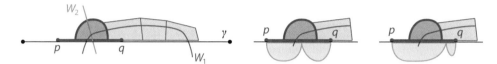

Figure 5.36. If W_1 double crosses γ then there is a nearby wall W_2 that single crosses γ.

C_1 and C_2, to obtain a path δ. We refer to the two rightmost diagrams in Figure 5.36. If $\{\{\gamma'\}\}_Y < \frac{\pi}{2}$ then $\{\{\delta\}\}_Y < \pi$ since the subpath of γ' contributes a defect of $< \frac{\pi}{2}$, and γ'' contributes 0, and there is at most a $\frac{\pi}{2}$ defect at the transition between these subpaths since there are no acute corners. But δ crosses $W_1 \cap Y$ in two distinct hyperplanes which contradicts Condition 5.1(4).

By the $\frac{\pi}{2}$-strong separation hypothesis, there exists a wall W_2 in Y that separates p, q. Let e_2 be the edge between p, q in γ that W_2 is dual to. To see that W_2 cannot double cross γ, suppose W_2 was dual to another edge e_2' of γ, and let γ_2 be the subpath starting with e_2 and ending with e_2', and let $L_2 \to N(W_2)$ be a minimal ladder such that γ_2 is a path in L_2. We examine how L_1 and L_2 overlap with γ and reach a contradiction. We refer to Figure 5.37. All possible situations lead to either p or q lying in the same piece as one of the hyperplanes of $W_2 \cap Y$, or lead to a path μ with $\{\{\mu\}\}_Y < \pi$ that passes through distinct hyperplanes in the same wall of a cone-cell, thus contradicting Condition 5.1.(4).

Since e_2 and e_1 lie on the same cone, they are uniformly close in γ. $\qquad \square$

5.1 Obtaining Proper Actions on the Dual

Given a group G acting on a wallspace, there is a corresponding action of G on the associated dual CAT(0) cube complex (see Chapter 7). Lemma 5.42 gives a simple criterion to ensure that G acts with small stabilizers on the dual cube

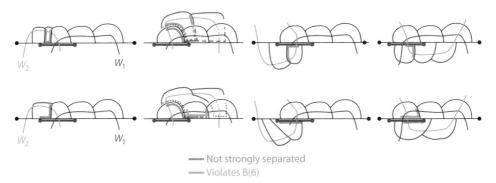

W_2 W_1

W_2 W_1

━━━━━ Not strongly separated
━━━━━ Violates B(6)

Figure 5.37. There are a variety of cases of how γ travels in the W_2-ladder. In each case, there is a violation of either $\frac{\pi}{2}$-separation of p, q or a violation of the $B(6)$ condition.

complex. Theorem 5.44 then gives a condition ensuring this criterion in our context where G acts on the wallspace $\widetilde{X}^*$.

Let g be an automorphism of a wallspace. A wall W *cuts* g if there is a g-invariant subspace $\mathbb{R}_g$ that is a copy of $\mathbb{R}$ such that $g^n W \cap \mathbb{R}_g = \{n\}$ for each $n \in \mathbb{Z}$. We moreover assume that the two components of $\mathbb{R}_g - W$ lie in distinct halfspaces of W. (There is an equally effective version of this notion allowing an odd number of points in the intersection.) The following is a variant of the properness criterion given in [Wis14, Lem 2.1] and we refer to [Wis12, Lem 7.16] for the version below.

Lemma 5.42. *Let G act on a wallspace. If each infinite order $g \in G$ is cut by a wall, then G acts with torsion stabilizers on the associated dual CAT(0) cube complex.*

A hyperplane U is m-*proximate* to a 0-cube v if there is a path $P = P_1 \cdots P_m$ such that each P_i is either a single edge or a path in a piece, and v is the initial vertex of P_1 and U is dual to an edge in P_m. A wall is m-*proximate* to v if it has a hyperplane that is m-proximate to v.

Let Y be a cone of a cubical presentation X^*. A hyperplane w of Y is *piecefully convex* if the following holds: For any path $\xi\rho \to Y$ with endpoints on $N(w)$, if ξ is a geodesic and ρ is trivial or lies in a piece of Y containing an edge dual to w, then $\xi\rho$ is path-homotopic in Y to a path $\mu \to N(w)$. Note that if w is piecefully convex then $N(w) \to Y$ is convex.

Remark 5.43. Let M be the maximal diameter of any piece of Y_i in X^*. Then a hyperplane H of Y_i is piecefully convex provided its carrier $N = N(H)$ satisfies: $\mathsf{d}_{\widetilde{Y}_i}(g\widetilde{N}, \widetilde{N}) > M$ for any translate $g\widetilde{N} \neq \widetilde{N} \subset \widetilde{Y}_i$.

Consequently, if each Y_i is compact, and each $\pi_1 H \subset \pi_1 Y_i$ is separable, there are finite covers $\widehat{Y}_i$ satisfying this injectivity radius condition (for any further covers). Hence pieceful convexity holds for all hyperplanes of cones of $\langle X \mid \{\widehat{Y}_i\}\rangle$.

In the following theorem we allow a technical generalization where we designate a set of the hyperplanes of X as *preferred* and where the wallspace structure on each Y_i has the property that if one hyperplane in a wall is preferred (i.e., the preimage of a preferred wall) then they all are, and hence each wall of $\widetilde{X}^*$ consists entirely of hyperplanes that are preferred or entirely of hyperplanes that are not preferred. A natural example of such a structure arises when X contains some isolated edges, and these are the preferred hyperplanes, and each wall of each Y_i consists of a collection of hyperplanes dual to isolated edges. The reader can ignore this technicality by assuming all hyperplanes are preferred when reading the following theorem.

Theorem 5.44. *Suppose $\langle X \mid \{Y_i\}\rangle$ satisfies the following hypotheses:*

(1) *X^* satisfies the $B(6)$ condition and has short innerpaths.*
(2) *Each [preferred] hyperplane w of each cone Y_i is piecefully convex.*
(3) *Let $\kappa \to Y \in \{Y_i\}$ be a geodesic with endpoints p, q. Let w_1 and w'_1 be distinct hyperplanes in the same [preferred] wall of Y. Suppose κ traverses a 1-cell dual to w_1, and either w'_1 is 1-proximate to q or κ traverses a 1-cell dual to w'_1. Then there is a [preferred] wall w_2 in Y that separates p, q but is not 2-proximate to p or q.*

Then each $g \in \pi_1 X^*$ has one of the following properties:

(1) *$g \in \operatorname{Aut}(Y)$ for some lift of Y to $\widetilde{X}^*$.*
(2) *g is cut by a [preferred] wall of $\widetilde{X}^*$.*
(3) *g is the image of an element $\tilde{g} \in \pi_1 X$ that is not cut by any [preferred] hyperplane.*
(4) *g has finite order in $\pi_1 X^*$.*

We refer to Figure 5.38 for a depiction of the notation in Hypothesis (3).

Corollary 5.45. *In the case of Theorem 5.44 where all hyperplanes are preferred, and where for each Y_i, each infinite order element of $\operatorname{Aut}(Y_i)$ is cut by a wall, then the action of $\pi_1 X^*$ on its associated dual cube complex has torsion stabilizers.*

Proof. If the stabilizer of a 0-cube is not a torsion group, then it has an infinite order element g, for which Property (4) cannot hold. Property (3) cannot hold since all hyperplanes are preferred. If Property (1) holds then g has an axis $\mathbb{R}_g$ in Y that is cut by a wall w of Y, by hypothesis. The wall w extends to a wall W of $\widetilde{X}^*$ cutting g since $(W \cap \mathbb{R}_g) \subset (W \cap Y_i) = (w \cap Y_i)$ by Theorem 5.18. Thus $\langle g \rangle$

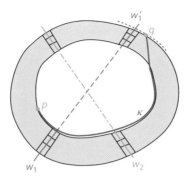

Figure 5.38. A scenario arising in Hypothesis 5.44(3). The dashed path is a piece showing that w_1' is 1-proximate to q. We have not illustrated several other similar scenarios.

acts freely on the dual cube complex by Lemma 5.42, which is impossible. Similarly, Property (2) cannot hold since then $\langle g \rangle$ acts freely by Lemma 5.42.   □

Remark 5.46. If we assume that the wall w_2 of Y separating p, q cuts κ in an edge lying within a uniformly bounded distance of $\kappa \cap w_1$, then the proof applied to the convex hull $\widetilde{S}$ of a geodesic $\widetilde{\gamma} \subset \widetilde{X}$ mapping to an arbitrary geodesic $\gamma \subset \widetilde{X}^*$ shows that $\widetilde{X}^*$ satisfies the linear separation property, and hence $\pi_1 X^*$ acts metrically properly on the CAT(0) cube complex dual to the wallspace on $\widetilde{X}^*$. In particular, this holds when there is a uniform upperbound on $\mathrm{diam}(Y_i)$.

Remark 5.47. For applications of Theorem 5.44 it can be useful to add each cube of X as a relator. We note as well that Theorem 5.44 can have interesting recubulation applications for artificial cubical presentations where each $\widetilde{Y}_i$ is a convex subcomplex equipped with an interesting wallspace structure.

Proof of Theorem 5.44. By possibly subdividing we can assume that $\pi_1 X$ acts without inversions on the hyperplanes of $\widetilde{X}$, and so each nontrivial element of $\pi_1 X$ has a combinatorial geodesic axis [Hag07]. If $g \in \pi_1 X^*$ has finite order then Property (4) holds, so we assume g has infinite order. We choose $\tilde{g} \in \pi_1 X$ that maps to g, and moreover choose $\tilde{g}$ so that its translation length is minimal among all such choices. Let $\tilde{\gamma}$ be an axis for $\tilde{g}$. Let $\gamma \to \widetilde{X}^*$ be the projection of $\tilde{\gamma}$ and note that $\gamma \to \widetilde{X}^*$ is embedded, since otherwise there would be a shorter choice for $\tilde{g}$.

If some Y_i contains a ray of γ, and W double crosses γ at e, e', then there exists n such that $g^n Y_i$ contains the subgeodesic $\gamma_\circ$ starting with e and ending with e'. Moreover, at least one of $g^{n \pm 1} Y_i$ also contains $\gamma_\circ$. If $g \notin \mathrm{Aut}(Y_i)$ then $g^n Y_i \neq g^{n \pm 1} Y_i$ but $\gamma_\circ$ lies in the piece between $g^n Y_i$ and $g^{n \pm 1} Y_i$, so $\{\{\gamma_\circ\}\}_{Y_i} = 0$, and we contradict that $\gamma_\circ$ is a geodesic since e, e' traverse the same hyperplane using Condition 5.1.(4). Thus if Property (1) does not hold, then no ray of γ

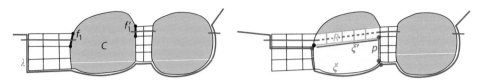

Figure 5.39. If $w_1 = w_1'$ then Hypothesis (2) allows us to find a shorter W_1-ladder.

lies in a cone Y_i of $\widetilde{X}^*$. Note that each cone is convex by Theorem 5.31, so γ intersects each cone in an interval. We thus proceed under the assumption that $\gamma \cap Y_i$ is finite for each cone.

If no edge of γ is dual to a preferred hyperplane of $\widetilde{X}$ then Property (3) holds. So we consider an edge e_1 of γ dual to a preferred hyperplane that extends to a wall W_1 of $\widetilde{X}^*$. If W_1 does not cross γ in a second edge then Property (2) holds.

Suppose an edge e_1 of γ is dual to a [preferred] wall W_1 of $\widetilde{X}^*$ intersecting γ in another edge that is dual to a second [preferred] hyperplane. By Theorem 5.31, the subtended subpath of γ is square-homotopic to a geodesic λ that lies on a *gridded ladder* $L_1 \to T(W_1)$. This is a ladder that is allowed to start and/or end with a grid instead of a cone-cell. We choose L_1 above so that it has as few cone-cells as possible. Note that L_1 must contain at least one cone-cell, for otherwise γ is not a geodesic. The square diagram between λ and the subpath of γ is contained in the convex hull of each of these two paths. Accordingly, L_1 begins with a (possibly degenerate) rectangle that starts at e_1 and terminates at a 1-cell f_1 and is followed by a cone-cell C mapping to a cone Y, such that f_1 maps to Y and is dual to a hyperplane w_1 of Y, and w_1 lies in the same wall of Y as a hyperplane w_1' of Y. And w_1' is dual to an edge f_1' that lies in a cone-piece or wall-piece with the next part of L_1.

We now show that $w_1 \neq w_1'$. By possibly performing a square-homotopy in $T(W_1)$, we can assume that λ has a subpath $\xi \to Y$ that starts at an endpoint of f_1, and such that ξ can be concatenated with a path $\rho \to Y$ ending at an endpoint of f_1', and either ρ is trivial or ρ lies in a piece of Y. See Figure 5.39. By Hypothesis (2), if $w_1 = w_1'$ then $\xi\rho$ is path-homotopic in Y to a path $\xi' \to R$ in a square ladder in $N(w_1)$. This induces a square-homotopy from λ to a path λ' that travels along a W_1-ladder with fewer cone-cells. In particular, the first cone-cell C of L_1 was unnecessary as it can be replaced by R, and this violates our assumption that the W_1-ladder L_1 has a minimal number of cone-cells.

Let κ be the subpath of γ contained in Y, and let p, q be the endpoints of κ. We refer to Figure 5.40. Let γ' be the geodesic obtained from γ by possibly homotoping it across one or two square grids so that $\gamma' \cap Y$ consists of a path κ' with endpoints p', q', and κ' traverses a 1-cell dual to w_1, and either κ' traverses a 1-cell dual to w_1' or q' lies on a 1-cell containing q' or contained in a piece containing q'. Note that γ' and γ are cut by precisely the same walls, so it suffices to show that γ' is cut by a wall separating its rays.

By Hypothesis (3), applied to $\kappa', p', q', w_1, w_1'$, there exists a [preferred] wall w_2 of Y that separates p' and q' but is not proximate to p' or q'. Let W_2 be the

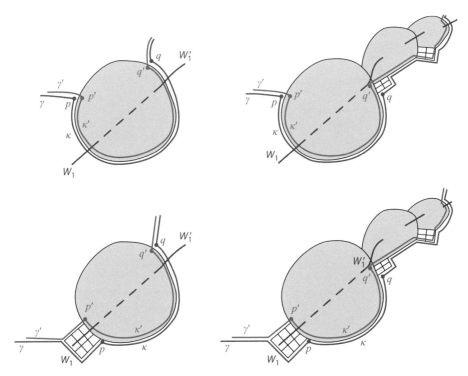

Figure 5.40. The diagrams illustrate scenarios where γ is double crossed by a wall W_1. The path γ travels in Y along a subpath κ with endpoints p, q. Either e_1 is a 1-cell of κ or e_1 lies in the p-ray of γ and there is a nontrivial square ladder at the beginning of L_1 from e_1 to the Y cone. The geodesic γ' is obtained from γ by pushing across at most two grids. The subpath $\kappa' = \gamma' \cap Y$ has endpoints p', q'. Note that κ' traverses a 1-cell dual to w_1, and either κ' traverses a 1-cell dual to w_1' or w_1' is 1-proximate to q'.

[preferred] wall of $\widetilde{X}^*$ corresponding to w_2. We will show below that W_2 cuts γ and so Property (2) holds.

Suppose W_2 crosses γ' outside κ'. Then Theorem 5.31 provides a W_2-ladder L_2 that starts at an edge of κ' and ends at an edge outside of κ' that is dual to W_2, and with a square diagram between the subpath of γ' between these edges, and a path on the extended carrier of W_2. See Figure 5.41. In the various cases that arise, the key point about L_2 is that γ' leaves Y at p', q', and hence if W_2 crosses the p'-ray of γ' then w_2 is 2-proximate to p', which is impossible. The analogous statement holds for a q'-ray and q'. $\qquad\square$

Corollary 5.48. *Let X be a compact nonpositively curved cube complex with $\pi_1 X$ hyperbolic. Let $z_1, \ldots, z_r$ be elements generating a collection $\{Z_1, \ldots, Z_r\}$ of cyclic subgroups. Suppose that conjugates of Z_p and Z_q have trivial intersection*

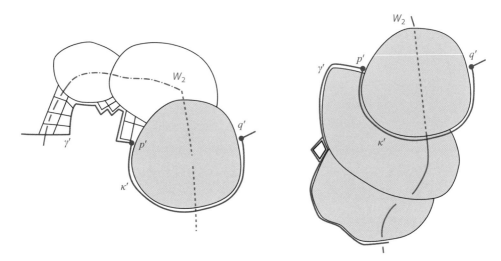

Figure 5.41. The wall w_2 of Y separates p', q' but is not 2-proximate to p' or q'. If the wall W_2 of $\widetilde{X}^*$ extending w_2 intersects γ' outside κ' then the diagram between γ' and a W_2-ladder L_2 shows that w_2 is 2-proximate to p' or q', which is impossible. This is illustrated where the p'-ray of γ crosses W_2 and hence w_2 is 2-proximate to p'. If the q'-ray of γ crosses W_2 then w_2 is 2-proximate to q', and the picture is analogous.

for $p \neq q$. Suppose $Z_i \cap \mathrm{Stab}(\widetilde{U})$ is either trivial or equal to Z_i for each $\widetilde{U}$. There exists N such that whenever $n_i \geq N$, the group $\pi_1 X / \langle\langle z_1^{n_1}, \ldots, z_r^{n_r} \rangle\rangle$ acts properly and cocompactly on a CAT(0) cube complex.

If we drop hyperbolicity and compactness, but add the hypothesis that each Z_i has a Z_i-cocompact superconvex core, then the proof below shows that we still get a proper action on a CAT(0) cube complex.

Proof. By Lemma 2.36, let $Y_1, \ldots, Y_r$ be complexes such that each $\widetilde{Y}_i$ is Z_i-cocompact and superconvex. Under the assumption that each $\mathrm{Stab}_{Z_i}(\widetilde{U}) = 1$ we see that each hyperplane of $\widetilde{Z}_i$ is finite and contractible. Hence there exists N, such $\langle z_i \rangle \backslash \widetilde{Z}_i$ has embedded 2-sided contractible hyperplanes when $n_i \geq N$. Subdividing X guarantees that each Y_i has an even number of finite CAT(0) hyperplanes that cut the generator of Y_i. A choice of local geodesic generator of $\pi_1 Y_i$ (which is then of even length) allows us to pair up "antipodal" hyperplanes of Y_i to form walls. We likewise use such an antipodal choice for any finite cover $\widehat{Y}_i$. The hyperplanes that do not cut such a generator already provide walls (an essential halfspace contains the generator and a frivolous part does not). Note that some of these hyperplanes are finite and contractible whereas others have the homotopy type of a circle.

Superconvexity and malnormality provide an upperbound M on the diameter of any wall-piece or cone-piece. Consequently, there exists N such that for any connected covers $\widehat{Y}_i \to Y_i$ each of whose degrees are $\geq N$, the cubical presentation $\langle X \mid \widehat{Y}_1, \ldots, \widehat{Y}_r \rangle$ is $C'(\frac{1}{24})$.

By choosing N a bit larger, the antipodal hyperplanes are sufficiently far that the wallspace satisfies the $B(6)$ condition and the conditions of Theorem 5.44 are satisfied. $\qquad\square$

The following analogue of Corollary 5.48 depends on results discussed later in the text. It can be generalized further to cover the situation where X is sparse.

Proposition 5.49 (Cubulating toral Dehn fillings). *Let X be a compact non-positively curved cube complex. Let $Y_i \to X$ be superconvex with Y_i compact for $1 \leq i \leq r$. Suppose each $\pi_1 Y_i \subset \pi_1 X$ is maximal virtually abelian. Suppose the nondiagonal part of $(\sqcup_i Y_i) \otimes_X (\sqcup_i Y_i)$ consists of contractible components. For any finite covers $\widehat{Y}_i^{\circ} \to Y_i$ there exist finite covers $\widehat{Y}_i$ factoring through $\widehat{Y}_i^{\circ}$, such that $X^* = \langle X \mid \{\widehat{Y}_i\} \rangle$ is $B(8)$ and $\pi_1 X^*$ acts properly on the cube complex dual to the wallspace on $\widetilde{X}^*$.*

The Y_i are provided by Lemma 2.38, and we can choose them to be superconvex if each $\pi_1 Y_i$ has trivial intersection with other abelian subgroups of rank ≥ 2, as happens when $\pi_1 X$ is hyperbolic relative to $\{\pi_1 Y_i\}$. When the $\{Y_i\}$ are also all superconvex, then distinct translates of the $\widetilde{Y}_i$ have uniformly bounded overlap so the nondiagonal components of $(\sqcup_i Y_i) \otimes_X (\sqcup_i Y_i)$ are contractible as in Corollary 8.10. We are then in the compact case of the sparse setting of Definition 7.21.

I expect the conclusion of Proposition 5.49 can be strengthened to actually generalize Corollary 5.48, by asserting that there exist finite covers $\widehat{Y}_i^{\circ} \to Y_i$ such that for any regular covers $\widehat{Y}_i \to Y_i$ factoring through $\widehat{Y}_i^{\circ}$, the group $\pi_1 X^*$ acts properly on the cube complex dual to the wallspace on $\widetilde{X}^*$.

Proof of Proposition 5.49. By Lemma 2.40, let M' be an upperbound on rectangular strips meeting each $\widetilde{Y}_i$. Let M'' be an upperbound on the diameters of nondiagonal components of $(\sqcup_i Y_i) \otimes_X (\sqcup_i Y_i)$. Let $M = \max(M', M'')$. By residual finiteness, let $\bar{Y}_i \to Y_i$ be finite regular covers so that $\|\bar{Y}_i\| > 14M$. This will ensure short innerpaths of X^* below by Lemma 3.70. Since a f.g. abelian group has separable subgroups, we may assume that the carrier of each hyperplane of $\bar{Y}_i$ is embedded. Remark 5.43 provides finite covers of Y_i so that pieceful convexity holds in the cubical presentation associated to any further covers, and we assume each $\bar{Y}_i$ factors through these covers. Moreover, we assume each $\bar{Y}_i$ factors through each $\widehat{Y}_i^{\circ}$.

Let $\widetilde{Z}_i$ be the subcomplex of $\widetilde{Y}_i$ that is the intersection of deep minor halfspaces of hyperplanes that are deep on exactly one side. Let R_3 be so that each $\widetilde{Y}_i \subset \mathcal{N}_{R_3}(\widetilde{Z}_i)$. Note that $\widetilde{Z}_i = K_i \times \prod_j \widetilde{L}_{ij}$ where K_i is compact and each $\widetilde{L}_{ij}$ is a quasiline. By passing to a further finite index subgroup, we may assume that $\bar{P}_i = \pi_1 \bar{Y}_i$ preserves the factors and acts trivially on K_i. Let $\bar{Z}_i = \bar{P}_i \backslash \widetilde{Z}_i$ and note that $\bar{Z}_i \cong K_i \times \prod_j \bar{L}_{ij}$.

We will apply Construction 9.1 to obtain a wallspace structure on a finite cover $\ddot{\bar{Z}}_i \to \bar{Z}_i$. This induces a finite cover $\ddot{\bar{Y}}_i$ and the wallspace structure on $\ddot{\bar{Z}}_i$ extends to a wallspace structure on $\ddot{\bar{Y}}_i$ as follows: If a hyperplane U of $\ddot{\bar{Y}}_i$ is disjoint from $\ddot{\bar{Z}}_i$ then we declare it to be an entire wall—note that U is separating in this case. The remaining hyperplanes of $\ddot{\bar{Y}}_i$ are partitioned into walls exactly the way their intersections with $\ddot{\bar{Z}}_i$ are partitioned. Note that this wallspace differs from the one obtained by applying Construction 9.1 directly to $\bar{Y}_i$, as inessential hyperplanes of $\ddot{\bar{Y}}_i$ are not combined with any others.

Verifying the $B(6)$ (or $B(8)$) structure is done as in the proof of Theorem 15.6, and depends on choosing $\bar{Y}_i$ so that the systoles of each $\bar{L}_{ij}$ is sufficiently large. In the remainder of the proof, we verify Hypotheses 5.44.(3). This depends on a special case of an argument proving Theorem 9.12. It requires choosing each $\bar{Y}_i$ so that the systoles of each L_{ij} factor is sufficiently large. As we will follow the same reasoning for each cone, to reduce notation we henceforth suppress the subscripts of cones and use the notation Y instead of Y_i, and $\widetilde{Z} \cong K \times \prod \widetilde{L}_j$ etc.

Observe that $\ddot{\bar{Z}} \cong K \times \prod L_j$ where each $L_j \to \bar{L}_j$ is a connected double cover. Indeed, the homomorphism inducing the cover in Construction 9.1 decomposes into a product of homomorphisms for each factor in the product structure.

Each quasiline $\widetilde{L}_j$ has finitely many $\text{Aut}(\widetilde{L}_j)$-orbits of hyperplanes, and each such hyperplane has compact carrier. Let R_1 be an upperbound on the number of hyperplanes in the carrier $N(U)$ of each hyperplane of each $\widetilde{L}_j$. By cocompactness, let R_2 be an upperbound on the number of hyperplanes that are 2-proximate to a 0-cell of Y. Assume the systole of each $\bar{L}_j$ strictly exceeds $R = 11R_1 + 2R_2 + 2R_3 + 4M$ in our choice of covers of cones above.

Let w_1, w_1' be distinct hyperplanes in the same wall of $\ddot{\bar{Y}}$. Let $\kappa \to \ddot{\bar{Y}}$ be a geodesic from p to q that traverses an edge dual to w_1 and such that either w_1' is 1-proximate to q or else κ traverses a 1-cell dual to w_1'. Since $w_1 \neq w_1'$ we see that w_1, w_1' intersect $\ddot{\bar{Z}}$ and moreover $w_1 \cap \ddot{\bar{Z}}$ and $w_1' \cap \ddot{\bar{Z}}$ project to the same hyperplane $\bar{w}_1$ of $\bar{Z}$. Note that $\bar{w}_1$ arises from a hyperplane of some $\bar{L}_\ell$ in the decomposition $K \times \prod_j \bar{L}_j$ of $\bar{Z}$.

Each of p, q is within R_3 of a closest point $\dot{p}, \dot{q}$ in $\ddot{\bar{Z}}$. Choose a minimal area disc diagram $D \to \ddot{\bar{Y}}$ with $\partial_{\mathsf{p}} D = \mu_p \sigma \mu_q \hat{\kappa}$, where μ_p, μ_q are geodesics with endpoints $p, \dot{p}$ and $q, \dot{q}$, where $\sigma \to \ddot{\bar{Z}}$ is a geodesic, and where $\hat{\kappa}$ is a geodesic that is path-homotopic to κ. Corollary 2.24 ensures that D is a grid with (possibly trivial) tails ending at p, q, but no tails at $\dot{p}, \dot{q}$ since μ_p, μ_q are shortest paths to $\ddot{\bar{Z}}$.

Let σ_ℓ be the path from p_ℓ to q_ℓ that is the projection of σ (and hence $\hat{\kappa}$) onto the L_ℓ factor, and regard L_ℓ as a subcomplex of $\ddot{\bar{Z}} \cong K \times \prod_j L_j$. Note that σ_ℓ is itself a geodesic in L_ℓ since the product structure ensures that σ is path-homotopic to a geodesic of the form $\sigma_\ell \hat{\sigma}'$.

We verify the following claim below: $\mathsf{d}(p_\ell, (w_1 \cap L_\ell))$ and $\mathsf{d}(q_\ell, (w_1' \cap L_\ell))$ are each bounded above by $2M + 5R_1$. Consequently σ_ℓ projects to a path $\bar{\sigma}$ in the factor $\bar{L}_\ell$ of $\bar{Z} = K \times \prod_j \bar{L}_j$, and $\bar{\sigma}_\ell$ can be augmented to a closed path $\bar{\sigma}'$ in $\bar{L}_\ell$ generating $\pi_1 \bar{L}_\ell$ by adding initial and terminal paths of length $\leq 2M + 5R_1$ to $\bar{\sigma}_\ell$.

Consider the hyperplanes dual to edges of $\bar{\sigma}'$. At most $2M + 5R_1$ such hyperplanes are dual to edges of each augmenting paths, at most R_2 such hyperplanes are 2-proximate to each of p, q, at most R_1 such hyperplanes cross $\bar{w}_1$, and at most R_3 such hyperplanes separate p, p_ℓ or q, q_ℓ. Since the systole of $\bar{L}_\ell$ is more than $R = 11R_1 + 2R_2 + 2R_3 + 4M$, we find that there is a hyperplane $\bar{w}_2 \cap \bar{L}_\ell$ dual to an edge of $\bar{\sigma}'$ with none of the above properties, and so it extends to a hyperplane $\bar{w}_2$ whose preimage w_2, w_2' in $\ddot{Y}$ separates p, q but is not 2-proximate to either of them.

We now prove the claim. Let $\bar{c}_\ell$ be a closed geodesic with $\pi_1 \bar{L}_\ell$ generated by $[\bar{c}_\ell]$. Then $\operatorname{diam}(L_\ell) \leq |\bar{c}_\ell| + 2R_1$. Indeed, the double cover c_ℓ of $\bar{c}_\ell$ has $\operatorname{diam}(c_\ell) = |\bar{c}_\ell|$, and each hyperplane of L_ℓ crosses c_ℓ, and hyperplanes have diameter $\leq R_1$.

If κ traverses w_1', then let $\sigma_\ell = \nu\alpha\beta$ where the first and last edges of α are dual to $w_1' \cap L_\ell$ and $w_1 \cap L_\ell$. The triangle inequality gives $|\nu| + |\beta| \leq 4R_1$. Consequently $|\nu| + |\beta| = |\sigma_\ell| - |\alpha| \leq (\operatorname{diam}(c_\ell) + 2R_1) - |\alpha| = 2R_1 + |c_\ell| - |\alpha| \leq 4R_1$.

If κ does not traverse w_1', then the endpoint of κ is within $2M$ of w_1', so the endpoint of σ_ℓ is joined to $w_1' \cap L_\ell$ by a geodesic β with $|\beta| \leq 2M$. Let $\sigma_\ell = \nu\alpha$ where the initial edge of α is dual to $w_1 \cap L_\ell$. The triangle inequality gives $|\bar{c}_\ell| - |\beta| \leq 2M + 2R_1$, and as σ_ℓ is a geodesic we have $|\nu| + |\beta| \leq \operatorname{diam}(L_\ell) \leq |c_\ell| + 2R_1$. Consequently $|\nu| \leq -|\beta| + |\bar{c}_\ell| + 2R_1 \leq 2M + 4R_1$. $\qquad\square$

Problem 5.50. Let G be a hyperbolic group that equals $\pi_1 X$ where X is a compact nonpositively curved cube complex. Let H be a quasiconvex subgroup. Under what conditions is there a finite subset $S \subset H - \{1\}$ such that $G/\langle\!\langle H' \rangle\!\rangle$ acts properly and cocompactly on a CAT(0) cube complex whenever $[H : H'] < \infty$ and $S \cap H' = \emptyset$?

5.m   Codimension-1 Subgroup Preserved

Let $\langle X \mid \{Y_i\} \rangle$ satisfy the $B(6)$ condition and suppose that X is compact but not contractible. Often, $\pi_1 X^*$ has a codimension-1 subgroup that is the stabilizer of a halfspace of a wall W of $\widetilde{X}^*$. We refer to Definition 7.3. However, $\pi_1 X^*$ might not have a codimension-1 subgroup. For instance, consider the cubical presentation $\langle X \mid Y \rangle$ where Y is a cover of X that satisfies the $B(6)$ properties, as obtainable by taking a regular cover $\widehat{X}_1 \to X$ whose hyperplanes are 2-sided and embedded, followed by a cover $Y = \widehat{X} \to \widehat{X}_1$ corresponding to the $\mathbb{Z}_2^W$ homomorphism of Section 9.a together with its induced wall structure. Then although X^* is $B(6)$, the group $\pi_1 X^*$ is finite and thus has no codimension-1 subgroup. It is interesting to ask for natural conditions on X^* guaranteeing that $\pi_1 X^*$ has a codimension-1 subgroup. The following criterion is often satisfied in practice:

Proposition 5.51 (Codimension-1 criterion). *Let $X^* = \langle X \mid \{Y_i\} \rangle$ be a $B(6)$ cubical presentation that is $C'(\alpha)$ for some $\alpha \leq \frac{1}{12}$, and has short innerpaths.*

Let γ be a closed geodesic in $\widetilde{X}$ so that $\widetilde{\gamma}$ and $\widetilde{U}$ intersect at a single point. Let $\widetilde{J} = \mathrm{hull}(\widetilde{\gamma})$ and let $J = \langle \gamma \rangle \backslash \widetilde{J}$ so there is a local-isometry $J \to X$.

Let W be the wall of $\widetilde{X}^*$ containing a hyperplane that is the image of $\widetilde{U}$. Then $\mathrm{Stab}(W) \subset \pi_1 X^*$ is a codimension-1 subgroup provided there exists β with $2\alpha + \beta \leq \frac{1}{2}$ such that the following holds.

(1) *Each intersection satisfies $\mathrm{diam}(\widetilde{Y}_i \cap \widetilde{J}) < \beta \|Y_i\|$.*
(2) *No $\widetilde{Y}_i \cap \widetilde{J}$ contains a path S projecting to a path in Y_i that starts and ends on pieces intersecting distinct hyperplanes in the same wall of $W \cap Y_i$.*

Condition (2) holds if $\mathrm{diam}(\widetilde{Y}_i \cap \widetilde{J}) < \beta \|Y_i\|$, and edges of Y_i dual to distinct hyperplanes in the same wall of Y_i are at distance greater than $(2\alpha + \beta)\|Y_i\|$.

Proof. Observe that the conditions of Lemma 3.74 are satisfied for X^* with $A = J$, and $A^* = A$, as we can ignore the contractible cones in the induced presentation A^*. Hence, by Lemma 3.74, there is an embedding $\widetilde{J} \to \widetilde{X}^*$ as a convex subcomplex. Consequently $\widetilde{\gamma} \subset \widetilde{X}$ maps to a geodesic that we also denote by $\widetilde{\gamma} \subset \widetilde{X}^*$. Suppose $\widetilde{\gamma}$ contains a second edge dual to a hyperplane of W, and let $\widetilde{\gamma}'$ be the smallest subgeodesic containing these two edges.

By Theorem 5.31, $\widetilde{\gamma}' \subset E(W)$, and moreover, there is a gridded ladder $L \to T(W)$ with γ' square-homotopic to a path $\lambda' \to L$. If L is a square ladder, then lifting to $\widetilde{X}$ we see that $\widetilde{\gamma}$ is crossed twice by the same hyperplane, and thus not a geodesic by Corollary 2.16. We may thus assume L contains a cone-cell C, and there is (possibly trivial) grid $I_m \times I_n$ such that $\{m\} \times I_n$ lies on C, and $\widetilde{\gamma}$ has an initial subpath $\widetilde{\gamma}'$ of the form vhc where $v = \{0\} \times I_n$ and $h = I_m \times \{n\}$ and $c \to C$ is a path in $\widetilde{J} \cap \widetilde{Y}_i$ that ends on a piece containing an edge dual to another hyperplane of W. As $\mathrm{diam}(c) < \beta \|Y_i\|$ by Condition (1), we obtain a contradiction of Condition (2). $\qquad \square$

5.n   Elliptic Annuli

Definition 5.52 (Elliptic Annulus). An element $g \in \pi_1 X^*$ is *elliptic* if $g\widetilde{Y}_i = \widetilde{Y}_i$ for some lift $\widetilde{Y}_i$ of some cone Y_i of X^*.

An annular diagram $A \to X^*$ is *elliptic* if the lift of its universal cover $\widetilde{A} \to \widetilde{X}^*$ actually lifts to a cone $\widetilde{Y}_i$. Note that when A is elliptic, both boundary paths P_+, P_- of A represent elliptic elements in $\pi_1 X^*$. The annulus A is *essential* if its boundary paths are essential in X^*. (Note that the trivial element is elliptic in $\pi_1 X^*$.)

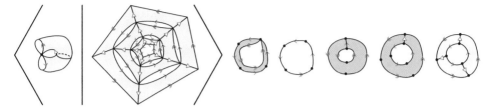

Figure 5.42. Elliptic annuli.

A typical elliptic annulus $A \to X^*$ consists of a single cone-cell that overlaps with itself at an internal path that is not a piece. As opposed to the typical such situation encountered in a disk diagram, it is impossible to "reduce" A by combining the cone-cell with itself. When a cone Y_i of X^* contains a square annulus that is essential in Y_i, then one obtains an elliptic annulus which is disguised by being built out of squares—though it could be replaced by a single self-overlapping cone-cell as above. There are two special cases worth mentioning: Any conjugacy class of an elliptic element yields an elliptic annulus that is isomorphic to a subdivided circle. If Y_i contains a hyperplane that is stabilized (without inversion) by some element $g \in \mathrm{Aut}(Y_i)$, then this data yields an interesting elliptic annulus that looks like a closed square annuladder. More generally, in the $B(6)$ case, we regard an elliptic annuladder carrying the dual curve of some wall as a *length* 0 annuladder. See Figure 5.42.

Lemma 5.53 (Elliptic subannuli merge). *Suppose X^* is a negatively curved small-cancellation presentation, $\widetilde{Y}$ is superconvex for each cone Y, and contiguous cone-pieces have trivial stabilizers in $\pi_1 X$.*

If an annular diagram A contains two or more essential elliptic annular subdiagrams, then they are associated to the same cone $\widetilde{Y}$.

Proof. Suppose these elliptic annuli intersect in A. Then the lift $\widetilde{A} \to \widetilde{X}^*$ shows that the intersection of the associated cones has a nontrivial stabilizer. However, this intersection, which is $\mathrm{CAT}(0)$ by Lemma 5.7, provides a contiguous cone-piece in $\widetilde{X}$ with infinite stabilizer, which contradicts our hypothesis unless the associated cones were the same.

We now assume that the elliptic annuli are disjoint. By repeatedly reducing we eventually arrive at a minimal complexity annular diagram B bounded by (boundary paths of) the two nontrivial elliptic annuli. We can assume there is no essential elliptic annular subdiagram between them (otherwise consider consecutive pairs).

Consider a cornsquare or nonnegatively curved shell along ∂B. The former would absorb into one of the two elliptic annuli, and likewise, the latter must be combinable or replaceable since the defect sum around it would be less than 2π. Note that even if the outerpath is at the transition in the elliptic annulus, we may shift this transition to occur in a different spot, and then absorb to reduce the complexity of B.

Thus B is a flat annulus by Lemma 2.25. If B contains a square, then the superconvexity of cones ensures that a width 1 square annuli carrying a closed dual curve within B is absorbable in the elliptic annuli, and so the complexity of B can be reduced further. Hence we may assume B has no square. But then $\widetilde{B}$ would be an infinite contiguous cone-piece between the two distinct cones associated to the elliptic annuli on either side of ∂B, thus these cones are identical as above, and the elliptic annuli are associated to the same cone-cell. □

We define a *reduced* annular diagram $B \to X^*$ analogously to the way we defined a reduced disk diagram in Definition (3.11). Specifically,

(1) There is no bigon in a square subdiagram of B.
(2) There is no cornsquare whose outerpath lies on a cone-cell of B.
(3) There do not exist a cancellable pair of squares in B.
(4) There is no square s in B that is absorbable into a cone-cell C of B.
(5) For each internal cone-cell C of B mapping to a cone Y, its boundary path $\partial_p C$ is essential in Y.
(6) There does not exist a pair of combinable cone-cells in B.

Excluding a cancellable pair of squares appears to be unnecessary in our arguments, but it seems best to avoid this pathology and the awkward examples it permits. We also caution that a finite cover of an annular diagram might not be reduced, since immersed square bigons can lift to embeddings.

We now prove a W-annuladder variant of Lemma 5.53. This is not an important point in the exposition and depends upon Definition 5.55.

Lemma 5.54. *Suppose $\langle X \mid \{Y_i\} \rangle$ is $B(8)$ with no acute corners, and no Y_i has a piece with infinite stabilizer. Let B be a reduced annuladder that deformation retracts to two annuladders L_1, L_2, where L_i is a W_i-annuladder for a wall W_i of $\widetilde{X}^*$. Suppose that L_1 is elliptic, and that L_2 is either elliptic or has no ejectable cone-cell. Then $L_1 = L_2$.*

Note that disallowing pieces with infinite stabilizer is essentially the same as hypothesizing that each $\widetilde{Y_i}$ is superconvex and $\{\pi_1 Y_i\}$ is malnormal.

Proof. If L_2 crosses L_1, then regarding L_1 as a single cone-cell, and considering the smallest subdiagram containing them both, we find that $L_1 = L_2$, since we obtain a reduced diagram after cutting along a cone-cell and then we can apply Theorem 3.43. We henceforth assume that L_2 does not cross L_1.

If L_2 is not elliptic, then consider the subdiagram A complementary to L_1. We now use that L_2 has no ejectable cone-cell to see that all cone-cells on the L_2 side are negatively curved, and since L_2 has a cone-cell, we can assume the squares in L_2 all lie in rectangles emerging from cones in the rectification of the diagram, and so there are no cornsquares with outerpath on the L_2 side. There are no cornsquares or nonnegatively curved shells with outerpath on the L_1 side, for then absorbing or combining with L_1 shows that B is not reduced. If some cone-cell of L_2 has an external arc on the L_1 side, then cutting along that cone-cell and applying Theorem 3.43 provides a ladder, and hence that cone-cell is ejectable in L_2, which is impossible. We thus find that there is no nonnegatively curved cone-cell in A, and so A is a flat annulus that is not a circle. This yields a piece with L_1 having nontrivial stabilizer, which is impossible.

Suppose L_2 is elliptic. Let A be the annular diagram between L_1, L_2, so the boundary paths of A are the boundary paths of L_1, L_2 that are internal to B. Note that A cannot have an outerpath of a cornsquare on its boundary since it would then be absorbable into L_1 or L_2. By minimality a cone-cell cannot be combinable with L_i, hence A has no nonnegatively curved shell with outerpath on L_i, for otherwise there would be diagram with a non-replaceable internal cone-cell having nonnegative curvature by Lemma 5.10. Similarly, if A has a cone-cell C with an external arc on each boundary path of A, then cutting along C shows that A is an annuladder by Theorem 3.43, and then Lemma 5.10 shows that C arises as a nonnegatively curved internal cone-cell of a reduced diagram, which is impossible as $\partial_{\mathsf{p}} C$ is the concatenation of four pieces. In summary, if there is a cone-cell then Theorem 3.23 leads us to the contradiction that $\chi(A) < 0$. Hence there is no cone-cell, and so A is a flat annulus by Lemma 2.25, and then either A has a square, in which case there is a nontrivially stabilized wall-piece arising from the universal cover of a square annulus S carrying a dual curve of A and $\tilde{L}_i$. Or else A contains no square, and there is a nontrivially stabilized cone-piece between $\tilde{L}_1, \tilde{L}_2$ (or more accurately, between the corresponding lifts of universal covers of cones). $\square$

5.o Annular Diagrams and the $B(8)$ Condition

The following condition strengthens the $B(6)$ condition by imposing certain negative curvature hypotheses. It will restrict the structure of certain annuli in a cubical presentation:

Definition 5.55 (Generalized $B(8)$ Condition). A cubical presentation $\langle X \mid \{Y_i\} \rangle$ satisfies the $B(8)$ *condition* if it satisfies the $B(6)$ condition and additionally:

(1) X^* is a negatively curved small-cancellation presentation.
(2) Let $S \to Y$ be a path whose first and last 1-cell are dual to hyperplanes of a wall W of Y. If $\{\{S\}\}_Y \leq \pi$ then S is path-homotopic into the carrier $N(W)$.

Remark 5.56 (Comparison with $B(6)$ conditions of Definition 5.1).
Condition 5.55.(1) strengthens Condition 5.1.(1).
Condition 5.55.(2) strengthens Condition 5.1.(4).

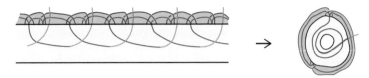

Figure 5.43. Self-crossing of a W-ladder in an annular diagram A surrounded by $N(W)$ implies self-crossing of W in $\widetilde{A}$.

Remark 5.57. When X^* is $C'(\frac{1}{12})$, Theorem 3.32 implies that Condition (1) holds. Moreover, in view of Lemma 3.70, Condition 5.55.(2) holds in the $C'(\frac{1}{12})$ case, provided the following holds for each path $P \to Y$ that is the concatenation of at most 8 pieces: If P starts and ends at endpoints of 1-cells dual to hyperplanes in the same wall of a cone Y then these hyperplanes are the same and P is path-homotopic into its carrier.

We define W-ladders in an annular diagram analogously to the disk diagram case given in Definition 5.27.

Lemma 5.58 (W-ladders cannot self-cross). *Let X^* be a $B(8)$ cubical presentation. Let $A \to X^*$ be an annular diagram whose outside boundary λ is collared by a wall W, in the sense that the lift $\widetilde{\lambda}$ lies on $N(W)$.*

Let $L \to A$ be a W-ladder that is dual to a 1-cell of λ. Then L cannot self-cross within A, in the sense that the wall carried by L passes through the same square in non-parallel 1-cells, or the same cone-cell in 1-cells that map to distinct walls in the corresponding cone.

Moreover, any 1-cell of λ that L ends on is dual to W with respect to $\widetilde{\lambda}$.

Proof. This is true for a disk diagram because of Theorem 5.18.

Let $\widetilde{A} \to A$ be the universal cover of A, and observe that $\widetilde{A}$ is collared by the universal cover of the collar of A. If L self-crosses in A then a pair of distinct lifts of $\widetilde{L}$ in $\widetilde{A}$ cross each other. Connecting them along the collar of $\widetilde{A}$, this gives a self-crossing W-ladder within a disk diagram, which is impossible by Theorem 5.18.

Similarly, if L enters A on a 1-cell on the outside boundary that is dual to W but exits at a 1-cell on the outside boundary that is not dual to W within the collar, then the lift of $\widetilde{L}$ in $\widetilde{A}$ does the same thing, which is impossible. We refer the reader to Figure 5.43. $\square$

Definition 5.59 (Annuladder). We use the term *annuladder* for an annular diagram $A \to X^*$ with the structure of a ladder as in Definition 3.42 but we now include the possibility of a single pseudo-grid with its left and right sides identified. (Hence a flat annulus is a type of annuladder.) Specifically, A does not start and end at cone-cells, but there is instead a cyclic ordering of its sequence of cone-cells and pseudo-grids. We use the term *W-annuladder* to mean that its

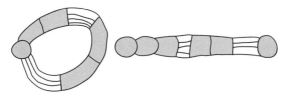

Figure 5.44. An annular diagram with a doubly-external cone-cell but no positively curved cells must be an annuladder. This is because when we cut along the cone-cell (and retain two copies of it) we obtain a ladder.

universal cover maps to $N(W)$ and it contains a "closed" immersed W-ladder whose fundamental group maps isomorphically to it.

Lemma 5.60 (Annuladder criterion). *Let X^* be a small-cancellation cubical presentation. Let $A \to X^*$ be a reduced annular diagram having no spurs, cornsquares, or shells. Suppose A contains a cone-cell C such that $\partial_p C$ intersects each boundary path of A. Then A is an annuladder.*

Consequently, if all internal cone-cells and θ-shells are negatively curved, then either A has no cone-cell or A is an annuladder.

Proof. We cut A along C to obtain a disk diagram D with a cone-cell C at each end. Since D has no other spurs, cornsquares, or shells, we see from Theorem 3.43 that D is a ladder, and hence A is an annuladder. ☐

We now use the tight innerpath condition described in Definition 3.69.

Theorem 5.61 (Conjugate into Wall). *Suppose X^* is $B(8)$ and has tight innerpaths and no inversions. Let $\gamma \to X^*$ be an immersed essential circle that has minimal length within its homotopy class in X^*. Suppose γ is homotopic to a closed path λ that lifts to the carrier of a wall W of $\widetilde{X}^*$. So $\widetilde{\lambda} \subset N = N(W)$, and so $\lambda \to X^*$ factors as $\lambda \to L \to \bar{N} \to X^*$, where $\bar{N} = \langle \gamma \rangle \backslash N$ and $L \to \bar{N}$ is an annuladder.*

Let A be an annular diagram between γ and λ, and suppose that A is of minimal complexity subject to $\lambda \to \bar{N}$ varying within its homotopy class in $\bar{N}$, and subject to $\widetilde{A}$ having one side equal to the fixed copy $\widetilde{\gamma} \subset \widetilde{X}^*$ of the universal cover of γ, and the other side lying on a fixed copy of $N \subset \widetilde{X}^*$.

Then A is a square annular diagram.

Moreover, we can choose A to be of the form $A = A_\gamma \cup_{\gamma'} A_\lambda$ where γ and λ are boundary paths of A_γ and A_λ respectively, and they meet along their other boundary path γ' where $|\gamma'| = |\gamma|$, and γ' lies in the local convex hull of γ in A.

Moreover, if L can be chosen to be a square W-annuladder then $A = A_\lambda$ is a flat annulus, and otherwise γ intersects each cone-cell of L, and $\widetilde{\gamma} \subset E(W)$.

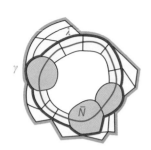
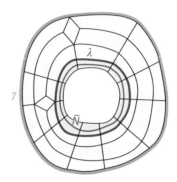

Figure 5.45. A minimal annulus between a closed curve γ and an immersed wall it is homotopic into. There are two cases depending on whether or not γ is homotopic to a path in a square annuladder $L \to \bar{N}(W)$.

We refer the reader to Figure 5.45 for two possible scenarios. In the second case, γ is homotopic to a path in a square annulus in $\bar{N}$. In the first case γ is homotopic to a more general annulus in $\bar{N}$ containing cone-cells.

Remark 5.62. Although we have not pursued this, the original intended role of Theorem 5.61 was to provide a route towards Theorem 5.68 by using a fiber-product computation as in Lemma 8.9. It serves the purpose of ensuring that the components of the fiber-product of the superconvex hull of W with itself correspond precisely to the nontrivial intersections of conjugates.

Proof. Let A be an annular diagram as in the hypothesis of the theorem. So the inside boundary path of A is γ which is of minimal length in its homotopy class in X^*, and the outside boundary path of A is a path $\lambda \to \bar{N}$, and A is reduced, and there does not exist a local complexity reduction achieved by pushing λ across some square or cone-cell that maps to $N(W)$. We note that all such complexity reductions preserve the "class" of A in the sense that the lift of $\widetilde{A}$ at $\widetilde{\gamma}$ has the property that the outside boundary path still lifts to the same fixed copy of N.

If γ has a common point with λ then the first part of the result holds by Theorem 5.31 which states that a geodesic starting and ending on $N(W)$ is square-homotopic into $T(W)$.

As suggested by the first and second diagrams in Figure 5.46, our choice of A implies that λ cannot pass through any dual 1-cell of W. For then, following a simple W-ladder emanating from this dual curve, we are able to find a lower complexity annulus of the same class, using a different choice of λ but the same γ. Note that this W-ladder cannot self-cross within A, as each of the various self-intersection possibilities are ruled out by Theorem 5.18 as established in Lemma 5.58. We note that pushing λ inwards to absorb a cell of A that maps to $\bar{N}$ reduces the complexity without affecting the class. Moreover, if we also

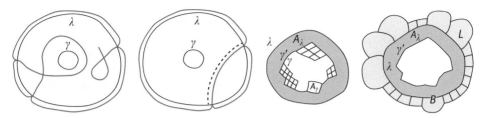

Figure 5.46. The two diagrams on the left indicate situations to consider where λ traverses a 1-cell dual to W, in which case there is a lower complexity annular diagram in the same class. The diagrams on the right depict $A = A_\gamma \cup_{\gamma'} A_\lambda$ and $B = A_\lambda \cup_\lambda L$.

choose A so that $(\mathsf{Comp}(A), |\lambda|)$ is minimal, then we can also assume that λ does not backtrack along a spur of A.

We will choose annular subdiagrams A_γ and A_λ with the following properties: A_γ will be a square annular diagram within A between γ and another path γ' of the same length. Specifically, there is a sequence of diagrams $\gamma = E_0$, $E_1, \ldots, E_t = A_\gamma$, where for each $0 \leq i < t$, the annular diagram E_{i+1} is obtained from E_i by adding a square s_i whose outerpath lies on ∂E_i. Such a square arises from a cornsquare in the complement of E_i. Hence A_γ is the final diagram in this sequence, and we note that the two boundary paths of A_γ are γ and γ', and that each is the local convex hull of the other within A_γ. We let A_λ the remaining annular diagram be the remaining diagram after following this procedure, so that (after shuffling) A equals the union of A_γ and A_λ along γ'. We refer the reader to the third diagram in Figure 5.46.

Let L be a W-annuladder having λ as one of its boundary paths. As illustrated in the fourth diagram of Figure 5.46, let $B = A_\lambda \cup_\lambda L$. This is analogous to what we did in Construction 5.19, except that we are collaring an annular diagram, and there is no corner. Finally, we chose $(B, A, A_\lambda, A_\gamma)$ such that among all possible such choices with γ and the class of A fixed, $(\#\mathsf{Cone\text{-}cells}(B), \#\mathsf{Squares}(B))$ is minimized.

We form the rectified annular diagram $\bar{B}$ from B as in Section 3.f, and then examine the possible positively curved cells in $\bar{B}$. We refer to Figure 5.47 for illustrations of some of the scenarios that we will exclude in the next few paragraphs. In each case, the scenario is illustrated in a diagram above, and the "reducing action" is illustrated directly below it.

There is no outerpath of a cornsquare in $\bar{B}$ along γ', for then we could push the square across γ' to reduce $\mathsf{Area}(A_\lambda)$ and hence $\#\mathsf{Squares}(B)$ at the expense of increasing $\mathsf{Area}(A_\gamma)$. See the first pair of diagrams in Figure 5.47.

There is no nonnegatively curved shell in $\bar{B}$ with outerpath Q along γ', for otherwise, denoting its boundary by QS, with Q outer and S inner, we would have $\{\{S\}\} \leq \pi$ and hence by tight innerpaths, either $|S| < |Q|$ so γ' and hence $\gamma \to X^*$ is not of minimal length in its homotopy class, or the shell bounded by QS can be replaced by a square diagram, thus reducing the complexity. See the second pair of diagrams in Figure 5.47.

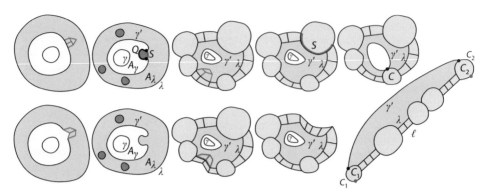

Figure 5.47. Scenarios leading to lower complexity diagrams illustrated below them.

We furthermore deduce that no cone-cell C of B intersects γ' in more than one arc, for then C would subtend a subdiagram of B, that is not a single cone-cell (since it contains more than C), and would therefore have at least two shells and/or cornsquares on the part of its boundary in γ' by Theorem 3.46, but we have excluded these above.

There is no outerpath of a cornsquare in $\bar{B}$ along the part of the boundary of B in ∂L. Indeed, the square itself would have to lie within A_λ, and the outerpath would necessarily lie along the external part of a rectangle in L, but then this square can be pushed across λ and outside the diagram to reduce the area with the path replacing λ remaining on $\bar{N}$. (In fact, because of the way rectangles are admitted, there can be no explicit cornsquare unless L contains a cone-cell.) See the third pair of diagrams in Figure 5.47.

There is no nonnegatively curved shell in L. Indeed, suppose a cone-cell C in L mapping to a cone Y is such a shell. The innerpath S of C traverses 1-cells dual to the same wall. Since $\{\{S\}\}_Y \leq \pi$, Condition 5.55.(2) asserts that these 1-cells are dual to the same hyperplane of Y, and a subpath of S is homotopic into the carrier of this hyperplane by a square diagram. We can therefore replace C by a rectangle and a diagram in Y providing the homotopy between this subpath of S and a side of the rectangle. This pushes λ to a path alongside it in $\bar{N}$, but reduces the number of cone-cells and hence violates minimal complexity of B. (This maneuver was performed earlier, and we refer to the second sequence in Figure 5.16.) We conclude that each cone-cell in L is negatively curved. See the fourth pair of diagrams in Figure 5.47.

If some cone-cell C of L intersects γ', then B is an annuladder by Lemma 5.60. Indeed, repeating that argument, we can cut B along C leaving copies C_1, C_2 of C at each end, to form a diagram D whose boundary path is a concatenation $\gamma' c_1 \ell c_2$ where ℓ is a path along $\partial L - \lambda$, and c_1, c_2 are paths on C. This is illustrated in the fifth pair of diagrams in Figure 5.47. The arguments above show that D has no positively curved cells except for C_1, C_2, and hence Theorem 3.43 shows that D is a ladder. This completes the proof in this case, as it follows that

γ' lies on $\bar{T} = \langle \gamma \rangle \backslash T(W)$, and hence γ lies in the local convex hull of λ within A after shuffling by Theorem 5.31.

Each cone-cell of B in L that doesn't intersect γ' has negative curvature and each cone-cell of B not in L with an arc on γ' has negative curvature, and there are no cornsquares, and internal cone-cells have negative curvature by hypothesis. Since $\chi(B) = 0$, we see from Theorem 3.23 that each cone-cell in B is a cone-cell in L that intersects γ', in which case $\gamma \subset T(W)$ as above.

We conclude that if B contains cone-cells, then these cone-cells lie in L, and γ' intersects each such cone-cell of L, and $\widetilde{\gamma}' \subset T(W)$ and $\widetilde{\gamma} \subset E(W)$.

But if B contains no cone-cell, then L is a width 1 square annulus consisting of the product of a 1-cube and a subdivided circle, and A_λ is a flat annulus by Lemma 2.25. It follows that $\gamma = \gamma'$ and so $A = A_\lambda$ as claimed, since if A_γ contains a square then A contains a cornsquare whose outerpath lies on λ and we can thus reduce the complexity as above. $\qquad\square$

5.p Doubly Collared Annular Diagrams

Theorem 5.63 (Doubly Collared Annulus). *Let $\langle X \mid \{Y_i\} \rangle$ be a $B(6)$ cubical presentation with no inversions. Let $A \to X^*$ be an annular diagram with boundary paths α_1, α_2 that are essential in X^*. Suppose that the lift of its universal cover $\widetilde{A} \to \widetilde{X}^*$ has the property that the induced lifts of $\widetilde{\alpha}_i \to \widetilde{X}^*$ lie on carriers of walls $N_1 = N(W_1)$ and $N_2 = N(W_2)$. And suppose $\text{Aut}(\widetilde{A})$ stabilizes W_1 and W_2.*

There exists a new annular diagram B that is reduced and satisfies the following:

(1) *B does not contain a cornsquare whose outerpath lies on one of its boundary paths.*

(2) *No cone-cell along the boundary of B is replaceable by a rectangle (with no adjustment of internal boundary path).*

Furthermore, B contains a pair of annuladders L_1, L_2 where L_i is a W_i-annuladder. Each $L_i \to B$ is an embedding and B deformation retracts to L_i. For each i, the path β_i is one of the boundary paths of L_i. And B has one of the following two structures:

(1) *Either B is thick in which case L_1, L_2 have disjoint interiors and lie along the boundary of B in the sense that β_1, β_2 are the boundary paths of B.*

(2) *Or B is thin and B is itself an annuladder. We emphasize that in this case β_1, β_2 might not equal the boundary paths of B.*

Finally, B lies in the same class as A in the sense that the lifts of $\widetilde{A}$ and $\widetilde{B}$ have: $\widetilde{\alpha}_i, \widetilde{\beta}_i$ lie in the same translate of N_i for each i, and α_i, β_i represent the same conjugacy class in each $\text{Stab}(N_i)$ (and hence in $\pi_1 X^*$ as well).

Without the no inversion hypothesis, one could deduce a similar statement with a Moebius strip possibly replacing the annulus.

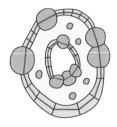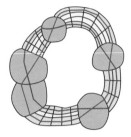

Figure 5.48. B is thick on the left, and B is thin on the right.

We refer the reader to Figure 5.48 for illustrations of the thick and thin cases of an annular diagram B together with the annuladders L_1, L_2 inside it.

As these two type of annular diagrams are occurring together here, we point out that an annuladder may have large nondegenerate pseudo-grids between cone-cells and an annuladder is not necessarily associated to a particular wall. In contrast, a W-ladder is narrow and associated to a particular wall, and consists of a cyclic sequence of squares and cone-cells, and its universal cover lifts to $N(W)$.

Proof of Theorem 5.63. **Preliminary note:** The thick case arises when the translates of W_1 and W_2 under consideration do not cross the same cone or square. The diagram B can then be chosen to be the union of a minimal $\big(\#\text{Cone-cells}, \#\text{Squares}\big)$ complexity annular diagram $A' \to X^*$ in the same class as A, together with minimal complexity annuladders $L_i \to \bar{N}_i$ each having a common boundary path with A'. Here $\bar{N}_i = \langle \gamma_i \rangle \backslash N_i$ where γ_i is the group element stabilizing $\widetilde{\alpha}_i$.

The thin case arises when W_1 and W_2 cross, or are equal to each other. Note that there is a degenerate case where $L_1 = L_2$ that arises in this situation.

A minimal annular diagram in the class: Let E be an annular diagram in the same class as A with boundary paths ε_i that represent elements conjugate to α_i in $\text{Stab}(N_i)$. Suppose moreover, that the complexity $\big(\#\text{Cone-cells}, \#\text{Squares}\big)$ of E is minimal among all possible such diagrams.

Properties conferred by minimality: The minimality ensures that E is reduced, but in addition, minimality ensures there are other properties related to "compressions into the boundary."

If E has a square or cone-cell with a 1-cell along ε_i and this square or cone-cell maps to N_i under the map $\widetilde{E} \to \widetilde{X}^*$, then we could push ε_i through this square or cone-cell and obtain a lower complexity diagram in the same class. Thus no such configuration exists.

If E has a cutpoint, then this cutpoint subtends a subpath of either ε_1 or ε_2 which bounds a disk diagram that is a subdiagram of E. Chopping off this disk diagram, and replacing this subpath by a point reduces the complexity but does not affect the class. We can therefore assume that no such configuration exists.

Accordingly, if e_i is a 1-cell on ε_i whose lift is dual to a 1-cell of W_i then e_i must be an isolated 1-cell of E and the other boundary path ε_j must also pass through e_i (which cannot be a cut-cell as above).

Finally, if ε_i contains the outerpath cd of a cornsquare of E, and this length-2 path lies along the external boundary of a length-2 square W_i-annuladder (i.e., there is a 1-cell e_i dual to W_i that forms squares with corners ce_i and $e_i^{-1}d$ in $\widetilde{X}^*$) then this cornsquare can be pushed out of the diagram, to decrease the complexity while maintaining the class.

The thick case: In case W_1 and W_2 don't cross the same square or cone and aren't equal, then ε_i cannot pass through a 1-cell e_i lifting to a dual 1-cell of W_i. Indeed, as above, such a 1-cell is forced to be an isolated 1-cell of E that also lies on ε_j (here $i \neq j$). A cone or square dual to W_j in N_j that contains e_i on its boundary is crossed by both W_i and W_j.

We shall now assume that no such 1-cells e_1, e_2 are traversed by $\varepsilon_1, \varepsilon_2$.

Minimal W_i-annuladders: Following a procedure similar to Construction 5.19, for each i we let $L_i \to \bar{N}_i$ be a W_i-annuladder having ε_i as one of its boundary paths, and moreover, assume that L_i is chosen to have minimal $(\#\text{Cone-cells}, \#\text{Squares})$ complexity among all such choices with ε_i fixed.

Forming B in the thick case: We now form the thick annuladder $B = L_1 \cup_{\varepsilon_1} E \cup_{\varepsilon_2} L_2$ by gluing L_1, L_2 along $\varepsilon_1, \varepsilon_2$ to E.

We are assuming here that the lifts $\widetilde{L}_1$ and $\widetilde{L}_2$ do not "cross" the same cone or the same square (possibly even along the same dual 1-cell).

We now verify that B is reduced and has the desired properties. Since E is reduced, we need only consider the interaction between cells in L_1, L_2, and the interaction between cells in E with L_1, L_2. A combinable or absorbable pair of cells between L_1, L_2 would lift to the same cone in $\widetilde{X}$ and hence $\widetilde{X}^*$, and would imply that W_1, W_2 both cross some cone. A combinable or absorbable pair of cells that are both in L_i would violate minimality if they are already adjacent in L_i. A combinable or absorbable pair of cells between L_i, L_i that are not already adjacent would mean that these cells are adjacent in E, and hence ε_i passes through a cutpoint which would violate the minimality of E as above. Minimality would also be violated if we could replace a cone-cell by a square diagram without affecting ε_i. A combinable pair between a cell in L_i and a cell in E would imply that there is a cell of N_i within E along the boundary of ε_i, contradicting the minimality of E, as above. Finally, a cornsquare along B would have to come from a square within E, since for any square in L_i, one of the dual curves of this square lies entirely in L_i and doesn't meet ∂B. But, such a cornsquare within E was ruled out by the minimality of E above.

Thin case: If W_1, W_2 cross, then by Corollary 5.37, they intersect a common cone $Y \subset \widetilde{X}^*$. Thus if W_1, W_2 cross or if $W_1 = W_2$, then we can choose a pair of annular diagrams A_1, A_2 that carry paths homotopic to α_1, α_2 in $\bar{N}_1, \bar{N}_2$, such that $\widetilde{A}_1, \widetilde{A}_2$ lift to N_1, N_2, and such that A_1, A_2 contain cone-cells C_1, C_2 such that in the lifts $\widetilde{A}_1, \widetilde{A}_2$, each cone-cell $\widetilde{C}_i \to \widetilde{X}^*$ factors through the same cone Y of $\widetilde{X}^*$.

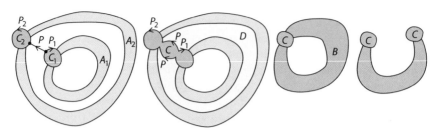

Figure 5.49. Since A_1, A_2 contain cone-cells C_1, C_2 that map to the same cone, we can glue them together and fill the result by a disk diagram D to form an annular diagram A as illustrated in the two diagrams on the left. After repeatedly reducing, A leads to an annular diagram B containing a cone-cell C that cuts B into a disk diagram, as illustrated in the right two diagrams. The only two positively curved features of this disk diagram are the two copies of C, and so it is a ladder, and hence B is an annuladder.

We glue A_1, A_2 together along a cone-cell C that combines with both C_1, C_2. That is, choose a path P in Y from the basepoint of $P_1 = \partial_{\mathsf{p}} C_1$ to the basepoint of $P_2 = \partial_{\mathsf{p}} C_2$, and let C be the cone-cell with $\partial_{\mathsf{p}} C = P_1 P P_2^{-1} P^{-1}$, and then form A' by attaching each $A_i - C_i$ to C along P_i. We then form the annular diagram A by filling in the interior boundary path of A' with a disk diagram, thus A contains W_i-annuladders which are variants of both A_i with C_i replaced by C. See Figure 5.49.

We then proceed to create a reduced annular diagram by combining and ejecting cone-cells while maintaining the W_i-annuladders. As the cone-cell C with respect to both boundary paths, our sequence of annular diagrams continues to contain a cone-cell that we continue to call C at each stage. At the end we arrive at a reduced annular diagram B with this doubly-external cone-cell C. We cut B along C, to obtain a disk diagram D with two copies of C. We then apply Theorem 3.43 to see that D is a ladder. Consequently, the end result B is an annuladder as claimed. $\qquad\square$

Corollary 5.64. *Suppose the $B(8)$ condition is added to the hypotheses of Theorem 5.63. Then either B is thin, or B has no cone-cells and is thus a square annular diagram $B \to X$, or at least one of L_1, L_2 is elliptic.*

Proof. Suppose B is a reduced thick ladder containing W-annuladders L_1 and L_2 on its inside and outside, such that L_1, L_2 have disjoint interiors and are not elliptic. Condition 5.55.(1) implies that any internal cone-cell is negatively curved and Condition 5.55.(2) implies that any cone-cell with a single external boundary arc is a negatively curved shell (since it lies on L_1 or L_2). There is hence no cone-cell by Theorem 3.23 since $\chi(B) = 0$. Thus B must be a square annular diagram in the thick non-elliptic case. $\qquad\square$

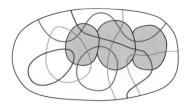

Figure 5.50. To verify no self-grazing: Each wall in a cone of $\langle X \mid \{Y_i\} \rangle$ maps to a single equivalence class of hyperplanes in X, but a single equivalence class has connected intersection with each piece in each cone.

Remark 5.65 (Elliptic Degeneracy). As alluded to in Corollary 5.64, the conclusion of Theorem 5.63 requires additional care when one or both of L_1, L_2 is an elliptic annuladder. Under additional assumptions, e.g., as in Lemmas 5.53 and 5.54, we can force L_1, L_2 to merge within B, or ensure their associated cones in $\widetilde{X}^*$ to be identical.

5.q Malnormality of Wall Stabilizers

We shall give a criterion for verifying that the stabilizer of a wall of $\widetilde{X}^*$ is almost malnormal when its associated hyperplanes form an almost malnormal collection in $\widetilde{X}$. We refer to Definition 8.1.

Definition 5.66 (Self-Grazing). The $B(6)$ cubical presentation X^* has *no self-grazing* walls if the following holds for each cone Y, wall W in $\widetilde{X}^*$, and element $g \in \pi_1 X^*$: If $Y \cap W$ and $Y \cap gW$ contain hyperplanes dual to 1-cubes lying in the same (contiguous or noncontiguous) cone-piece of Y, then $W = gW$. (Hence $gW \cap Y$ and $W \cap Y$ are the same wall of Y.)

Remark 5.67. No self-grazing states that for an immersed wall $W \to X^*$ and a cone Y, the distinct ways that W intersects Y are far from each other—i.e., they cannot pass through the same piece.

No self-grazing is a global condition that is difficult to check in practice, but is instead verified through the following scenario which represents a stronger condition. The hyperplanes of X are partitioned into equivalence classes. For each i, the hyperplanes of each wall of Y_i map to hyperplanes in X that are in the same equivalence class. The union of all hyperplanes of X in an equivalence class have connected intersection with (or more generally, preimage in) each piece of each Y_i. This is illustrated heuristically in Figure 5.50. We will later use this to verify no self-grazing in a situation where all hyperplanes of X embed, each wall of Y maps to the same hyperplane in X, and the injectivity radius of hyperplanes of X exceeds the diameter of the largest cone-piece or wall-piece in any Y.

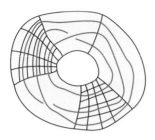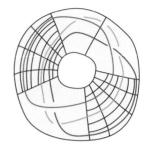

Figure 5.51. The first annular diagram shows that there is self-grazing. In contrast, the immersed walls traveling through pieces in the second annular diagram will not be in the same $\pi_1 X^*$ orbit.

No self-grazing implies that the dual curves illustrated in the first diagram of Figure 5.51 actually lie within the same hyperplane in each of the cones (receiving the cone-cells). If this is not the case for a pair of dual curves as in the second diagram of Figure 5.51, then that pair of dual curves does not have lifts contained in the same wall of $\tilde{X}^*$.

Theorem 5.68. *Let* $\langle X \mid \{Y_i\} \rangle$ *be a* $B(8)$ *cubical presentation with no inversions that satisfies the conditions below. Then* $\mathrm{Stab}(W)$ *is almost malnormal in* $\pi_1 X^*$ *for each wall* W.

(1) $\{\pi_1 H_1, \ldots, \pi_1 H_k\}$ *is a malnormal collection of subgroups of* $\pi_1 X$ *when* $\{H_1, \ldots, H_k\}$ *are distinct immersed hyperplanes of* X *that are images of hyperplanes in the same wall* W *of* $\tilde{X}^*$.
(2) X^* *has no self-grazing.*
(3) $\mathrm{Aut}(Y_i)$ *is finite for each* i.

Proof of Theorem 5.68. The idea of the proof is that if $h \in \mathrm{Stab}(W) \cap \mathrm{Stab}(gW)$ then one of three things happen: Either W, gW do not cross a common cone-cell, in which case by the $B(8)$ condition there is an h-periodic infinite square strip joining $N(W)$ and $N(gW)$, and this is impossible by Condition (1). Or W and gW pass through more than one common cone-cell, in which case $W = gW$ by no self-grazing and Theorem 5.34. Or W and gW pass through a unique common cone-cell Y, in which case $\mathrm{Stab}(W) \cap \mathrm{Stab}(gW) \subset \mathrm{Stab}(Y)$ which is finite by Condition 3.

As Γ_W is a tree by Theorem 5.20, any subgroup of $\mathrm{Stab}(W)$ either stabilizes a vertex of Γ_W and hence stabilizes a cone or hyperplane of W or else the subgroup stabilizes neither. Thus either $\mathrm{Stab}(W) \cap \mathrm{Stab}(gW)$ consists entirely of elliptic elements stabilizing a cone in which case $\mathrm{Stab}(W) \cap \mathrm{Stab}(gW)$ is finite by our hypothesis that each $\mathrm{Aut}(Y_i)$ is finite, or else $\mathrm{Stab}(W) \cap \mathrm{Stab}(gW)$ is not elliptic, in which case we can assume the chosen element $h \in \mathrm{Stab}(W) \cap \mathrm{Stab}(gW)$ is not an elliptic element stabilizing a cone.

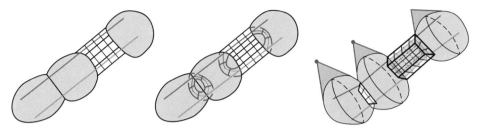

Figure 5.52. Since the annulus B is thin, Condition 5.68 implies that the two annuladders travel through the same immersed wall $\bar{W} \to X^*$, and hence $B \to X^*$ factors through the immersed carrier $\bar{N} \to X^*$ consisting of cones and carriers of hyperplanes.

Let $N = N(W)$ be the carrier of W and let α_1, α_2 be immersed combinatorial circles in X whose universal covers $\widetilde{\alpha}_1, \widetilde{\alpha}_2$ have lifts to N and gN that are stabilized by h. Let B be the annular diagram provided by Theorem 5.63. By Corollary 5.64, B is either thin or it has no cone-cells and is a square diagram.

In the latter case, B represents a homotopy in X between closed curves in distinct immersed hyperplanes $\bar{H}_1, \bar{H}_2$ of X where H_1, H_2 are components of their preimages in $\widetilde{X}^*$ that lie in the same wall W. Condition (1) on malnormality implies that $\bar{H}_1 = \bar{H}_2$ and that B can be homotoped into the carrier of $\bar{H}_1$ (relative to β_1, β_2). In particular, B is homotopic into $\bar{N} = \mathrm{Stab}(W) \backslash N$ relative to β_1, β_2 and we find that β_1, β_2 are conjugate in $\mathrm{Stab}(W)$ and hence the same holds for α_1, α_2.

The above argument deals with the case that B contains no cone-cell, and is independent of whether B is thick or thin. We now examine the case where B is thin and contains a cone-cell. Thus B contains a pair of annular W-ladders L_1, L_2 passing through each cone-cell of B and traveling around B so that each generates $\pi_1 B$. As h is not elliptic, B is not an elliptic annulus consisting of a single cone-cell so there are actual pieces between successive cone-cells. Hence no self-grazing implies that the dual curves of L_1, L_2 lie in the same hyperplane in each cone (at entrance and exit). We refer the reader to Figure 5.52. This gives us a homotopy between β_1 and β_2 within $\bar{N}$, and more importantly, considering $\widetilde{B}$, we find that $gW = W$. $\qquad\square$

We close this section by pointing out that $\mathrm{Aut}(W)$ is often equal to $\mathrm{Aut}(N)$.

Lemma 5.69. *Let W be a wall of $\widetilde{X}^*$. Suppose $N(W)$ does not consist of a single cone. If W consists of a single hyperplane, then suppose the associated immersed hyperplane embeds in X. Finally, assume X^* has no self-grazing. Then $\mathrm{Stab}(N) = \mathrm{Stab}(W)$.*

Note that Lemma 5.69 can fail when N consists of a single cone-cell Y, in which case $\mathrm{Aut}(N) = \mathrm{Aut}(Y)$.

Proof. It is obvious that $\text{Stab}(W) \subset \text{Stab}(N)$. To see that $\text{Stab}(N) \subset \text{Stab}(W)$ we suppose that $gN = N$ and deduce that $gW = W$. If W consists of a single hyperplane then we use that $\text{Stab}(W)\backslash W$ is a single embedded hyperplane of $\text{Stab}(N)\backslash N$ to see that $\pi_1 X$ automorphisms of W are in one-to-one correspondence with $\pi_1 X$ automorphisms of N. If there is a cone Y in N, then we compare the walls $W \cap gY$ and $g(W \cap Y)$, or translating by g^{-1} we compare $(g^{-1}W) \cap Y$ and $W \cap Y$. Let v be a 0-cube of $N(W) - Y$, and choose H so that v lies in the wallray $[Y, H)$. Since $v \subset g^{-1}N = N$ we see that v must lie in the wallray $[Y, H_1)$ for some hyperplane H_1 of $g^{-1}W$. Hence Theorem 5.34 implies that $Y \cap H$ and $Y \cap H_1$ have 1-cells dual to edges in the same piece. But then no self-grazing implies that $H = H_1$, so $g^{-1}W = W$ so $g \in \text{Stab}(W)$ as required. $\qquad \square$

5.r Artin Groups

Let $(x, y)^m$ denote the initial half of the word $(xy)^m$. A f.g. *Artin* group is presented by $\langle a_1, a_2, \ldots \,|\, \{(a_i, a_j)^{m_{ij}} = (a_j, a_i)^{m_{ji}}\} : i \neq j \rangle$ where M is a symmetric square matrix whose entries are natural numbers that are ≥ 2. We allow $m_{ij} = \infty = m_{ji}$, in which case there is no relation between a_i, a_j among the relators.

For each $i < j$, let Y_{ij} denote the 1-skeleton of the universal cover of the standard 2-complex of $\langle a_i, a_j \,|\, \{(a_i, a_j)^{m_{ij}} = (a_i, a_j)^{m_{ij}}\}\rangle$. When $m_{ij} = \infty$, we omit Y_{ij}.

Let $A = A(M)$ be an Artin group as above. Then A has the following cubical presentation satisfying the $C(6)$ condition:

$$\langle a_1, a_2, \ldots \,|\, Y_{ij} : i < j \text{ and } m_{ij} < \infty \rangle \qquad (5.1)$$

Theorem 5.70. *Suppose that $3 \leq m_{ij} \leq \infty$ for each i, j. Then the cubical presentation in Equation (5.1) satisfies the $C(6)$ property.*

More generally, suppose that there is no triangle of form $(2, 3, 3)$, $(2, 3, 4)$, $(2, 3, 5)$, or $(2, 2, n)$, in the graph associated to the Artin presentation. Then the cubical presentation satisfies the $C(6)$ conditions.

Proof. Observe that pieces between distinct translates of Y_{ij} and $Y_{k\ell}$ in $\widetilde{X}^*$ correspond to lines a_p^∞ where one of i, j equals p, and one of k, ℓ equals p.

In the first case, at least 6 pieces are needed to form an essential cycle in Y_{ij}.

In the more general case, use the angle grading. $\qquad \square$

Remark 5.71. Jon McCammond reported on a similar idea at a talk he gave in Albany around 2000. He grouped the relations of the same type in an Artin group within a disk diagram to obtain small-cancellation behavior under "large enough type" conditions. Perhaps the earliest occurrence of this idea is in the work of Appel and Schupp [AS83].

Remark 5.72 (Not $B(6)$). There are several natural wallspace structures on Y_{ij}. The first arises from the induced map to the associated Coxeter group. Namely, two edges are dual to the same wall if and only if they lie in the same equivalence class, where the equivalence relation is generated by opposite edges in the same $2m_{ij}$-gon. However, this wallspace is not Hausdorff, and in fact, corresponds to an action of the 2-generator Artin group on a cubulated copy of $\mathbb{E}^3$. A second structure arises from the map to $\mathbb{Z}$ induced by sending each generator $a_k \mapsto 1$. In this case, the walls don't even cross each other, and the cubulation gives an action on $\mathbb{E}^1$. In particular, we note that the above cubical presentations are not $B(6)$. To achieve the $B(6)$ condition it will be necessary to replace the bouquet of circles corresponding to the free group, by a cube complex X—that is preferably a compact pseudograph.

Chapter Six

Special Cube Complexes

In [HW08] we introduced "special cube complexes" and explored some of their properties. In Section 6.b we review the definition of special cube complexes in terms of illegal hyperplane pathologies, and we state the characterization in terms of local-isometries to the cube complex of a right-angled Artin group. In Section 6.d we review the definition of canonical completion and retraction. The hyperplane pathology definition of special cube complex arose originally from our desire to define canonical completion and retraction above dimension one, where it was originally used for graphs [Wis00, Wis02, Wis06]. Subsequently, we recognized this definition was equivalent to admitting a local-isometry to the cube complex of a right-angled Artin group. Many other aspects of special cube complexes are explored in [HW08], [HW12] and [HW10a] including various conditions which imply that a cube complex is virtually special. The material in Section 6.f is new.

6.a Immersed Hyperplanes

We now continue the discussion of Section 2.c by defining an immersed hyperplane in an arbitrary cube complex C. Let M denote the disjoint union of the collection of midcubes of cubes of C. Let D denote the quotient space of M induced by identifying faces of midcubes under the inclusion map. The connected components of D are the *immersed hyperplanes* of C.

6.b Hyperplane Definition of Special Cube Complex

The definition is crafted by prohibiting certain pathologies related to immersed hyperplanes.

An immersed hyperplane D *crosses itself* if it contains two different midcubes from the same cube of C.

An immersed hyperplane D is *2-sided* if the map $D \to C$ extends to a map $D \times I \to C$ that is a combinatorial map of cube complexes.

A 1-cube of C is *dual* to D if its midcube is a 0-cube of D. When D is 2-sided, it is possible to consistently orient its dual 1-cubes so that any two dual

Figure 6.1. Immersed hyperplane pathologies.

1-cubes lying (opposite each other) in the same 2-cube are oriented in the same direction.

An immersed 2-sided hyperplane D *self-osculates* if for one of the two choices of induced orientations on its dual 1-cells, some 0-cube v of C is the initial 0-cube of two distinct dual 1-cells of D.

A pair D, E of distinct immersed hyperplanes *cross* if they contain midcubes lying in the same cube of C. We say D, E *osculate* if they have dual 1-cubes which contain a common 0-cube, but do not lie in a common 2-cube. Finally, a pair of distinct immersed hyperplanes D, E *inter-osculate* if they both cross and osculate.

A cube complex is *special* if it is nonpositively curved and the following hold:

(1) No immersed hyperplane crosses itself.
(2) Each immersed hyperplane is 2-sided.
(3) No immersed hyperplane self-osculates.
(4) No two immersed hyperplanes inter-osculate.

Example 6.1. Any graph is a 1-dimensional cube complex that is special: This is immediate since hyperplanes are the midpoints of edges.

Any CAT(0) cube complex is special: Indeed, it follows from Lemma 2.14.(1) that hyperplanes do not self-cross, are 2-sided, and do not self-osculate. It follows from Lemma 2.14.(4) that no pair of hyperplanes can inter-osculate.

The cube complex $C(\Gamma)$ associated to a right-angled Artin group $G(\Gamma)$ is special: Indeed, each hyperplane is dual to a single 1-cube. As the 2-cubes are attached by commutators we see that hyperplanes are 1-sided, and since these commutators are associated to distinct generators we see that hyperplanes do not self-cross. There is no self-osculation or inter-osculation since each hyperplane is dual to exactly one 1-cube.

6.c   Right-Angled Artin Group Characterization

We give the following surprisingly simple characterization of special cube complexes in [HW08]:

Proposition 6.2. *A nonpositively curved cube complex X is special if and only if it admits a combinatorial local-isometry $X \to C(\Gamma)$ to the cube complex of a right-angled Artin group.*

An immediate consequence of Proposition 6.2 is that if X is special and $Y \to X$ is a local-isometry of nonpositively curved cube complexes, then Y is special.

As the following proof shows, when X has finitely many hyperplanes, Γ is finite, and hence $C(\Gamma)$ is compact.

Proof. When X is special we define Γ to be the graph whose vertices are the immersed hyperplanes of X, and whose edges correspond to intersecting hyperplanes. Let $C(\Gamma)$ be the nonpositively curved cube complex associated to Γ that was discussed in Section 2.b.

We label each 1-cube by the hyperplane dual to it, and we use that hyperplanes are 2-sided to orient all 1-cubes so that the attaching map of each 2-cube looks like a commutator. As we may likewise label and orient the edges of $C(\Gamma)$, there is a natural label-preserving and orientation-preserving combinatorial map $X^2 \to C(\Gamma)$. This extends to each n-cube of X, since n pairwise intersecting hyperplanes in X would correspond to an n-clique in Γ. The map is an immersion since no two outgoing or incoming 1-cubes have the same label, as X has no self-osculating hyperplane. The map is a local-isometry since if two 1-cubes a, b at a 0-cube of X map to 1-cubes of $C(\Gamma)$ that bound the corner of a square, then the hyperplanes associated to a, b cross in X, whence a, b form the corner of a square in X as there is no inter-osculation.

The converse holds since for a local-isometry $X \to C$, a hyperplane pathology in X would project to a hyperplane pathology in C. Thus if C is special (and in particular, if $C = C(\Gamma)$), we see that a local-isometry $X \to C$ implies that X is special.   □

6.d   Canonical Completion and Retraction

We refer to [HW08, HW12] for more details about the following fundamental property of special cube complexes:

Proposition 6.3 (Canonical completion and retraction). *Let $f : Y \to X$ be a local-isometry from a compact nonpositively curved cube complex to a special cube complex. There exists a finite degree covering space $C(Y \to X) \to X$ called the* canonical completion *of f such that $f : Y \to X$ lifts to an embedding $\widehat{f} : Y \to C(Y \to X)$, and there is a retraction map $C(Y \to X) \to Y$ called the* canonical retraction.

A map $A \to B$ between cube complexes is a *cubical map* if each cube of A maps to a cube of B by a map modeled on $I^{n+m} \to I^n$ that collapses any extra dimensions.

A local-isometry $\phi : Y \to X$ of nonpositively curved cube complexes is *segregated* if for edges e_1, e_2 at a vertex of Y, their images $\phi(e_1), \phi(e_2)$ are not dual to the same immersed hyperplane of X.

Note that every local-isometry $Y \to X$ is segregated provided that X is special and $N(H) \to X$ is an embedding for each hyperplane H. The latter condition holds when X has no indirectly self-osculating hyperplanes (see [HW08]). When X is special, the cubical subdivision of X has this property, and the (natural bipartite) double cover of X has this property.

The following is a variation of [HW12, Lem 3.8] where it was proven that $\mathsf{C}(Y \to X) \to Y$ is cubical after subdividing the domain. It follows directly from the definition of the canonical completion and retraction construction, since when ϕ is segregated, then in $\mathsf{C}(Y^1 \to R^1)$ all new edges are either completions of loops or bigons retracting to vertices or edges (see [HW12, Def 3.1]).

Proposition 6.4. *In the context of Proposition 6.3: Suppose, moreover, that f is segregated. Then $\mathsf{C}(Y \to X) \to Y$ is a cubical map.*

A subgroup $H \subset G$ is *separable* if for each $g \notin H$ there is a finite quotient $G \to \bar{G}$ such that $\bar{g} \notin \bar{H}$. Equivalently, H is separable if it is the intersection of the finite index subgroups of G that contain it. It follows that if $[G : G'] < \infty$ and $H \subset G'$ is separable then $H \subset G$ is separable. From another viewpoint, H is separable if and only if it is a closed subset in the *profinite topology* of G, which is the topology generated by the basis consisting of cosets of finite index subgroups. A group G is *residually finite* if its trivial subgroup is separable, and this is equivalent to the profinite topology being Hausdorff.

Corollary 6.5. *Let X be a special cube complex and let $Y \to X$ be a local-isometry with Y compact. Then Y is a virtual retract. Hence $\pi_1 Y$ is separable in $\pi_1 X$.*

Proof. The first claim holds by applying Proposition 6.3 to obtain the virtual retraction $\mathsf{C}(Y \to X) \to Y$. The second claim holds since a retract of a Hausdorff topological space is closed in the profinite topology.

Note that $\pi_1 X$ is residually finite since $\pi_1 C(\Gamma)$ is residually finite where $X \to C(\Gamma)$ is the local-isometry of Proposition 6.2. It is easy to prove that raags are residually finite, and we note that they are even linear [Hum94] and in fact subgroups of $SL_n(\mathbb{Z})$ when they are f.g. (see, e.g., [HW99]). One can prove this along the lines of this discussion as follows: Let σ be a closed path containing a nontrivial element of $\pi_1 X$. Then σ lies in a compact convex subcomplex (its hull) Y of $\widetilde{X}$. Hence $\mathsf{C}(Y \to X)$ provides a finite cover of X where the lift of σ is not closed. $\square$

As explained in [HW08] we have the following consequence, and note that analogous results hold in the sparse case as discussed in Theorem 15.13.

Corollary 6.6. *Let X be a compact special cube complex with $\pi_1 X$ word-hyperbolic. Let $H \subset \pi_1 X$ be a quasiconvex subgroup. Then H is a virtual retract, and hence H is separable.*

Proof. By Proposition 2.31, there is a based local-isometry $Y \to X$ with $\pi_1 Y = H$. The result thus follows from Corollary 6.5.   □

We frequently use the following additional consequence of Proposition 6.3 which is a consequence of Corollary 6.5 when X is compact.

Corollary 6.7. *Let U be a hyperplane of a special cube complex X. Then $\pi_1 U \subset \pi_1 X$ is separable.*

Proof. Let $X \to R$ be the local-isometry of Proposition 6.2. Let V be the hyperplane of R that U maps to. Observe that $\pi_1 U = \pi_1 X \cap \pi_1 V$ and so it suffices to show that $\pi_1 V$ is separable in $\pi_1 R$. This follows since V is a retract of R. Algebraically, we collapse the generator v dual to V as well as those generators that do not commute with v.   □

Corollary 6.8. *Let G be a f.g. group that is virtually special. Every virtually abelian subgroup A of G is separable.*

Proof. Let $G' \subset G$ be a special finite index subgroup. Then A is the union of finitely many cosets of $A \cap G'$. Hence it suffices to show that $A \cap G'$ is separable in G', and hence in G. We thus prove the statement under the assumption that G is special, and $G \subset \pi_1 R$ where R is the Salvetti complex of a raag, which we can assume is f.g. since G is. As the preimage of a separable subgroup is separable, it suffices to show that $A \subset \pi_1 R$ is separable. Note that virtually abelian subgroups of special groups are actually abelian (see [WW17, Lem 5.3]). Let $M \subset \pi_1 R$ be a maximal abelian subgroup containing A. By Lemma 2.38, there is a compact nonpositively curved cube complex Y and a local-isometry $Y \to R$ such that $\pi_1 Y$ maps to M. A finite covering space $\widehat{Y}$ shows the same holds for any finite index subgroup $\widehat{M}$. Each $\widehat{M}$ is thus separable by Corollary 6.5. Finally, A is separable in $\pi_1 R$ since it is the intersection of separable subgroups. Alternately, one can complete the argument by observing that A is a virtual retract of G since A is a virtual retract of M which is a virtual retract of $\pi_1 R$.   □

6.e   Double Cosets and Virtual Specialness

The following provides a characterization of virtual specialness in terms of separability and hyperplanes. We refer the reader to [HW08, Prop 9.7] for the case when X is compact, and to [HW10a, Thm 4.1] for the general case.

Proposition 6.9 (Double coset criterion). *Let X be a nonpositively curved cube complex. Suppose $\widetilde{X}$ has finitely many $\pi_1 X$-orbits of hyperplanes, and finitely many orbits of pairs of hyperplanes that cross or osculate. Then X is virtually special if (and only if) for each pair of immersed hyperplanes U_1, U_2 and choice of basepoint $x \in U_1 \cap U_2$, the double coset $\pi_1 U_1 \pi_1 U_2$ is separable in $\pi_1 X$.*

The following consequence of Proposition 6.9 was proven in [HW08]:

Corollary 6.10. *Let X be a compact nonpositively curved cube complex. Suppose $\pi_1 X$ is word-hyperbolic and has a finite index subgroup that is the fundamental group of a compact special cube complex. Then X is virtually special.*

6.f   Extensions of Quasiconvex Codimension-1 Subgroups

Definition 6.11 (K-Partitions and K-Walls). Let G be a f.g. group with Cayley graph $\Gamma = \Gamma(G, S)$ and let K be a subgroup of G. The subset $A \subset \Gamma$ is K-*shallow* if $A \subset \mathcal{N}_s(K)$ for some $s \geq 0$. We say A is K-*deep* if A is not K-shallow.

A K-*partition* of G is a collection of subsets $\{G_1, \ldots, G_m\}$ with $G = G_1 \cup \cdots \cup G_m$ such that there exists $r \geq 1$ with the following properties:

(1) $\mathcal{N}_r(G_i) \cap \mathcal{N}_r(G_j)$ is connected and K-shallow for $i \neq j$.
(2) $\{kG_1, \ldots, kG_n\} = \{G_1, \ldots, G_n\}$ for each $k \in K$. So k stabilizes the partition but might permute the parts.

We emphasize that a K-partition might not be a genuine partition as the parts can intersect.

A K-*wall* is a K-partition with exactly two parts. Of greatest interest is the case of a K-wall each of whose parts is K-deep and K-invariant, and the f.g. subgroup $K \subset G$ is *codimension*-1 precisely if there is such a K-wall. We consider this notion again in Definition 7.3, and discuss the applicability of K-walls in Chapter 7.

Let G' be a finite index subgroup of G with $K \subset G'$. Given a K-partition $\{G_1, \ldots, G_m\}$ of G, we obtain a K-partition $\{G'_1, \ldots, G'_m\}$ of G' by setting $G'_i = G_i \cap G'$ for each i. If each part of G is K-deep then so is each part of G'. Conversely, let $\{G'_1, \ldots, G'_m\}$ be a K-partition of G' and suppose $G \subset \mathcal{N}_r(G')$ for some r. Then $\{G_1, \ldots, G_m\}$ is a K-partition of G, where $G_i = \mathcal{N}_r(G'_i) \cap G$.

Definition 6.12 (Extension of K-Partitions). Let $K \subset H$ and $H \subset G$ be subgroups, and let $\{H_1, \ldots, H_m\}$ be a K-partition of H. Let $K' \subset G$ be a subgroup with $K' \cap H = K$. We say $\{H_1, \ldots, H_m\}$ *extends* to a K'-partition $\{G_1, \ldots, G_m\}$ of G if there is an inclusion $(H_1, \ldots, H_m) \subset (G_1, \ldots, G_m)$ that is K-equivariant with respect to the left action on the parts, and such that G_i is K'-deep when H_i is K-deep.

Remark 6.13. When $[G : H] < \infty$, any K-partition of H extends to a K-partition of G. In particular a K-wall of H extends to a K-wall of G, and a codimension-1 subgroup of H has codimension-1 in G. Indeed, if $\mathcal{N}_r(H)$ separates the deep parts in $\Gamma(H)$ then $\mathcal{N}_{r'}(H)$ separates them in $\Gamma(G)$ for sufficiently large r'. For each deep part H_i, let G_i consist of the elements of G that lie in the components of $\Gamma(G) - \mathcal{N}_{r'}(H)$ containing points of $H_i - \mathcal{N}_{r'}(H)$ for sufficiently large r'. We can extend each shallow H_i to $G_i = G \cap \mathcal{N}_{r'}(H)$.

The subgroup $H \subset G$ has the *extension property for K-partitions* if each K-partition of H extends to a K'-partition of G, for some $K' \subset G$ with $K' \cap H = K$. Note that K' itself depends on the chosen K-partition of H.

We now describe a variation of the definitions above. The reader can skip to Theorem 6.19 on a first reading.

Although the above definitions are expressed in terms of a Cayley graph Γ for G, they can be voiced in terms of a group acting properly and cocompactly on a geodesic metric space X. Moreover, the notion of K-partition and K-wall can be analogously defined as a decomposition of the space X that G is acting on. One can likewise use any proper left-invariant metric on G instead of using a Cayley graph.

A *geometric wall* in a topological space X is a pair of subspaces $\overleftarrow{W}, \overrightarrow{W}$ such that $X = \overleftarrow{W} \cup \overrightarrow{W}$ and the intersection $W = \overleftarrow{W} \cap \overrightarrow{W}$ is connected. Often $X - W$ consists of two components whose closures are $\overleftarrow{W}$ and $\overrightarrow{W}$, and so we refer to W itself as the wall. This is revisited in Definition 7.2. When the group G acts on X, we say W is a *geometric K-wall* if both $\overleftarrow{W}$ and $\overrightarrow{W}$ are K-invariant, and W is K-cocompact.

The following stricter version of a geometric wall supports the application in [HW10b] that is described in Theorem 7.59:

Definition 6.14. Let A be a subgroup of a f.g. group H acting on a geodesic metric space Γ_H. An *A-wall in Γ_H* consists of an A-cocompact connected subspace $W \subset \Gamma_H$ together with a decomposition of Γ_H as the union of two A-invariant subspaces $\Gamma_H = \overleftarrow{A} \cup \overrightarrow{A}$ with $\overleftarrow{A} \cap \overrightarrow{A} = W$.

Let H be a subgroup of a f.g. group G, and let $\phi : \Gamma_H \to \Gamma_G$ be an H-equivariant map between geodesic metric spaces that H and G act on. An A-wall in H *extends* to a B-wall in G if $A = H \cap B$ and one of the following holds for some $t \geq 0$:

(1) $\phi(\overleftarrow{A}) \subset \mathcal{N}_t(\overleftarrow{B})$ and $\phi(\overrightarrow{A}) \subset \mathcal{N}_t(\overrightarrow{B})$,

(2) $\phi(\overleftarrow{A}) \subset \mathcal{N}_t(\overrightarrow{B})$ and $\phi(\overrightarrow{A}) \subset \mathcal{N}_t(\overleftarrow{B})$.

Let H be a quasiconvex subgroup of G. We say that G has the *extension property for quasiconvex walls relative to* H if it has a finite index subgroup G' such that for any quasiconvex $A \subset G' \cap H$, each codimension-1 A-wall in H extends to a B-wall in G where B is quasiconvex. Note that we may choose G' to be normal without loss of generality. This has the advantage that any A^h-wall extends to a B-wall whenever A^h is a conjugate of A by $h \in H$.

Remark 6.15. A natural setting is where Γ_H is a Cayley graph of H, and A is a codimension-1 subgroup, and $W = \mathcal{N}_r(W)$ is a connected neighborhood. Likewise, for Γ_G. Then $\phi : \Gamma_H \to \Gamma_G$ is induced from a map between Cayley graphs by sending edges of Γ_H to associated paths in Γ_G.

In our case of main interest, $\phi : \Gamma_H \to \Gamma_G$ is an embedding of a convex subcomplex of a CAT(0) cube complex. However, the A-wall W will not be a hyperplane whose halfspaces are stabilized by A. It arises from a more complicated convex subspace associated to an alternate CAT(0) cube complex for H.

The map ϕ was omitted in [HW10b, Def 7.1] but the notion employs ϕ in [HW10b, Def 6.1].

Remark 6.16 (Geometric Walls versus Partitions)**.** Let G act properly and cocompactly on a length space X. Given a geometric H-wall $\overleftarrow{W}, \overrightarrow{W}$ in X, let $\overleftarrow{G} = \{g \in G : gx \in \overleftarrow{W}\}$ and let $\overrightarrow{G} = \{g \in G : gx \in \overrightarrow{W}\}$.

Observe that when W is connected, then $\mathcal{N}_q(W)$ is connected for $q \geq 0$. Hence if $\{\overleftarrow{W}, \overrightarrow{W}\}$ is a geometric H-wall for X, then so is $\{\mathcal{N}_q(W) \cup \overleftarrow{W}, \mathcal{N}_q(W) \cup \overrightarrow{W}\}$.

Conversely, suppose there is an H-partition $\{\overleftarrow{G}, \overrightarrow{G}\}$, so $\bar{G} = \mathcal{N}_r(\overleftarrow{G}) \cap \mathcal{N}_r(\overrightarrow{G})$ is connected in the Cayley graph Γ for some r. Let $W = \mathcal{N}_m(\bar{G}x)$, and note that W is connected when $m \geq \max_i(\mathsf{d}(x, h_i x))$ where $H = \langle\{h_i\}\rangle$. Choose n so that $X = \mathcal{N}_n(Gx)$. For each m, let $\overleftarrow{X}_m = \mathcal{N}_n(\overleftarrow{G}x) \cup \mathcal{N}_m(\bar{G}x)$ and let $\overrightarrow{X}_m = \mathcal{N}_n(\overrightarrow{G}x) \cup \mathcal{N}_m(\bar{G}x)$. Note that for sufficiently large m, we have $\mathcal{N}_n(\overleftarrow{G}x) \cap \mathcal{N}_n(\overrightarrow{G}x) \subset \mathcal{N}_m(\bar{G}x)$ and hence $\overleftarrow{X}_m \cap \overrightarrow{X}_m = \mathcal{N}_m(\bar{G}x)$. Define $W = \mathcal{N}_m(\bar{G}x)$, and define $\overleftarrow{W} = \overleftarrow{X}_m$ and $\overrightarrow{W} = \overrightarrow{X}_m$.

Lemma 6.17. *Let $H \hookrightarrow G$ be a subgroup. Suppose H acts properly and cocompactly on Y, and let G act properly and cocompactly on X. Let $\phi : Y \to X$ be an H-equivariant continuous map.*

If every quasiconvex wall in H extends to a quasiconvex wall in G, then every quasiconvex geometric wall in H extends to a quasiconvex geometric wall in G.

Proof. Let W_A be an A-wall of H, and let $\overleftarrow{W}_A, \overrightarrow{W}_A$ be its halfspaces in Y. Let $y \in W_A$ and let $x = \phi(y)$. Following Remark 6.16, there is an associated A-partition $\overleftarrow{H}, \overrightarrow{H}$. By hypothesis, this extends to a B-partition $\overleftarrow{G}, \overrightarrow{G}$ where

$B \subset G$ is a quasiconvex subgroup such that $B \cap H = A$. By Remark 6.16, there is a geometric B-wall W_B whose halfspaces in X are $\{\overleftarrow{W}_B, \overrightarrow{W}_B\}$. The result now follows by considering the following sequence of inclusions, and the analogous sequence for $\overrightarrow{W}_A, \overrightarrow{W}_B$. The inclusions are all either elementary or exist via the discussion in Remark 6.16.

$$\phi(\overleftarrow{W}_A) \subset \phi(\mathcal{N}_a(\overleftarrow{H}y)) \subset \mathcal{N}_b(\overleftarrow{H}x) \subset \mathcal{N}_b(\mathcal{N}_c(\overleftarrow{G}x)) = \mathcal{N}_{b+c}(\overleftarrow{G}x) \subset \mathcal{N}_d(\overleftarrow{W}_B). \qquad \Box$$

Remark 6.18. Lemma 6.17 ensures that the statements of Corollary 6.24 and Theorem 6.25 apply to geometric walls. Note that the proof of Theorem 6.25 depends on a local-isometry $X \to R$ where X and R are both compact, and Lemma 6.17 is used at that stage.

The goal of this section is to prove the following form of the extension property:

Theorem 6.19 (Quasiconvex Extension Property)**.** *Let* $G = \pi_1 X$ *where* X *is a compact special cube complex. Let* H *be a subgroup represented by a based segregated local-isometry* $Y \to X$ *with* Y *compact. Let* K *be a subgroup represented by a based local-isometry* $Z \to Y$ *with* Z *compact. Then any* K-partition *of* H *extends to a* K'-partition *of* G *such that* K' *is represented by a local-isometry* $Z' \to X$ *with* Z' *compact.*

Remark 6.20. It is conceivable that Theorem 6.19 might hold under the weaker assumption that H is a virtual retract of G without assuming that G is virtually special, but some effort would be required to produce K' that is actually quasiconvex.

The proof of Theorem 6.19 depends on the following convenient property:

Lemma 6.21 (Locally-convex canonical preimage)**.** *Let* $Z \subset Y$ *be a locally-convex subcomplex of a nonpositively curved cube complex. Let* $\phi: W \to Y$ *be a cubical map of nonpositively curved cube complexes. Then* $\phi^{-1}(Z)$ *is a locally-convex subcomplex of* W.

Remark 6.22 (The Extension Z^+)**.** We will apply Lemma 6.21 where Y is compact and X is special, and $Y \to X$ is a segregated local-isometry, $W = \mathsf{C}(Y \to X)$, and the cubical map is $\phi: \mathsf{C}(Y \to X) \to Y$. We will use the notation Z^+ to mean the component of $\phi^{-1}(Z)$ in $\mathsf{C}(Y \to X)$ that contains Z.

Proof of Lemma 6.21. Suppose two 1-cubes e_1, e_2 are adjacent along a 0-cube v with $e_1, e_2, v \in Z^+$. Suppose e_1, e_2 form the corner of a 2-cube c at v in W. If $\phi(e_1), \phi(e_2)$ form the corner of a 2-cube at $\phi(v)$ then it must be $\phi(c)$, and moreover $\phi(c)$ lies in Z by local convexity of Z in Y, and so c is in $\phi^{-1}(Z)$. The alternative is that one of $\phi(e_1)$ or $\phi(e_2)$ collapses to $\phi(v)$, or perhaps that both collapse to $\phi(v)$. In the former case $\phi(c)$ must collapse to $\phi(e_2)$ or $\phi(e_1)$, and in the latter case $\phi(c)$ collapses to $\phi(v)$. Either way, $\phi(c) \subset Z$ so $c \subset \phi^{-1}(Z)$. $\quad\square$

Remark 6.23. Choose a basepoint in $Z \subset Z^+ \subset \mathsf{C}(Y \to X)$. Then $\pi_1 Z^+ \cap \pi_1 Y = \pi_1 Z$. This holds by the algebra of retractions: Indeed $(\pi_1 Z^+ \cap \pi_1 Y) \supset \pi_1(Z^+ \cap Y) = \pi_1 Z$. And $(\pi_1 Z^+ \cap \pi_1 Y) = \phi(\pi_1 Z^+ \cap \pi_1 Y) \subset \phi(\pi_1 Z^+) = \pi_1 Z$.

Another explanation arises because both Z^+ and Y are locally-convex subcomplexes of $\mathsf{C}(Y \to X)$. In general, if $A \subset C$ and $B \subset C$ are locally-convex based subcomplexes of the nonpositively curved cube complex C, then $\pi_1 A \cap \pi_1 B = \pi_1 D$ where D denotes the component of $A \cap B$ containing the basepoint $c \in C$. Indeed, let $\tilde{c} \in \widetilde{C}$ be a lift of the basepoint, and let $\widetilde{A}, \widetilde{B}, \widetilde{D}$ be the based lifts of universal covers in $\widetilde{C}$. Then $\widetilde{A} \cap \widetilde{B} = \widetilde{D}$. Each $g \in \pi_1 C$ is represented by a geodesic γ from $\tilde{c}$ to $g\tilde{c}$. Thus if $g \in \pi_1 A \cap \pi_1 B$ then γ lies in both $\widetilde{A}$ and $\widetilde{B}$ by their convexity, as in Lemma 2.11. Hence γ lies in their intersection $\widetilde{D}$.

We refer to Figure 6.2 for a depiction of the objects in the following proof which establishes the main goal of this section:

Proof of Theorem 6.19. Let Γ_H be a Cayley graph of H. Let $\{H_1, \ldots, H_m\}$ be a K-partition of H, and let r be so that each $\mathcal{N}_r(H_i) \cap \mathcal{N}_r(H_j)$ is connected and K-shallow, so $\mathcal{N}_r(H_i) \cap \mathcal{N}_r(H_j) \subset \mathcal{N}_d(K)$.

Let $\tilde{z}$ be a lift of the basepoint z of Z to the universal cover $\widetilde{Y}$, and let $\widetilde{Z} \subset \widetilde{Y}$ be the based lift of the universal cover of Z. Let $\psi : \Gamma_H \to \widetilde{Y}$ be an H-equivariant continuous map with $1 \mapsto \tilde{z}$ and each edge mapping to a geodesic in $\widetilde{Z}$.

Choose q so that $H_i \tilde{z}$ and $H_j \tilde{z}$ do not intersect the same component of $\widetilde{Y} - \mathcal{N}_q(\widetilde{Z})$ for $i \neq j$. Moreover, assume $H_i \tilde{z} \subset \mathcal{N}_q(\widetilde{Z})$ for any shallow part H_i. The following routine computation verifies that q exists: As $\psi : \Gamma_H \to \widetilde{Y}$ is a quasi-isometry, choose a, b, c so that $\widetilde{Y} = \mathcal{N}_a(\psi(\Gamma_H))$, and $\mathsf{d}(p, q) \leq b$ when $\mathsf{d}(\psi(p), \psi(q)) \leq 2a + 1$, and $\mathsf{d}(\psi(p), \psi(q)) \leq c$ when $\mathsf{d}(p, q) \leq 1$. Choose a path in $\widetilde{Y}$ with vertices $y_1, \ldots, y_\ell$, and with $y_1 \in H_i \tilde{z}$ and $y_\ell \in H_j \tilde{z}$. Choose $u_k \in \Gamma_H$ with $\mathsf{d}(y_k, \psi(u_k)) \leq a$, and where $u_1, u_\ell \in H$. Observe that $\mathsf{d}(u_k, u_{k+1}) \leq b$ since $\mathsf{d}(\psi(u_k), \psi(u_{k+1})) \leq 2a + 1$. As y_1, y_ℓ lie in distinct components of $\Gamma_H - \mathcal{N}_d(K)$, we have $u_k \in \mathcal{N}_d(K)$ for some k, so $\psi(u_k) \in \mathcal{N}_{cd}(K\tilde{z})$. Hence $y_k \in \mathcal{N}_q(\psi(K\tilde{z}))$ where $q = cd + b$. Choosing q a bit larger, we may assume $H_i \tilde{z} \subset \mathcal{N}_q(\widetilde{Z})$ when H_i is shallow.

By Proposition 2.33, $\mathcal{N}_q(\widetilde{Z})$ lies in a K-cocompact convex subcomplex $\widetilde{Z}_\diamond \subset \widetilde{Y}$. Let $Z_\diamond = K \backslash \widetilde{Z}_\diamond$.

Pass to a finite based cover $\widehat{Y} \to Y$ such that $Z_\diamond \to Y$ lifts to a based embedding in $\widehat{Y}$. Such a cover exists by separability of $\pi_1 Z_\diamond \subset \pi_1 Y$, but we could also simply let $\widehat{Y} = \mathsf{C}(Z_\diamond \to Y)$. Let $\widehat{H} = \pi_1 \widehat{Y}$, and note that $K \subset \widehat{H}$ since $Z_\diamond \hookrightarrow \widehat{Y}$. We similarly pass to a finite based cover $\widehat{X} \to X$ such that $\widehat{Y} \to X$ lifts to an embedding $\widehat{Y} \hookrightarrow \widehat{X}$.

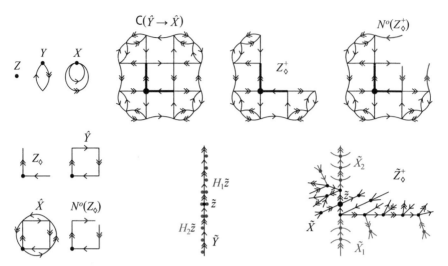

Figure 6.2. The graphs above illustrate the proof of Theorem 6.19 in a very simple case. The K-wall is depicted in $\widetilde{Y}$ by the sets $\{H_1, H_2\}$. The intersection $H_1 \cap H_2$ is contained in $Z_\diamond$ which equals $\mathcal{N}_1(\tilde{z})$ where $Z = \tilde{z}$ consists of the base vertex, and K is trivial in this case. $Z_\diamond$ embeds in a finite cover $\widehat{Y}$ which itself embeds in a finite cover $\widehat{X}$ of X which is a bouquet of circles. The retraction map $\mathsf{C}(\widehat{Y} \to \widehat{X}) \to \widehat{Y}$ maps four edges to edges of $\widehat{Y}$ they are parallel to, and maps all other edges of $\mathsf{C}(\widehat{Y} \to \widehat{X})$ to vertices of $\widehat{Y}$. The vertices in $H_1 \tilde{z}$, $H_2 \tilde{z}$ that are not within $\widetilde{Z}_\diamond \subset \widetilde{Y}$ are associated to the components of $N^o(Z_\diamond) - Z_\diamond$. These pull back to corresponding components in $N^o(Z_\diamond^+) - Z_\diamond^+$ via the retraction map. That determines the K'-partition of $\widetilde{X}$.

Consider the canonical completion $\mathsf{C}(\widehat{Y} \to \widehat{X})$ and canonical retraction $\mathsf{C}(\widehat{Y} \to \widehat{X}) \to \widehat{Y}$. It lifts to a based retraction map $\widetilde{X} \to \widetilde{Y}$ that is $\pi_1 \mathsf{C}(\widehat{Y} \to \widehat{X})$-equivariant. Here $\pi_1 \mathsf{C}(\widehat{Y} \to \widehat{X})$ acts on $\widetilde{Y}$ via the homomorphism $\pi_1 \mathsf{C}(\widehat{Y} \to \widehat{X}) \to \pi_1 \widehat{Y}$ and the action of $\pi_1 \widehat{Y}$ on $\widetilde{Y}$.

Let $Z_\diamond^+$ denote the base component of the preimage of $Z_\diamond$ in $\mathsf{C}(\widehat{Y} \to \widehat{X})$ under the retraction map, and let $K' = \pi_1 Z_\diamond^+$. (The complex Z' that is claimed in the conclusion of the theorem is equal to $Z_\diamond^+$, but we will retain the notation $Z_\diamond^+$ within this proof.) Note that $Z_\diamond^+$ is a subcomplex of $\mathsf{C}(\widehat{Y} \to \widehat{X})$ since the map $\mathsf{C}(\widehat{Y} \to \widehat{X}) \to Y$ is a cubical map by Proposition 6.4.

Let $\widetilde{Z}_\diamond^+ \subset \widetilde{X}$ be the component of the preimage of $Z_\diamond^+$ containing $\tilde{z}$. Note that $Z_\diamond^+$ is locally-convex by Lemma 6.21, and so $\widetilde{Z}_\diamond^+ \subset \widetilde{X}$ is convex by Lemma 2.11. Observe that $K' \cap \widehat{H} = K$ by Remark 6.23 (with $\widehat{Y}$ playing the role of Y).

The open cubical neighborhood $N^o(\widetilde{Z}_\diamond^+)$ in $\widetilde{X}$ retracts to the open cubical neighborhood $N^o(\widetilde{Z}_\diamond)$ in $\widetilde{Y}$. Moreover, the retraction restricts to a map $\left(N^o(\widetilde{Z}_\diamond^+) - \widetilde{Z}_\diamond^+ \right) \to \left(N^o(\widetilde{Z}_\diamond) - \widetilde{Z}_\diamond \right)$. This is because the analogous statement holds under the map $C(\widehat{Y} \to \widehat{X}) \to \widehat{Y}$ restricted to $\left(N^o(Z_\diamond^+) - Z_\diamond^+\right) \to \left(N^o(Z_\diamond) - Z_\diamond\right)$.

Indeed, for any cube c adjacent to $Z_\diamond^+$, if c maps to $Z_\diamond$ then $c \subset Z_\diamond^+$ by definition, and otherwise c maps to a cube outside $Z_\diamond$ but adjacent to $Z_\diamond$. There is a bijection between components of $\widetilde{X} - \widetilde{Z}_\diamond^+$ and components of $N^o(\widetilde{Z}_\diamond^+) - \widetilde{Z}_\diamond^+$, for otherwise $H_1(\widetilde{X}) \neq 0$. We similarly, have a bijection between components of $\widetilde{Y} - \widetilde{Z}_\diamond$ and components of $N^o(\widetilde{Z}_\diamond) - \widetilde{Z}_\diamond$.

Thus the map $\left(N^o(\widetilde{Z}_\diamond^+) - \widetilde{Z}_\diamond^+\right) \to \left(N^o(\widetilde{Z}_\diamond) - \widetilde{Z}_\diamond\right)$ allows us to *associate* each component of $\widetilde{X} - \widetilde{Z}_\diamond^+$ to a component of $\widetilde{Y} - \widetilde{Z}_\diamond$. For each i, let $\widetilde{X}_i$ be the union of the components of $\widetilde{X} - \widetilde{Z}_\diamond^+$ that are associated to components of $\widetilde{Y} - \widetilde{Z}_\diamond$ that nontrivially intersect $H_i \tilde{z}$.

For each i we let $G_i = \{g \in G : g\tilde{z} \in \widetilde{X}_i \cup \widetilde{Z}_\diamond^+\}$.

This yields a K'-partition of G that extends the K-partition of H. To see that $H_i \subset G_i$ for each i, note that for $g \in H_i$, either $g\tilde{z} \in \widetilde{Z}_\diamond \subset \widetilde{Z}_\diamond^+$ or $g\tilde{z}$ lies in a component of $\widetilde{X} - \widetilde{Z}_\diamond^+$ that is associated to H_i by definition.

The K'-action on $\{G_1, \ldots, G_m\}$ agrees with the K-action on $\{H_1, \ldots, H_m\}$ as follows: First note that each G_i is either K'-deep or equal to $\{g \in G : g\tilde{z} \in \widetilde{Z}_\diamond^+\}$. If H_i is K-deep, and $kH_i = H_j$ then $kG_i = G_j$ since we can choose $h_i \in H_i$ such that $h_i \tilde{z} \in \widetilde{X}_i$ but $kh_i \tilde{z} \in k\widetilde{X}_i$ which must then be $\widetilde{X}_j$. The action on the K'-shallow parts (which are identical to each other) is defined by using the retraction $K' \to K$ to induce the action of K' from the action of K (which is only visible on the indices).

Finally, the construction extends deep sets of the K-partition to deep sets of the K'-partition. Indeed, the cubical map $\widetilde{X} \to \widetilde{Y}$ is distance non-increasing, but fixes the subcomplex $\widetilde{Y}$. Therefore, if $G_i \subset \mathcal{N}_s(K')$ in the Cayley graph of G for some $s > 0$, then $G_i \tilde{z} \subset \mathcal{N}_t(\widetilde{Z}_\diamond^+)$ for some $t > 0$, and so $H_i \tilde{z} \subset \mathcal{N}_t(\widetilde{Z}_\diamond)$ using the retraction map, in which case $H_i \subset \mathcal{N}_u(K)$ in the Cayley graph of H for some $u > 0$. $\qquad\square$

The wall case of the following consequence of Theorem 6.19 is an important ingredient in the hypotheses of Theorem 7.59.

Corollary 6.24. *Let G be a word-hyperbolic group with a finite index subgroup $\bar{G}$ that is the fundamental group of a compact special cube complex. Let H be a quasiconvex subgroup of G and let K be a quasiconvex subgroup of $H \cap \bar{G}$. Then any K-partition of H extends to a K'-partition of G such that K' is quasiconvex.*

Proof. Let $\bar{G} = \pi_1 X$ where X is a compact nonpositively curved cube complex. By subdividing X, we may assume that any local-isometry to X is segregated.

Let $\bar{H} = \bar{G} \cap H$, and note that the K-partition $\{H_1, \ldots, H_m\}$ of H induces a K-partition $\{\bar{H}_1, \ldots, \bar{H}_m\}$ of $\bar{H}$ where $\bar{H}_i = \bar{H} \cap H_i$.

Theorem 6.19 provides a K'-partition $\{\bar{G}_1, \ldots, \bar{G}_m\}$ of $\bar{G}$ extending the K-partition of $\bar{H}$. For any sufficiently large s, the neighborhoods $\{\mathcal{N}_s(\bar{G}_i), \ldots,$

$\mathcal{N}_s(\bar{G}_i)\}$ in the Cayley graph of G provide a K'-partition of G extending the K-partition of H.   □

The following result presumes familiarity with sparse cube complexes as treated in Definition 7.21. It is used in the proof of Theorem 15.1.

Theorem 6.25 (Extending Walls in Sparse Case). *Let G act freely and cosparsely on $\widetilde{X}$ and suppose there is a torsion-free finite index subgroup G' such that $X' = G'\backslash\widetilde{X}$ is special. Let $H \subset G$ be a relatively quasiconvex subgroup that is aparabolic in G. Let K be a quasiconvex subgroup of $H \cap G'$. Then every K-wall of H extends to a K'-wall of G for some relatively quasiconvex subgroup $K' \subset G$. Moreover, if each part of the former is K-invariant, then each part of the latter is K'-invariant.*

Remark 6.26. The proof operates under the following slightly more general hypotheses that $H \subset G'$ is full and relatively quasiconvex, and $K \subset H$ is represented by a local-isometry with compact domain.

Proof. By either subdividing $\widetilde{X}$ or by replacing X' by a bipartite double cover, we may assume that each local-isometry to X' is segregated. Note that subdivision preserves cosparseness.

Let $X' \to R$ be the local-isometry to the special cube complex associated to the hyperplanes of X' as described in Proposition 6.2. Since X' is sparse, X' has finitely many hyperplanes and so R is compact.

By Proposition 2.31, there is a based local-isometry $Y \to X'$ with $\pi_1 Y = H$ and Y compact, and there is a based local-isometry $Z \to Y$ with $\pi_1 Z = K$ and Z compact.

Apply Theorem 6.19 to $Z \to Y \to R$, and extend the K-wall in H to a $\pi_1 J$-wall $\{A_1, A_2\}$ of $\pi_1 R$, where $J \to R$ is a based local-isometry with J compact. We can moreover assume that left-invariance is preserved in the sense that if K stabilizes the parts of H then $\pi_1 J$ stabilizes the parts of $\pi_1 R$. Let $K' = G' \cap \pi_1 J$. Let $\{G' \cap A_1, G' \cap A_2\}$ be the induced K'-wall of G'. We obtain a K'-wall $\{\mathcal{N}_s(G' \cap A_1), \mathcal{N}_s(G' \cap A_2)\}$ in G by taking finite neighborhoods in the Cayley graph of G. These last two steps also preserve left-invariance.

As it is cosparse, $\widetilde{X}$ has the isometric core property by Lemma 7.37, and so there is a G'-cocompact convex subspace $\widetilde{S} \subset \widetilde{X}$. Quasiconvexity of K' holds since $\widetilde{S} \cap \widetilde{J}$ is a nonempty K'-cocompact convex subspace. Indeed, it is the universal cover of a compact subspace of the fiber-product $S \otimes_R J \subset X' \otimes_R J$. See Section 8.b.   □

The following observation requires familiarity with the dual cube complex construction discussed in Chapter 7. It is of interest because the failure of cocompactness of the cube complex dual to a collection of quasiconvex walls in a relatively hyperbolic group is pinpointed to the failure of cocompactness of the dual cube complexes of the parabolic subgroups. When P is virtually $\mathbb{Z}^n$, if P acts properly on the dual, then the dual is not cocompact precisely when the number of distinct commensurability classes of codimension-1 walls exceeds n.

Remark 6.27 (No New Parabolic Walls). Let G be hyperbolic relative to virtually abelian subgroups. Suppose that G has a finite index subgroup G' such that $G' = \pi_1 X$ where X is a compact (or more generally sparse) special cube complex. Let P be a parabolic subgroup of G, and observe that the hyperplanes of $\widetilde{X}$ determine a wallspace structure on $P \cap G'$ and hence on P.

Let W be a K'-wall of G arising from a K-wall in a quasiconvex subgroup $H \subset \pi_1 X$ as produced by Theorem 6.19 (or Theorem 6.25). Then W does not contribute genuinely new walls on P in the following sense: Each $P \cap (K')^g$ is either commensurable with P, or commensurable with $\cap_{i=1}^{k} \mathrm{Stab}_P(W_i)$ where each W_i is a hyperplane that cuts an orbit $P\widetilde{x}$ essentially. The reason for this is as follows: Note that subdivision does not effect the commensurability classes of hyperplane stabilizers. By the construction given in Theorem 6.19, the wall is represented by a local-isometry $Z' \to X$ with Z' compact, and hence by a convex $\pi_1 Z'$-cocompact subcomplex $\widetilde{Z'} \subset \widetilde{X}$. The claim therefore follows from Corollary 7.24 and Remark 7.25.

After developing the material in this section, I found that Masters had produced related results for certain right-angled hyperbolic Coxeter groups. Specifically, Masters showed that every f.g. free subgroup of a 3-dimensional closed right-angled hyperbolic Coxeter group lies in a quasifuchsian surface subgroup [Mas08]. We can use the material developed here to give a variation on his result, which we state in the simplest case. Similar statements hold in the virtually compact special and virtually sparse special settings.

Theorem 6.28. *Let X be a compact special cube complex with $\pi_1 X$ word-hyperbolic. Let K be an infinite index quasiconvex subgroup of $\pi_1 X$. Then K lies in a codimension-1 quasiconvex subgroup K' of $\pi_1 X$.*

Proof. Since K is quasiconvex and has infinite index in $\pi_1 X$ we can choose an infinite order element a such that $\langle K, a \rangle \cong K * \mathbb{Z}$ (see [Arz01, Git99b]). Let $H = \langle K, a \rangle$. By Proposition 2.31, let $Z \to X$ be a local-isometry with $\pi_1 Z = K$ and Z compact. Again, by Proposition 2.31, let $Y \to X$ be a based local-isometry with $\pi_1 Y = H$ and Y compact, and such that $Z \to X$ factors as $Z \to Y \to X$. By Corollary 6.24, extend the codimension-1 quasiconvex subgroup $K \subset H$ to a codimension-1 quasiconvex subgroup $K' \subset G$. □

Problem 6.29. Does every quasiconvex free subgroup of a higher dimensional word-hyperbolic right-angled Coxeter group lie in a surface subgroup?

Is every quasiconvex free subgroup of a nonelementary indecomposable word-hyperbolic compact special group contained in a surface subgroup?

Does every free subgroup of a (virtually special) hyperbolic 3-manifold group lie in a surface subgroup?

Problem 6.30. Let G be virtually sparse special. Does G have the extension property with respect to any quasiconvex H-wall in an arbitrary quasiconvex subgroup K of G? In particular, does this hold when K is a parabolic subgroup?

6.g The Malnormal Combination Theorem

In [HW12] we prove the following:

Proposition 6.31. *Let Q be a compact nonpositively curved cube complex with an embedded 2-sided hyperplane H. Suppose that $\pi_1 Q$ is word-hyperbolic. Suppose that $\pi_1 H \subset \pi_1 Q$ is malnormal. Let $N^o(H)$ denote the open cubical neighborhood of H. Suppose that each component of $Q - N^o(H)$ is virtually special.*

Then Q is virtually special.

Here is a more general formulation of the hypothesis that permits torsion: Let G act properly and cocompactly on a CAT(0) cube complex $\widetilde{Q}$. Suppose there is a hyperplane $\widetilde{H}$ such that:

(1) $\operatorname{Stab}(\widetilde{H})$ *is almost malnormal in* G.
(2) $g\widetilde{H} \cap \widetilde{H} = \emptyset$ *for each* $g \in G - \operatorname{Stab}(\widetilde{H})$.
(3) $\operatorname{Stab}(\widetilde{H})$ *preserves each component of* $N^o(\widetilde{H}) - \widetilde{H}$.
(4) *For each component $\widetilde{X}$ of $\widetilde{Q} - GN^o(\widetilde{H})$, the group $\operatorname{Stab}(\widetilde{X})$ has a finite index torsion-free subgroup J such that $J \backslash \widetilde{X}$ is special.*

Note that Condition (3) holds in a finite index subgroup of G provided Condition (2) holds.

Chapter Seven

Cubulations

7.a   Wallspaces

Definition 7.1. Haglund and Paulin introduced the notion of a wallspace to abstract a property that arises in many natural scenarios, and especially for Coxeter groups [HP98]. A *wallspace* is a set X together with a collection of *walls* each of which is a partition $X = \overleftarrow{N} \sqcup \overrightarrow{N}$ into *halfspaces*, and such that moreover, $\#(p, q) < \infty$ for each $p, q \in X$ where $\#(p, q)$ equals the number of walls separating p, q.

The fundamental example of a wallspace is the 0-skeleton of a CAT(0) cube complex, together with a system of walls associated to the hyperplanes.

In [HW14] we gave a slightly more general version of the definition of wallspace, by relaxing the requirement that the two halfspaces be disjoint. This is in agreement with the language used in Section 6.f. We also describe the following special case of this variant:

Definition 7.2. A *geometric wallspace* is a pair $(X, \mathcal{W})$, where X is a metric space, and each wall W is path-connected and arises from a pair of connected halfspaces with $X = \overleftarrow{W} \cup \overrightarrow{W}$ and $W = \overleftarrow{W} \cap \overrightarrow{W}$. A wall *separates* two sets if they lie in distinct *open halfspaces* $X - \overleftarrow{W}$ and $X - \overrightarrow{W}$.

In many natural cases the halfspaces in a geometric wallspace are actually determined by the subspace W. Generally speaking, the geometric wallspace hypothesis merely simplifies certain proofs, and most statements that hold for geometric wallspaces can be suitably reinterpreted for general abstract wallspaces. One useful aspect of a geometric wallspace is that we can naturally refer to neighborhoods of walls when we are working in a metric space. Moreover, given a wallspace structure on a metric space, one can often replace the original walls by finite thickenings of their halfspaces to obtain a geometric wallspace. The result on the dual cube complex (defined below) is that the dual is thickened in the sense that it has more higher cubes.

7.b   Sageev's Construction

Definition 7.3. Let G be a f.g. group with Cayley graph Γ. A subgroup $H \subset G$ is *codimension*-1 if it has a finite neighborhood $\mathcal{N}_r(H)$ such that $\Gamma - \mathcal{N}_r(H)$ contains at least two H-orbits of components that are *deep* in the sense that they do not lie in $\mathcal{N}_s(H)$ for any $s \geq 0$.

For instance any $\mathbb{Z}^n$ subgroup of $\mathbb{Z}^{n+1}$ is codimension-1, and any infinite cyclic subgroup of a closed orientable surface subgroup is as well. An edge group of a nontrivial splitting of a group is codimension-1. If the coset diagram $H\backslash\Gamma$ has more than one topological end, then H is codimension-1. There is a closely related notion: H is *divisive* if $\Gamma - \mathcal{N}_r(H)$ has two or more deep components for some $r > 0$. Every codimension-1 subgroup is divisive, however there are divisive subgroups that are not codimension-1. The difficulty is that the action of H on Γ might permute the deep components of $\Gamma - \mathcal{N}_r(H)$. When $\Gamma - \mathcal{N}_r(H)$ has finitely many deep components, there is a finite index subgroup $H' \subset H$ whose action stabilizes each of these components, and one obtains a multi-ended coset diagram $H'\backslash\Gamma$, which is equivalent to H' being codimension-1 in G.

Given a finite collection of codimension-1 subgroups $H_1, \ldots, H_k$ of G, Michah Sageev introduced a simple but powerful construction that yields an action of G on a CAT(0) cube complex C that is *dual* to a system of walls associated to these subgroups [Sag95].

For each i, let $N_i = \mathcal{N}_{r_i}(H_i)$ be a neighborhood of H_i that separates Γ into at least two deep components. The *wall* associated to N_i is a fixed partition $\{\overleftarrow{N}_i, \overrightarrow{N}_i\}$ consisting of one of these deep components $\overleftarrow{N}_i$ together with its complement $\overrightarrow{N}_i = \Gamma - \overleftarrow{N}_i$, and more generally, the translated *wall* associated to gN_i is the partition $\{g\overleftarrow{N}_i, g\overrightarrow{N}_i\}$. The two parts of the wall are *halfspaces*.

We presume a certain degree of familiarity with the details of Sageev's construction here, but hope that any interested reader will mostly be able to follow the arguments. We shall not describe the structure of the dual cube complex C here but will describe its 1-skeleton. A 0-cube of C is a choice of one halfspace from each wall such that each element of G lies in all but finitely many of these chosen halfspaces. A wall is thought of as "facing" the points in its chosen halfspace. Two 0-cubes are joined by a 1-cube precisely when their choices differ on exactly one wall. See Figure 7.1 for two particularly simple dual cube complexes.

The walls in Γ are in one-to-one correspondence with the hyperplanes of the CAT(0) cube complex C produced by Sageev's construction, and the stabilizer of each such hyperplane equals the codimension-1 subgroup that stabilizes the associated translated wall: The stabilizer of the hyperplane corresponding to a translated wall associated to gN_i is commensurable with gH_ig^{-1}.

Sageev's construction naturally decomposes into two separate ideas. The first is that a collection of codimension-1 subgroups yields a wallspace. The second is that a wallspace yields a dual cube complex (see [CN05, Nic04] for more details on the latter).

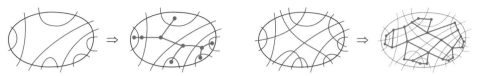

Figure 7.1. The dual cube complex on the left is a tree, the one on the right is 2-dimensional. The n-cubes correspond to certain n-fold collections of pairwise crossing walls.

7.c   Finiteness Properties of the Dual Cube Complex

Cocompactness properties of the action of G on the CAT(0) cube complex dual to a wallspace associated to a collection of codimension-1 subgroups was analyzed in [Sag97] where Sageev proved that:

Proposition 7.4. *Let G be a word-hyperbolic group, and let $\{H_1, \ldots, H_k\}$ be a collection of quasiconvex codimension-1 subgroups. Then the action of G on the dual cube complex is cocompact.*

Note that Proposition 7.4 requires a choice of Cayley graph as well as a chosen H_i-wall $\{\overleftarrow{W}_i, \overrightarrow{W}_i\}$ for each i. But although the resulting dual cube complex depends on these choices, it is always G-cocompact. We refer to [HW14] for a more elaborate discussion of the finiteness properties of the action obtained from Sageev's construction, as well as for background and an account of the literature.

Definition 7.5. Let $(X, \mathcal{W})$ be a wallspace with X a proper metric space. Say $(X, \mathcal{W})$ has *Ball-Wall separation* if for each finite ball $B \subset X$ there exists $r > 0$ such that if $\mathsf{d}(B, W) > r$ for some $W \in \mathcal{W}$, then there exists $W' \in \mathcal{W}$ such that W' separates B, W. Here $\mathsf{d}(B, W) = \min(\mathsf{d}(B, \overleftarrow{W}), \mathsf{d}(B, \overrightarrow{W}))$. Here W' *separates* B, W if a halfspace of W' contains W and the complement of that halfspace contains B.

The following is proven in [HW14, Thm 5.7]:

Lemma 7.6 (Local finiteness). *Let $(X, \mathcal{W})$ be a wallspace where X is a proper metric space. Suppose that for each finite ball B there are finitely many walls $W \in \mathcal{W}$ which separate two points of B. If $(X, \mathcal{W})$ has Ball-Wall separation then the associated dual cube complex is locally finite.*

Example 7.7 (Not locally finite). Figure 7.2 illustrates the cautionary example of a wallspace obtained from the universal cover $\widetilde{X}$, where $X = T^2 \cup (S^1 \times I)$ is obtained by attaching a cylinder to a torus. The walls in $\widetilde{X}$ arise from three immersed walls in X. The first of these is a circle in the cylinder: $S^1 \times \{\frac{1}{2}\} \subset$

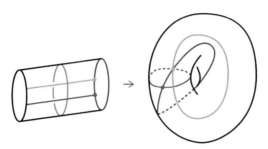

Figure 7.2. The universal cover of the space above provides a wallspace with three orbits of walls whose dual is $\mathbb{R} \times \mathbb{R} \times T_\infty$.

$S^1 \times I$. Each of the other two consists of the union of a circle in T^2 together with an arc in $S^1 \times I$. The dual of the associated wallspace for the universal cover is isomorphic to $\mathbb{R} \times \mathbb{R} \times T_\infty$ where $\mathbb{R}$ is cubulated in the usual way, and T_∞ is an ∞-star consisting of the wedge of infinitely many edges along a central vertex.

Note that the Ball-Wall separation property fails in this wallspace for any ball intersecting $\widetilde{T}^2$, and any wall that is the universal cover of $S^1 \times \{\frac{1}{2}\}$.

We now state a properness criterion (see, e.g., [HW14] for a proof). We use the notation $\#(p, q)$ for the number of walls separating p, q.

Proposition 7.8. *If $\#(1, g) \to \infty$ as $\mathsf{d}_\Gamma(1, g) \to \infty$ then G acts properly on C.*

We now describe a form of "relative cocompactness" that arises for dual CAT(0) cube complexes in the relatively hyperbolic case. The first explicit appearance of such actions was for the CAT(0) cube complexes dual to the wallspaces on $B(6)$ small-cancellation groups [Wis04] (they were called "cofinite" there, but we now use the term "cosparse"). However the notion reflects a general phenomenon associated to relative hyperbolicity. Proposition 7.4 generalizes to a relatively hyperbolic context through the following whose two parts are simplified restatements of [HW14, Thm 7.12 and Prop 8.1]:

Theorem 7.9 (Relatively Cocompact Dual). *Let G be a f.g. group that is hyperbolic relative to a collection of subgroups $P_1, \ldots, P_s$. Let $\{H_1, \ldots, H_k\}$ be a collection of quasi-isometrically embedded codimension-1 subgroups of G, and for each i let $W_i = \{\overleftarrow{W}_i, \overrightarrow{W}_i\}$ be an H_i-wall in the Cayley graph. Let C denote the CAT(0) cube complex dual to the G-translates of $W_1, \ldots, W_k$. There is a compact subcomplex $K \subset C$ with GK connected, and for $1 \leq i \leq s$ there is a nonempty P_i-invariant convex subcomplex $C_i \subset C$, such that:*

(1) $C = GK \cup \bigcup_{i=1}^s GC_i$.
(2) $g_i C_i \cap g_j C_j \subset GK$ unless $i = j$ and $g_j^{-1} g_i \in \mathrm{Stab}(C_i)$.
(3) For each i there exists a compact subcomplex K_i' such that $C_i \cap GK = P_i K_i'$.

Additionally, if G acts properly on C, then the following also hold:

(4) *Each 0-cube lies in uniformly finitely many $g_j C_i$.*
(5) *C is [uniformly] locally finite if each C_i is [uniformly] locally finite.*
(6) *Each C_i is superconvex.*

We note that local finiteness for C can fail, and indeed it can fail for an individual C_i. We refer to Lemma 7.6 and Example 7.7.

Remark 7.10. The subcomplexes C_i above are described in [HW14, Thm 7.12] with more precision: Specifically, we can choose each C_i to be the CAT(0) cube complex dual to the wallspace consisting of those walls that *slice* $\mathcal{N}_r(P)$ in the sense that both halfspaces have infinite diameter intersection with $\mathcal{N}_r(P)$ for some r. This corresponds to the notation there. The CAT(0) cube complex C_i embeds in C as a convex subcomplex by orienting the walls not slicing $\mathcal{N}_r(P)$ towards P. We refer to the notion of *hemiwallspace* in [HW14].

When G is hyperbolic relative to virtually abelian groups $\{P_i\}$, cocompactness is provided by Theorem 7.9 when each C_i is P_i-cocompact. Lemma 7.12 provides the criterion for detecting cocompactness in this case.

Remark 7.11. When a group G acts on a wallspace $(X, \mathcal{W})$, the typical way of detecting that the dual C is G-cocompact, is to verify that there are finitely many G-orbits of maximal cubes in C. This is equivalent to showing that there are finitely many G-orbits of maximal collections of pairwise crossing walls in $\mathcal{W}$. We refer to [HW14, Cor 3.13]. When X is δ-hyperbolic and the walls are κ-quasiconvex (i.e., their halfspaces are κ-quasiconvex) a collection $\{W_1, \ldots, W_m\}$ is pairwise crossing if and only if there is a point at distance $\mu = \mu(\delta, \kappa)$ from all the walls (i.e., all the halfspaces). Hence local finiteness of the collection implies cocompactness. This was proven in [Sag97] and clarified in [NR03], and then led to the bounded packing notion and a relatively hyperbolic generalization in [HW09, HW14].

We now describe conditions that ensure that the dual to $(X, \mathcal{W})$ is P-cocompact when P is virtually $\mathbb{Z}^d$. As cocompactness is preserved by passage to a finite index subgroup, we assume $P = \mathbb{Z}^d$.

When P acts on the wallspace $(X, \mathcal{W})$, the wall W is P-*essential* if no orbit Px lies in a finite neighborhood of either halfspace of W. We sometimes say a wall W is *essential* if both of its halfspaces are deep, and the group P is understood.

Lemma 7.12 ($\mathbb{Z}^d$-cocompact dual)**.** *Let $P = \mathbb{Z}^d$ act on a wallspace $(Y, \mathcal{W})$ where Y is a P-cocompact proper geodesic metric space. Then P acts properly and cocompactly on the dual C to $(Y, \mathcal{W})$ if and only if the following conditions hold:*

(1) *There are d distinct commensurability classes of stabilizers of P-essential walls. Let $W_1, \ldots, W_d$ be representatives of these basic walls.*

(2) *The corresponding* $\mathrm{Stab}(W_i)$*-invariant codimension-1 subspaces* $E_i \subset \mathbb{R}^d$ *are orthogonal to vectors* v_i *such that* $\{v_1, \ldots, v_d\}$ *form a basis for* $\mathbb{R}^d$.

(3) *For each* $U \in \mathcal{W}$, *the subgroup* $\mathrm{Stab}(U)$ *is commensurable with* $\cap_{j \in J} \mathrm{Stab}(W_j)$ *for some subset* $J \subset \{1, \ldots, d\}$.

Regard $\mathbb{Z}^d$ as a subgroup of $\mathbb{R}^d$ in Condition (2) by extending a $\mathbb{Z}$-basis to an $\mathbb{R}$-basis. Regard P as being the intersection of an empty collection of subgroups in Condition (3), and note that the finiteness condition of a wallspace ensures that there are only finitely many walls whose stabilizer is commensurable with P.

In practice, P is a subgroup of a group G acting on a wallspace $(X, \mathcal{W}_X)$, and $Y \subset X$ is a P-cocompact subspace, and the walls of $\mathcal{W}$ arise from a subset of walls $W \in \mathcal{W}_X$ having the property that $\overleftarrow{W} \cap Y \neq \emptyset$ and $\overrightarrow{W} \cap Y \neq \emptyset$ and such that two walls of $\mathcal{W}$ cross in X if and only if they cross in Y. The dual to $(Y, \mathcal{W})$ will embed as a convex subcomplex in the dual to $(X, \mathcal{W}_X)$ by Lemma 7.41.

Proof. In view of Remark 7.11, we shall determine whether or not there are finitely many P-orbits of maximal collections of pairwise crossing walls. Let Υ be a Cayley graph with respect to a finite generating set for $\mathbb{Z}^d$. We will use the $\mathbb{Z}^d$-equivariant quasi-isometry $\Upsilon \to X$ induced by choosing a basepoint $p \in X$. Each wall U is associated with a coset of $\mathrm{Stab}(W)$.

Cocompactness implies that for each $\nu > 0$ there exists $\beta = \beta(\nu)$ such that for each i if $g\mathcal{N}_\nu(W_i) \cap \mathcal{N}_\nu(W_i) \neq \emptyset$, then $g\mathcal{N}_\nu(W_i) \subset \mathcal{N}_\beta(W_i)$. By Condition (3), there exists $\nu > 0$ so that each wall $U \in \mathcal{W}$ lies in the intersection of translates of neighborhoods $\cap_{j \in J} g_j \mathcal{N}_\nu(W_j)$ of a set of basic walls, and moreover U is coarsely equal to this intersection. Furthermore, each such intersection contains at most ρ walls in the commensurability class of U, where $\rho = \rho(\nu)$ can be chosen uniformly as there are 2^d distinct such intersections.

For a maximal collection $\{g_k U_k\}$ of pairwise crossing walls, as above $U_k \subset \cap_{j \in J_k} g_{kj} \mathcal{N}_\nu(W_{kj})$. For each basic wall W_i, there exists k and $j \in J_k$ so that $W_i = W_{kj}$, as otherwise we could add W_i to the collection, which violates maximality.

Thus for each $g_k U_k$ in our maximal collection, we have $g_k U_k \subset \cap_{i \in I_k} g_i \mathcal{N}_\beta(W_i)$ for some $I_k \subset \{1, \ldots, d\}$. As above, there are finitely many walls in the commensurability class of U_k within this intersection. Finally, there are finitely many distinct P-orbits of collections $\{g_i \mathrm{Stab}(W_i)\}$, since the image of P in $\prod_{i=1}^d P/\mathrm{Stab}(W_i)$ has finite index.

For a collection $\{U_i\}$ of subspaces of X, its *inradius* is the smallest integer r such that there is a point p with $\mathcal{N}_r(p) \cap U_i \neq \emptyset$ for each i. If there are $d' > d$ commensurability classes of P-essential walls, then the dual is not cocompact, since we can produce a sequence $\{g_{in} W_i : 1 \leq i \leq d'\}_{n \in \mathbb{N}}$ of collections of translates of basic walls with inradius $r(n)$ and $\lim_{n \to \infty} r(n) = \infty$. To see this, first observe that since $\{v_1, \ldots, v_{d'}\}$ are not linearly independent, by reordering, we may assume that $v_1 \in \mathrm{span}\{v_i\}_{i>1}$. Consequently, $\cap_{i>1} v_i^\perp \subset v_1^\perp$. Let $g \in P$ be so that $\langle g \rangle \cap \mathrm{Stab}(W_1) = \{1_P\}$, e.g., $g = sv_1$ for some $s > 0$. Thus $\cap_{i>1} \mathrm{Stab}(W_i)$

is commensurable with a subgroup of $\text{Stab}(W_1)$. Then $\{W_1, g^n W_2, \ldots, g^n W_{d'}\}$ has linearly growing inradius since the coset $g^n \bigcap_{i>1} \text{Stab}(W_i)$ is increasingly far from $\text{Stab}(W_1)$.

If there are $d' < d$ commensurability classes of P-essential walls then $\cap_{i=1}^{d'} \text{Stab}(W_i)$ is infinite, and provides a nontrivial element that fixes a 0-cube.

If $\text{Stab}(U)$ is not commensurable with the intersection of P-essential walls for some $U \in \mathcal{W}$ then the action cannot be both cocompact and free, since the dual would not be locally finite by Lemma 7.23. $\square$

We specify an important case strengthening Theorem 7.9 in Theorem 7.28 after describing sparse complexes.

7.d   Virtually Cubulated

We now describe how a finite index subgroup of a CAT(0) cube complex induces an action of the ambient group on a different CAT(0) cube complex. We begin with an observation that is devoid of geometric hypotheses, and then turn to conclusions that utilize hyperbolic assumptions.

Lemma 7.13. *Suppose the group G has a finite index subgroup H that acts either properly or with finite stabilizers on a CAT(0) cube complex $\widetilde{Y}$. Then G acts properly or with finite stabilizers respectively on a CAT(0) cube complex $\widetilde{X}$ that is isomorphic to the cartesian product of finitely many copies of $\widetilde{Y}$.*

Proof. Fix a point $p \in \widetilde{Y}$. For each hyperplane U of $\widetilde{Y}$, let S_U be the stabilizer of its halfspace containing p. Consider the full collection $\{hS_U U\}$ where we vary over both the hyperplanes U and the cosets of S_U in H. We extend this collection by allowing the cosets to lie in G. Declare $g'S_{U'}U'$ and $gS_U U$ to *cross* if either $g^{-1}g' \notin H$ or else $g^{-1}g' \in H$ and $g^{-1}g'U'$ crosses U. Multiplication of the original collection by the distinct left cosets representatives $\{f_i\}$ of H in G shows that our collection is partitioned into pairwise crossing subcollections, and so the cube complex $\widetilde{X}$ dual to the entire collection is isomorphic to $\widetilde{Y}^d$ where $d = [G:H]$. The factor that a hyperplane belongs to is determined by its left coset representative, and this is preserved by the group action, since for a subgroup $S \subset H$ all elements of gS belong to the same gH coset. Finally, the action of the subgroup H on the first factor (corresponding to cosets with representatives in H) is precisely the original action of H on $\widetilde{Y}$.

Alternate viewpoint on $\widetilde{X}$: Consider the product $G \times \widetilde{Y}$ which is topologized with the discrete topology for G. Form the quotient where we identify $(g, \widetilde{y})$ with $(gh^{-1}, h\widetilde{y})$ for $h \in H$. The left action by G on the first factor of $G \times \widetilde{Y}$ induces an action of G on the quotient. The quotient consists of $[G:H]$ components, as two elements are in the same component precisely if their first coordinates lie in the same left coset of H in G. Moreover, H stabilizes the *trivial component* consisting of elements whose first coordinate is an element of H. As $h(1, \widetilde{Y}) = (h, \widetilde{Y}) = (1, h\widetilde{Y})$, the H-action on the trivial component is identical to the H-action on $\widetilde{Y}$. Choosing coset representatives $f_1, \ldots, f_d$ with f_1 trivial, each point in the quotient is represented uniquely by $(f_i, \tilde{y})$ for some i and $\tilde{y} \in \widetilde{Y}$. Thus each point of $\widetilde{Y}^d$ corresponds to a d-tuple consisting of a point in each component of the quotient. Specifically, the i-th coordinate of $\tilde{y} = (\tilde{y}_i)_{1 \le i \le d}$ is identified with $\{(f_i, \tilde{y}_i) : 1 \le i \le d\}$. Hence the action of G on the quotient determines an action of G on the product $\widetilde{X} = \widetilde{Y}^d$.

Suppose H acts properly on $\widetilde{Y}$. Then H acts properly on $\widetilde{X}$. Indeed, an orbit Hx has finitely many points in a finite ball in $\widetilde{X}$, since these project to orbit points in a finite ball of the same radius in the first factor. Thus G acts properly since if $J \subset G$ is an infinite set with Jx in a finite ball, then there is an infinite subset $K \subset J$ with $K \subset fH$, and so $f^{-1}K \subset H$ is an infinite subset with $f^{-1}Kx$ in a finite ball of $\widetilde{X}$.

Suppose the action of H on $\widetilde{Y}$ has finite 0-cube stabilizers. Then H acts on $\widetilde{X}$ with finite stabilizers as well, since if $J \subset H$ stabilizes a 0-cube of $\widetilde{X}$, then it stabilizes the 0-cube that is its first coordinate in the product. Consequently, if $J \subset G$ stabilizes a 0-cube then it must be finite. Indeed $(J \cap H) \subset H$ and is hence finite, but then J itself is finite since $[J : (J \cap H)] < \infty$. $\square$

Lemma 7.14. *Let G be word-hyperbolic and suppose G has a finite index subgroup G' that acts properly and cocompactly on a CAT(0) cube complex $\widetilde{Y}$. Then G acts properly and cocompactly on a CAT(0) cube complex $\widetilde{X}$.*

Moreover, for any quasiconvex subgroup H and H-wall of G, we can assume that H stabilizes a hyperplane $\widetilde{V}$ of $\widetilde{X}$. Furthermore, the two parts of the H-wall map to finite neighborhoods of the two halfspaces of $\widetilde{V}$ under an orbit map $G \to G\tilde{x}$ with $\tilde{x} \in \widetilde{X}$.

Proof. By possibly passing to a subcomplex, we can assume that $\widetilde{Y}$ does not contain a nonempty G'-invariant convex subcomplex. Let $U_1, \ldots, U_r$ denote representatives of the G'-orbits of hyperplanes in $\widetilde{Y}$. The $G'\tilde{y}$ orbit determines a $\mathrm{Stab}(U_i)$-wall for G', where $G'\tilde{y}$ and hence G' is partitioned by the halfspaces of U_i. This extends to a $\mathrm{Stab}(U_i)$-wall for G, as noted in Remark 6.13.

Consider the full collection of G-translates of such walls, and observe that G acts properly on the resulting dual cube complex $\widetilde{X}$ since G' does. Specifically, G' acts properly since there is a quotient to the cube complex with respect to the original walls. Thus G acts properly since if $\{g_i x\}$ lies in a finite ball then $\{f_i g_i x\}$ lies in a finite ball that is slightly larger, where $\{f_i\}$ are the finitely many left coset representatives for G' in G. Proposition 7.4 implies that G acts cocompactly, as $\mathrm{Stab}(U_i)$ is quasiconvex since $N(U_i)$ is a convex subcomplex.

The moreover conclusion holds by adding the given H-wall to the collection of walls given above. $\square$

Remark 7.15. A relatively hyperbolic generalization is given in Lemma 7.34, but that situation is more delicate. For instance, the Coxeter group $P = \langle a_1, a_2, a_3 \mid a_i^2, (a_i a_j)^3 : 1 \leq i, j \leq 3 \rangle$ has an index 6 subgroup isomorphic to $\mathbb{Z}^2$ but doesn't act properly and cocompactly on a CAT(0) cube complex. It follows that if G is hyperbolic relative to P, then G cannot act properly and cocompactly on a CAT(0) cube complex.

Definition 7.16 (Compatible). Let P', P'' be finite index subgroups of a group P. Suppose P' and P'' act properly and cocompactly on CAT(0) cube complexes $\widetilde{Y}', \widetilde{Y}''$. These two actions are *compatible* if each P'-essential hyperplane stabilizer is commensurable with a P''-essential hyperplane stabilizer, and vice versa. Compatibility is an equivalence relation, and we refer to a collection of such actions as *compatible* if each two are compatible.

We now describe how compatibilty will arise in the statement of Lemma 7.17 below. Let G be hyperbolic relative to virtually abelian subgroups, let $G' \subset G$ be a finite index subgroup, and for simplicity assume G' is normal. Given a cocompact action of G' on $\widetilde{X}$, and a peripheral subgroup P of G, let $P' = P \cap G'$. For each $g_i \in G$, there is a cocompact action of $(P')^{g_i}$ on a CAT(0) cube complex $\widetilde{Y}_i \subset \widetilde{X}$ by Lemma 2.38, and hence there is an action of P' on $\widetilde{Y}_i$ through the composition $P' \to (P')^{g_i} \subset G'$. To say the various cubulations of $(P^g \cap G')$ are compatible means that these actions are compatible.

When $G = P$, the hypothesis implies that the set of commensurability classes of essential hyperplane stabilizers is invariant under conjugation by G.

Lemma 7.17. *Let G be hyperbolic relative to virtually abelian subgroups $\{P_i\}$. Suppose G has a finite index subgroup G' that acts properly and cocompactly on a CAT(0) cube complex $\widetilde{Y}$. Suppose that for each P_i and $g \in G$, the cubulations of $P_i \cap G'$ and $P_i^g \cap G'$ are compatible. Then G acts properly and cocompactly on a CAT(0) cube complex $\widetilde{X}$.*

Proof. The argument uses the exact same construction as in the proof of Lemma 7.14, so we describe the extra explanation to obtain the cocompactness conclusion.

As explained in Remark 7.10, the relative cocompactness conclusion of Theorem 7.9 actually yields cocompactness since compatibility is satisfied for the parabolic cubulations and so the conditions of Lemma 7.12 are satisfied. $\square$

7.e Sparse Complexes

Definition 7.18 (Quasiflat). A *quasiflat* $\widetilde{F}$ is a locally finite CAT(0) cube complex with a proper action by a f.g. virtually abelian group P such that there are finitely many P-orbits of hyperplanes. A *strong quasiflat* has the additional

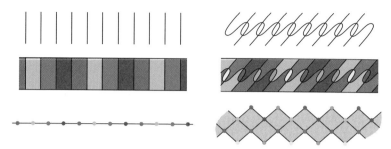

Figure 7.3. Jiggled walls: The wallspace on an infinite strip on the left has dual cube complex homeomorphic to $\mathbb{R}$. If we "jiggle" the walls so that they cross their neighbors, then one obtains a thicker cube complex consisting of the quasiline on the right. Similar "thickenings" occur if we jiggle typical wallspaces on $\mathbb{R}^n$.

property that there are finitely many P-orbits of pairs of osculating hyperplanes, and finitely many P-orbits of pairs of crossing hyperplanes.

Local finiteness is a delicate point here, and we refer to Example 7.7 for a cubulation that is not locally finite, but which arises from a seemingly reasonable wallspace with a cocompact $\mathbb{Z}^2$ action.

A *quasiline* is a locally finite CAT(0) cube complex $\widetilde{L}$ with a cocompact $\mathbb{Z}$ group action. Many quasiflats arise as products $\widetilde{F} = \prod_{i=1}^r \widetilde{L}_i$ for some r, where each $\widetilde{L}_i$ is a quasiline. It is then the case that $\mathrm{Aut}(\widetilde{F})$ acts cocompactly on $\widetilde{F}$.

The plainest example of a strong quasiflat $\widetilde{F}$ is isomorphic to the universal cover of the standard n-torus T^n for some n, that is dual to a collection of n parallelism classes of hyperplanes in $\mathbb{R}^m$ with a $P = \mathbb{Z}^m$ action. The general case is not much more complicated: The typical situation is obtained by "jiggling" these hyperplanes somewhat so that they aren't convex. See Figure 7.3. See Example 7.47 for examples of strong quasiflats arising from "overcubulated" Euclidean groups.

Consider the action of $\mathbb{Z}^n = \pi_1 T^n$ on $\widetilde{T}^n$. Let $\mathbb{Z}^m \subset \mathbb{Z}^n$ be an arbitrary rank m subgroup. Then the pair $(\widetilde{T}^n, \mathbb{Z}^m)$ is a strong quasiflat if and only if there is a $\mathbb{Z}^m$-invariant convex flat $\mathbb{E}^m \subset \widetilde{T}^n$ such that each codimension-1 hyperplane and each codimension-2 hyperplane of $\widetilde{T}^n$ intersects $\mathbb{E}^m$. Thus $\widetilde{T}^n$ is encoded within the $\mathbb{Z}^m$-cocompact wallspace $\mathbb{E}^m$, whose walls arise from intersections with hyperplanes. We revisit this in Lemma 7.42.

A diagonal action of $\mathbb{Z}$ on $\widetilde{T}^2$ shows that local finiteness does not imply finitely many P-orbits of crossing hyperplanes. In contrast, typical $\mathbb{Z}^2$ actions on $\widetilde{T}^n$ are strong quasiflats, as a $\mathbb{Z}^2$-invariant flat typically intersects non-parallel hyperplanes of $\widetilde{T}^n$ in non-parallel lines, and all the crossings thus appear within the flat.

Consider $\widetilde{T}^2$ with coordinates so that vertices are at $\mathbb{Z}^2 \subset \mathbb{R}^2$. Let $J \subset \widetilde{T}^2$ be the "jagged half-flat" subcomplex consisting of all cubes that lie entirely in the subset $\{(x, y) : x + y \geq 0\}$. The cyclic group $\langle (1, -1) \rangle$ acts freely on J but there are infinitely many orbits of crossing hyperplanes. This type of example rarely arises in practice but is often the object to be excluded in theory.

A quasiflat that is not a strong quasiflat arises in Example 7.36.

A strong quasiflat with infinitely many P–orbits of hyperplanes at distance 2 is described in Example 7.32.

Remark 7.19 (rank 2 Is Often Strong). As described above, there are examples of (rank 1 and) rank 2 quasiflats that are not strong, for instance let $\widetilde{F} = \widetilde{T}^3$ and $P = \langle (1,0,0),(0,1,0) \rangle$.

However, quasiflats are often strong in practice when $\operatorname{rank}(P) = 2$. In particular, this is the case when the quasiflat $\widetilde{F}$ arises from a locally finite $\widetilde{C}$ cubulation of a geometric wallspace $\widetilde{X}$ with quasiconvex walls. Indeed, when $\widetilde{C}$ is locally finite then by Lemma 7.23, each hyperplane of $\widetilde{F} \subset \widetilde{C}$ has stabilizer that is either commensurable with P, finite, or commensurable with a P-essential hyperplane of E. Hence by Lemma 7.31, Ball-WallNbd separation and WallNbd-WallNbd separation hold with respect to a P-cocompact subspace $\widetilde{Y}$ of $\widetilde{X}$ (see Definition 7.27). Consequently, $\widetilde{F}$ will actually be a strong quasiflat, following Theorem 7.28. We revisit this point in Lemma 7.44 where we show how to replace a cosparse cubulation by a strongly cosparse cubulation in this context.

Lemma 7.20 (Strong quasiflats are virtually special). *Let* $(\widetilde{F}, P)$ *be a strong quasiflat. Then* P *has a finite index torsion-free subgroup* P' *such that* $P' \backslash \widetilde{F}$ *is special.*

Proof. Let P_o be a finite index free-abelian subgroup. Note that all subgroups of a f.g. abelian group, and hence all double cosets of P_o are separable. The result follows directly from [HW10a, Thm 4.1]. ☐

Definition 7.21 (Sparse). Let G be hyperbolic relative to f.g. virtually abelian groups $\{P_i\}$. We say G acts [*strongly*] *cosparsely* on a $\mathrm{CAT}(0)$ cube complex $\widetilde{X}$ if there is a compact subcomplex K, and [strong] quasiflats $\widetilde{F}_i$ with $P_i = \operatorname{Stab}(\widetilde{F}_i)$ for each i, such that:

(1) $\widetilde{X} = GK \cup \bigcup_i G\widetilde{F}_i$.
(2) For each i we have $(\widetilde{F}_i \cap GK) \subset P_i K_i$ for some compact K_i.
(3) For i,j and $g \in G$, either $(\widetilde{F}_i \cap g\widetilde{F}_j) \subset GK$ or else $i = j$ and $\widetilde{F}_i = g\widetilde{F}_j$.

It follows that translates of quasiflats are either equal or are coarsely isolated in the sense that $\operatorname{diam}(g_i\widetilde{F}_i \cap g_j\widetilde{F}_j)$ is bounded by some uniform constant. See Figure 7.4 for a "picture" of a cosparse cube complex.

We will be especially interested in the case when G acts both properly and [strongly] cosparsely. In particular, when G acts freely and [strongly] cosparsely on $\widetilde{X}$, then the quotient $X = G \backslash \widetilde{X}$ is said to be [*strongly*] *sparse*. In this case, X is the finite union $K \cup \bigcup_i F_i$ where K is compact and $(F_i \cap F_j) \subset K$ for $i \neq j$,

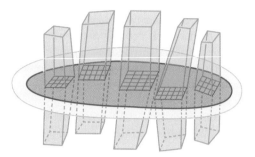

Figure 7.4. Heuristic picture of a cosparse cubulation.

and each F_i equals $P_i \backslash \widetilde{F_i}$ where P_i is a f.g. virtually abelian group acting freely on a [strong] quasiflat $\widetilde{F_i}$. We summarize this by saying that X is *sparse relative to* the subcomplexes $\{F_i\}$.

When G has torsion, the quotient X is a "[strongly] sparse orbihedron" that is still quasi-isometric to the wedge of finitely many flats and half-flats, but we shall avoid treating this situation.

The finiteness property hypothesized on the quasiflats propagates to all of $\widetilde{X}$:

Lemma 7.22. *If G acts properly and cosparsely on $\widetilde{X}$ then $\widetilde{X}$ is locally finite and has finitely many G-orbits of hyperplanes. Moreover, if G acts strongly cosparsely then there are finitely many G-orbits of pairs of crossing hyperplanes and of pairs of osculating hyperplanes.*

Proof. We first examine the claims under the cosparse hypothesis. First note that local finiteness holds for each 0-cube of $\widetilde{X} - GK$ since each $\widetilde{F_i}$ is locally finite. Each 0-cube v of GK is adjacent to finitely many $g_j \widetilde{F_i}$ since each $(\widetilde{F_i} \cap GK) \subset P_i K_i$ for some compact K_i. Hence the degree of v is bounded by the sum of its degrees in the finitely many $g_j \widetilde{F_i}$ containing it together with the degree of v in GK, which is also finite since G acts properly and cocompactly on GK. There are finitely many G-orbits of hyperplanes, since each hyperplane is either dual to a 1-cube in GK or a 1-cube of some $\widetilde{F_i}$. The former are bounded by the number of 1-cubes in K, the latter are bounded by the number of P-orbits of hyperplanes in each $\widetilde{F_i}$.

We now assume strong cosparseness. Any crossing pair of hyperplanes is either dual to a 2-cube in GK and there are finitely many of these in K, or is dual to a 2-cube in some $\widetilde{F_i}$, but there are finitely many P_i-orbits of crossing hyperplanes in $\widetilde{F_i}$ since it is a strong quasiflat. Any osculating pair of hyperplanes are either dual to adjacent 1-cubes in $\mathcal{N}_1(G(K))$ and hence the pair can be translated to be dual to adjacent 1-cubes in $\mathcal{N}_1(K)$ which is compact since $\widetilde{X}$ is locally finite, or else the osculating pair are dual to osculating hyperplanes in some $\widetilde{F_i}$, and there are finitely many P_i-orbits of such pairs, since $\widetilde{F_i}$ is a strong quasiflat. $\qquad\square$

Lemma 7.23 (Essential walls conspire). *Let P be a virtually abelian group that acts on a wallspace $(X, \mathcal{W})$. Suppose there are finitely many P-orbits of walls and that the dual CAT(0) cube complex is locally finite, and P acts properly on it.*

Then for each wall W, we have $\mathrm{Stab}(W)$ is commensurable with $\cap_{i=1}^{k} \mathrm{Stab}(W_i)$ where each W_i is P-essential.

Note that $\cap_{i=1}^{k} \mathrm{Stab}(W_i) = P$ when $k = 0$.

As exhibited in Example 7.26, it is possible that $\mathrm{Stab}(W)$ is not equal to the intersection $\cap_{i=1}^{k} \mathrm{Stab}(W_i)$ of stabilizers of P-essential hyperplanes. So the commensurability softening is necessary.

Proof. We may assume without loss of generality that P is free-abelian and acts without wall-inversions.

There is a transitive relation $(\mathcal{W}, \preceq)$ on the walls, where $W \preceq W'$ if a translate of W' separates aW, bW for some $a, b \in P$. This factors to a relation on orbits, in the sense that for all $g_1, g_2 \in G$ we have $(W_1 \preceq W_2) \Leftrightarrow (g_1 W_1 \preceq g_2 W_2)$.

We use the symbol $W \prec W'$ if $W \preceq W'$ but $W' \not\preceq W$. It follows from the transitivity of $\preceq$ that $\prec$ is also transitive. Note that $W \not\prec W$. Consequently the length of a sequence $W_1 \prec W_2 \prec \cdots \prec W_m$ is bounded by the number of P-orbits of walls.

If $W \prec W'$ then $\mathrm{Stab}(W)$ is commensurable with a subgroup of $\mathrm{Stab}(W')$. Indeed, suppose W' separates aW, bW, and consider the finitely many walls separating aW, bW. Since $\mathrm{Stab}(aW) = \mathrm{Stab}(W) = \mathrm{Stab}(bW)$ must permute these walls, a finite index subgroup of $\mathrm{Stab}(W)$ must stabilize each of them.

The proof of the lemma is executed by downward induction on $\prec$. If $W \preceq W$ then W is P-essential, and we are done with $k = 1$. We therefore assume that $W \not\preceq W$. Consider a maximal set $\{W_1, \ldots, W_\ell\}$ of walls in distinct P-orbits such that $W \prec W_i$. Let $K = \cap_{i=1}^{\ell} K_i$ where $K_i = \mathrm{Stab}(W_i)$. By induction, each K_i is commensurable with an intersection of stabilizers of finitely many P-essential walls, and so the same holds for K.

Since each $W \prec W_i$, we see that $\mathrm{Stab}(W)$ is commensurable with a subgroup of each K_i, and hence $\mathrm{Stab}(W)$ is commensurable with a subgroup H of K. If $[K : H] < \infty$ then we are done: Indeed, as $W \prec W_i$, the claim holds for each W_i by downward induction, hence each K_i is the intersection of stabilizers of P-essential walls, and so $K = \cap K_i$ is as well. Otherwise, $[K : H] = \infty$ and we will obtain contradiction with local finiteness as follows: Let $g \in K$ be an element whose image in K/H has infinite order. Consider $\{g^n W : n \in \mathbb{Z}\}$, and note that for each W_i, there is a halfspace of W_i that intersects each $g^n W$. If some wall V separates two elements of $\{g^n W : n \in \mathbb{Z}\}$, then $W \preceq V$. However, $W \prec V$ would contradict the maximality of $\{W_1, \ldots, W_\ell\}$. Thus $V \preceq W$ holds, and consequently $W \preceq W$ which is a contradiction.

We now use $\{g^n W : n \in \mathbb{Z}\}$ to exhibit a 0-cube x in the dual, such that local finiteness does not hold at x. Indeed, we can orient all walls in $\mathcal{W}$ towards $\{g^n W : n \in \mathbb{Z}\}$ as each wall has a halfspace that intersects $g^n W$ for each n. For each $m \in \mathbb{Z}$, we let x_m be the 0-cube adjacent to x which differs from x by orienting $g^m W$ away from each $g^n W$ with $n \neq m$.   □

Corollary 7.24. *Let G act properly on a locally finite CAT(0) cube complex $\widetilde{X}$. Let P be a f.g. virtually abelian group stabilizing a superconvex subcomplex $\widetilde{F}$ of $\widetilde{X}$. And suppose $\widetilde{F}^{+r}$ has finitely many orbits of hyperplanes for each $r \geq 0$.*

For each hyperplane U of $\widetilde{X}$, there exist (a possibly empty set of) P-essential hyperplanes $\{U_i\}_{i=1}^k$ of $\widetilde{F}$ such that $\cap_{i=1}^k \operatorname{Stab}_P(U_i)$ is commensurable with $\operatorname{Stab}_P(U)$.

Let $\widetilde{Y} \subset \widetilde{X}$ be a convex subcomplex, and let $H = \operatorname{Stab}(\widetilde{Y})$. Suppose the collection $\{g\widetilde{Y} : gH \in G/H\}$ is locally finite. Then there exist P-essential hyperplanes $\{U_i\}_{i=1}^k$ of $\widetilde{F}$ such that $\cap_{i=1}^k \operatorname{Stab}_P(U_i)$ is commensurable with $\operatorname{Stab}_P(\widetilde{Y})$.

Remark 7.25. The hypothesis that $\widetilde{F}^{+r}$ has finitely many P-orbits of hyperplanes for each r holds when $\widetilde{F}$ is P-cocompact. It also holds when $\widetilde{F}$ is a quasiflat and G acts cosparsely on $\widetilde{X}$ and $\widetilde{F}$ is a quasiflat stabilized by P.

The hypothesis that the collection $\{g\widetilde{Y} : gH \in G/H\}$ is locally finite holds when $\widetilde{Y}$ is H-cocompact. It also holds when $\widetilde{X}$ is the universal cover of a nonpositively curved cube complex X, and $\widetilde{Y}$ is the universal cover of a locally-convex subcomplex $Y \subset \widehat{X}$ where $\widehat{X} \to X$ is a finite cover.

Proof of Corollary 7.24. Choose r such that $U' = U \cap \widetilde{F}^{+r} \neq \emptyset$. Note that $\operatorname{Stab}_P(U') = \operatorname{Stab}_P(U)$ by Lemma 8.9 since we can assume $\widetilde{F}$ and hence $\widetilde{F}^{+r}$ is superconvex by Remark 7.10. The result now follows from Lemma 7.23.

We now show how to deduce the second statement from the first statement. Let $\widetilde{X}'$ be the CAT(0) cube complex formed by attaching a copy of $g\widetilde{Y} \times [0,1]$ along $g\widetilde{Y}$ for each gH. Note that $\widetilde{X}'$ is locally finite if $\widetilde{X}$ is and that the action of G on $\widetilde{X}$ extends to an action of G on $\widetilde{X}'$ where G acts on the added part according to the left coset representation on H. Note that a hyperplane is P-essential in $\widetilde{X}$ if and only if it extends to a P-essential hyperplane in $\widetilde{X}'$. Let U be the hyperplane at $\widetilde{Y} \times \{\frac{1}{2}\}$ within $1_G \widetilde{Y} \times [0,1]$, and observe that $\operatorname{Stab}(U) = \operatorname{Stab}(\widetilde{Y})$. The second statement now follows from the first statement applied to $\widetilde{X}, U, P$.   □

Example 7.26. We now describe a quasiflat $\widetilde{E}$ with a hyperplane $\widetilde{U}$ such that $\operatorname{Stab}(U) \not\subset \operatorname{Stab}(W)$ for any essential hyperplane W. The simplest example consists of the CAT(0) square complex obtained from the universal cover of the square complex obtained from a Klein bottle with a cylinder attached along a circle associated to a glide reflection element. The hyperplane U associated to the universal cover of the cylinder exhibits the desired behavior. Of course, the Klein bottle group does have an action (with inversions) on a CAT(0) square complex where $\operatorname{Stab}(U)$ equals the stabilizer of an essential hyperplane.

We can form a more troublesome example as follows: Let E be formed from a 3-torus $S^1 \times S^1 \times S^1$ by attaching a cylinder $S^1 \times I$ along the third factor. Thus $\widetilde{E}$ is a copy of $\widetilde{T}^3$ with a $\mathbb{Z}^2$-invariant family of attached strips. Regard $\widetilde{T}^3$ as $\mathbb{E}^2 \times \mathbb{R}$. Let $P = \langle a, b, c \rangle$ where $\langle a, b \rangle \cong \mathbb{Z}^2$ is the group acting by translations on $\mathbb{E}^2$ and acting trivially on $\mathbb{R}$, and where c translates along $\mathbb{R}$, and acts on $\mathbb{E}^2$ by a $\frac{\pi}{2}$ rotation fixing a vertex v and stabilizes the strip $e \times \mathbb{R}$ attached along $v \times \mathbb{R}$. Let U be the hyperplane dual to this strip. Then $\mathrm{Stab}(U)$ is not properly contained in the stabilizer of any essential hyperplanes. Indeed, any subgroup properly containing $\mathrm{Stab}(U)$ must be commensurable with P.

Building upon the previous example, we cubulate P with an additional two families of walls, namely those obtained from the original walls by a $\frac{\pi}{4}$ rotation of the $\mathbb{E}^2$ factor. The resulting dual CAT(0) cube complex is a 5-dimensional strongly cosparse quasiflat that is not closeable. A similar 4-dimensional example is available with $\mathrm{rank}(P) = 2$ but with torsion. One simply quotients the $\mathbb{R}$ factor.

Recall from Definition 7.2 that a wall in a geometric wallspace separates two sets if they lie in distinct open halfspaces.

Definition 7.27. Let $\{Y_i\}$ be a collection of subspaces. We say X has *WallNbd-WallNbd separation* with respect to $\{Y_i\}$ if the following holds: For each r there exists s, such that for each Y_i and osculating walls $W, W' \in \mathcal{W}$: if $\mathsf{d}\big((\mathcal{N}_r(W) \cap Y_i), (\mathcal{N}_r(W') \cap Y_i)\big) > s$ then there is a wall W'' separating $(\mathcal{N}_r(W) \cap Y_i)$ from $(\mathcal{N}_r(W') \cap Y_i)$.

We say X has *Ball-WallNbd separation* with respect to $\{Y_i\}$ if the following holds for each Y_i and each $x \in X$ and $W \in \mathcal{W}$: For each r there exists s, such that if $\mathsf{d}\big((\mathcal{N}_r(x) \cap Y_i), (\mathcal{N}_r(W) \cap Y_i)\big) > s$ then there is a wall W' separating $(\mathcal{N}_r(x) \cap Y_i)$ from $(\mathcal{N}_r(W) \cap Y_i)$.

The following strengthening of Theorem 7.9 is the primary source of cosparse actions on CAT(0) cube complexes. It is a simplified version of [HW14, Thm 8.13]:

Theorem 7.28 (Cosparse Criterion)**.** *Let G be hyperbolic relative to f.g. virtually abelian groups $\{P_i\}$. Suppose G acts properly and cocompactly on a geometric wallspace $(X, \mathcal{W})$ where each $\mathrm{Stab}(W)$ is quasi-isometrically embedded in G. Let Y_i be a P_i-cocompact nonempty subspace for each i. Suppose G acts properly on the dual cube complex $C = C(X, \mathcal{W})$.*

If $(X, \mathcal{W})$ has $[$WallNbd-WallNbd and$]$ Ball-WallNbd separation relative to each Y_i then G acts $[$strongly$]$ cosparsely on C.

Lemma 7.29 (Ball-WallNbd criterion)**.** *Let G be hyperbolic relative to f.g. virtually abelian subgroups $\{P_i\}$. Consider a proper and cocompact action of G on a wallspace $\widetilde{X}$. Suppose $\widetilde{X}$ is a geodesic metric space, and each wall W is* *the intersection of its two halfspaces so* $W = \overleftarrow{W} \cap \overrightarrow{W}$. *Suppose* $\mathrm{Stab}(\overleftarrow{W})$ *is relatively quasiconvex and acts cocompactly on* W. *Suppose that for each maximal parabolic subgroup* P_i, *and for each wall* W, *the subgroup* $\mathrm{Stab}_{P_i}(W)$ *is commensurable with* $\cap_{j=1}^{m} \mathrm{Stab}_{P_i}(W_j)$ *where each* W_j *is a* P-*essential wall. Let* $\widetilde{Y}_i$ *be a nonempty* P_i-*cocompact subspace. Then Ball-WallNbd separation holds for* P_i *acting on* $\widetilde{Y}_i$.

Proof. Consider a geodesic σ in $\widetilde{Y}_i$ from $\mathcal{N}_r(p) \cap \widetilde{Y}_i$ to $\mathcal{N}_r(W) \cap \widetilde{Y}_i$. For $1 \le j \le m$ there are finitely many P_i-translates of W_j that intersect $\mathcal{N}_r(p)$ or $\mathcal{N}_r(W) \cap \widetilde{Y}_i$, where finiteness of the former holds by local finiteness of the collection of walls, and finiteness of the latter holds because of our hypothesis that $\mathrm{Stab}_{P_i}(W)$ is commensurable with a subgroup of $\mathrm{Stab}_{P_i}(W_j)$. Let n_r denote an upperbound on the number of walls that intersect an r-ball.

By hypothesis, the P-translates of the $\{W_j \cap P_i\}_{j=1}^{m}$ cut P_i into pieces, each of which is coarsely equal to a translate of $W \cap P_i$, in the sense that it lies in the s-neighborhood of such a translate. By restricting to a finite index subgroup of P_i, at the expense of increasing s, we may consider only translates of the $\{W_j\}_{j=1}^{m}$ that are disjoint from $W \cap \widetilde{Y}_i$. There is thus an s upperbound on the length of a subpath of σ disjoint from the P_i-translates of the $\{W_j\}_{j=1}^{m}$, for otherwise we could shorten the terminal length s subpath of σ. Similarly, a terminal length ns subpath of σ would intersect at least n walls. We thus find that if $\mathsf{d}(p, W) > n_r s$ then there is a wall separating $\mathcal{N}_r(S) \cap \widetilde{Y}_i$ and $\mathcal{N}_r(W) \cap \widetilde{Y}_i$. $\square$

Lemma 7.30. *Let G act properly on a CAT(0) cube complex $\widetilde{X}$. Let $G' \subset G$ be a finite index normal subgroup. Then G acts properly on a CAT(0) cube complex $\widetilde{X}'$ such that $\mathrm{Stab}(\widetilde{U}) \subset G'$ for each hyperplane $\widetilde{U}$ of $\widetilde{X}'$.*

Specifically, for each hyperplane $\widetilde{V}$ of $\widetilde{X}$, there is a hyperplane $\widetilde{V}'$ of $\widetilde{X}'$ such that $\mathrm{Stab}_G(\widetilde{V}') = G' \cap \mathrm{Stab}_G(\widetilde{V})$, and their halfspaces coarsely contain the same G orbits. Furthermore, each hyperplane $\widetilde{V}'$ is associated in this fashion with some hyperplane $\widetilde{V}$.

Moreover, if G acts cocompactly on $\widetilde{X}$ then G acts cocompactly on $\widetilde{X}'$. And if G acts [strongly] cosparsely on $\widetilde{X}$ then G acts [strongly] cosparsely on $\widetilde{X}'$.

Proof. For each hyperplane $\widetilde{V} \subset \widetilde{X}$, let $H = \mathrm{Stab}_G(\widetilde{V})$ and let $H' = \mathrm{Stab}_{G'}(\widetilde{V})$. The wall W in $\widetilde{X}$ corresponding to $\widetilde{V}$ is the pair of closed halfspaces meeting at $\widetilde{V}$. Our new wallspace for G will have underlying set $\widetilde{X}$, and will have $m = [H : H']$ distinct walls $W_1, \ldots, W_m$ that are copies of W. These walls correspond to the left cosets of H' in H, and we let H act on them using the left coset representation. More generally, the action of G on the orbit GW lifts to an action of G on $G\{W_1, \ldots, W_m\}$ using this left coset representation. We emphasize that W_i, W_j are identical partitions, and distinguishable only through their indices.

Having done this for a representative of each G-orbit of hyperplane of $\widetilde{X}$, we obtain a new wallspace with an action by G. Let $\widetilde{X}'$ be the cube complex dual to this new wallspace.

Note that G acts properly on $\widetilde{X}'$, since letting $x \in \widetilde{X}$ be a 0-cube, we find that the number of walls separating $g_1\widetilde{x}, g_2\widetilde{x}$ is bounded below by the number of hyperplanes of $\widetilde{X}$ separating them. (In fact, there is a G-equivariant quasi-isometry between $\widetilde{X}$ and $\widetilde{X}'$.)

Observe that $\mathrm{Stab}(kW_i) = \mathrm{Stab}(kg_iH') = (kg_i)H'(kg_i)^{-1} \subset G'$ since $H' \subset G'$ and $G' \subset G$ is normal.

If G acts cocompactly on $\widetilde{X}$, then G acts cocompactly on $\widetilde{X}'$. Indeed, a maximal cube of $\widetilde{X}'$ corresponds to a maximal collection of pairwise crossing walls. This is associated to a corresponding maximal collection of pairwise crossing hyperplanes of $\widetilde{X}$. Note that for each hyperplane of the latter, there is a finite collection of identical corresponding walls in the former. The cardinality of each such collection is bounded above by $\max[\mathrm{Stab}_G(\widetilde{U}) : \mathrm{Stab}_{G'}(\widetilde{U})]$ as $\widetilde{U}$ varies. Hence the cardinality of the maximal pairwise crossing collection is bounded. Note that identical walls W_i, W_j cross since each halfspace is closed, and hence contains the associated hyperplane $\widetilde{V}$.

Now suppose that G is relatively hyperbolic and acts cosparsely on $\widetilde{X}$. We will apply the relative cocompactness conclusion of Theorem 7.9. Note that there are finitely many P-orbits of hyperplanes in each $C_\star(P)$ by construction, since there are finitely many P-orbits of walls slicing P from each of the conjugates of G' in P. For each parabolic subgroup P, the P-stabilizer of each hyperplane of $\widetilde{X}$ is commensurable with the intersection of essential P-wall stabilizers by Lemma 7.23. Ball-WallNbd separation holds for each P by Lemma 7.29. Consequently, local finiteness of the quasiflats (and in fact, all of $\widetilde{X}'$) holds by Lemma 7.6. Lemma 7.33 ensures that osculating hyperplanes $\widetilde{V}_1, \widetilde{V}_2$ of $\widetilde{X}$ have the property that $\langle \mathrm{Stab}'_P(\widetilde{V}_1), \mathrm{Stab}'_P(\widetilde{V}_2) \rangle$ whenever P' is a maximal abelian subgroup of a peripheral subgroup of the relatively hyperbolic structure of G. A pair of walls osculate in the new wallspace if and only if their associated hyperplanes osculate in the original cube complex (note that new walls associated to the same hyperplane actually cross each other since we use pairs of closed halfspaces). Hence WallNbd-WallNbd separation holds by Lemma 7.31. Thus strong cosparseness holds by Theorem 7.28. $\square$

The following revisits (and generalizes) Lemma 7.29.

Lemma 7.31. *Let G act properly and cocompactly on a geometric wallspace X. Suppose G is hyperbolic relative to virtually abelian subgroups, and that all walls have relatively quasiconvex stabilizers. Let $P \subset G$ be a maximal parabolic subgroup that is virtually abelian and let Y be a nonempty P-cocompact subspace.*

WallNbd-WallNbd separation holds with respect to Y provided that for each pair of osculating walls W_1, W_2, the subgroup $\langle \mathrm{Stab}_P(W_1), \mathrm{Stab}_P(W_2) \rangle$ is commensurable with the intersection of the stabilizers of finitely many P-essential walls.

Ball-WallNbd separation holds with respect to Y provided that each wall W has the property that $\mathrm{Stab}_P(W)$ is commensurable with the intersection of the stabilizers of finitely many P-essential walls.

Proof. It suffices to assume the wallspace is a Cayley graph $\Gamma = \Gamma(G, S)$. And we identify each wall with its corresponding coset for the coarse computations below. The computation below is simplified if we assume that P is free-abelian. However, there is no loss of generality, since passing to a finite index abelian subgroup just increases the number of orbits of walls.

Let $K_1 = \mathrm{Stab}_P(W_1)$ and $K_2 = \mathrm{Stab}_P(W_2)$. The subgroup $\langle K_1, K_2 \rangle$ is commensurable with the group $K = \cap_{j=1}^{m} \mathrm{Stab}(W_j')$ for some finite set of walls. Observe that the K cosets in P are separated by $\{pW_j : p \in P\}$ in the sense that the number of these walls separating aK, bK is bounded below by a linear function of $\mathsf{d}(aK, bK)$.

Either aK_1, bK_2 lie within nearby translates aK, bK, and hence are close to each other since $\mathsf{d}(aK_1, bK_2) \leq \mathsf{d}(aK_1, aK_2) + \mathsf{d}(aK_2, bK_2)$ where the first distance is uniformly small since K is commensurable with $\langle K_1, K_2 \rangle$, and the second distance is small since the orbits are nearby.

Or as above, the number of $\{pW_j : p \in P\}$ walls separating aK, bK is proportional to $\mathsf{d}(aK, bK)$, and these walls separate aK_1, bK_2 as well.

Finally, taking finite neighborhoods only excludes finitely many separating walls, and so WallNbd-WallNbd separation holds for G with respect to each P.

Note that relative hyperbolicity and relative quasiconvexity of the wall stabilizers ensures that there are finitely many ways that a wall can coarsely intersect $\widetilde{Y}$, and hence finitely many distinct subgroups $\mathrm{Stab}_P(gW)$ arise. This allows us to choose s depending on r independently of the walls W, W' in the definition of WallNbd-WallNbd separation.

The proof for Ball-WallNbd separation is similar, except that we let the trivial subgroup play the role of K_1.   □

Example 7.32. Consider the following wallspace with a $\mathbb{Z}^3$ action. Firstly, let $\mathbb{R}^3 = \langle a, b, c \rangle_{\mathbb{R}}$ and let $\mathbb{Z}^3 \subset \mathbb{R}^3$ be the subgroup $\langle a, b, c \rangle$. Here $\langle v, \dots \rangle_{\mathbb{R}}$ denotes $\mathrm{span}\{v, \dots\}$. Consider the hyperplanes in $\mathbb{R}^3 = \langle a, b, c \rangle_{\mathbb{R}}$ that are $\mathbb{Z}^3$-translates of the following four hyperplanes: $\{\langle a, b \rangle_{\mathbb{R}}, \langle b, c \rangle_{\mathbb{R}}, \langle ac, bc \rangle_{\mathbb{R}}, \langle a, c \rangle_{\mathbb{R}}\}$. Extend this wallspace to $\mathbb{R}^3 \times [0, 1]$ by extending each hyperplane H above to $H \times [0, 1] \subset \mathbb{R}^3 \times [0, 1]$. We attach strips $\mathbb{R} \times [0, 1]$ along $\mathbb{Z}^3$-translates of $\langle b \rangle_{\mathbb{R}} \times \{0\}$ by identifying $\mathbb{R} \times \{0\}$ with $\langle b \rangle_{\mathbb{R}} \times \{0\}$. We similarly attach strips $\mathbb{R} \times [0, 1]$ along $\mathbb{Z}^3$ translates of the lines $\langle c \rangle_{\mathbb{R}} \times \{1\}$. We add a wall at $\mathbb{R} \times \{\frac{1}{2}\}$ for the above two types of strips.

The dual to the resulting wallspace is a strong quasiflat quasi-isometric to $\mathbb{R}^4$. It has infinitely many distinct $\mathbb{Z}^3$-orbits of pairs of hyperplanes that are separated by a single hyperplane. Specifically, there are infinitely many $\mathbb{Z}^3$-orbits of pairs of strip walls corresponding to $\langle b \rangle_{\mathbb{R}}, a^n \langle ac \rangle_{\mathbb{R}}$, and they are separated by $\mathbb{R}^3 \times \{\frac{1}{2}\}$.

Lemma 7.33. *If G acts strongly cosparsely on $\widetilde{X}$ then for each parabolic subgroup P with finite index abelian subgroup P', and each pair of osculating hyperplanes U_1, U_2, the subgroup $\langle \mathrm{Stab}_{P'}(U_1), \mathrm{Stab}_{P'}(U_2) \rangle$ is commensurable with $\cap_{j=1}^{k} \mathrm{Stab}_{P'}(W_j)$ for some finite set of P-essential hyperplanes.*

Proof. Let K_1, K_2 denote the P' stabilizers of U_1, U_2. Let K be the intersection of stabilizers of P-essential hyperplanes that virtually contain $\langle K_1, K_2 \rangle$. Let $K' = K \cap \langle K_1, K_2 \rangle$.

We claim that $[K : K'] < \infty$. Suppose otherwise, and let $Z = \langle z \rangle$ be an infinite cyclic subgroup of K with $Z \cap K' = 1_G$. We show below that $\{U_1, z^n U_2\}_{n \in \mathbb{Z}}$ form an infinite set of osculating pairs of hyperplanes in distinct P-orbits, and this violates our hypothesis.

Let V be a hyperplane separating $U_1, z^n U_2$ for some $n \in \mathbb{Z}$. Observe that V cuts Z since U_2 is on the same side of V as U_1 but $z^n U_2$ is on the other side of V. Note that both $\mathrm{Stab}(U_1)$ and $\mathrm{Stab}(z^n U_2) = \mathrm{Stab}(U_2)$ are virtually in $\mathrm{Stab}(V)$. Indeed, consider the orbits $K_1 z^n U_2$ and $K_2 U_1$, and note that they are both K_1 and K_2 invariant. If V cuts gU_1 then V cuts each $g'U_1$ as they are parallel, and likewise, if V cuts $gz^n U_2$ then V cuts each $g'z^n U_2$ as they are parallel. There are finitely many hyperplanes separating $K_1 z^n U_2$ and $K_2 U_1$. These are stabilized by K_1 and K_2, and hence a finite index subgroup of $K_1 K_2$ stabilizes each hyperplane separating $K_1 z^n U_2$ and $K_2 U_1$. $\square$

Lemma 7.34 (Virtually cosparse is cosparse). *Let G be hyperbolic relative to f.g. virtually abelian subgroups $\{P_1, \ldots, P_k\}$. Suppose G has a finite index subgroup G' that acts properly and cosparsely on a CAT(0) cube complex $\widetilde{X}'$. Then G acts properly and cosparsely on a CAT(0) cube complex $\widetilde{X}$. Moreover, the stabilizers of hyperplanes of $\widetilde{X}$, are equal to conjugates of stabilizers of hyperplanes of $\widetilde{X}'$.*

Furthermore, if G' acts strongly cosparsely on $\widetilde{X}'$ then G acts strongly cosparsely on $\widetilde{X}$ provided the following additional condition holds:

For each P_i and pair of hyperplanes U_1, U_2 of $\widetilde{X}'$ and elements $g_1, g_2 \in G$: The subgroup $\langle P_i \cap g_1 \mathrm{Stab}_{G'}(U_1)g_1^{-1}, P_i \cap g_2 \mathrm{Stab}_{G'}(U_2)g_2^{-1} \rangle$ is commensurable with $P_i \cap \bigcap_j h_j \mathrm{Stab}(W_j)h_j^{-1}$ for some finite set $\{W_j\}$ of hyperplanes where each W_j is an essential hyperplane with respect to the action of $h_j^{-1} P_i h_j \cap G'$ on $\widetilde{X}'$ for some $h_j \in G$.

Remark 7.35. The final hypothesis of Lemma 7.34 (i.e., the hypothesis enabling WallNbd-WallNbd separability in Lemma 7.31) holds when for each P_i and U, the subgroup $P_i \cap g \mathrm{Stab}_{G'}(U)g^{-1}$ is either finite, or of finite index in P_i, or a codimension-1 subgroup of P_i. In particular, this holds when $\mathrm{rank}(P_i) \leq 2$ for each parabolic subgroup P_i, since by local finiteness, Lemma 7.23 provides the necessary essential wall in that case.

Consequently, in the rank 2 case, G has a finite index subgroup that acts strongly cosparsely if and only if G acts strongly cosparsely.

Proof of Lemma 7.34. This follows the proof of Lemma 7.14. Fix a Cayley graph $\Gamma = \Gamma(G, S)$. Without loss of generality, we may assume G' is normal, which simplifies some conjugation below. Each wall of G' induces a wall for G as in Remark 6.13. We choose one representative for each G'-orbit of hyperplane of $\widetilde{X}'$ to induce a wall for G, and extend this to a G-invariant collection of walls for G by translating, so each wall for G' provides $[G:G']$ walls for G.

The hyperplane stabilizers of $\widetilde{X}'$ are relatively quasiconvex in G since they are relatively quasiconvex in G'. Indeed, the cocompact convex subspace of Lemma 7.37 shows that hyperplane stabilizers are quasi-isometrically embedded, and being relatively quasiconvex is equivalent to being quasi-isometrically embedded for the case of f.g. groups that are hyperbolic relative to virtually abelian subgroups [Hru10].

Properness of the action on the dual holds just as in the hyperbolic case.

As G' and hence $(g^{-1}P_i g \cap G')$ acts properly on $\widetilde{X}'$, Lemma 7.23 ensures that the $g^{-1}P_i g$-stabilizer of each wall is commensurable with the intersection of $g^{-1}P_i g$-essential walls. Thus Ball-WallNbd holds by Lemma 7.29 (or Lemma 7.31). [The additional hypothesis implies that WallNbd-WallNbd separation holds by Lemma 7.31.]

As [WallNbd-WallNbd and] Ball-WallNbd separation hold in the wallspace on $\Gamma(G)$ with respect to each maximal parabolic subgroup P of G, the action on the dual cube complex $\widetilde{X}$ is [strongly] cosparse by Theorem 7.28. $\square$

Example 7.36. The group $G = \mathbb{Z} * \mathbb{Z}^3$ has an index 2 subgroup $G' = \mathbb{Z}^3 * \mathbb{Z} * \mathbb{Z}^3$ with a strongly cosparse cubulation, but where the construction of Lemma 7.34 does not produce a strongly cosparse cubulation for G. The idea is to use incompatible cubulations of the two $\mathbb{Z}^3$ factors of G', together with accompanying inessential $\mathbb{Z}$ walls whose cyclic subgroups do not generate the stabilizer of any hyperplane of the $\mathbb{Z}^3$ factors.

Let G be generated as $G = \langle t \rangle * \langle a, b, c \rangle$, and G' be generated by $\langle a, b, c \rangle * \langle t^2 \rangle * \langle a', b', c' \rangle$ where $a' = a^t, b' = b^t, c' = c^t$, and the inclusion $G' \subset G$ is according to these generators.

We now describe the cocompact cubulation of G', which can be described as the wedge of a circle and two 3-tori: For $\langle a, b, c \rangle$ we use essential walls with stabilizers $\langle a, b \rangle, \langle b, c \rangle, \langle c, a \rangle$ together with an inessential wall with stabilizer $\langle a \rangle$. For $\langle a', b', c' \rangle$ we use essential walls with stabilizers $\langle ab, c \rangle, \langle bc, a \rangle, \langle ab^{-1}, ac^{-1} \rangle$ together with an inessential wall with stabilizer $\langle abc \rangle$. We use a wall for t^2 with trivial stabilizer.

The construction of Lemma 7.34 leads to a cubulation of G that is locally finite by Lemma 7.29, but does not have finitely many $\mathbb{Z}^3$-orbits of osculating walls. The problem is that each pair of $\langle a \rangle, \langle abc \rangle$ walls in the cubulation of $\mathbb{Z}^3$ osculate, but there are infinitely many $\mathbb{Z}^3$ orbits of such pairs, since the corresponding pairs of lines can be chosen arbitrarily distant.

The seemingly artificial inessential walls can be upgraded to essential walls in a natural example by amalgamating with some other group (e.g., a surface group $\pi_1 S$) along the cyclic subgroup.

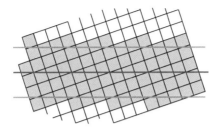

Figure 7.5. The smallest subcomplex containing a convex subspace is isometrically embedded.

7.e.1   Isometric core property

The CAT(0) cube complex $\widetilde{F}$ has the *isometric core property* with respect to the action of a group H, if each H-cocompact subcomplex $\widetilde{K} \subset \widetilde{F}$ lies in an H-cocompact isometrically embedded subcomplex $\widetilde{K}'$. Similarly, $\widetilde{F}$ has the *convex core property* if each H-cocompact subcomplex $\widetilde{K} \subset \widetilde{F}$ lies in an H-cocompact convex subspace $\widetilde{K}'$.

Finally, $\widetilde{F}$ has the *convex subcomplex core property* if each H-cocompact subcomplex $\widetilde{K} \subset \widetilde{F}$ lies in an H-cocompact convex subcomplex $\widetilde{K}'$.

The first two of these properties hold when H is a f.g. virtually abelian group acting properly on a locally finite CAT(0) cube complex $\widetilde{F}$. To see this, note that there exists an H-cocompact CAT(0) convex subspace $\widetilde{U} \subset \widetilde{F}$, and moreover, $\mathcal{N}_r(U)$ is likewise H-cocompact and CAT(0) convex and any $\widetilde{K}$ lies in some $\mathcal{N}_r(U)$. The smallest subcomplex $\widetilde{K}'$ containing $\mathcal{N}_r(U)$ is our desired isometric core. See Figure 7.5. To see this, let $a, b \in \widetilde{K}^0$, so a, b lie on cubes whose interiors contain points α, β which lie in $\mathcal{N}_r(U)$. Note that a, b lie in $N(\alpha\beta) \subset \widetilde{K}'$, but $N(\alpha\beta)$ is isometrically embedded by Corollary 7.39. It follows that $\widetilde{K}' \subset \widetilde{X}$ is isometrically embedded by Lemma 7.38. The last property holds whenever there is some H-compact convex subcomplex and $\widetilde{F}$ is locally finite. Indeed, larger H-compact convex subcomplexes are available by using $\widetilde{K}^{+r}$ as in Proposition 2.33.

Let G act properly and cocompactly on a CAT(0) cube complex $\widetilde{X}$, and let $H \subset G$ be a f.g. subgroup that is not quasi-isometrically embedded. Then $\widetilde{X}$ does not have any of these three core properties with respect to H.

Lemma 7.37. *Let G act on the CAT(0) cube complex $\widetilde{X}$ that is a union $GK \cup_{i=1}^r GF_i$ with K compact, and where $(g_i\widetilde{F}_i \cap g_j\widetilde{F}_j) \subset GK$ unless $g_i\widetilde{F}_i = g_j\widetilde{F}_j$, and each $\widetilde{F}_i \cap GK$ is P_i-cocompact. If each $\widetilde{F}_i$ has the isometric core property with respect to its stabilizer, then $\widetilde{X}$ has the isometric core property. If each $\widetilde{F}_i$ has the convex core property with respect to its stabilizer, then $\widetilde{X}$ has the convex core property.*

The proof that $\widetilde{X}$ has the convex core property is given in [HW14, Thm 9.3]. We focus on the isometric core property and the convex subcomplex core property.

Proof of the isometric core property. Let $\widetilde{B}$ be a G-cocompact subcomplex of $\widetilde{X}$, and assume without loss of generality that $GK \subset \widetilde{B}$. Let $\widetilde{B}_i = \widetilde{B} \cap \widetilde{F}_i$, and observe that $\widetilde{B}_i$ is P_i-cocompact. By the isometric core property of $\widetilde{F}_i$, the subcomplex $\widetilde{B}_i$ extends to a $\mathrm{Stab}(\widetilde{F}_i)$-cocompact subcomplex $\widetilde{B}'_i \subset \widetilde{F}_i$.

Let $\widetilde{X}^\square = \widetilde{B} \cup_i G\widetilde{B}'_i$. We now show that $\widetilde{X}^\square \subset \widetilde{X}$ is isometrically embedded (though perhaps not convex). Indeed, for any geodesic γ in $\widetilde{X}$ joining points in $\widetilde{X}^\square$, consider a subpath γ_i where γ departs $\widetilde{X}^\square$ and travels within $g\widetilde{F}_i$. Then γ_i has the same length as a path γ'_i in $g\widetilde{B}_i$ with the same endpoints. Successively replacing γ_i by γ'_i, we pull γ' to a path in $\widetilde{X}^\square$ with the same endpoints such that $|\gamma'| = |\gamma|$.   □

Proof of the convex subcomplex core property. By the convex subcomplex property, for each i, let $\widetilde{D}_i \subset \widetilde{F}_i$ be a convex P_i-cocompact subcomplex that contains all cubes intersecting $\widetilde{F}_i \cap GK$.

Let $\widetilde{Y} = GK \cup \bigcup_i G\widetilde{D}_i$, and note that $\widetilde{Y}$ is cocompact since each $\widetilde{D}_i$ is P_i-cocompact. We now show $\widetilde{Y} \subset \widetilde{X}$ is convex. Let e, e' be edges meeting at a vertex v of a square s of $\widetilde{X}$. Suppose e and e' lie in $\widetilde{Y}$. If $s \subset GK$, then $s \subset \widetilde{Y}$. Otherwise, $s \subset g\widetilde{F}_i$ in a unique $g\widetilde{F}_i$. If e and e' both lie in some $g\widetilde{D}$, then $s \subset \widetilde{D}$ by convexity of $g\widetilde{D} \subset g\widetilde{F}$. Suppose $e \subset g\widetilde{D}$ but $e' \subset g'\widetilde{D}'$ with $g\widetilde{D} \neq g'\widetilde{D}'$. Then $v \in g\widetilde{D} \cap g'\widetilde{D}' \subset GK$. Consequently $s \subset g\widetilde{D}_i \subset g\widetilde{F}_i$, since we assumed above that each $\widetilde{D}_i \subset \widetilde{F}_i$ was chosen so that it contains each cube of $\widetilde{F}_i$ that intersects GK.   □

Lemma 7.38. *A connected subcomplex $\widetilde{Y} \subset \widetilde{X}$ of a CAT(0) cube complex is isometrically embedded in the combinatorial metric if and only if its intersection $N(V) \cap \widetilde{Y}$ with each hyperplane carrier is isometrically embedded (and thus connected).*

Proof. If $\widetilde{Y}$ is isometrically embedded, then each $N(V) \cap \widetilde{Y}$ is isometrically embedded since $N(V)$ is convex by Lemma 2.14.(2).

Consider a geodesic $\gamma \to \widetilde{X}$ whose endpoints p, q lie on $\widetilde{Y}$ but such that $\mathrm{d}_{\widetilde{Y}}(p, q) > |\gamma|$. Consider a disk diagram $D \to \widetilde{X}$ between γ and a geodesic $\sigma \to \widetilde{Y}$. If each dual curve that starts on σ ends on γ, then $|\sigma| \leq |\gamma|$ so σ is also a geodesic. Therefore, there is a dual curve α in D that starts and ends on edges e_1, e_2 where $e_1 \sigma' e_2$ is a subpath of σ. Let p', q' be the endpoints of $e_1 \sigma' e_2$. Then p', q' lie in $N(V) \cap \widetilde{Y}$ where V is the hyperplane carrying σ. However, $e_1 \sigma' e_2$ is not a geodesic in $\widetilde{X}$, and this contradicts that $N(V) \cap \widetilde{Y}$ is isometrically embedded.   □

Corollary 7.39. *Let $\widetilde{X}$ be a CAT(0) cube complex. Let $\gamma \subset \widetilde{X}$ be a geodesic in the CAT(0) metric. Let $N(\gamma)$ be the subcomplex consisting of all closed cubes whose interiors intersect γ. Then $N(\gamma)$ is isometrically embedded in the combinatorial metric.*

Proof. We give the proof in the finite dimensional case (note that this also handles the case when γ is finite, since then $N(\gamma)$ is compact). We will prove that $N(U) \cap N(\gamma)$ is isometrically embedded for each hyperplane U, and so $N(\gamma)$ is isometrically embedded by Lemma 7.38. The proof is by induction on $\dim(\widetilde{X})$. The statement is trivial when $\dim(\widetilde{X}) = 0$. Suppose it holds for $\dim \leq d$. Consider $N(U) \cap N(\gamma)$. If it lies in a frontier of U consisting of a component U' of $N(U) - N^o(U)$, then $\gamma \cap N(U)$ lies in U', and so $N(U) \cap N(\gamma) = N(U' \cap \gamma)$, which is isometrically embedded in U' by induction on dimension. Otherwise, $N(U) \cap N(\gamma)$ equals $N_U(\bar{\gamma}) \times [0,1]$ where $\bar{\gamma}$ is the projection of $N(U) \cap \gamma$ onto a geodesic $\bar{\gamma} \subset U$, and $N_U(\bar{\gamma})$ is its carrier in U, so $N_U(\bar{\gamma}) \subset U$ is isometrically embedded by induction, and so $N_U(\bar{\gamma}) \times [0,1] \subset N(U)$ is isometrically embedded. $\qquad\square$

An *isocore* $X^\square \subset X$ is a subcomplex such that $\pi_1 X^\square \to \pi_1 X$ is an isomorphism and $\widetilde{X^\square} \to \widetilde{X}$ is an isometric embedding. Note that Lemma 7.37 provides a compact isocore $X^\square \subset X$ when X is cosparse, but we emphasize that the subcomplex $X^\square$ is not necessarily nonpositively curved.

Remark 7.40. Given a local-isometry $A \to X$, and an isocore $X^\square$ of X, there is an induced isocore $A^\square$ of A. Indeed, the intersection of a convex subcomplex and an isometrically embedded subcomplex is isometrically embedded. Hence we can let $A^\square$ be the quotient of $\widetilde{A} \cap \widetilde{X^\square}$ by the action of $\pi_1 A$.

7.f   Useful Subwallspaces

This subsection focuses on cocompact subwallspaces that can be useful as geometric wallspaces. Lemma 7.42 explains that strongly cosparse cube complexes are dual to cocompact geometric wallspaces. Lemma 7.44 explains how to get from a cosparse cubulation to a strongly cosparse cubulation when the peripheral subgroups are small.

Lemma 7.41. *Let $f : (S, W) \to (S', W')$ be a map between wallspaces, with $f : S \to S'$ and $f : W \to W'$ injective, and such that the preimages of the halfspaces of $f(W)$ are the halfspaces of W for each $W \in W$. Suppose that for $W' \in W' - f(W)$, there do not exist walls $W_1, W_2 \in W$ such that W' separates $f(W_1), f(W_2)$. (More formally, we do not allow $f(W_1) \neq W \neq f(W_2)$ where distinct halfspaces of W contain halfspaces of $f(W_1)$ and $f(W_2)$.) Suppose there does not exist $W' \in W' - f(W)$ with the property that W' crosses each element of $f(W)$.*

Then there is an isometric embedding $C(S, W) \subset C(S', W')$. Moreover it embeds as a convex subcomplex provided that walls of W cross (if and) only if their images cross.

Furthermore, if G acts on W' and preserves W, then the inclusion is G-equivariant.

Proof. We map a 0-cube $v \in C(\mathcal{W})$ to the 0-cube $f(v) \in C(\mathcal{W}')$ by orienting each $W \in f(\mathcal{W})$ as it is oriented by v, and orienting each $W \notin f(\mathcal{W})$ towards all the walls of $f(\mathcal{W})$. Equivariance holds since a halfspace containing some wall in $f(\mathcal{W})$ is translated to a halfspace with the same property. The map is an isometric embedding since the distance between 0-cubes is precisely the number of walls whose orientation they disagree upon. Convexity holds by Lemma 2.11.   □

Lemma 7.42 (Cocompact subwallspace). *Let G act properly and strongly cosparsely on a CAT(0) cube complex $\widetilde{X}$. Then there is a convex G-cocompact subspace $S \subset \widetilde{X}$ such that $\widetilde{X}$ is isomorphic to the cube complex dual to $(S, \mathcal{W})$ where $\mathcal{W}$ is the set of walls arising from the intersection with hyperplanes of $\widetilde{X}$.*

In particular, if $(\widetilde{F}_i, P_i)$ is a strong quasiflat then there is a convex P_i-cocompact subspace $S_i \subset \widetilde{F}_i$ such that $\widetilde{F}_i$ is isomorphic to the cube complex dual to $(S_i, \mathcal{W}_i)$ where $\mathcal{W}_i$ is the set of walls arising from the intersection with hyperplanes of $\widetilde{F}_i$.

Proof. For each i, let $\mathbb{E}_i$ be a P_i-invariant P_i-cocompact flat in the quasiflat $\widetilde{F}_i$. Since $\widetilde{F}_i$ has finitely many P_i-orbits of hyperplanes and finitely many P_i-orbits of pairs of crossing hyperplanes, we can choose r such that $S_i = \mathcal{N}_r(\mathbb{E}_i)$ has the property that each hyperplane of $\widetilde{F}_i$ intersects S_i, and each codimension-2 hyperplane of $\widetilde{F}_i$ intersects S_i. Moreover, we can arrange that distinct hyperplanes have distinct intersections. We apply Lemma 7.37 to choose the convex G-cocompact subspace S such that $S_i \subset S$ for each i. The dual of $(S, \mathcal{W})$ is isomorphic to $\widetilde{F}$ by Lemma 7.41.   □

Elementary peripheral subgroups can be eliminated from a relatively hyperbolic structure, and likewise, the following explains how to avoid rank 1 quasiflats. It is employed in the proof of the cosparse case of Theorem 15.1.

Lemma 7.43 (Omitting rank 1 quasiflats). *Let G act cosparsely on a CAT(0) cube complex $\widetilde{X}$. Then G acts cosparsely on a CAT(0) cube complex $\widetilde{Y}$, where the relatively hyperbolic structure omits the peripheral subgroups of rank < 2, and where each quasiflat of $\widetilde{Y}$ is equivariantly isomorphic to a quasiflat of $\widetilde{Y}$.*

Proof. We will use the following claim: Let $(\widetilde{F}, P)$ be a quasiflat with P virtually cyclic. Let J be a compact subcomplex of $\widetilde{F}$. There exists a P-cocompact CAT(0) complex $\widetilde{E} \subset \widetilde{F}$ with $PJ \subset \widetilde{E}$.

We now prove the claim. Let $\widetilde{R} \subset \widetilde{F}$ be a P-cocompact line that embeds as a convex subspace, and note that $\widetilde{R}$ exists by the Flat Torus Theorem. Choose $r > 0$ so that $PJ \subset S = \mathcal{N}_r(R)$ and such that each hyperplane of $\widetilde{F}$ intersects S. Let

$(S, \mathcal{W})$ be the wallspace whose walls arise from intersections with hyperplanes of $\widetilde{F}$. Let $\widetilde{E}$ be the CAT(0) cube complex dual to $(S, \mathcal{W})$, and note that $\widetilde{E}$ is P-cocompact by Proposition 7.4. Observe that there is a P-invariant embedding $\widetilde{E} \subset \widetilde{F}$ by Lemma 7.41, and moreover, $PK \subset \widetilde{E}$, since each 0-cube of PK is a canonical cube of the dual.

We now prove the lemma. Let $\widetilde{X} = GK \cup \bigcup_i G\widetilde{F}_i \cup \bigcup_j G\widetilde{F}'_j$ where K is the compact subcomplex of $\widetilde{X}$ of Definition 7.21, and each $\widetilde{F}_i$ is associated to a peripheral subgroup of rank ≥ 2, and each $\widetilde{F}'_j$ is associated to a peripheral subgroup of rank < 2. For each j, let $K_j \subset \widetilde{F}'_j$ be a compact subcomplex so that $(GK \cap \widetilde{F}'_j) \subset P_j K_j$, as in Definition 7.21.(2). By the claim, there exists a CAT(0) subcomplex $\widetilde{E}_j \subset \widetilde{F}'_j$ so that $P_j K_j \subset \widetilde{E}_j$. Let $\widetilde{E} = GK \cup \bigcup_i G\widetilde{F}_i \cup \bigcup_j G\widetilde{E}_j$. Note that $\widetilde{E}$ is nonpositively curved, since for each $v \in \widetilde{E}^0$, either v has a neighborhood entirely within some $\widetilde{F}_i$ or $\widetilde{E}_j$, or else $v \in GK$ in which case its neighborhood is the same as its neighborhood in $\widetilde{X}$. Hence $\widetilde{E}$ is CAT(0) since it is simply-connected. $\qquad\square$

Lemma 7.44. *Let G be hyperbolic relative to virtually abelian subgroups $\{P_i\}$. Suppose G acts cosparsely on $\widetilde{X}$. If each $\mathrm{rank}(P_i) \leq 2$ then G acts strongly cosparsely on a CAT(0) cube complex $\widetilde{C}$.*

Proof. Let $\widetilde{S} \subset \widetilde{X}$ be a G-cocompact convex subspace provided by Lemma 7.37. Note that we can choose $\widetilde{S}$ so that each hyperplane of $\widetilde{X}$ intersects $\widetilde{S}$, but it is possible that there are crossing hyperplanes of $\widetilde{X}$ whose intersections with $\widetilde{S}$ do not cross. Let $\widetilde{C}$ be the CAT(0) cube complex dual to this induced wallspace on $\widetilde{S}$, and note that $\widetilde{S}$ is a cocompact geometric wallspace. Note that G acts properly by Lemma 5.42 since each nontrivial g has an axis in $\widetilde{S}$, and that axis is cut by the same hyperplanes cutting any combinatorial axis of g in $\widetilde{X}$, and the number of $\langle g \rangle$-orbits of such hyperplanes is the period of the combinatorial axis.

Observe that S has Ball-WallNbd separation with respect to P_i-invariant flats by Lemma 7.23 and Lemma 7.29 and has WallNbd-WallNbd separation by Lemma 7.31 which applies by Remark 7.35, since peripheral subgroups have rank ≤ 2. Finally G acts strongly cosparsely on $\widetilde{C}$ by Theorem 7.28. $\qquad\square$

Note that recubulating using the geometric wallspace can also be used to repair individual low rank quasiflats that fail to be strong.

7.f.1   Virtually abelian extensions

The goal of this subsection is to prove a result about extending codimension-1 subgroups in virtually abelian groups. This is obtained in Lemma 7.45 and Remark 7.46.

Lemma 7.45. *Let P be a [torsion-free] f.g. virtually abelian group. Then P is a subgroup of a [torsion-free] virtually $\mathbb{Z}^n$ group P' that acts freely and cocompactly on the standard n-dimensional cube complex.*

Proof. Let $m = \mathrm{rank}(P)$. Then P acts properly and cocompactly on $\mathbb{E}^m$ as proven by Zassenhaus [Rat94, Thm 7.4.5]. Choose finitely many isometric copies $\{W_j\}_{j \in J}$ of $\mathbb{E}^{m-1} \subset \mathbb{E}^m$ that have cocompact stabilizers. Suppose moreover that each bi-infinite geodesic in $\mathbb{E}^m$ is transverse to at least one $\{W_j\}$. Translating by the action of P, we obtain a wallspace with underlying set $\mathbb{E}^m$ and whose walls are of the form $\{pW_j : p \in P, j \in J\}$. The wallspace has n parallelism classes of walls for some $n \geq m$. Any two isometrically embedded copies of $\mathbb{E}^{m-1}$ in $\mathbb{E}^m$ are either parallel or cross. Consequently, applying Sageev's construction of Section 7.b, the dual cube complex is the standard cubulation $\widetilde{X}$ of $\mathbb{E}^n$, and moreover, P acts properly on $\widetilde{X}$ by Lemma 5.42.

Observe that $\mathrm{Aut}(\widetilde{X})$ acts properly and cocompactly on $\widetilde{X}$. We have thus shown that an arbitrary virtually $\mathbb{Z}^m$ group P has a finite kernel quotient that is a subgroup of the group $\mathrm{Aut}(\widetilde{X})$ acting properly and cocompactly on the standard cubulation of $\mathbb{E}^n$ for some n. In particular, as P is torsion-free it embeds in $\mathrm{Aut}(\widetilde{X})$.

It was proven in [BL77] that each maximal torsion-free subgroup of a polycyclic-by-finite group is of finite index. We complete the proof by letting P' be a maximal torsion-free subgroup of $\mathrm{Aut}(\widetilde{X})$ that contains P.   □

Remark 7.46 (Virtually $\mathbb{Z}^m$ Extension Property). The construction in the proof of Lemma 7.45 shows the following: Let P be a f.g. [torsion-free] virtually abelian group. Let $\{H_j\}_{j \in J}$ be a finite collection of codimension-1 subgroups of P, with associated H_j-walls. And suppose P acts properly on the dual CAT(0) cube complex. There exists a f.g. [torsion-free] virtually abelian group P', and codimension-1 subgroups $\{H'_j\}_{j \in J}$ such that each H_j-wall extends to an H'_j-wall in P', and moreover P' acts properly and cocompactly on the associated dual CAT(0) cube complex.

Indeed, for each j, we let $W_j \subset \mathbb{E}^m$ be an H_j-cocompact isometric copy of $\mathbb{E}^{m-1}$. Then, applying the construction to $\{W_j\}_{j \in J}$, we obtain a proper cocompact action of P' on $\widetilde{X}$ that extends the action of P on $\widetilde{X}$. Each H_j stabilizes a hyperplane U_j of $\widetilde{X}$. We then use separability of $H_j \subset P'$ to separate H_j from its cosets in $H'_j \cap P$ to choose a finite index subgroup $H'_j \subset \mathrm{Stab}_{P'}(U_j)$ with $H'_j \cap P = H_j$.

In an earlier version of this text, I had asked whether Lemma 7.45 can be strengthened to choose P' so that $P' \cong P \times \mathbb{Z}^d$ for some d. This was subsequently answered negatively by Hagen in [Hag14] who used a simplicial structure on the sphere at infinity associated to Example 7.47.

Example 7.47 (Overcubulated Euclidean groups). Allowing torsion, the $(3, 3, 3)$ and $(2, 3, 6)$ triangle groups are virtually $\mathbb{Z}^2$ but do not act on the standard cubulation of $\mathbb{E}^2$. On the other hand, the $(2, 4, 4)$ triangle group does act properly on the standard cubulation of $\mathbb{E}^2$.

I am grateful to Bill Dunbar for explaining the following torsion-free example to me: Let $G = \mathbb{Z}^2 \rtimes_{\phi} \mathbb{Z}$ be the group where $\phi : \mathbb{Z}^2 \to \mathbb{Z}^2$ is an order-6 automorphism. So $G \cong \langle a, b, t \mid [a, b], a^t = b, b^t = a^{-1}b \rangle$. Then G is virtually $\mathbb{Z}^3$, but G does not act freely on the standard cubulation of $\mathbb{E}^3$. Indeed, in an action of G on $\mathbb{E}^3$, the element t acts by a screw-motion rotating by $\frac{\pi}{3}$ about its axis. Now, G acts by automorphisms on the sphere at ∞, and $\langle t \rangle / \langle\!\langle t^6 \rangle\!\rangle$ would act faithfully there, but this action also gives rise to an action on a 3-cube which does not admit an order-6 orientation-preserving automorphism.

Consequently G cannot act freely and cocompactly on any CAT(0) cube complex $\widetilde{X}$, for then by Lemma 7.48, G would act freely and cocompactly on the standard cubulation of $\mathbb{E}^3$ which is impossible.

Lemma 7.48. *Suppose that G is virtually $\mathbb{Z}^n$ and suppose that G acts properly and cocompactly on a CAT(0) cube complex $\widetilde{X}$. Then G acts properly and cocompactly on the standard cubulation of $\mathbb{E}^n$.*

Note that $\widetilde{X}$ might not contain a convex subcomplex isomorphic to $\widetilde{T}^n$. For instance, let $\widetilde{X}$ denote the cartesian product of three copies of the cube complex in Figure 7.3 with the associated $\mathbb{Z}^3$ action.

Proof. The flat torus theorem of [BH99] shows that $\widetilde{X}$ contains a G-invariant isometric copy F of $\mathbb{E}^n$. There is a wallspace structure on F arising from the intersections of hyperplanes $\widetilde{H} \subset \widetilde{X}$ with F. We henceforth ignore those hyperplanes $\widetilde{H}$ with the property that either $\widetilde{H} \cap F = F$ or $\widetilde{H} \cap F = \emptyset$. Each intersection $\widetilde{H} \cap F$ is an isometric copy of $\mathbb{E}^{n-1}$ and if several hyperplanes of $\widetilde{X}$ intersect F in the same wall then we simply identify these from the viewpoint of the wallspace, so a single wall of F could correspond to several hyperplanes of $\widetilde{X}$. We emphasize that crossing hyperplanes of $\widetilde{X}$ might intersect F in parallel hyperplanes.

Suppose there are exactly m parallelism classes of walls in F, and observe that $m \leq n$ for otherwise $\widetilde{X}$ would coarsely contain a copy of $\mathbb{E}^m$ which is impossible since $\widetilde{X}$ is quasi-isometric to $\mathbb{Z}^n$. Applying Sageev's construction of Section 7.b, we see that G acts on the cube complex $\widetilde{T}^m$ dual to the wallspace of F. Note that $\widetilde{T}^m$ is the standard cubulation of $\mathbb{E}^m$ since there are m-parallelism classes of walls. The action is proper by Proposition 7.8. The dimension $m \geq n$ since $\mathbb{Z}^n \subset G$ cannot act freely on $\widetilde{T}^m$ with $m < n$. Finally G acts cocompactly since $\mathbb{Z}^n$ acts cocompactly on $\widetilde{T}^n$ as it is of finite index in $\mathrm{Aut}(\widetilde{T}^n)$. $\qquad\square$

7.f.2 Closing up quasiflats

The goal of this subsection is to present Theorem 7.54 which is sometimes helpful for providing a shortcut via a cocompact interpolation. In particular, it can be used to simplify the proofs of Theorems 14.2 and 15.10.

Definition 7.49 (Closeable Quasiflat). Let $(\widetilde{F}, P)$ be a quasiflat with a proper action by a f.g. virtually abelian group P. We say it is *closeable* if there is a quasiflat $(\widetilde{F}', P')$ such that P' acts properly and cocompactly on $\widetilde{F}'$, and such that there is an inclusion $P \hookrightarrow P'$, and an equivariant convex embedding $\widetilde{F} \to \widetilde{F}'$.
Note that if $\mathrm{Aut}(\widetilde{F})$ acts cocompactly on $\widetilde{F}$ then $\widetilde{F}$ is closeable.

There are several reasons for expecting quasiflats to be closeable. We refer to Remark 7.50, Lemma 7.51, Remark 7.52 and Lemma 7.53.

Remark 7.50. When $\widetilde{Y}$ is a cosparse cube complex that arises from applying Proposition 2.31 to a relatively hyperbolic subgroup H of a group G that is hyperbolic relative to virtually abelian subgroups and acts properly and cocompactly on a CAT(0) cube complex, then we can assume each quasiflat of $\widetilde{Y}$ is closeable.

Lemma 7.51 (Virtually special implies closeable). *Let Y be sparse. If Y is virtually special then each quasiflat $\widetilde{F}$ of $\widetilde{Y}$ is closeable.*
Consequently, strong quasiflats are closeable.

A converse to Lemma 7.51 in the presence of embedded hyperplanes is stated in Corollary 15.12.

Proof. Let $\widehat{Y}$ be a finite special cover of Y, and note that $\widehat{Y}$ has finitely many hyperplanes. Let $\widehat{Y} \to R$ be the local-isometry to a compact Salvetti complex provided by Proposition 6.2. Let $P = \pi_1 \widehat{Y} \cap \mathrm{Stab}(\widetilde{F})$. Let M be a maximal abelian subgroup of $\pi_1 R$ that contains P. By Lemma 2.38, there is a compact nonpositively curved cube complex E and a local-isometry $E \to R$ such that $\pi_1 E$ maps to M. As explained below, we can assume E is thick enough, so that choosing the appropriate translate of $\widetilde{E} \subset \widetilde{R}$, we have $\widetilde{F} \subset \widetilde{E}$, and have thus shown that $\widetilde{F}$ is a convex subcomplex of an M-cocompact subcomplex. To see that we may assume that $\widetilde{F} \subset \widetilde{E}$ when $P \subset M$, observe that by possibly replacing $\widetilde{E}$ by its cubical thickening $\widetilde{E}^{+r}$ for sufficiently large r (see Proposition 2.33), we can ensure that no hyperplane separates $\widetilde{F}, \widetilde{E}$, and each hyperplane of $\widetilde{F}$ intersects $\widetilde{E}$ (there are finitely many P-orbits of such hyperplanes). By Lemma 2.19, both $\widetilde{F}$ and $\widetilde{E}$ are the intersections of the minor halfspaces containing them. However, each minor halfspace containing $\widetilde{E}$ must contain $\widetilde{F}$, since its hyperplane cannot separate $\widetilde{E}, \widetilde{F}$, and its hyperplane cannot intersect $\widetilde{F}$ since it doesn't intersect $\widetilde{E}$.
  Combining the first statement with Lemma 7.20 yields the second statement. $\qquad\square$

A nonempty CAT(0) cube complex $\widetilde{M}$ is *minimal* with respect to an action by a group G if $\widetilde{M}$ has no proper nonempty G-invariant convex subcomplex.

Remark 7.52 (Quasiflats Often Have Cocompact Minimum). Quasiflats typically arise as $C_\star(Ps)$ where P is a parabolic subgroup of the relatively hyperbolic group G acting on a wallspace $(S, \mathcal{W})$ and $s \in S$ is some basepoint.

Let $\widetilde{M}$ be the P-minimal subcomplex of $C_\star(Ps)$. We show that $\widetilde{M}$ has a proper cocompact automorphism group in this case. The hyperplanes of $\widetilde{M}$ correspond to walls that essentially cut Ps. There are finitely many commensurability classes of (essential) wall stabilizers. Since essential walls with non-commensurable stabilizers must cross each other, we see that $\widetilde{M} = \prod_{i=1}^d \widetilde{L}_i$ is a product, each factor $\widetilde{L}_i$ corresponds to a commensurability class, and is dual to the wallspace consisting of the walls whose stabilizers are commensurable with that class.

We now show that each $\widetilde{L}_i$ is actually a quasiline. By [HW14, Lem 8.11], there is a uniform constant μ such that for two walls that cross Ps, if they cross each other then their μ-neighborhoods both contain a common element of Ps. Thus the local finiteness of the system of walls (and P-cocompactness of Ps) ensures there are finitely many P-orbits of crossing walls in $\widetilde{L}_i$. Moreover, there are finitely many orbits of maximal cubes in a quasiline $\widetilde{L}_i$, since each such maximal cube corresponds to a collection of walls whose neighborhoods intersect Ps in translates of parallel pairwise uniformly close cosets of our commensurability class in P. Finally, let $p \in P$ be an element not having a power in these (commensurable) codimension-1 subgroups, then $\langle p \rangle$ acts cocompactly on $\widetilde{L}_i$. Thus $\prod \mathrm{Aut}(\widetilde{L}_i)$ acts properly and cocompactly on $\prod \widetilde{L}_i = \widetilde{M}$.

The action of P on $\prod \widetilde{L}_i$ is factor preserving since P acts on its Euclidean flat by translations. Consequently $P \hookrightarrow \prod \mathrm{Aut}(\widetilde{L}_i)$. We now show that P extends to a free-abelian subgroup $P' \subset \prod \mathrm{Aut}(\widetilde{L}_i)$ that acts properly and cocompactly on $\prod \widetilde{L}_i$. Let P_i be the projection of P onto $\mathrm{Aut}(\widetilde{L}_i)$. Note that P_i is a f.g. abelian group that acts cocompactly on $\widetilde{L}_i$. We thus have an embedding $P \subset \prod P_i$. Let $R_o = P$. Starting with $m = 0$, for each $m \leq d$, if R_m contains an element r_{m+1} that acts hyperbolically on the $\widetilde{L}_i$ factor and acts elliptically on all other factors, then we let $R_{m+1} = R_m$. Otherwise, let $r_{m+1} \in \prod P_i$ be an element with this property, and let $R_{m+1} = \langle R_m, r_{m+1} \rangle$. Then R_{m+1} acts freely on $\prod \widetilde{L}_i$ and contains a subgroup whose projection acts properly and cocompactly on $\prod_{i=1}^{m+1} \widetilde{L}_i$. The result follows with $P' = R_d$.

Remark 7.52 shows that $C_\star(Ps)$ has cocompact minimum, so if P is abelian then $(C_\star(Ps), P)$ is closeable by Lemma 7.53. We caution that Lemma 7.53 is not useful without local finiteness of the entire dual, and this can fail as in Example 7.7.

Lemma 7.53 (Closeability from cocompact minimum). *Let $(\widetilde{E}, P)$ be a quasiflat with P free-abelian. Let $\widetilde{M} \subset \widetilde{E}$ be a minimal P-invariant subcomplex, and suppose P extends to a free-abelian group P' that acts properly and cocompactly on $\widetilde{M}$. Then $(\widetilde{E}, P)$ is closeable to a quasiflat $(\widetilde{E}', P')$.*

We use that P is free-abelian at one point of the construction to ensure that each codimension-1 subgroup H of P extends to a codimension-1 subgroup H' of P'. We refer to Example 7.26.

Proof. Let U be a hyperplane of $\widetilde{E}$ and let $H = \mathrm{Stab}_P(U)$. When U is a P-essential hyperplane, we define $H' \subset P'$ by $H' = \mathrm{Stab}_{P'}(U)$. More generally, by Lemma 7.23, H is commensurable with the intersection $\bigcap_{j \in J} H_j$ of the stabilizers of finitely many P-essential walls. We let $\{H'_j\}_{j \in J}$ be the corresponding subgroups of P' declared above, so $P \cap \bigcap_{j \in J} H'_j$ is commensurable with H.

We then let H' equal a subgroup of P' such that H' is commensurable with $\bigcap_{j \in J} H'_j$ and such that $H' \cap P = H$. It is for this last property that we use that $P \subset P'$ is free-abelian, as we can let $K = \bigcap_{j \in J} H'_j$ and then let $K' = HK$ which is commensurable with K (since HK is a subgroup and K contains a finite index subgroup of H), and then apply separability to let $H' \subset K'$ be the desired finite index subgroup containing H but not containing any nontrivial coset of H in $H' \cap P$, and so we can ensure that $H' \cap P = H$. Note that when H is commensurable with P, we have $\bigcap_j H_j = P$, as the intersection is vacuous.

We make our choices P-equivariantly, so that for each of the finitely many representatives of P-orbits of hyperplanes U of $\widetilde{X}$, we choose H' associated to $H = \mathrm{Stab}_P(U)$ as above, and thus pU has stabilizer pHp^{-1} which extends to $pH'p^{-1}$ in H'.

The action of P on $\widetilde{E}$ extends to an action of P' on $\sqcup_{p' \in P'/P} \, p'\widetilde{E}$. Indeed, choose a basepoint v in a minimal P-invariant subcomplex of $\widetilde{E}$, and consider its orbit $Pv \subset \widetilde{E}$. Regard Pv as being the vertices of a Cayley graph $\Gamma(P)$, and extend this to a Cayley graph $\Gamma(P')$, so that the translates of Pv are in correspondence with the cosets of P in P'.

For each hyperplane U of $\widetilde{E}$ with $H = \mathrm{Stab}_P(U)$, let H' be the extension of H, and note that $H' = \sqcup_{p' \in H'/H} \, p'H$. Accordingly, H' stabilizes the wall $W = H'U$ and we define $\overrightarrow{W} = H'\overrightarrow{U}$, and $\overleftarrow{W} = H'\overleftarrow{U}$. Let $\mathcal{W}$ denote the set of all such walls with its associated P' action.

Let $\widetilde{E}'$ be the CAT(0) cube complex dual to the wallspace $(\sqcup_{p'} \, p'\widetilde{E}, \mathcal{W})$. We will apply Lemma 7.41 to see that there is an equivariant embedding $\widetilde{E} \subset \widetilde{E}'$ as a convex subcomplex. We must thus verify that W_1, W_2 cross if and only if U_1, U_2 cross.

Consider two halfspaces with nonempty intersection: $\overrightarrow{W}_1 \cap \overrightarrow{W}_2 \neq \emptyset$, so $h'_1 \overrightarrow{U}_1 \cap h'_2 \overrightarrow{U}_2 \neq \emptyset$ for some $h'_1 \in H'_1$ and $h'_2 \in H'_2$. Note that $p'\overrightarrow{U}_1 \cap p''\overrightarrow{U}_2 = \emptyset$ when p', p''

lie in distinct cosets of P. Hence $h_1' = p'h_1$ and $h_2' = p'h_2$ where $h_1 \in H_1$ and $h_2 \in H_2$, and $p' \in P'$. Left multiplying by $(p')^{-1}$ we see that $h_1 \overrightarrow{U}_1 \cap h_2 \overrightarrow{U}_2 \neq \emptyset$, and hence $\overrightarrow{U}_1 \cap \overrightarrow{U}_2 \neq \emptyset$.

Ball-WallNbd separation holds by Lemma 7.29, as wall stabilizers were arranged to be commensurable with the intersections of essential wall stabilizers. (Note that we are applying Lemma 7.29 in the trivial case where $\widetilde{Y}_i = \widetilde{X}$.) Consequently, Ball-Wall separation holds. Thus local finiteness of $\widetilde{E}'$ holds by Lemma 7.6. Note that to apply Lemma 7.6 we must ensure that $\bigsqcup_{p'} p'\widetilde{E}$ is a proper metric space, and that finitely many walls cross each finite ball. To this end, we regard it as a subspace of the metric space $\Gamma(P') \cup \bigsqcup_{p'} p'\widetilde{E}$ described above by adding a Cayley graph. The metric associated to the resulting nonpositively curved cube complex is hence proper by local finiteness. A wall is declared to cut an additional edge precisely if it separates its endpoints. By construction, exactly one wall cuts an edge of $p'\widetilde{E}$. An additional edge is cut by finitely many walls since the same is true for the Cayley graph $\Gamma(P')$, as our walls coarsely correspond to walls obtained by removing a finite neighborhood of a f.g. subgroup. Each ball is crossed by finitely many walls since balls have finitely many edges, and finitely many walls cross each edge.

To see that P' acts properly on $\widetilde{E}'$, observe that each infinite order element of P' is cut by an essential wall in the action of P' on $\widetilde{T}$ provided by Lemma 7.45. Each such essential wall has stabilizer that is a codimension-1 subgroup H' of P' which extends a codimension-1 subgroup H of P. And being cut by such a wall is unaffected by passage to a finite index subgroup of H'. Hence properness follows by Lemma 5.42.

Consider the inclusions $\widetilde{M} \subset \widetilde{E} \subset \widetilde{E}'$ of subcomplexes. There are finitely many P'-orbits of hyperplanes in $\widetilde{E}'$ that do not intersect $\widetilde{M}$, since all hyperplanes have translates that intersect $\widetilde{E}$, but there are finitely many P-orbits of hyperplanes in $\widetilde{E}$ that don't intersect $\widetilde{M}$. Consequently, since $\widetilde{E}'$ is locally finite, and $\widetilde{M} \subset \widetilde{E}'$ is P'-cocompact, and since there are finitely many additional P'-orbits of hyperplanes, we see that $\widetilde{E}'$ is also P'-cocompact.   □

An r-*star* is a tree formed by gluing r edges together along a vertex.

Theorem 7.54. *Let X be a sparse cube complex with fundamental group G. Suppose each quasiflat of $\widetilde{X}$ is closeable. There exists a compact nonpositively curved cube complex X' and group $G' = \pi_1 X'$ such that:*

(1) *G' splits as a tree T_r of groups where T_r is an r-star.*
(2) *The central vertex group of T_r is G.*
(3) *Each leaf vertex group P_i' is f.g. and virtually abelian.*
(4) *Each edge group of T_r is a peripheral subgroup P_i of G.*
(5) *There is a local-isometry $X \to X'$ inducing $G \subset G'$.*
(6) *Each 2-sided embedded hyperplane of X extends to a 2-sided embedded hyperplane in X'.*

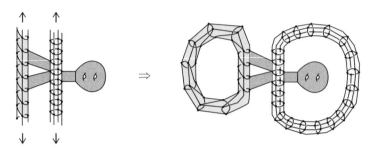

Figure 7.6. Closing up two infinite cylinders in base space.

The motivating case is where the cube complexes F_i^n associated to the peripheral subgroups have $\widetilde{F}_i^n$ isomorphic to standard cubulations of $\mathbb{E}^n$, though the peripheral subgroup P_i might only be virtually $\mathbb{Z}^m$ with $m < n$. In this case, it is easy to imagine quotienting each such F_i^n by a $\mathbb{Z}^{n-m}$ subgroup, to obtain cocompactness, and with a large enough fundamental domain so that the new space looks locally like the old—so nonpositive curvature is maintained. A simple example of this is illustrated in Figure 7.6.

Proof. By closeability, for each quasiflat $\widetilde{E}_i$ of $\widetilde{X}$ with stabilizer P_i, there is a quasiflat $\widetilde{E}_i'$ with a cocompact group action by a virtually abelian group P_i', and an embedding $P_i \hookrightarrow P_i'$, and an equivariant convex embedding $\widetilde{E}_i \subset \widetilde{E}_i'$.

Building X'. Let K be a compact subcomplex such that $\widetilde{X} = GK \cup \bigcup G\widetilde{E}_i$, and $g_i \widetilde{E}_i \cap g_j \widetilde{E}_j \subset GK$ unless $g_i \widetilde{E}_i = g_j \widetilde{E}_j$. For each i, let $K_i' \subset \widetilde{E}_i$ be a compact subcomplex such that $\widetilde{E}_i \cap GK \subset P_i K_i'$. By separability of $P_i \subset P_i'$, we may rechoose P_i' to ensure that $p_i'(P_i K_i') \cap P_i K_i' = \emptyset$ whenever $p_i' \in P_i' - P_i$.

Having extended $(P_i, \widetilde{E}_i)$ to $(P_i', \widetilde{E}_i')$ for each i, we form X' from $X \cup \bigsqcup E_i'$ by identifying the images $X \leftarrow E_i \rightarrow E_i'$ for each i. Here $E_i = P_i \backslash \widetilde{E}_i$ and $E_i' = P_i' \backslash \widetilde{E}_i'$ for each i. (A similar construction works equivariantly, in the more general case where there is torsion.) The resulting cube complex is nonpositively curved since E_i maps by a local-isometry on each side. And one sees the splitting along each P_i, since $\widetilde{X}'$ splits as a tree of spaces as for each i, the $(P_i' - P_i)$ translates of $\widetilde{X}$ are disjoint from each other outside of $\widetilde{E}_i'$.

Extending good hyperplanes: We now verify that 2-sided embedded hyperplanes of X extend to 2-sided embedded hyperplanes in X'. For simplicity, we prove the stronger statement that if all hyperplanes of X are embedded and 2-sided then they extend to embedded 2-sided hyperplanes in X'. The proof for individual hyperplanes is similar. By Lemma 7.55, by passing to a finite cover, we may assume that each E_i' has the property that 2-sided/embedded/non-crossing hyperplanes of E_i extend to 2-sided/embedded/non-crossing hyperplanes of E_i'. Let U be an embedded 2-sided hyperplane of X. The preimage of U under $E_i \rightarrow X$ consists of finitely many embedded, 2-sided, pairwise non-crossing hyperplanes in E_i. These extend to distinct 2-sided, embedded, pairwise non-crossing hyperplanes in each E_i'. Consequently, Conclusion (6) holds since U extends to a hyperplane U' of X' that contains U together with these disjoint hyperplanes in each E_i', and hence U' is embedded and 2-sided. $\qquad\square$

Lemma 7.55. *Let $E \to E'$ be a local-isometry where $(\widetilde{E}, P)$ and $(\widetilde{E}', P')$ are quasiflats, with $P = \pi_1 E$ and $P' = \pi_1 E'$, and suppose E' is compact. Suppose all hyperplanes of E are 2-sided and embedded. Then there is a finite cover $\widehat{E}' \to E$ such that all hyperplanes of E' are embedded and 2-sided, E lifts to $\widehat{E}'$, distinct hyperplanes of E map to distinct hyperplanes of $\widehat{E}'$, and non-crossing hyperplanes of E map to non-crossing hyperplanes of $\widehat{E}'$.*

We cannot force all hyperplanes of a cover of E' to be embedded. For instance, let E' the union of two squares carrying a self-crossing hyperplane, and let $E \subset E'$ be a locally-convex embedded circle consisting of a single edge and vertex. Then $\pi_1 E \to \pi_1 E'$ is an isomorphism, so the extraneous self-crossing is unavoidable.

Proof. We use the separability of multiple cosets in a f.g. virtually abelian group. Multiple cosets in a f.g. abelian group are separable since they are cosets of normal subgroups with f.g. abelian quotient. Hence separability also holds in the virtually abelian case.

Let U and V be hyperplanes of $\widetilde{E}$ and let U' and V' be the hyperplanes they map to in $\widetilde{E}'$. Note that U and V have different images in E if and only if $pU \neq V$ for $p \in P$. Let $H = \mathrm{Stab}_{P'}(U')$ and $K = \mathrm{Stab}_{P'}(V')$. Suppose $gU' = V'$ for some $g \in P'$. The multiple coset $KgH \subset P'$ consists of all elements that translate U' to V'. Since $pU' \neq V'$ for $p \in P$, we see that $P \cap KgH = \emptyset$, or equivalently, $1 \notin PKgH$. By separability of multiple cosets in P', there is a finite quotient $\phi : P' \to \overline{P'}$ with $\bar{1} \notin \bar{P}\bar{K}\bar{g}\bar{H}$. Letting $N = \ker(\phi)$ we have $1 \notin PNKgH$, so PN is a finite index subgroup of P' and $V' \neq p'U'$ for any $p' \in PN$. Consequently, U and V extend to distinct hyperplanes in $(PN)\backslash\widetilde{E}'$, and likewise for any finite index subgroup of PN containing P. The analogous argument applied to halfspaces of U shows that if no element of P translates one halfspace of U to its other halfspace, then there is a finite index subgroup of P' containing P having the same property, and so a 2-sided hyperplane of E will extend to a 2-sided hyperplane in a finite cover of E'. Suppose the images of U and V do not cross in E, i.e., no P-translate of U crosses V. Let $\{KgH\}_{g \in F}$ be the finitely many double cosets in P' consisting of elements that translate U' to cross V'. Note that $1 \notin PKgH$ for $g \in F$, since if $pU' \cap V' \neq \emptyset$, then $pU' \cap V' \cap \widetilde{E} \neq \emptyset$ by Lemma 2.10, and so $pU \cap V = p(U' \cap \widetilde{E}) \cap (V' \cap \widetilde{E}) \neq \emptyset$, which is impossible. By multiple coset separability, there is a finite index subgroup $P_* \subset P'$ with $P \subset P_*$ but $1 \notin P_*KgH$ for any $g \in F$. Note that there are finitely many P-orbits of hyperplanes, and we find such finite index subgroups for each pair U, V of representatives of P-orbits. Taking the intersection of the finite index subgroups above, we obtain an associated finite cover $\widehat{E}'$ so that E lifts to $\widehat{E}'$, and $E \to \widehat{E}'$ maps distinct hyperplanes to distinct hyperplanes, maps 2-sided hyperplanes to 2-sided hyperplanes, and maps non-crossing hyperplanes to non-crossing hyperplanes (and in particular, each embedded hyperplane maps to an embedded hyperplane). $\qquad\square$

The following special case of Theorem 7.54 is often sufficient in practice.

Lemma 7.56. *Let $X \to R$ be a local-isometry to a compact special cube complex. If X is sparse then there is a local-isometry $X \to X'$ where X' is compact and $\pi_1 X'$ is hyperbolic relative to abelian subgroups $\{P_i'\}$ that contain the corresponding parabolic subgroups $\{P_i\}$ of the relatively hyperbolic structure of $\pi_1 X$. And there is a local-isometry $X' \to R$ such that $X \to R$ factors as $X \to X' \to R$.*

Remark 7.57. As each P_i (and P_i') is free-abelian, we may also assume $\pi_1 X'$ splits over a tree T_r, whose central vertex group is $\pi_1 X$, whose edge groups are the $\{P_i\}$, and whose leaf vertex groups P_i' are of the form $P_i \times \mathbb{Z}^{m_i}$ for some m_i. This follows immediately from the proof, by interpolating an additional separability step to ensure each $P_i \subset P_{i\#}$ is a direct factor.

The proof functions in the generality where Definition 7.21 is satisfied without the assumption that $\pi_1 X$ is hyperbolic relative to the stabilizers of the quasiflats.

Proof. Let $\widetilde{F}$ be a P-quasiflat. Let A be the maximal abelian subgroup of $\pi_1 R$ containing $P \subset \pi_1 R$. By Lemma 2.38, the group A acts properly and cocompactly on a convex subcomplex $\widetilde{E}$ that is the product of cube and finitely many quasilines. We may assume $\widetilde{F} \subset \widetilde{E}$ by replacing $\widetilde{E}$ with $\widetilde{E}^{+s}$ for some s.

Let $\widetilde{F}_m \subset \widetilde{F}$ be a minimal subcomplex. Observe that $\widetilde{F}_m$ is P'-cocompact for some $P' \subset A$ since $\widetilde{F}_m$ is a subproduct of a minimal P-invariant subcomplex $\widetilde{E}_m$ of $\widetilde{E}$. Note that $\widetilde{F}_m^{+r}$ is P'-cocompact for each r. Choose r so that $\widetilde{F} \subset \widetilde{F}_m^{+r}$, and let $\widetilde{F}' = \widetilde{F}_m^{+(r+1)}$, so that $\widetilde{F}'$ contains an entire cubical neighborhood of $\widetilde{F}$.

Let $K \subset \widetilde{X}$ be a compact subcomplex such that $\widetilde{X} = \pi_1 X K \cup \bigcup_i \pi_1 X \widetilde{F}_i$, and the intersection of distinct quasiflats lies in $\pi_1 X K$. Let $K^{+1} \subset \widetilde{X}$ be a compact subcomplex containing all cubes that intersect K. Let K_i be a compact subcomplex such that $\widetilde{F}_i \cap \pi_1 X K^{+1} = P K_i$. By Lemma 7.58, let $\widetilde{F}_{\#} \subset \widetilde{F}'$ be a $P_{\#}$-cocompact quasiflat with $P \subset P_{\#}$ and $\widetilde{F} \cap \pi_1 X K^{+1} = \widetilde{F}_{\#} \cap \pi_1 X K^{+1}$. Follow the above process for each i to obtain a $P_{\#i}$-cocompact quasiflat $\widetilde{F}_{\#i}$ with $\widetilde{F}_{\#i} \cap \pi_1 X K^{+1} = \widetilde{F}_i \cap \pi_1 X K^{+1}$.

Let $F_i = P_i \backslash \widetilde{F}_i$ and let $F_{i\#} = P_{i\#} \backslash \widetilde{F}_{i\#}$ and note that there is a map $\phi_i : F_i \to F_{i\#}$ and that $F_i \to R$ factors as $F_i \to F_{i\#} \to R$. Let $X' = (X \sqcup \bigsqcup_i F_{i\#}) / \{x \sim \phi_i(x) : x \in F_i\}$. Then $X' \to R$ is a local-isometry.

To see this, observe that at each 0-cube $p \in X'$, the map to R looks either like $X \to R$ or $F_{i\#} \to R$. Indeed, if $p \in \pi_1 X \backslash (\pi_1 X K)$ then no new cells are attached along p, so p has a neighborhood entirely in X. If $p \notin \pi_1 X \backslash (\pi_1 X K)$ then p has a neighborhood entirely in some $F_{i\#}$. $\qquad\square$

Lemma 7.58 (Controlled closeability). *Let $(\widetilde{F}, P)$ be a quasiflat that extends to a cocompact quasiflat $(\widetilde{F}', P')$. Let $K \subset \widetilde{F}'$ be a compact subcomplex. There exists a finite index subgroup $P_\# \subset P'$ with $P \subset P_\#$ and a convex $P_\#$-cocompact convex subcomplex $\widetilde{F}_\# \subset \widetilde{F}'$ such that $\widetilde{F} \subset \widetilde{F}_\#$ and $\widetilde{F}_\# \cap PK = \widetilde{F} \cap PK$.*

Proof. Let $\widetilde{F}'_m$ be a minimal subcomplex of $\widetilde{F}'$. Then $\widetilde{F}'_m \cong K \times \prod_i \widetilde{L}_i$ is a product of a compact complex and finitely many quasilines L_i, where each hyperplane of each $\widetilde{L}_i$ is essential. Observe that $\widetilde{F}'_m$ is P'-cocompact. By the flat torus theorem, let $\mathbb{E} \subset \widetilde{F}$ be a P-cocompact flat. Let $\widetilde{E} = \text{hull}(\mathbb{E})$. Observe that $\widetilde{E}$ is a P-invariant convex subcomplex of $\widetilde{F}'_m \cap \widetilde{F}$. Moreover, $\widetilde{E} = K' \times \prod_i \widetilde{L}'_i$ where for each i, either $\widetilde{L}'_i = \widetilde{L}_i$ or $\widetilde{L}'_i$ is compact, and $K' \subset K$ is a subcomplex. Let $P_e = \text{Stab}_{P'}(\widetilde{E})$ and note that $P \subset P_e$ and that P_e acts cocompactly on $\widetilde{E}$. Furthermore, $\widetilde{F} \subset \widetilde{E}^{+r}$ for some r, and P_e cocompactly stabilizes $\widetilde{E}^{+r}$. Let U be a hyperplane of $\widetilde{E}^{+r}$ that is not a hyperplane of $\widetilde{E}$. Then U does not separate $\mathbb{E}, z\mathbb{E}$ for any $z \in P_e$, as $z\mathbb{E} \subset \widetilde{E}$, so $\mathbb{E}$ and $z\mathbb{E}$ lie in the same halfspace of U. Since $\widetilde{E} = \text{hull}(\mathbb{E})$, each hyperplane U of $\widetilde{E}$ intersects $\mathbb{E}$ and hence U cannot separate $\mathbb{E}$ from $z\mathbb{E}$.

By replacing $(\widetilde{F}', P')$ with $(\widetilde{E}^{+r}, P_e)$ we may assume that each hyperplane U of $\widetilde{F}'$ has the property that either $U \cap \mathbb{E} \neq \emptyset$ or U does not separate $\mathbb{E}, z\mathbb{E}$.

Let $v \in PK^0 - \widetilde{F}$. By Lemma 2.18, let U be a hyperplane separating $v, \widetilde{F}$. Let $C \subset \widetilde{F}$ be a compact subcomplex such that $\text{hull}(PC) = \widetilde{F}$. Let $P'_v \subset P'$ be a finite index subgroup such that $P \subset P'_v$ but $P'_v U \cap PC = \emptyset$, and hence $U \cap P'_v C = \emptyset$. To see that P'_v exists, let V be a hyperplane intersecting PC, let $H = \text{Stab}_{P'}(V)$, and let $g \in P'$ be such that $U = gV$. Since $P \cap gH = \emptyset$, we have $1 \notin PgH$, hence since PgH is closed in P', there is a finite index subgroup $P'_v \subset P'$ so that $1 \notin P'_v gH$. Moreover, as there are finitely many P-orbits of such hyperplanes V, we can do this so that $1 \notin P'_v g \, \text{Stab}(V)$ for each such pair (g, V).

Do this for a representative of each of the finitely many P-orbits of such 0-cubes v. Let $P_\# = \cap_v P'_v$. We claim that $\widetilde{F}_\# = \text{hull}(P_\# C)$ has the desired property. Indeed, if U separates two points of $P_\# C$ then either $U \cap P_\# C \neq \emptyset$ or else U separates PC, zPC and hence $\mathbb{E}, z\mathbb{E}$ for some $z \in P_\#$. Both of these were excluded above. □

7.g Cubulating Amalgams

We now summarize the results from [HW15]. As the main statement is rather technical, we first give an imprecise simplification that disregards certain facilitating conditions.

Quasi-Theorem. *Let G split as $A *_C B$ or $A*_{C^t = C'}$. Suppose G is hyperbolic relative to virtually abelian subgroups, and that C is malnormal and quasiconvex in G. If A, B act properly and cocompactly on CAT(0) cube complexes then G*

acts properly on a CAT(0) cube complex dual to a system of walls associated to quasi-isometrically embedded subgroups.

When G is hyperbolic relative to subgroup $\{P_1, \ldots, P_k\}$ we call each conjugate of each P_i a *parabolic* or *peripheral* subgroup of G. A subgroup $H \subset G$ of a relatively hyperbolic group is *aparabolic* if it has finite intersection with each (noncyclic) parabolic subgroup of G.

A slightly simplified version of the main result from [HW15] is:

Theorem 7.59. *If G has all the following properties then G acts properly on a CAT(0) cube complex, and the stabilizers of the hyperplanes are quasi-isometrically embedded and hence relatively quasiconvex.*

(1) *G is hyperbolic relative to f.g. virtually abelian subgroups.*
(2) *G splits as an amalgamated product $G \cong A *_C B$.*
(3) *A and B act properly and cosparsely (e.g., cocompactly) on CAT(0) cube complexes.*
(4) *Let C_- be the image of $C \hookrightarrow A$ and C_+ be the image of $C \hookrightarrow B$. Then $C_- \subset A$ and $C_+ \subset B$ are relatively quasiconvex in their vertex groups.*
(5) *$C_- \subset A$ and $C_+ \subset B$ are almost malnormal.*
(6) *$C_- \subset A$ and $C_+ \subset B$ are aparabolic.*
(7) *C has separable quasiconvex subgroups.*
(8) *There are quasiconvex subgroups $H_1, \ldots, H_r$ of C and H_i-walls in C such that C acts properly on the resulting cube complex.*
(9) *Each H_i-wall of C extends to an H_i^A-wall of A and an H_i^B-wall of B, where H_i^A, H_i^B are relatively quasiconvex subgroups of A, B.*

Alternatively, we can assume the following slightly more flexible possibility:

(5') *and* (6') *C_+ is almost malnormal in B and C_+ is aparabolic in B.*
(8') *and* (9') *There is a system of walls for A so that A acts properly and cosparsely on the resulting cube complex and each induced H_i-wall extends to an H_i^B-wall in B.*

In the HNN case we have the following adjusted statements:

$(\overline{2})$ *G splits as $A*_C$.*
$(\overline{4})$ *C is quasi-isometrically embedded in G.*
$(\overline{5})$ *$\{C_+, C_-\}$ are an almost malnormal pair of subgroups of A.*
$(\overline{6})$ *$\{C_+, C_-\}$ are aparabolic in A.*
$(\overline{8})$ *There are quasiconvex subgroups $H_1, \ldots, H_r$ and H_i-walls of C, so that C acts properly on the resulting cube complex.*
$(\overline{9})$ *Each H_i-wall of C_+ and each H_i-wall of C_- extends to an H_i^A-wall.*

Remark 7.60. The CAT(0) cube complex of Theorem 7.59 is dual to a wallspace $\widetilde{X}$ having the structure of a *tree of wallspaces*. Briefly, G acts on a wallspace $\widetilde{X}$, and there is a G-equivariant map $\widetilde{X} \to T$ where T is the Bass-Serre tree of the splitting of G. The preimage of each vertex v of T is a wallspace that we call a vertex space $\widetilde{X}_v$ of $\widetilde{X}$. Each wall of $\widetilde{X}$ either projects to the center of an edge of T, or projects to a subtree of T, and has the property that its intersection with each vertex space is either empty or consists of a wall of that vertex space. In our setting the underlying spaces of these vertex spaces are CAT(0) cube complexes. However, they differ in two ways from the cube complexes that are provided as input in the hypotheses of Theorem 7.59. Firstly, they are new CAT(0) cube complexes produced using the "recubulation" procedure of [HW15, Thm 5.4]. Secondly, the walls of these vertex spaces are of two types: They are either hyperplanes, or they are extension-walls produced via Theorem 6.25 or 6.19.

In the more flexible adjusted version of Theorem 7.59 applied to an amalgam $A *_C B$, a new CAT(0) cube complex is produced for a vertex space of $\widetilde{X}$ via recubulation only for vertices associated to the B factors, and likewise, the extension-walls only arise within the B-wallspaces. The wallspaces of the A-vertices are the original hypothesized CAT(0) cube complexes and their walls are their hyperplanes.

Remark 7.61. We caution that "cosparse" in [HW15] has a weaker meaning than we require here, as it is missing certain useful finiteness properties that were discussed in Section 7.e.

Chapter Eight

Malnormality and Fiber-Products

8.a Height and Virtual Almost Malnormality

Definition 8.1 (Malnormal Collection). A collection of subgroups $\{H_1, \ldots, H_r\}$ of G is *malnormal* provided that $H_i^g \cap H_j = \{1_G\}$ unless $i = j$ and $g \in H_i$. Similarly, the collection is *almost malnormal* if intersections of nontrivial conjugates are finite (instead of trivial). Note that this condition implies that $H_i \neq H_j$ (unless they are finite in the almost malnormal case).

Definition 8.2 (Height). Consider a collection $\{H_1, \ldots, H_k\}$ of subgroups of G. We use the notation $H^g = g^{-1}Hg$. We say $H_{m_i}^{g_i}$ and $H_{m_j}^{g_j}$ are *distinct conjugates* unless $m_i = m_j$ and $H_{m_i}g_i = H_{m_j}g_j$. We emphasize that each "conjugate" corresponds to a value of $(m_i, H_{m_i}g_i)$ and not just a subgroup $H_{m_i}^{g_i}$.

We say the collection has *height* $0 \leq h \leq \infty$ if h is the largest number so that there are h distinct conjugates of these subgroups whose intersection $H_{m_1}^{g_1} \cap H_{m_2}^{g_2} \cap \cdots \cap H_{m_h}^{g_h}$ is infinite. If each H_i is finite, then the height of the collection is $h = 0$. If there is an infinite collection of distinct conjugates whose intersection is an infinite subgroup, then the height of the collection is $h = \infty$.

For instance, if G is infinite and $[G : H]$ is finite then the height of H in G equals $[G : H]$.

The notion of height was introduced and studied in [GMRS98] where it was shown that quasiconvex subgroups of word-hyperbolic groups have finite height. It follows that a finite collection of quasiconvex subgroups also has finite height. We have explored the notion a bit further for relatively hyperbolic groups in [HW09].

A collection of subgroups is almost malnormal as in Definition 8.1 precisely if its height is ≤ 1.

We will later make use of the following closely related result which also hinges on the key point implying the height finiteness. It holds because for conjugates whose intersection contains an infinite order element, the corresponding cosets must lie within a uniformly bounded distance of an axis, and because there are finitely many A-translates of gB cosets within a finite distance of A.

The following was first proven in [GMRS98]:

Proposition 8.3. *Let G be a hyperbolic group and A, B be quasiconvex subgroups. There are finitely many double cosets AgB such that $A^g \cap B$ is infinite.*

Definition 8.4 (Commensurator). The *commensurator* $\mathbb{C}_G(H)$ of H in G is defined by: $\mathbb{C}_G(H) = \{g \in G : [H : H^g \cap H] < \infty\}$. It is shown in [KS96] that $[\mathbb{C}_G(H) : H] < \infty$ for any infinite quasiconvex subgroup H of the word-hyperbolic group G. Consequently, in this case $\mathbb{C}_G(H)$ is itself quasiconvex in G.

The notion of height might have been better crafted in terms of intersections up to commensurability. This can be remedied by using the simple observation in Lemma 8.5 below. The subgroup H is *virtually almost malnormal* in G if for each $g \in G$ either $H^g \cap H$ is finite, or $g \in \mathbb{C}_G(H)$.

Lemma 8.5. *If $H \subset G$ is virtually almost malnormal then $\mathbb{C}_G(H)$ is almost malnormal in G.*

Proof. Suppose $\mathbb{C}_G(H)^g \cap \mathbb{C}_G(H)$ is infinite. Since $[\mathbb{C}_G(H) : H] < \infty$ we see that $H^g \cap H$ has finite index in $\mathbb{C}_G(H)^g \cap \mathbb{C}_G(H)$. Thus $H^g \cap H$ is infinite, and so $g \in \mathbb{C}_G(H)$ by virtual almost malnormality. $\square$

Lemma 8.6. *Let $H_1, \ldots, H_r$ be a collection of quasiconvex subgroups of a word-hyperbolic group G. Let $K_1, \ldots, K_s$ be representatives of the finitely many distinct conjugacy classes of subgroups consisting of intersections of collections of distinct conjugates of $H_1, \ldots, H_k$ in G that are maximal with respect to having infinite intersection. Then $\{\mathbb{C}_G(K_1), \ldots, \mathbb{C}_G(K_s)\}$ is an almost malnormal collection of subgroups of G.*

Note that this is indeed a finite collection by [GMRS98, HW09].

Proof. By maximality, for each K_i, H_j, and g_ℓ we have either:

(1) $K_i \cap H_j^{g_\ell}$ is finite.
(2) $K_i \subset H_j^{g_\ell}$.

Suppose $\mathbb{C}_G(K_s) \cap \mathbb{C}_G(K_t)^g$ is infinite for some $g \in G$. Since each $[\mathbb{C}_G(K) : K] < \infty$ we see that $K_s \cap K_t^g$ is infinite as well. Now $K_t = \bigcap_{i=1}^m H_{i_t}^{g_{i_t}}$ so:

$$K_s \cap K_t^g = K_s \cap \bigcap_{i=1}^m H_{i_t}^{g_{i_t}g} = \bigcap_{i=1}^m \left(K_s \cap H_{i_t}^{g_{i_t}g}\right).$$

Since the intersection is infinite, we must have alternative (2) that $K_s \subset H_{i_t}^{g_{i_t}g}$ for each i. But then $K_s \subset K_t^g$. Likewise $K_t^g \subset K_s$. Since we have chosen representatives of *distinct* conjugacy classes of maximal intersections (that is, a maximal number of factors) we see that $s = t$, and that $K_s^g = K_s$, and so $g \in \mathbb{C}_G(K_s)$. $\square$

Lemma 8.7 (Malnormal intersection). *Let $\{B_1, \ldots, B_k\}$ be an almost malnormal collection of subgroups of G. For each i, let $\{B_i g_{ij} A : j \in J_i\}$ denote a collection of distinct double cosets such that $B_i^g \cap A$ is infinite if and only if $g \in B_i g_{ij} A$ for some ij. For each ij, let $M_{ij} = B_i^{g_{ij}} \cap A$. Then $\{M_{ij} : 1 \le i \le k, j \in J_i\}$ is an almost malnormal collection of subgroups of A.*

Proof. Suppose $M_{ij}{}^a \cap M_{pq}$ is infinite for some $a \in A$. Then $B_i^{g_{ij} a} \cap B_p$ is infinite and so by almost malnormality of $\{B_1, \ldots, B_k\}$ in G, we must have $i = p$. Thus $M_{ij} = B_i^{g_{ij}} \cap A$ for some ij, and $M_{iq} = B_i^{g_{iq}} \cap A$ for some iq. Since $((B_i^{g_{ij}})^a \cap B_i^{g_{iq}}) \supset ((B_i^{g_{ij}} \cap A)^a \cap (B_i^{g_{iq}} \cap A)) = (M_{ij}^a \cap M_{iq})$, we see that $(B_i^{g_{ij}})^a \cap B_i^{g_{iq}}$ is infinite, and so $g_{ij} a g_{iq}^{-1} \in B_i$ by almost malnormality. Thus $B_i g_{ij} A = B_i g_{iq} A$ and hence $j = q$. Finally, $(B_i^a \cap B_i) \supset (M_{ij}^a \cap M_{ij})$ which is infinite and so $a \in B_i$ by almost malnormality. $\qquad\square$

8.b Fiber-Products

Definition 8.8 (Fiber-Product). Given a pair of combinatorial maps $A \to X$ and $B \to X$ between cube complexes, we define their *fiber-product* $A \otimes_X B$ to be a cube complex, whose i-cubes are pairs of i-cubes in A, B that map to the same i-cube in X. There is a commutative diagram:

$$
\begin{array}{ccc}
A \otimes_X B & \to & B \\
\downarrow & & \downarrow \\
A & \to & X
\end{array}
$$

Note that $A \otimes_X B$ is the subspace of $A \times B$ that is the preimage of the diagonal $D \subset X \times X$ under the map $A \times B \to X \times X$. For any cube Q, the diagonal of $Q \times Q$ is isomorphic to Q by either of the projections, and this makes D into a cube complex isomorphic to X. We thus obtain an induced cube complex structure on $A \otimes_X B$.

Our description of $A \otimes_X B$ as a subspace of the cartesian product $A \times B$ endows the fiber-product $A \otimes_X B$ with the property of being a universal receiver in the following sense: Consider a commutative diagram as below. Then there is an induced map $C \to A \otimes_X B$ such that the following diagram commutes:

$$
\begin{array}{ccccc}
C & \longrightarrow & & & B \\
& \searrow & & \nearrow & \\
\downarrow & & A \otimes_X B & & \downarrow \\
& \swarrow & & \searrow & \\
A & & \longrightarrow & & X
\end{array}
$$

When the target space X is understood, we will simply write $A \otimes B$. When A, B, and X have basepoints and the maps are basepoint preserving, then $A \otimes B$ has a basepoint, and it is often natural to consider only the base component.

When $A \to X$ and $B \to X$ are covering maps, then so is $A \otimes B \to X$. Moreover, let $(a, b) \in A \times B$ reflect choices of the preimage of the basepoint $x \in X$, then the component of $A \otimes B$ containing (a, b) is the covering space of X corresponding to $\pi_1(A, a) \cap \pi_1(B, b)$. This leads to the following:

Lemma 8.9. *Let $A \to X$ and $B \to X$ be local-isometries of connected nonpositively curved cube complexes. Suppose the induced lift of universal covers $\tilde{A} \subset \tilde{X}$ is a superconvex subcomplex. Then the noncontractible components of $A \otimes B$ correspond precisely to the nontrivial intersections of conjugates of $\pi_1(A, a)$ and $\pi_1(B, b)$ in $\pi_1 X$.*

Let $Y \to Z$ be a map and let $\hat{Z} \to Z$ be a covering map. We generalize the notion of a lift as follows: An *elevation* $\hat{Y} \to \hat{Z}$ of $Y \to Z$ is a map where the composition $\hat{Y} \to Y \to Z$ equals $\hat{Y} \to \hat{Z} \to Z$, and such that choosing basepoints so the above maps are basepoint preserving, we have $\pi_1 \hat{Y}$ equals the preimage of $\pi_1 \hat{Z}$ in $\pi_1 Y$.

Proof. Let $\hat{X}_A$ and $\hat{X}_B$ denote the based covers of X corresponding to $\pi_1(A, a)$ and $\pi_1(B, b)$, and note that there are locally-isometric embeddings $A \subset X_A$ and $B \subset X_B$. The universal property of the fiber-product gives a map $A \otimes_X B \to \hat{X}_A \otimes_X \hat{X}_B$. This map is a local-isometric embedding, since $A \to X$ and $B \to X$ are local-isometries and hence so is $A \otimes_X B \to X$. Hence $A \otimes_X B \to \hat{X}_A \otimes_X \hat{X}_B$ is a π_1-injection on each component by Corollary 2.12.

We now show that each noncontractible component of $\hat{X}_A \otimes_X \hat{X}_B$ contains a component of $A \otimes_X B$. Let $\gamma \to \hat{X}_A \otimes_X \hat{X}_B$ be a closed geodesic (i.e., a local-isometry from a circle) in a component $\hat{X}$ of $\hat{X}_A \otimes_X \hat{X}_B$, and consider the composition $\gamma \to \hat{X} \to \hat{X}_B$ induced by composing with the right-projection $\hat{X}_A \otimes_X \hat{X}_B \to \hat{X}_B$. Since $B \to \hat{X}_B$ is a π_1-isomorphism and a local-isometry, there exists a closed geodesic $\gamma' \to B \subset \hat{X}_B$ such that γ' is homotopic to γ in $\hat{X}_B$. Since $A \to X$ is superconvex, so is the elevation $\hat{A} \to \hat{X}_A \otimes_X \hat{X}_B$ to $\hat{X}$. Hence γ' lies in $\hat{A}$, and so $\gamma' \to X$ factors as $\gamma' \to A \to X$. Thus γ' lies in a component K of $A \otimes_X B$ contained in $\hat{X}$.

We now use that K is nonempty to show that the map $K \to \hat{X}$ is π_1-surjective. Consider the components $\tilde{A}, \tilde{B}$ of the preimages of $\hat{A} \subset \hat{X}$ and $\hat{B} \subset \hat{X}$ in $\tilde{X}$ that contain a component of $\tilde{K}$ (which acts as our basepoint). Note that $\tilde{A} \cap \tilde{B} \neq \emptyset$ since it contains $\tilde{K}$, and $\tilde{A} \cap \tilde{B}$ is clearly $(\pi_1 A \cap \pi_1 B)$-invariant. Finally, $\tilde{A} \cap \tilde{B}$ is connected since it has the stronger property of being convex, as it is the intersection of convex subcomplexes. The quotient $(\pi_1 A \cap \pi_1 B) \backslash (\tilde{A} \cap \tilde{B})$ is thus the component K of $A \otimes_X B$ that is the desired locally-convex π_1-isomorphic subcomplex of our component $\hat{X} \subset \hat{X}_A \otimes_X \hat{X}_B$. $\square$

The following is a simple consequence of Lemma 8.9.

Corollary 8.10. *Let $A \to X$ be a superconvex local-isometry of connected non-positively curved cube complexes. Then $\pi_1 A$ is malnormal if and only if $A \otimes_X A$ consists entirely of contractible components except for its diagonal component which is a copy of A.*

More generally, when $A = \sqcup A_i$ consists of a collection of components, and each $A_i \to X$ is superconvex, the malnormality of $\{\pi_1 A_i\}$ corresponds to contractibility of all nondiagonal components of $A \otimes A$.

The height of $\{\pi_1 A_i\}$ is the maximal number of factors in a multiple fiber-product $A \otimes_X A \otimes_X \cdots \otimes_X A$ such that there is a noncontractible component outside the large diagonal.

Consider a finite collection $\{Y_i \to X\}$ of superconvex local-isometries with each Y_i connected, and consider the local-isometry $Y \to X$ where $Y = \sqcup_i Y_i$. For each n, let $\{Y_{nj}\}$ denote the various nondiagonal components of the n-fold fiber-product $Y \otimes_X \cdots \otimes_X Y$. We say Y_{nj} has *degree* n.

When each Y_i is compact, there is number h such that each $Y_{(h+1)j}$ is contractible. As above, this is the *height* of $\{\pi_1 Y_i\}$. To see that h exists, note that it is bounded above by the sum $\sum |Y_i^0|$ of the numbers of 0-cells in the various Y_i. It is natural to henceforth ignore contractible components in our case of interest where each Y_i is compact, and in particular, we will not consider Y_{nj} for $n > h$.

Finally, when $\{Y_i\}$ has height h, the collection $\{Y_{hj}\}$ of degree h components are symmetric, and their fundamental groups are normal finite index subgroups in their commensurators, which are malnormal subgroups of $\pi_1 X$.

Definition 8.11 (Symmetric). A local-isometry $Y \to X$ with Y connected is *symmetric* if for each component K of $Y \otimes_X Y$, either K maps isomorphically to each copy of Y, or $[\pi_1 Y : \pi_1 K] = \infty$.

Lemma 8.12. *Let $Z \to X$ be a superconvex local-isometry with Z compact. Then Z is symmetric if and only if $\mathbb{C}_{\pi_1 X}(\pi_1 Z) = \mathrm{Stab}(\widetilde{Z})$ and $\pi_1 Z \subset \mathrm{Stab}(\widetilde{Z})$ is a finite index normal subgroup.*

Proof. Suppose $\mathbb{C}_{\pi_1 X}(\pi_1 Z) = \mathrm{Stab}(\widetilde{Z})$. Let K be a component of $Z \otimes_X Z$. Then $\widetilde{K} = g\widetilde{Z} \cap \widetilde{Z}$ for some g. If $[\pi_1 Z : \pi_1 K] < \infty$ then $g \in \mathbb{C}_{\pi_1 X}(\pi_1 Z)$. Since $\mathbb{C}_{\pi_1 X}(\pi_1 Z) = \mathrm{Stab}(\widetilde{Z})$ we have $g \in \mathrm{Stab}(\widetilde{Z})$, so $g\widetilde{Z} = \widetilde{Z}$, so K projects to Z.

Suppose Z is symmetric and not contractible. If $g \in \mathbb{C}_{\pi_1 X}(\pi_1 Z)$ then $g\pi_1 Z g^{-1} \cap \pi_1 Z$ acts cocompactly on $g\widetilde{Z} \cap \widetilde{Z}$ with quotient a component K of $Z \otimes_X Z$ with $[\pi_1 Z : \pi_1 K] < \infty$, and hence $K \to Z$ is an isomorphism since Z is symmetric, but then $g\widetilde{Z} = \widetilde{Z}$, so $g \in \mathrm{Stab}(\widetilde{Z})$. The normality of $\pi_1 Z \subset \mathrm{Stab}(\widetilde{Z})$ holds since if $g \in \mathrm{Stab}(\widetilde{Z})$ but $g\pi_1 Z g^{-1} \neq \pi_1 Z$ then the quotient

$(g\pi_1 Z g^{-1} \cap \pi_1 Z) \backslash (g\widetilde{Z} \cap \widetilde{Z})$ is a component of $Z \otimes_X Z$ that is a proper finite cover of Z. We have $\mathrm{Stab}(\widetilde{Z}) \subset \mathbb{C}_{\pi_1 X}(\pi_1 Z)$ since $\widetilde{Z}$ is $\pi_1 Z$-cocompact. The latter implies that $[\mathrm{Stab}(\widetilde{Z}) : \pi_1 Z] < \infty$. $\qquad\square$

8.c Graded Systems

The following material will be useful in organizing statements and proofs in Section 12.c.

Definition 8.13 (Graded System). Let $\{Y_i \to X\}$ be a finite collection of local-isometries. A *graded system* $\{Y_{jk}\}$ associated to $\{Y_i\}$ consists of local-isometries $Y_{jk} \to X$ with the following properties:

(1) $\pi_1 Y_{jk} = \mathbb{C}_{\pi_1 X}(\pi_1 Y_{jk})$ for each Y_{jk}.
(2) Each $\pi_1 Y_{jk}$ is commensurable in $\pi_1 X$ to a nontrivial subgroup H that is the intersection of finitely many conjugates of the various $\pi_1 Y_i$.
(3) For each such H there is a unique ab such that H is commensurable with a conjugate of $\pi_1 Y_{ab}$.
(4) If $g\pi_1 Y_{ab} g^{-1} \subset \pi_1 Y_{cd}$ for some $g \in \pi_1 X$ then $g\widetilde{Y}_{ab} \subset \widetilde{Y}_{cd}$.

The graded system is *compact* if $\sqcup Y_{jk}$ is compact, and it is *superconvex* if each Y_{jk} is superconvex.

We declare $Y_{ab} \preceq Y_{cd}$ if a translate of $\widetilde{Y}_{ab}$ lies in $\widetilde{Y}_{cd}$. The *height* h of an element p of a poset is the length of a longest chain of the form $p_1 \prec p_2 \prec \cdots \prec p_{h+1} = p$. We assume each element of the poset $\{Y_{jk}\}$ has finite height, and the notation is assigned so that the height of Y_{jk} is j. We declare the *grade* of Y_{jk} to be j.

Construction 8.14 (Producing graded system). Let $\{Y_i \to X\}$ be a finite collection of local-isometries where X is compact and $\pi_1 X$ is word-hyperbolic, and each Y_i is compact. We describe how to associate a graded system $\{Y_{jk}\}$ to $\{Y_i\}$.

By Lemma 2.36, we may assume that each Y_i is superconvex. Let m be the total number of 0-cubes in $Y = \sqcup Y_i$. The components of $\otimes_m Y$ are compact and superconvex. For each component A of $\otimes_m Y$, let $H = \mathbb{C}_{\pi_1 X}(\pi_1 A)$. Let $\widetilde{B} = \cap_{h \in H} h\widetilde{A}$ and let $B = H \backslash \widetilde{B}$.

Given B and B' produced as above, we declare $B \preceq B'$ if $\pi_1 B \subset g^{-1} \pi_1 B' g$ for some $g \in \pi_1 X$. Equivalently, $g\widetilde{B} \subset \mathcal{N}_r(\widetilde{B}')$ for some $g \in \pi_1 X$ and $r \geq 0$. We discard multiple copies representing conjugate commensurability classes, so we may assume that $B \preceq B'$ and $B' \preceq B$ implies $B = B'$. The *grade* of B is the length j of the longest chain $B_1 \prec B_2 \prec \cdots \prec B_{j+1} = B$.

Let $r_1 = 0$. And for each $i \geq 1$, choose r_i so that if $C \prec B$ and $j = \mathrm{grade}(C)$, then $C^{+r_j} \to X$ factors as $C^{+r_j} \to B^{+r_i} \to X$. Let $\{Y_{jk}\}$ be copies of $\{B^{+r_j}\}$

where the notation Y_{jk} has a first subscript to indicate that B has grade j, and a second subscript k to distinguish between the various elements with the same grade. We continue to regard Y_{jk} as having grade j.

When we choose the superconvex hull of a core for a quasiconvex subgroup H, it should be chosen to be invariant under the action of $\mathbb{C}_G(H)$. This cautionary point is especially noteworthy in the graded case. Indeed, even though $\mathrm{Stab}(\widetilde{Y_i}) = \mathbb{C}_{\pi_1 X}(\pi_1 Y_i)$, it is not necessarily the case that $\mathrm{Stab}(\widetilde{Y_1} \cap \widetilde{Y_2}) = \mathbb{C}_{\pi_1 X}(\pi_1 Y_1 \cap \pi_1 Y_2)$. This explains some of the extra details in Construction 8.14.

Chapter Nine

Splicing Walls

9.a Finite Cover That Is a Wallspace

Construction 9.1 (Splicing). Let Y be a connected nonpositively curved cube complex whose hyperplanes are 2-sided and embedded. Let $\Lambda(Y)$ denote the set of hyperplanes of Y. Let $q : \Lambda(Y) \to S$ be a map with the property that for each $s \in S$, no two hyperplanes in $q^{-1}(s)$ cross each other.

Consider the homomorphism $\#_q : \pi_1 Y \to \mathbb{Z}_2^S$ induced by $\#_q(e) = v_{q(\Lambda_e)}$ where Λ_e is the hyperplane dual to the 1-cube e and where $\mathbb{Z}_2^S$ has basis $\{v_s : s \in S\}$.

Let $\ddot{Y}$ denote the cover of Y corresponding to the kernel of $\#_q$. Let $s \in S$ lie in the image of q. Let W_s denote the collection of hyperplanes of $\ddot{Y}$ that map to hyperplanes of Y which map to s. See Figure 9.1.

Remark 9.2. We have in mind the following situation: $Y \hookrightarrow X$ is an embedded locally-convex subcomplex of a nonpositively curved cube complex X whose hyperplanes are embedded and 2-sided. The set S equals the collection $\Lambda(X)$ of hyperplanes of X. The map $q : \Lambda(Y) \to \Lambda(X)$ sends a hyperplane of Y to the hyperplane of X containing it. The map $\#_q : \pi_1 X \to \mathbb{Z}_2^{\Lambda(X)}$ sends each path σ to the $\mathbb{Z}_2$-vector whose v_s coordinate is the number of times (modulo 2) that σ passes through the hyperplane $s \in \Lambda(X)$. The map $\#_q : \pi_1 Y \to \mathbb{Z}_2^{\Lambda(X)}$ is induced by composition with the natural map $\pi_1 Y \to \pi_1 X$.

Lemma 9.3. *For each $s \in q(\Lambda(Y))$, the collection of hyperplanes W_s separates $\ddot{Y}$.*

The motivation is to produce a collection of walls in $\ddot{Y}$ each of which corresponds to some hyperplane of X.

Proof. Consider a closed edge path $\ddot{\sigma}$ in $\ddot{Y}$. We must show that $\ddot{\sigma}$ cannot pass through hyperplanes of W_s an odd number of times. If this were the case, then the image σ of $\ddot{\sigma}$ in Y would pass through hyperplanes of $q^{-1}(s)$ an odd number of times. But then σ would not be in the kernel of $\pi_1 Y \to \mathbb{Z}_2^S$, for $\#_q(\sigma)$ would take the value 1 on the v_s coordinate. Hence $\ddot{\sigma}$ would not be closed. $\square$

We also employ the following simple method of inducing a wallspace on a cover.

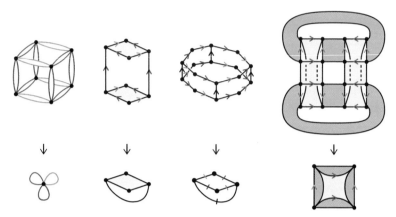

Figure 9.1. The preimages in $\ddot{Y} \to Y$ of hyperplanes in Y are walls in $\ddot{Y}$. We have indicated some walls with corresponding labelled edges. An interesting case is on the right, where $\#_W \neq H_1(-; \mathbb{Z}_2)$.

Construction 9.4 (Cover induced wallspace). Let Y be a nonpositively curved cube complex that is a wallspace. Let $Y' \to Y$ be a covering space. There is an *induced wallspace structure* on Y' where each wall of Y' is the preimage of a wall of Y, and each halfspace of that wall in Y' is the preimage of a halfspace in Y.

9.b Preservation of Small-Cancellation and Obtaining Wall Convexity

Lemma 9.5 (Obtaining local convexity of walls). *Let $\langle X \mid Y_1, \ldots, Y_k \rangle$ be a cubical presentation with an angling rule satisfying Condition (4) of Section 3.o, and with the property that there is a uniform lowerbound α on its nonzero assigned angles. Suppose that X has finitely many immersed hyperplanes D, and that for each D and $r > 0$, there are finitely many distinct translates gD with $d_{\tilde{X}}(D, gD) \leq r$. Suppose that $\pi_1 D$ is separable in $\pi_1 X$ for each hyperplane D, and suppose that there is a uniform upperbound M on the diameters of wall-pieces and cone-pieces. Then for each θ there is a finite regular cover $\widehat{X} \to X$ whose induced covers $\widehat{Y}_1, \ldots, \widehat{Y}_k$ have the following property: Consider $\langle \widehat{X} \mid g\widehat{Y}_i : 1 \leq i \leq k, g \in \mathrm{Aut}(\widehat{X}) \rangle$. For any path $S \to \widehat{Y}_i$ that starts and ends on 1-cells of a hyperplane $\widehat{D}$ of $\widehat{X}$ but is not homotopic into $\widehat{D}$, we have $\{\{S\}\}_{Y_i} \geq \theta$.*

The angling rule for $\langle \widehat{X} \mid \{gY_i\} \rangle$ is *induced* by the angling rule for X^* in the obvious way: Any diagram $D \to \widehat{X}^*$ projects to a diagram $D \to X^*$ and so a rectification and angling of the latter provides one for the former.

The conditions on the angling rule for X^* hold when it is small-cancellation. The finiteness condition on the hyperplanes obviously holds when X is compact, and the separability condition holds when X is special.

Proof. Choose $r \geq \frac{M\theta}{\pi - \alpha}$. Consider a path $S \to Y_i$ such that $|S|_X \geq r$ and such that the projection $S \to X$ starts and ends on a hyperplane D. Our choice of r, together with the hypothesized upperbounds and lowerbounds on pieces and transition angles, ensures that $\{\{S\}\}_{Y_i} > \frac{r}{M}(\pi - \alpha) \geq \theta$.

Our finiteness condition on translates of $\widetilde{D}$ ensures that for each hyperplane D, there are finitely many nontrivial cosets $\pi_1 D g_j \pi_1 D$ such that $g_j \widetilde{D}$ and $\widetilde{D}$ are joined by a path S with $|S|_X < r$. By separability of $\pi_1 D$, we can choose a cover $\widehat{X}_D$ such that $\pi_1 \widehat{X}_D$ contains $\pi_1 D$ but is disjoint from each $\pi_1 D g_j \pi_1 D$. Hence there is no path $S \to Y_i$ that lifts to a path $\widehat{S}$ in $\widehat{X}_D$ that starts and ends on D, and has $\{\{S\}\}_{Y_i} < \theta$, unless S is homotopic into D.

Let $\widehat{X} \to X$ be a finite regular cover factoring through each $\widehat{X}_D$. Then $\widehat{X}$ has the desired property. $\qquad\square$

Lemma 9.6 (Obtaining and preserving wall convexity). *Let $\ddot{Y}$ be obtained as in Construction 9.1 from $Y \to X$.*

Suppose that for each path $P \to Y$ whose endpoints are on 1-cells dual to the same hyperplane D of X, either $\{\{P\}\}_Y \geq \pi$ or P is path-homotopic in X into the carrier of this hyperplane (and hence, path-homotopic in Y to the carrier of a hyperplane mapping to D). Then the same holds for paths $\ddot{P} \to \ddot{Y}$.

Suppose $\{\{P\}\}_Y \geq \theta$ whenever $P \to Y$ is a path whose first and last edges are dual to hyperplanes in the same wall of Y. Then the same holds for paths $\ddot{P} \to \ddot{Y}$ whose first and last edges are dual to the same wall of $\ddot{Y}$.

Proof. In each case, a path $\ddot{P} \to \ddot{Y}$ with the above start-end property, projects to a path $P \to Y$ with the same property. The defect $\{\{\ddot{P}\}\}_{\ddot{Y}} = \{\{P\}\}_Y$. $\qquad\square$

Lemma 9.7. *Let $\{Y_i \to X\}$ be local-isometries of cube complexes. Let $\widehat{Y}_i \to Y_i$ be regular covers with the following symmetry property: Each automorphism ϕ of $Y_i \to X$ lifts to an automorphism $\widehat{\phi}$ of $\widehat{Y}_i \to X$.*

For each condition listed in Definition 5.1, if $\langle X \mid \{Y_i\} \rangle$ satisfies this condition then so does $\langle X \mid \{\widehat{Y}_i\} \rangle$.

Proof. For Condition 5.1.(2), we utilize Construction 9.4 to produce the wallspace structure on each $\widehat{Y}_i$, and so if walls of Y_i agree with walls of X then so do walls of $\widehat{Y}_i$. The other conditions are almost immediate. The main point is that $\widehat{Y}_i$ has fewer essential paths than Y_i, their lengths are the same, and for a path P, we have $\{\{P\}\}_{Y_i} = \{\{P\}\}_{\widehat{Y}_i}$. $\qquad\square$

9.c   Obtaining the Separation Properties for Pseudographs

Definition 9.8 (Pseudograph). A nonpositively curved cube complex X is a *pseudograph* if each hyperplane of X is a compact $\mathrm{CAT}(0)$ space.

Lemma 9.9. *Suppose X is a compact pseudograph. Then $\pi_1 X$ is free.*

I don't know if Lemma 9.9 holds when X is a nonpositively curved cube complex whose immersed hyperplanes are simply-connected (without a compactness hypothesis).

Proof. Let H be a hyperplane in $\widetilde{X}$.

If neither component of $X - H$ is deep, then $\pi_1 X$ stabilizes H and so $\pi_1 X$ is trivial.

If exactly one component of $X - H$ is not deep, then $\pi_1 X$ leaves invariant the intersection of the deep halfspaces, and so the theorem is true by induction on the number of cells in X.

If each component of $X - H$ is deep, then we see that $\widetilde{X}$ and hence X has more than one end, and hence $\pi_1 X$ is infinite cyclic or splits as a free product $A * B$. We will show below that each factor A, B cocompactly stablizes a convex subcomplex of $\widetilde{X}$. Consequently each factor is again the fundamental group of a pseudograph. Hence each factor is free, by induction on rank.

We now show that if A is a f.g. subgroup of $\pi_1 X$ then A cocompactly stabilizes a convex subcomplex of $\widetilde{X}$. First observe that there is a nonempty A-invariant connected cocompact subcomplex $\widetilde{Y}_o \subset \widetilde{X}$. Indeed, we let $\widetilde{Y}_o$ be the image of a combinatorial A-invariant map $\Upsilon(A) \to \widetilde{X}$. Let D be an upperbound on the diameter of the carrier of each hyperplane of $\widetilde{X}$. Then we claim that the convex hull of $(\widetilde{Y}_o)$ lies in $\mathcal{N}_{D+2}(\widetilde{Y}_o)$, and so it is cocompact, in addition to obviously being A-invariant and connected. To see this, suppose that $\mathsf{d}_{\widetilde{X}}(p, \widetilde{Y}_o) \geq D+2$ for some 0-cube $p \in \widetilde{X}$. Let γ be a geodesic from a 0-cube of $\widetilde{Y}_o$ to p. Let H be the hyperplane dual to the $(D+2)$-th edge of γ. Then H is disjoint from $\widetilde{Y}_o$ since $N(H)$ is. But H separates p from $\widetilde{Y}_o$, and thus p is not in the convex hull of $\widetilde{Y}_o$. $\qquad\square$

A simple source of pseudographs arises as follows:

Definition 9.10 (Complete). Let Y be a nonpositively curved cube complex and let $\Lambda(Y)$ be its set of hyperplanes. A subset $\Lambda' \subset \Lambda(Y)$ is *complete* if it has the following properties: each hyperplane in Λ' is embedded and 2-sided; no two elements of Λ' cross; and for each nontrivial element $g \in \pi_1 Y$ there is a hyperplane $\widetilde{V} \subset \widetilde{Y}$ that cuts g and such that $\widetilde{V}$ projects to an element of Λ'.

Note that if Y has a complete set of hyperplanes, then the homomorphism $\pi_1 Y \to \mathsf{H}^1(Y)$ has the same kernel as the homomorphism counting traversals of oriented edges dual to hyperplanes of Λ'. Moreover, this property is hereditary under passage of covers.

The following example of a pseudograph Y does not have a complete subset of hyperplanes. Moreover, neither does any finite cover $\widehat{Y} \to Y$.

Example 9.11 (No complete subset). Let Y be the square complex obtained by identifying two squares along their vertices as in the rightmost diagram in Figure 9.1. Then no finite cover $\widehat{Y}$ has a complete subset. To see this, note that $\pi_1 Y \cong F_3$, and Y has two pairs of crossing hyperplanes. For any cover $\widehat{Y} \to Y$, at most one hyperplane in each square of $\widehat{Y}$ could be in a set of complete hyperplanes. Hence for a degree d cover, at most $2d$ hyperplanes could lie in a complete set. However, $\pi_1 \widehat{Y}$ has rank $2d + 1$.

Lemma 9.12. *Let $\langle X \mid Y_1, \ldots, Y_r \rangle$ be a cubical presentation. Suppose each Y_i is a compact pseudograph. Suppose all pieces in each Y_i have finite diameter. There exist finite regular covers $\widehat{Y}_i \to Y_i$ such that $\langle X \mid \widehat{Y}_1, \ldots, \widehat{Y}_r \rangle$ satisfies the hypotheses of Theorem 5.44.*

We first prove Lemma 9.12 in the special case where each Y_i is a graph, and then explain it for a general pseudograph. It is an interesting problem to generalize Lemma 9.12 to non-pseudograph setting.

Proof of graph case. For simplicity, we will use the notation Y for a cone Y_i. First observe that the pieces of Y are collections of finite trees $\{T_j\}$ in a cone Y. Let $A \geq 1$ be an upperbound on the diameter of all such pieces, and let $B = 2A$. By residual finiteness, there exists a finite regular cover $\bar{Y} \to Y$, such that $\|\bar{Y}\| \geq 4B + 1$.

Let $\ddot{Y} \to \bar{Y}$ be the cover of Construction 9.1 where S consists of the set of hyperplanes of $\bar{Y}$, and the map $\Lambda(\bar{Y}) \to S$ is the identity. Thus $\ddot{Y}$ corresponds precisely to the cover associated to $\mathsf{H}_1(\bar{Y}, \mathbb{Z}_2)$.

Hypothesis (2) of Theorem 5.44 holds since a geodesic that starts on an edge and ends within a piece of that edge has length at most one more than a piece.

We now verify Hypothesis (3) of Theorem 5.44. Consider a geodesic $\kappa \to \widetilde{X}^*$ that lies in a cone $\ddot{Y}$, and let p, q be the endpoints of κ. Let $\ddot{w}_1, \ddot{w}_1'$ be distinct edges dual to the same wall $[w_1]$ of $\ddot{Y}$ so $\ddot{w}_1, \ddot{w}_1'$ project to the same edge w_1 of $\bar{Y}$, and suppose $\ddot{w}_1$ separates the endpoints of κ and that $\ddot{w}_1'$ is 1-proximate to q.

Let $\bar{\kappa} \to \bar{Y}$ be the projection of $\kappa \to \ddot{Y}$, and let $\bar{p}, \bar{q}$ be its endpoints.

If $\mathsf{d}(\bar{p}, \bar{q}) \geq 2B + 1$, then regarding $\bar{\kappa}$ as a $\mathbb{Z}_2$ 1-chain, we see that $\bar{\kappa}$ oddly traverses an edge w_2 outside $\mathcal{N}_B(p) \cup \mathcal{N}_B(q)$, and the preimage of w_2 provides a wall $W_2 = [\ddot{w}_2]$ in $\ddot{Y}$ that separates p, q but is not 2-proximate with either.

If $\mathsf{d}(\bar{p}, \bar{q}) \leq 2B$, then letting $Z = \mathcal{N}_{2B}(p)$ and letting $\bar{Z} \subset \bar{Y}$ be the image of Z, we note that $\mathsf{H}_1(\bar{Y}, \bar{Z}; \mathbb{Z}_2) = \mathsf{H}_1(\bar{Y}; \mathbb{Z}_2)$ since $\bar{Z} \cong Z$ is contractible. The path

$\kappa \to \overset{..}{\widetilde{Y}}$ represents an element $[\bar{\kappa}] \in \mathsf{H}_1(\bar{Y}, \bar{Z}; \mathbb{Z}_2)$. Observe that $[\bar{\kappa}] \neq 0$, for otherwise κ would start and end in the same component of the preimage of Z in $\overset{..}{\widetilde{Y}}$, and thus $\mathsf{d}_{\overset{..}{\widetilde{Y}}}(p, q) \leq \mathrm{diam}(Z) = B$, which contradicts that κ is a geodesic since $|\kappa| \geq \|\bar{Y}\|$ as $\ddot{w}_1, \ddot{w}_1'$ are distinct edges of κ with the same image in $\bar{Y}$.

Since $[\bar{\kappa}] \neq 0$ we see that $\bar{\kappa}$ oddly traverses an edge w_2 lying outside $\mathcal{N}_B(\bar{p}) \cup \mathcal{N}_B(\bar{q})$. Indeed, $\|\bar{Y}/\bar{Z}\| \geq \|\bar{Y}\| - 2B \geq 2B + 1$, and hence we may choose an edge $w_2/\bar{Z}$ in the quotient cycle κ/Z at distance at least B from the basepoint $\bar{Z}/\bar{Z}$ of the graph $\bar{Y}/\bar{Z}$. This provides the desired edge w_2 of $\bar{Y}$ yielding a wall in $\overset{..}{\widetilde{Y}}$ that separates p, q but is not 2-proximate to p or q.

The claim will follow for $X^* = \langle X \mid \widehat{Y}_i \rangle$ by letting $\widehat{Y}_i = \overset{..}{\widetilde{Y}}_i$ for each i. However, we also arrange that the $C'(\frac{1}{12})$ condition holds by assuming that each $\overset{..}{\widetilde{Y}}_i \to Y_i$ has $\|\overset{..}{\widetilde{Y}}\| \geq 12A + 1$, which follows if we assume each $\|\bar{Y}_i\| \geq 6A + 1$. Note that since wall-pieces are trivial in this case, short innerpaths also holds at $C'(\frac{1}{12})$ and there is no need to use $C'(\frac{1}{14})$ and invoke Lemma 3.70. In fact, one obtains the desired conclusion at $C'(\frac{1}{6})$ which hold when $\|\bar{Y}\| \geq 3A + 1$. $\square$

Proof of general case. Simplifying notation, let $Y = Y_i$ be a cone. Any compact pseudograph Y has a finite cover $\bar{Y}_1 \to Y$ that is special. Indeed, $\pi_1 Y$ is free by Lemma 9.9, and hence subgroup separable, and hence has separable double cosets of f.g. subgroups. Thus Y has a finite special cover by Proposition 6.9.

Choose a finite graph Γ and a map $Y \to \Gamma$ inducing an isomorphism $\pi_1 Y \to \pi_1 \Gamma$. We may also assume the carrier of each hyperplane of $\widetilde{Y}$ map to finite trees in $\widetilde{\Gamma}$.

Let A be an upperbound on the diameter of pieces occurring in $\langle X \mid \{Y_i\} \rangle$. By residual finiteness we can pass to a finite cover $\bar{Y} \to Y$ so that $\|\bar{Y}\| > 14A$, and so that for any hyperplane H of $\bar{Y}$, its carrier $N = N(H)$ has the property that N^{+A} is convex in $\bar{Y}$, and any path $P \to \bar{Y}$ that is the concatenation of fewer than 8 pieces and that starts and ends on N is path-homotopic into N. These choices enable most of the small-cancellation parts of the $B(6)$ condition, as well as Hypothesis 5.44.(2). The condition that each cone have a wallspace structure will arise by applying Construction (9.1) to each $\bar{Y}$.

Hypothesis 5.44.(3) will be enabled by two additional properties: Firstly, an embedded path $\bar{\kappa} \to \bar{Y}$ wh ose endpoints $\bar{p}, \bar{q}$ are quite far must traverse many hyperplanes. Secondly, a (nearly)-closed essential path $\bar{\kappa}$ that oddly traverses a hyperplane must oddly traverse many hyperplanes. Local finiteness and the upperbound A on piece diameter ensures that there is a uniform upperbound M on the number of walls that are 2-proximate with p or q.

Assume moreover that $\bar{Y}$ is chosen so that $\mathrm{girth}(\bar{\Gamma}) > 4M + 2$, where $\bar{\Gamma} \to \Gamma$ is the cover associated with $\bar{Y} \to Y$. Note that $\mathsf{H}_1(Y) \to \mathsf{H}_1(\Gamma)$ is an isomorphism and $\pi_1 Y \to \pi_1 \Gamma$ is the homomorphism counting the number of times a path traverses edges of Γ with multiplicity. The analogous statement holds for covers $\bar{Y}$ and $\bar{\Gamma}$. Let $\overset{..}{\widetilde{Y}} \to \bar{Y}$ and $\overset{..}{\widetilde{\Gamma}} \to \bar{\Gamma}$ be the covers induced by $\mathsf{H}_1(-, \mathbb{Z}_2)$ with wallspace structures as in Construction 9.1.

Let κ be a geodesic in $\ddot{Y}$ from p to q. Let $\bar{\kappa} \to \bar{Y}$ and $\bar{p}, \bar{q}$ be the images in $\bar{Y}$, and let $\dot{\kappa}, \dot{p}, \dot{q}$ be the images in $\dot{\Gamma}$. Now we follow the same plan as in the graph case: If $\mathsf{d}(\dot{p}, \dot{q}) \geq 2M + 1$, then regarding $\dot{\kappa} \to \dot{\Gamma}$ as a $\mathbb{Z}_2$-chain, we see that $\dot{\kappa}$ oddly traverses at least $2M + 1$ edges of $\dot{\Gamma}$ and hence $\bar{\kappa}$ oddly traverses at least $2M + 1$ hyperplanes associated to $2M + 1$ walls in $\ddot{Y}$. These walls all separate p, q, so at least one of them is not 2-proximate with p, q. If $\mathsf{d}(\dot{p}, \dot{q}) \leq 2M$, then $\dot{\kappa}$ represents a cycle in $\mathsf{H}(\Gamma, Z; \mathbb{Z}_2)$ where $Z = \mathcal{N}_{2M}(\dot{p})$. Moreover $[\dot{\kappa}] \neq 0$ since its endpoints do not lie in the same component of the preimage of Z in $\dot{\Gamma}$. Quotienting by Z, and regarding the image of $\dot{\kappa}$ as a $\mathbb{Z}_2$-chain, we see that it contains a closed cycle σ, whose length is at least $2M + 1$. Each edge of σ is traversed oddly by $\dot{\kappa}$, and each edge corresponds to a hyperplane of $\bar{Y}$ that is associated to a wall of $\ddot{Y}$ that is traversed an odd number of times by κ, and hence separates p, q. At most M of these walls are 2-proximate with p, and at most M are 2-proximate with q. Hence at least one of these walls separates p, q but is not 2-proximate with either. $\qquad\square$

Chapter Ten

Cutting X^*

The goal of this chapter is to give a geometric interpretation of how $\pi_1 X^*$ splits along a wall stabilizer, when the wall does not cross its translates. Ultimately, we will apply this to hierarchies of cubical presentations as described in Section 10.a. However, we set up the splitting by inflating conepoints in Section 10.b, we examine the persistence of conditions that guarantee the splitting have desired properties in Section 10.c, and we describe the two-steps of algebra that take us from a cubical presentation to the next cubical presentations in Section 10.d.

10.a   Hierarchies of Cubical Presentations

Definition 10.1. A *hierarchy* for a nonpositively curved cube complex X is a sequence of cube complexes $X_0, X_1, \ldots, X_r$ where $X_0 = X$ and $X_r = X^0$ is a set of 0-cells, and for $0 \leq i < r$ we have $X_{i+1} = X_i - N^o(D_i)$ where D_i is the union of a set of disjoint 2-sided embedded hyperplanes in the cube complex X_i, and where $N^o(D)$ denotes the open cubical neighborhood of D.

Every compact nonpositively curved cube complex whose hyperplanes are 2-sided and embedded has a hierarchy, so in particular this holds for any compact special cube complex. In the compact setting, if there is a hierarchy, then there is a (possibly longer) hierarchy where each D_i consists of a single hyperplane. While the original hierarchy for $\pi_1 X$ has a graph of groups for each stage, in contrast, with single hyperplanes, there would be an amalgamated product or HNN extension at each stage. However, for organizational purposes it is natural to allow D_i to consist of more than one hyperplane. Furthermore, permitting multiple components in D_i allows us to accommodate infinite situations, where this flexibility might be the only way for the hierarchy to be of finite length. Finally, we note that if X' has a hierarchy, and there is a local-isometry $X \to X'$ then there is an induced hierarchy for X.

There are also relevant generalizations where each terminal component of X_r has some property, and we would say the hierarchy terminates with this property. In particular, we will occasionally mention a *hierarchy terminating at cube complexes with virtually abelian* π_1.

Definition 10.2 (Hierarchies of Cubical Presentations). Let X^* be a cubical presentation. Suppose X has a hierarchy with $X = X_0, X_1, \ldots, X_r$ where

$X_{i+1} = X_i - N^o(D_i)$. Let $X_0^* = X^*$. For each $i \geq 0$, let X_{i+1}^* be the cubical presentation whose cones are the complementary components of the preimage of $N^o(D_i)$ in cones of X_i^*. Note that X_{i+1} might not be connected, so strictly speaking we are dealing with disjoint unions of cubical presentations.

Suppose that each cone of X^* is a wallspace as in Definition 2. A *hierarchy* for X^* is a hierarchy for X with the following additional property for each i: For each cone Y of X_i^*, the preimage of D_i in Y consists of a union of non-crossing walls of Y. (Two walls *cross* if all four intersections of halfspaces are nonempty.) The wallspace structure on a cone of X_{i+1}^* is induced by intersection in the sense that if Y_{i+1} is a component of the preimage of the complement of $N^o(D_i)$ in Y_i, then each wall of Y_{i+1} is a (nonempty) intersection with a wall of Y_i.

A primary source of examples arises where X^* is $B(6)$ and the hyperplanes of X are partitioned into collections of pairwise disjoint embedded 2-sided hyperplanes "base-walls," and for each base-wall, any nonempty preimage in a cone Y of X^* consists of a single wall of Y. In Constructions 10.10 and 10.11 we describe the hierarchy of $\pi_1 X^*$ that results from a hierarchy of X^*. This is very similar to the hierarchy for $\pi_1 X$ induced by a hierarchy for X, but there are some additional splittings arising from disconnected halfspaces in cones, and further complications arising when $\mathrm{Aut}(Y_i)$ is nontrivial.

10.b Inflations

Construction 10.3 (Inflating a single cone with respect to wall). Let Y be a nonpositively curved cube complex. Let w be a wall consisting of the union of a separating collection of disjoint 2-sided embedded hyperplanes that do not self-osculate or osculate with each other.

Let $N^o(w)$ be the open cubical neighborhood of w, and note that $N^o(w) \cong w \times (-\frac{1}{2}, \frac{1}{2})$ by assumption. There is a natural map $w \times [-\frac{1}{2}, \frac{1}{2}] \to Y$. Let Y^- and Y^+ denote the parts of $Y - N^o(w)$ on opposite sides of the wall w so that $Y^{\pm}$ contains the image of $w \times \{\pm\frac{1}{2}\}$ respectively.

We define the *inflated cone* $C_w(Y)$ to be the union of the ordinary cones $C(Y^+)$ and $C(Y^-)$ glued together with $C(w) \times [-\frac{1}{2}, \frac{1}{2}]$ by identifying $C(w) \times \{-\frac{1}{2}\}$ with its image in $C(Y^-)$ and identifying $C(w) \times \{+\frac{1}{2}\}$ with its image in $C(Y^+)$. We identify $C(w)$ with $C(w) \times \{0\} \subset C(w) \times (-\frac{1}{2}, \frac{1}{2})$. We refer the reader to Figure 10.1.

There is a natural map $C_w(Y) \to C(Y)$ induced by quotienting the cone-edge to the conepoint.

Construction 10.4 (Inflated cone with respect to a wall). We now consider $\langle X \mid \{Y_i\} \rangle$, and suppose that each Y_i is a wallspace. Let w be a *base-wall* of X^* consisting of a collection of disjoint hyperplanes with the property that $w_i = w \wedge Y_i$ is a wall in the wallspace structure of Y_i for each i. We use the notation $w \wedge Y_i$ for the preimage of w under the map $Y_i \to X$. (In our later application

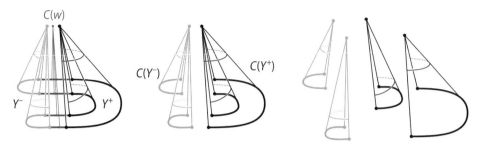

Figure 10.1. On the left is $C_w(Y)$ which is a variant of the cone $C(Y)$ that combinatorially enables a geometric splitting along a wall w of Y. The complex $C_w(Y)$ is the union of $C(Y^-)$ and $C(Y^+)$ which are illustrated at the center, together with $C(w) \times [-\frac{1}{2}, +\frac{1}{2}]$. On the right are the cones obtained by cutting $C(Y^\pm)$ along their conepoints.

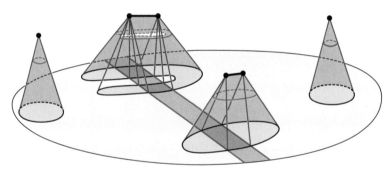

Figure 10.2. $C_w(\langle X \mid Y_1, Y_2, Y_3, Y_4 \rangle)$ has $Y_i \cap w$ with $0, 2, 1, 0$ components respectively.

of this construction, w is the image of $\mathrm{Stab}(W) \backslash W$ where W is a wall of $\widetilde{X}^*$.) We shall *inflate* the coned-off space X^* which equals the union $X \cup \bigcup_i C(Y_i)$ as follows: Let $C_{w_i}(Y_i)$ denote the inflated cone space of Y_i with respect to w_i. When $w_i = \emptyset$ we simply let $C_{w_i}(Y_i) = C(Y_i)$. The *base* of $C_{w_i}(Y_i)$ is Y_i as usual. Define $C_w(\langle X \mid \{Y_i\} \rangle)$ to be $X \cup \bigcup_i C_{w_i}(Y_i)$ where for each i we attach $C_{w_i}(Y_i)$ to X along its base using the map $Y_i \to X$. See Figure 10.2. The map $C_w(\langle X \mid \{Y_i\} \rangle) \to X^*$ that sends each cone-edge to a conepoint and each inflated cone to a cone is a homotopy equivalence.

Observe that the base-wall w has a natural geometric extension w^* in $C_w(\langle X \mid \{Y_i\} \rangle)$ consisting of the union of hyperplanes in w, together with the cone $C(w_i)$ on each w_i in Y_i, which lies in $C(w_i) \times \{0\} \subset C(w_i) \times (-\frac{1}{2}, \frac{1}{2})$. In particular, w^* has an open neighborhood $N^o(w^*)$ in $C_w(\langle X \mid \{Y_i\} \rangle)$ consisting of the union of open cubical neighborhoods of hyperplanes in w together with the open neighborhoods $C(w_i) \times (-\frac{1}{2}, \frac{1}{2})$ of $C(w_i)$ in each $C_{w_i}(Y_i)$ with $w_i \neq \emptyset$.

Finally, the base-walls of the components obtained from $C_w(\langle X \mid \{Y_i\} \rangle) - N_o(w^*)$ by cutting along conepoints are the intersections of base-walls of X with the components of $X - N_o(w)$.

We will examine the image of $\pi_1 w^*$ below by interpreting $N(w^*) = w^* \times I$, as a π_1-injective subcomplex and use this to understand the splitting of $\pi_1 X^*$.

The splitting of $\pi_1 X^*$ is represented geometrically by cutting $C_w(\langle X \mid \{Y_i\} \rangle)$ along w^*. In particular, $C_w(\langle X \mid \{Y_i\} \rangle) - w^*$ deformation retracts to $C_w(\langle X \mid \{Y_i\} \rangle) - N^o(w^*)$ which equals the space obtained from $X - N^o(W)$ by coning off the various Y_i^+ and Y_i^-. More precisely:

$$C_w(\langle X \mid \{Y_i\} \rangle) - N^o(w^*) = C\big(\langle X - N^o(W) \mid \{Y_i\}_{w_i = \emptyset}, \ \{Y_i^+, Y_i^-\}_{w_i \neq \emptyset} \rangle\big).$$

It is often the case that Y_i^+ or Y_i^- is not connected, but is treated as one unit and receives a single conepoint in $C_w(\langle X \mid \{Y_i\} \rangle)$. See Figure 10.1.

By utilizing additional splittings over conepoints with disconnected base we can pass to a space which is the coned-off space of the complex (which may have two components) consisting of $\langle X - N^o(w) \mid Y_j, Y_i^\pm \rangle$. The wallspace structures on Y_i^+ and Y_i^- are induced by the original wallspace Y_i. Note that a wall in $Y_i^\pm$ might have more hyperplanes than in Y_i.

10.c   Some Persistent Properties

Lemma 10.5 (Persistence after cutting). *Let $X^* = \langle X \mid \{Y_i\} \rangle$ be a cubical presentation. Let w be a base-wall of X^*, and consider the cubical presentations obtained from $C_w(\langle X \mid \{Y_i\} \rangle) - N(w^*)$ after cutting along conepoints, as described in Construction 10.4. Then each of the following conditions persist in the sense that if it holds for X^* then it holds for each cubical presentation after cutting:*

(1) *The $B(8)$ condition.*
(2) *Malnormality of the collection of hyperplanes in a base-wall.*
(3) *Short innerpaths.*
(4) *No self-grazing.*

Proof. Each component of $Y_i - N^o(w)$ is a nonpositively curved cube complex, and continues to map to $X - N^o(w)$ by a local-isometry.

If Y_i is a wallspace, then the walls of a component Z of $Y_i - N^o(w)$ are defined to be the intersection of Z with walls of Y_i.

Lemma 10.6 verifies the persistence of $B(8)$.

Lemma 2.30 verifies that malnormality of a collection of hyperplanes persists.

Lemma 10.8 verifies that short innerpaths persist.

Lemma 10.9 verifies the persistence of no self-grazing. □

Lemma 10.6. *The $B(8)$ conditions persist.*

Proof. We verify the persistence of Condition 5.55.(2) as the other properties are proven similarly. Let X_1^* be obtained from X^* by cutting along w and then along conepoints, and let Y_1 be a cone of X_1^* obtained in this way from a cone Y of X^*. Let S be a path in Y_1 that starts and ends on a wall v_1 of Y_1 and suppose $\{\{S\}\}_{Y_1} < \pi$. Then S is a path in Y starting and ending on a wall v inducing v_1 such that $\{\{S\}\}_Y < \pi$. Consequently, by Condition 5.55.(2) we see that S is homotopic into the carrier of a single hyperplane u of v.

Consider a minimal area square diagram D between S and a path $P \to N(u)$. Obviously, no 1-cell of S is dual to w. Hence no 1-cell of P is dual to w, for otherwise, we could reduce the complexity of D. Indeed, a dual curve in D that is dual to w must start and end on 1-cells of P, and so we could push P past it to obtain a smaller diagram D' between S and a new path P' that is still on $N(u)$. (This is essentially an application of Lemma 2.3.) As D is disjoint from w, it is a diagram in Y_1 and hence P' lies on the carrier of a single hyperplane $u_1 \subset u$ of Y_1. □

Remark 10.7. As opposed to the other parts of Lemma 10.5, the proof of Lemma 10.6 uses that each $Y_i \to X$ is embedded. More generally, it holds under the hypothesis that no Y_i contains a subcomplex with a nontrivial automorphism. The difficulty is that a large piece between $Y_i^{\pm}$ and $Y_j^{\pm}$ might not come from a piece before cutting since its lift to X might extend to a morphism between Y_i and Y_j. Consequently, the small-cancellation is preserved only under the assumption that Construction 10.4 preserves each $\mathrm{Aut}(Y_i)$. The reader can examine $\langle X \mid Y \rangle$ where X is a 4-cycle whose walls are opposite pairs of edges and where Y is a finite nontrivial cover of $\ddot{X}$.

Lemma 10.8. *Short innerpaths persists.*

Proof. Suppose that $QS \to Y_1$ is essential and $\{\{S\}\}_{Y_1} < \pi$. Let Y denote the cone that Y_1 is derived from after cutting X^* along the base-wall w. Regarding S as a path in Y we have $\{\{S\}\}_Y \leq \{\{S\}\}_{Y_1}$ and hence $\{\{S\}\}_Y < \pi$. Since QS is essential in Y_1 which is locally-convex in Y, we see that QS is essential in Y by Corollary 2.12. Since X^* has short innerpaths we see that $|S|_Y < |Q|$. Finally $|S|_{Y_1} = |S|_Y$ by local convexity, since $\widetilde{Y_1} \subset \widetilde{Y}$ is a convex subcomplex by Lemma 2.11. □

Lemma 10.9 (No self-grazing persists). *Suppose X^* has no self-grazing. Let w be a base-wall in X^*, and let $X_\circ^*$ be a cubical presentation obtained by cutting along w and then along conepoints. Then $X_\circ^*$ also has no self-grazing.*

Proof. Let Y_1 be a cone of $X_\circ^*$, and let h_1, k_1 be hyperplanes in the same wall of Y_1, and suppose h_1 and k_1 are dual to 1-cells in the same cone-piece of Y_1,

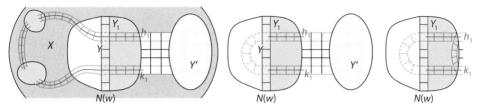

Figure 10.3. Persistence of no self-grazing: If h, k are dual to 1-cells in the same piece and lie in the same base-wall of Y_1, then they lie in the same base-wall of Y as on the left. By no self-grazing they lie in the same wall of Y as in the center, and consequently, are equal to the same hyperplane in Y_1 since they are connected by a hyperplane in the piece that they travel through.

say associated to a cone Y_1' along some rectangular diagram between Y_1, Y_1'. We refer the reader to Figure 10.3. By construction, Y_1 is a component of a cone Y obtained by cutting along $N(w)$, so Y_1 is either Y^+ or Y^-. And likewise Y_1' arises from some Y', and the piece K between Y_1, Y_1' arises from a cone-piece between Y and Y' accordingly. In particular, h_1, k_1 arise from hyperplanes h, k in Y that are dual to 1-cells e_h, e_k in this cone-piece. Since h_1, k_1 lie in hyperplanes in the same base-wall v_1 of $X^*_\circ$, and v_1 is induced from a base-wall v of X^*, we see that h, k lie in hyperplanes in the same base-wall v. No self-grazing applied to h, k, v in X^* implies that h, k are in the same hyperplane U of Y. Consider a path $P \to K$ from e_h to e_k, and note that $\{\{P\}\}_Y = 0 < \pi$. Condition 5.1.(3) implies that $P \to K$ is path-homotopic to the carrier of the hyperplane dual to e_h and e_k. The path-homotopy lifts to $\widetilde{Y}$ and shows that $\widetilde{K}$ intersects $\widetilde{U}$ in a hyperplane of $\widetilde{K}$. Thus h_1 and k_1 are equal to the same hyperplane of Y_1. $\square$

10.d   Additional Splitting along Conepoints

We now describe the two-step splitting of $\pi_1 X^*$ that results from cutting X^* along a base-wall.

The following requires that each cone is compact and has trivial automorphism group. We will explain the general case allowing torsion in Construction 10.11. Suiting our application, for simplicity we work under the assumption that the negatively curved small-cancellation structure for X^* arises from the $C'(\frac{1}{12})$ condition.

Construction 10.10 (Algebraic splitting). Under appropriate small-cancellation conditions, the geometric splitting of $C_w(\langle X \mid \{Y_i\} \rangle)$ along w^* induces a splitting of $\pi_1 X^*$ along the image of $\pi_1 w^*$.

Adding dummy squares to form $\underline{X}^*$: It is inconvenient that $w \wedge Y_i$ might be disconnected for some values of i. We remedy this by adding *dummy squares* as follows: Let $w_{i1}, \ldots, w_{ip_i}$ denote the hyperplanes that are the components of

Figure 10.4. Each $T_i \times I$ added on the right connects the wall in a cone.

$w_i = w \wedge Y_i$. Choose 1-cells $e_{i1}, \ldots, e_{ip_i}$ in Y_i that are dual to these. Let T_i be a p_i-star having base-vertex v_i of valence p_i, and p_i distinct edges that meet at v_i and end at leaves $v_{i1}, \ldots, v_{ip_i}$. (We shall later subdivide these edges to presume they have some minimal length, but for now the reader can assume each arc is a single edge.) Consider the square complex $T_i \times I$, and now attach a copy of $T_i \times I$ to Y_i with the edge $v_{ik} \times I$ attached to e_{ik} (so that the orientations are consistent), and likewise attach a copy of $T_i \times I$ to X with the edge $v_{ik} \times I$ attached to the image of e_{ik} under the map $Y_i \to X$. We do not perform this procedure when $p_i = 0$ in which case $w \wedge Y_i = \emptyset$.

Following the above procedure we obtain $\underline{Y}_i = Y_i \cup (T_i \times I)$ for each i, and $\underline{X} = X \cup_i (T_i \times I)$. Let $\underline{Y}_i = Y_i$ when $w_i = \emptyset$. Let $\underline{w}$ denote the hyperplane of $\underline{X}$ containing w, so the carrier $N(\underline{w})$ contains $N(w)$ and each $T_i \times I$. The advantage of $\underline{w}$ over w is that a wall $\underline{W}$ in $\widetilde{\underline{X}}^*$ corresponding to $\underline{w}$ has a unique hyperplane. Note that $\underline{W} \cap \widetilde{X}^* = W$, a wall mapping to the base-wall w.

Observe that $\langle \underline{X} \,|\, \{\underline{Y}_i\} \rangle$ deformation retracts to $\langle X \,|\, \{Y_i\} \rangle$ by collapsing along free faces. Moreover, there is a deformation retraction of $C_{\underline{w}}(\langle \underline{X} \,|\, \{\underline{Y}_i\} \rangle)$ onto $C_w(\langle X \,|\, \{Y_i\} \rangle)$ that pulls each base-edge of $T_i \times I$ upwards to the cone-edge of $C_{\underline{w}_i}(Y_i)$.

Dummy squares preserve small-cancellation: Suppose each $\text{Aut}(Y_i)$ is trivial (e.g., each $Y_i \to X$ is an embedding). If $\langle X \,|\, \{Y_i\} \rangle$ satisfies $C'(\alpha)$ then so does $\langle \underline{X} \,|\, \{\underline{Y}_i\} \rangle$. To see this, first note that 1-cells of $T_i \times I$ that are parallel to the T_i-factor do not lie in any piece, and if e is a 1-cell parallel to the I-factor, then if e lies in a new wall-piece with some $T_j \times I$, then e already lies in cone-piece with the Y_j. Finally, let M be an upperbound on the pieces in X^*. We can subdivide each edge of each T_i so that it has length $12M$. Hence $C'(\frac{1}{12})$ continues to hold for $\underline{X}^*$ if it holds for X^*, since any cycle in $\underline{Y}_i$ that was not already a cycle in Y_i has length exceeding $12M$. (Generalizing the supporting results would make this subdivision unnecessary.) Thus $\underline{X}^*$ has short innerpaths by Lemma 3.70.

Examining the splitting: Let $\underline{A} = N(\underline{w}) \cong \underline{w} \times [-\frac{1}{2}, \frac{1}{2}]$, and consider the map $\underline{A} \to \underline{X}$. As in Definition 3.63, let $\underline{A}^* = \langle\ \underline{A}\ |\ \{\underline{A} \otimes_X Y_i\}\ \rangle$ denote the induced presentation. Our $T_i \times I$ additions have made the cones connected, whereas $A \otimes_X Y_i$ has components in one-to-one correspondence with components of $w \otimes_X Y_i$.

Since an innerpath of a shell in $\underline{X}^*$ is already an innerpath of a shell in X^*, Condition 5.1.(3) implies that $\underline{A}^* \to \underline{X}^*$ has no missing shells, and is thus

Figure 10.5. Objects considered in the hierarchy for a cubical presentation of $\mathbb{Z}_3$: From left to right are X^*, $\widetilde{X}^\star$, $\overline{w}^\star \subset \overline{X}^\star$, $\overline{X}^\star_\pm$, $X^*_\pm$, X^*_+, and X^*_- expressed as cubical presentations.

π_1-injective by Theorem 3.68. Moreover, $\pi_1\underline{A}^* \subset \pi_1\underline{X}^*$ is quasi-isometrically embedded by Corollary 3.72. (We don't use that the cones embed here, so the same argument works within Construction 10.11.)

Let $\underline{w}^*$ denote the space obtained from $\underline{w}$ by coning off the various $\underline{w}_i$, so it effectively has the structure of a cubical presentation. Observe that $\underline{A}^*$ is isomorphic to a product $\underline{w}^* \times [-\frac{1}{2}, \frac{1}{2}]$ and that the deformation retraction map $C_{\underline{w}}(\langle \underline{X} \,|\, \{\underline{Y}_i\} \rangle) \to C_w(\langle X \,|\, \{Y_i\} \rangle)$ sends $\underline{w}^*$ to a space whose fundamental group is the same as $\mathrm{Stab}(W)$ where W is the wall in $\widetilde{X}^*$ corresponding to w, which in turn corresponds to the wall corresponding to the universal cover of $\underline{w}^*$.

We thus see that $\pi_1\underline{A}^* = \mathrm{Stab}(W)$, and that $\pi_1 X^* = \pi_1\underline{X}^*$ splits as an HNN extension or AFP over the subgroup $\mathrm{Stab}(W) = \mathrm{Stab}(\underline{W}) = \pi_1\underline{A}^*$.

The reader might prefer to consider the two maps from $\underline{w} \times \{\pm\frac{1}{2}\}$ to $\underline{X} - N^o(\underline{w})$, and verify that there are no missing shells, and that the induced presentations of $\underline{w} \times \{+\frac{1}{2}\}$ and $\underline{w} \times \{-\frac{1}{2}\}$ are the same.

Construction 10.11 (When cones have nontrivial automorphisms). When each $\mathrm{Aut}(Y_i)$ is trivial, Construction 10.10 explains the group theoretical splittings corresponding to the passage from X^* to $X^*_\pm$ by first splitting along the stabilizer of a wall and then along trivial subgroups at conepoints. When some of the $\mathrm{Aut}(Y)$ are nontrivial then we explain this transition using the action on the universal cover. A simple example is indicated explicitly in Figure 10.5. A heuristic sketch of the steps we will follow is illustrated in Figure 10.6.

Let $X^\star$ denote $C_w(X^*)$ which is the inflation of X^* along a base-wall w of X. Recall from Construction 10.4 that w^* denotes the coned-off copy of w within X^*. Let $X^\star_+$ and $X^\star_-$ denote the spaces obtained from $X^\star$ by removing $N^o(w^*)$. Let $X^*_\pm$ denote the spaces obtained from $X^\star_\pm$ by cutting along the conepoints, i.e., by "unwedging." Observe that $X^*_+ = \sqcup_k X^*_{+k}$ is the disjoint union of cubical presentations and likewise $X^*_- = \sqcup_k X^*_{-k}$, and that when the base-wall has multiple components it is possible for the cubical parts to be separated. (Our notation considers the case of an amalgamated free product—where w separates X. The non-separating case is similar.)

For a cone Y_i in $\widetilde{X}^*$, it is natural to identify the cones $C(Y_i)$ that lie in the same $\mathrm{Aut}(Y_i)$ orbit. Then $\pi_1 X^*$ acts with $\mathrm{Aut}(Y_i)$ stabilizers at conepoints. In a corresponding fashion, we can identify cones in $\widetilde{X}^\star$ and $\pi_1 X^*$ acts freely except at the cone-edges, which again have $\mathrm{Aut}(Y_i)$ stabilizers. Let $\overline{X}^\star$ denote the space

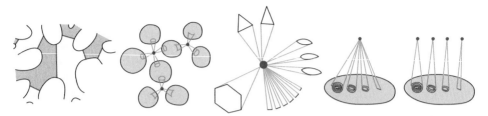

Figure 10.6. $\pi_1 X^* \overline{w}^\star \subset \overline{X}^\star$ is depicted in the first diagram. Each component of $\overline{X}^\star - \pi_1 X^* N^o(\overline{w}^\star)$ is a translate of $\overline{X}^\star_\pm$ that itself has a tree structure. This tree structure is indicated for $\overline{X}^\star_+$ in the second diagram. Its vertex spaces are conepoints and cube complexes and its edge spaces are cones. The stabilizers of the conepoint vertex spaces are the various $\mathrm{Aut}(Y_i)$ where Y_i is cut by $\overline{w}$. The stabilizers of the cone-edge spaces are the subgroups $\mathrm{Aut}(Y_{ij})$, where Y_{ij} are the parts of Y_i after cutting along $\overline{w}$. A neighborhood of a minor vertex space is depicted in the third diagram and its vertex group $\mathrm{Aut}(Y_i)$ permutes the orbits of the various Y_{ij} each of which corresponds to a cone-edge space attached to a distinct $\overline{X}_{+k}$ translate. The "unwedging" procedure to get from $X^\star_+$ to $X^\star_{+1}$ is suggested by the last two figures.

obtained by equivariantly identifying inflated cones in the same $\mathrm{Aut}(Y_i)$ orbit. (Warning: $X^\star$ is not usually a quotient of $\overline{X}^\star$.)

Let $\widetilde{w}^*$ denote a lift of the universal cover of w^* in $\widetilde{X}^*$, where the "cubical base" of $\widetilde{w}^*$ is a wall W of the type studied in Chapter 5. We use the notation $\widehat{w}^\star$ for its isomorphic image within $\widehat{X}^\star$, and use the notation $\overline{w}^\star$ for its quotient within $\overline{X}^\star$.

$\overline{X}^\star$ is simply-connected since $\widehat{X}^\star$ is simply-connected since $\widetilde{X}^*$ is. As $\overline{X}^\star$ is simply-connected, and the coned-off wall $\overline{w}^\star$ in $\overline{X}^\star$ is simply-connected (as explained below), we see that the complementary components of $\overline{X}^\star - \pi_1 X^* N^o(\overline{w}^\star)$ are simply-connected.

Let $\overline{X}^\star_\pm$ denote one of these components. We now describe how the stabilizer of $\overline{X}^\star_\pm$ splits as a graph of groups. Indeed, $\overline{X}^\star_\pm$ has a treelike structure: Its *minor vertex spaces* are the conepoints and its *major vertex spaces* are its maximal cubical subcomplexes (disjoint from conepoints). Each major vertex space is the cubical part of some $\widehat{X}^*_{\pm k}$. Indeed, the closure of a component of $\overline{X}^\star_\pm -$ {conepoints} is equal to a translate of some $\overline{X}^*_{\pm k}$ which is the quotient $\widehat{X}^*_{\pm k} \to \overline{X}^*_{\pm k}$ induced by $\widehat{X}^* \to \overline{X}^\star$. Note that we identify isomorphic cones with the same base that arises from the same cone of X^* and are hence in the same $\pi_1 X^*$ orbit.

Its *edge spaces* are cones consisting of lifts of cones from within $X^*_{\pm k}$. The underlying graph is a tree since the vertex and edge spaces are connected and $\overline{X}^\star_\pm$ is simply-connected. The stabilizers of edge spaces naturally embed in the major vertex groups regarded as $\mathrm{Aut}(\widehat{X}^*_{\pm k})$. The stabilizers of edge spaces naturally

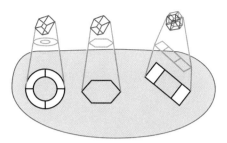

Figure 10.7. Cubical conepoints.

embed in the minor vertex groups using $\mathrm{Aut}(Y_{ij}) \subset \mathrm{Aut}(Y_i)$. Here Y_{ij} denote the various complementary components of the preimage of $N^o(w)$ in the Y_i. The cones that are not intersected by $\overline{w}$ can be ignored in the graphical structure and subsumed within the major vertex space parts. If we had attended to these cones and conepoints, they would have contributed spurs to the underlying tree.

Each $\overline{w}^\star$ is simply-connected: For a hyperplane $v \to X$, let $N = N(v)$ denote its carrier, and observe that the induced presentation $N^* \to X^*$ has no missing shells, and hence $\widetilde{N}^* \to \widetilde{X}^*$ lifts to an embedding. After adding dummy squares, we utilize this argument to quickly understand the simple-connectivity of $\overline{w}^*$.

By (equivariantly) adding dummy squares to X^*, we form a new cubical presentation $\langle \underline{X} \mid \{\underline{Y}_i\} \rangle$ satisfying the same small-cancellation conditions but with the further property that $\underline{w}$ now consists of a single hyperplane containing the components of the base-wall w and having the property that each $\underline{w} \wedge \underline{Y}_i$ consists of a single hyperplane containing the hyperplanes in the wall $w \wedge Y_i$ (which denotes the preimage of w in Y_i with respect to $Y_i \to X$). We do this *equivariantly* for each Y_i in the following sense: As in Construction 10.10, when $w \wedge Y_i$ has p_i components, we let T_i denote a p_i-star and add a copy of $T_i \times I$ to X for each i to form $\underline{X}$, but now add $\mathrm{Aut}(Y_i)$ copies of $T_i \times I$ to form $\underline{Y}_i$. Note that $\mathrm{Aut}(\underline{Y}_i) \cong \mathrm{Aut}(Y_i)$ and that $Y_i \subset \underline{Y}_i$ is an $\mathrm{Aut}(Y_i)$-equivariant inclusion.

Let $N = N(\underline{w})$ and as above, note that $N^* \to \underline{X}^*$ has no missing shells so $\widetilde{N}^*$ embeds in $\underline{\widetilde{X}}^*$. Consequently, the simply-connected $\overline{N}^*$ also embeds in $\underline{\overline{X}}^*$. Note that $\overline{N}^* \cong \overline{\underline{w}}^* \times I$ and so $\overline{\underline{w}}^*$ is simply-connected. The $\pi_1 X^*$-equivariant deformation retraction from $\underline{\overline{X}}^\star$ to $\overline{X}^\star$ collapses $\overline{\underline{w}}^\star$ to $\overline{w}^\star$.

Remark 10.12 (Cubes Instead of Conepoints)**.** An intriguing viewpoint that we have not pursued is to extend the idea of Construction 10.4 to a more complete setting. As each cone Y_i is a wallspace, there is an embedding $Y_i \to \mathsf{C}(Y_i)$ where $\mathsf{C}(Y_i)$ is the dual CAT(0) cube complex. (Note that it might not be a cube.) For each i, let M_i be the associated mapping cylinder, and then instead of coning off the cones, we attach a copy of M_i to X along its base $X \leftarrow Y_i \to M_i$. See Figure 10.7.

Chapter Eleven

Hierarchies

Definition 11.1. The class $\mathcal{MQH}$ of groups with an *almost malnormal quasiconvex hierarchy* is the smallest class of groups closed under the following conditions:

(1) If $|G| < \infty$ then $G \in \mathcal{MQH}$.
(2) If $G \cong A *_C B$ and $A, B \in \mathcal{MQH}$ and B is quasiconvex and almost malnormal in G, then $G \in \mathcal{MQH}$.
(3) If $G \cong A*_C$ and $A \in \mathcal{MQH}$ and B is quasiconvex and almost malnormal in G, then $G \in \mathcal{MQH}$.

It follows from the Bestvina-Feighn Combination Theorem that every group in $\mathcal{MQH}$ is word-hyperbolic [BF92].

In conjunction with Definition 11.1, a *hierarchy* for G is a specific sequence of splittings leading to terminal groups, and the *length* of this hierarchy is the longest maximal (nontrivial) subsequence. Depending on the context we might allow the splittings to be general graphs of groups, or we might insist that each is either an HNN extension or AFP. The hierarchy is *malnormal, quasiconvex*, etc. if each of its constituent splittings has this property. The *terminal* groups could be trivial as in $\mathcal{QH}$, or they could be finite as for $\mathcal{MQH}$. Other possibilities are considered in Chapter 15.

Theorem 11.2 ($\mathcal{MQH}$ Is Virtually Special). *Each group in $\mathcal{MQH}$ is virtually special and hyperbolic. More generally, if G has an almost malnormal quasiconvex hierarchy terminating at virtually compact special hyperbolic groups then G is virtually compact special and hyperbolic.*

Proof. This holds by induction on the "length" of a hierarchy using Theorem 11.3.

Note that every compact special cube complex with word-hyperbolic fundamental group has a finite cover with a malnormal quasiconvex hierarchy [HW12]. Thus such a hierarchy can be extended further to a virtual malnormal quasiconvex hierarchy. $\square$

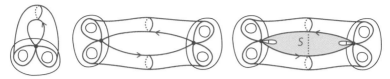

Figure 11.1. $X \leftarrow \widehat{X} \hookrightarrow \widehat{X}'$.

Theorem 11.3. *Let $G = A *_C B$ or $G = A *_C$ where C is almost malnormal and quasiconvex and A, B (respectively A) is virtually compact special and hyperbolic. Then G is virtually compact special.*

Proof. This holds by Proposition 6.31 together with Theorem 7.59 and the wall case of Corollary 6.24. For the amalgamated product $A *_C B$ one lets A', B' be special finite index subgroups of A, B, and then let $C' = A' \cap C \cap B'$. Then one chooses a collection $\{H_1, \ldots, H_r\}$ of quasiconvex codimension-1 subgroups cubulating C' (and hence cubulating C). By Corollary 6.24, this collection is extendible to quasiconvex codimension-1 subgroups $\{H_1^A, \ldots, H_r^A\}$ in A and $\{H_1^B, \ldots, H_r^B\}$ in B. The HNN case is similar but also follows from the AFP case as described in Remark 11.4. $\square$

Remark 11.4. Revising the notation, let X be a graph of spaces with a vertex space A and edge space $C \times I$ corresponding to the HNN extension, so $G = \pi_1 X$ splits as $\pi_1 A *_{\pi_1 C}$. Let $\widehat{X}$ denote the double cover of X corresponding to the map sending the stable letter t of the HNN extension to the generator of $\mathbb{Z}_2$ and sending $\pi_1 A$ to the identity element. Let S be a "dummy square" such that $\partial_p S$ has label $a_1 t_1 a_2 t_2$. We add S to $\widehat{X}$ to form $\widehat{X}'$ by identifying t_1 and t_2 with the two copies of the t-edge in $\widehat{X}$. We do this so that their orientations agree. See Figure 11.1. Finally, there is a splitting of $\widehat{X}'$ as a graph of spaces: each vertex space equals $\widehat{A}_i \cup a_i$ where $\widehat{A}_1, \widehat{A}_2$ are the two components of the preimage of A in $\widehat{X}$, and a_1, a_2 are the other two 1-cells on ∂S. And the sole edge space consists of the two copies $(C \times I)_1, (C \times I)_2$ of the preimage of $C \times I$, joined together with the square S. Note that $\pi_1 \widehat{X}' \cong \pi_1 \widehat{X} * \mathbb{Z}$ is word-hyperbolic if $\pi_1 X$ is. The reader can check that our splitting as a graph of spaces induces a splitting of $\pi_1 \widehat{X}'$ as an amalgamated free product, whose vertex groups are each isomorphic to $\pi_1 A * \mathbb{Z}$ and hence virtually special, and whose edge group is isomorphic to $\pi_1 C * \pi_1 C$ and is easily seen to be quasiconvex.

The height of the pair of inclusions of $\pi_1 C$ into $\pi_1 A$ inside the HNN extension equals the height of each edge group $\pi_1 C * \pi_1 C$ in each vertex group of the amalgamated free product. In particular, almost malnormality of the splitting is preserved.

If the amalgamated free product is virtually compact special, then the original group is as well, since it has a finite index subgroup with this property. (Here one needs word-hyperbolicity to guarantee the compactness. See Remark 7.15.)

The amalgamated free product case of Theorem 13.1 thus implies that $\pi_1 \widehat{X}'$ is virtually compact special, and hence the same holds for the quasiconvex subgroup $\pi_1 \widehat{X}$, and hence $\pi_1 X$.

Theorem 11.2 generalizes to the class $\mathcal{QMVH}$ of groups with a *quasiconvex malnormal virtual hierarchy*, which is defined by adding the following to the closure conditions indicated in Definition 11.1:

(4) If $[G:H] < \infty$ and $H \in \mathcal{QMVH}$ then $G \in \mathcal{QMVH}$.

We aim to analyze the class of groups that are hyperbolic relative to tori, and that have quasiconvex hierarchies.

Each compact special cube complex X has $\pi_1 X \in \mathcal{QH}$. It is therefore interesting to examine possible converses, but it is unclear exactly what extra ingredients are needed to ensure that a group with a hierarchy is virtually special.

Definition 11.5. Let $\mathcal{QH}$ denote the smallest class $\mathcal{G}$ of groups that is closed under the following three operations, and let $\mathcal{QVH}$ denote the related class obtained by adding the fourth operation.

(1) $\{1\} \in \mathcal{G}$.
(2) If $G = A *_B C$ and $A, C \in \mathcal{G}$ and B is f.g. and embeds by a quasi-isometry, then G is in $\mathcal{G}$.
(3) If $G = A*_B$ and $A \in \mathcal{G}$ and B is f.g. and embeds by a quasi-isometry, then G is in $\mathcal{G}$.
(4) Let $H \subset G$ with $[G:H] < \infty$. If $H \in \mathcal{G}$ then $G \in \mathcal{G}$.

Remark 11.6. The class of groups with a hierarchy such that each edge group is f.g. (and free) is nastier than one might expect, and certainly contains groups that are not virtually special. For instance it is known that such groups can have undecidable word problem.

Even among 2-dimensional cubical complexes whose fundamental groups exhibit such a hierarchy, there are examples that are not virtually special [BM97, Wis07]. The simplest examples with a hierarchy that are not virtually special are certain Baumslag-Solitar groups (see Problem 11.8).

One of the main results in this text is that every word-hyperbolic group in $\mathcal{QVH}$ is virtually special. It is unclear what hypothesis (besides hyperbolic relative to abelian subgroups) could replace hyperbolicity in a variant of the above statement.

Conjecture 11.7 ($\mathcal{SQVH}$ is virtually special). *Let us restrict the third operation as follows:*

(3') *If $G = A *_B$ and $A \in \mathcal{G}$ and B is f.g. and embeds by a quasi-isometry and B is separable in G, then G is in $\mathcal{G}$.*

Let $\mathcal{SQVH}$ denote the class $\mathcal{G}$ of groups that is closed under the four operations. These are the groups with a separable quasi-isometric virtual hierarchy. Then for every group $G \in \mathcal{SQVH}$ there is a finite index subgroup $H \subset G$ such that H is the fundamental group of a special cube complex.

Problem 11.8. A good test case is the class of one-relator groups. The strongest possible conjecture one could hope for here would be that a one-relator group is virtually special if and only if it contains no subgroup $BS(n, m)$ with $n \neq \pm m$ nonzero, and where $BS(n, m) = \langle a, t \mid (a^n)^t = a^m \rangle$.

Other important test cases are free-by-cyclic groups and more generally ascending HNN extensions of free groups that are hyperbolic relative to virtually abelian subgroups.

A special case of Conjecture 11.7 is:

Conjecture 11.9. *Let X be a (compact) nonpositively curved cube complex. If $\pi_1 D$ is separable in $\pi_1 X$ for each hyperplane D of X, then X is virtually special.*

Chapter Twelve

Virtually Special Quotient Theorem

The goal of this chapter is to prove the following theorem:

Theorem 12.1 (Virtually Special Quotient). *Let G be a word-hyperbolic group with a finite index subgroup that is the fundamental group of a compact special cube complex. Let $H_1, \ldots, H_r$ be quasiconvex subgroups of G. Then there are finite index subgroups $H'_1, \ldots, H'_r$ such that the quotient: $G' = G/\langle\langle H'_1, \ldots, H'_r \rangle\rangle$ is a word-hyperbolic group with a finite index subgroup that is the fundamental group of a compact special cube complex.*

12.a Malnormal Special Quotient Theorem

Theorem 12.1 will be a consequence of the following special case which is our main focus in this section, and probably the crux of this venture:

Theorem 12.2 (Malnormal Virtually Special Quotient). *Let G be a word-hyperbolic group with a finite index subgroup J that is the fundamental group of a compact special cube complex. Let $\{H_1, \ldots, H_r\}$ be an almost malnormal collection of quasiconvex subgroups of G. Then there are finite index subgroups $\ddot{H}_1, \ldots, \ddot{H}_r$ such that: For any finite index subgroups $H'_1, \ldots, H'_r$ contained in $\ddot{H}_1, \ldots, \ddot{H}_r$ the quotient $G' = G/\langle\langle H'_1, \ldots, H'_r \rangle\rangle$ is a word-hyperbolic group with a finite index subgroup J' that is the fundamental group of a compact special cube complex.*

The statement will be proven using subgroups H'_i and $\ddot{H}_i$ that are normal in H_i, but consequently holds (as stated) without assuming these subgroups are normal. We also note that the statement allows us to choose the H'_i to lie in any pre-chosen finite index subgroups H°_i of H_i.

Remark 12.3. There is a tricky point in the algebraic statement. When passing to a nonpositively curved cube complex X for a finite index subgroup J of G, each subgroup H_i appears in several ways as a subgroup $H_{ij} = H^{g_j}_i \cap J$. We could keep the subgroups H_{ij} abstractly isomorphic by assuming that J is normal in G. However, they might not have the same representative geometry when we pass to a nonpositively curved cube complex X with $\pi_1 X \cong J$. This can be remedied by recubulating with a cube complex containing an action of G/J. One way to do this is to use an action of G on $X^{[G:J]}$. Another way uses Lemma 7.14.

Proof of Theorem 12.2. By Lemma 7.14, there is a proper cocompact action of G on a CAT(0) cube complex $\widetilde{X}_0$.

By Corollary 6.10 the cube complex $X_0 = J \backslash \widetilde{X}_0$ is virtually special. So by replacing J with a finite index subgroup, we may assume X_0 is special. Note that specialness is preserved by arbitrary further covers.

By Lemma 2.36, for each i, let $\widetilde{Z}_i \subset \widetilde{X}_0$ denote an H_i-cocompact superconvex subcomplex. Let $H_{i0} = H_i \cap J$ for each i. Let $Z_i = H_{i0} \backslash \widetilde{Z}_i$ so $Z_i \to X_0$ is a compact superconvex local-isometry representing $H_{i0} \subset J$. (We can ignore the basepoint here as only the conjugacy class in J interests us.)

A regular cover $X \to X_0$ determines a cubical presentation in the following way: Let $\widehat{Z}_i$ denote an elevation of Z_i to X, so $\widehat{Z}_i$ is the cover of Z_i corresponding to $\pi_1 Z_i \cap \pi_1 X$. We shall choose X so that $\pi_1 X$ is normal in G. For $1 \leq i \leq r$ we let $Y_i = \widehat{Z}_i$, and employing the G-action on X to obtain other elevations, we extend this correspondence to let $\{Y_1, \ldots, Y_j\} = \{g\widehat{Z}_i : 1 \leq i \leq r, g \in G\}$ denote the collection of all possible such elevations, and we obtain the cubical presentation:

$$\langle X \mid Y_1, \ldots, Y_j \rangle$$

Our initial goal below is to show how to choose a finite regular cover $X \to X_0$ so that using the above notation, the cubical presentation $\langle X \mid Y_1, \ldots, Y_j \rangle$ satisfies the following properties:

(1) Each hyperplane M of X has $\pi_1 M$ malnormal in $\pi_1 X$.
(2) Each cone $Y_i \to X$ is an embedding.
(3) There exists D such that each cone-piece and wall-piece is a CAT(0) subcomplex of X with diameter $\leq D$.
(4) $\|Y_i\| > 14D$ for each i.
(5) Twice the injectivity radius of each hyperplane of X is greater than D.
(6) For each hyperplane $M \subset X$ and path $S \to Y_i$ starting and ending on 0-cells at endpoints of 1-cells dual to M if $\{\{S\}\}_{Y_i} \leq \pi$ then S is homotopic into $N(M) \cap Y_i$.

The main conditions above are *stable* in the sense that they are preserved by passage to a further finite cover, this facilitates achieving the conditions, as we can resolve each specific issue at some finite cover, and then conclude by taking a finite cover factoring through all of these whose fundamental group is normal in G.

Malnormality: Since X_0 is special, for each hyperplane M_ℓ of X_0, the subgroup $\pi_1 M_\ell$ is separable in $\pi_1 X_0$. By Proposition 8.3, there are finitely many cosets $\pi_1 M_\ell g_i$ such that $\pi_1 M_\ell^{g_i} \cap \pi_1 M_\ell$ is infinite. We can therefore separate $\pi_1 M_\ell$ from each g_i in a finite index subgroup to make $\pi_1 M_\ell$ malnormal there. We obtain a finite cover $X_{M_\ell} \to X_0$ for each hyperplane, and any finite regular cover of X_0 factoring through all of these has the desired property. The above sketch is explained in [HW09] in a relatively hyperbolic context.

Cone injectivity: Since X_0 is special we can pass to a finite cover $X_i \to X_0$ such that Z_i embeds. (E.g., let $X_i = \mathsf{C}(Z_i \to X_0)$.) Any finite regular cover of X_0 factoring through each X_i has the desired property.

Piece size and embedding: The almost malnormality of $\{H_1, \ldots, H_r\}$ and superconvexity of $\{\widetilde{Y}_1, \ldots, \widetilde{Y}_j\}$ bounds the cone-pieces and wall-pieces in each $g\widetilde{Y}_i$ since each such piece lies in a diameter D CAT(0) subcomplex of $\widetilde{X} = \widetilde{X}_0$. (See Lemma 3.52, and note that each $\widetilde{Y}_p$ is a translate of a $\widetilde{Z}_q$.) By residual finiteness of $\pi_1 X_0$, and local finiteness of $\widetilde{X}$, we can pass to a finite regular cover of X_0 such that all diameter $\leq D$ subcomplexes of $\widetilde{X}$ embed.

Small-cancellation and negative curvature: For each i, by residual finiteness, we can choose a finite cover of $\bar{Z}_i \to Z_i$ such that $\|Z_i\| > 14D$. We then choose a regular cover $\bar{X}_i$ such that each elevation of Z_i to $\bar{X}_i$ factors through $\bar{Z}_i$. Any further cover of $\bar{X}_i$ has the desired property for Z_i, and thus any cover X_0 factoring through each $\bar{X}_i$ has the desired property.

Injectivity radius $> \frac{1}{2}$ cone-piece diameter: This is a simple consequence of the separability of hyperplane subgroups. The *injectivity radius* of a hyperplane $M \subset X$ is half the length of the shortest path σ starting and ending on 1-cells of M such that σ is not homotopic into M. This equals half the minimal distance between lifts $g_1 \widetilde{M}, g_2 \widetilde{M}$ in $\widetilde{X}$. We will use this below to ensure no self-grazing.

π-wall separation: Like the malnormality claim and the injectivity radius claim, this is a consequence of separability of hyperplanes. Any path $S \to Y_i$ with $\{\{S\}\}_{Y_i} \leq \pi$ is the concatenation of at most B pieces where B is a uniform constant. (For instance, Lemma 3.70 shows that $B = 8$ works.) There are thus finitely many such situations to dispose of as above.

Splicing the cones to obtain $B(8)$: We now pass to a finite cover $\ddot{Y}_i \to Y_i$ that is induced by a cover $\ddot{X} \to X$ such that $\langle X \mid \ddot{Y}_i \rangle$ satisfies the conditions above (except (2)) as well as the following:

(7) The $B(8)$ small-cancellation condition.

Let $\mathbb{W} = \mathbb{W}(X)$ denote the family of hyperplanes of X. Consider the map $\#_{\mathbb{W}} : \pi_1 X \to \mathbb{Z}_2^{\mathbb{W}}$ induced by counting (modulo 2) the number of times a path passes through a hyperplane. As in Section 9.a, let $\ddot{X}$ denote the cover of X associated to $\#_{\mathbb{W}}$. Let $\ddot{Y}_i \to Y_i$ denote the cover induced by $\ddot{X} \to X$ for each i. According to Construction 9.1, each $\ddot{Y}_i$ is a wallspace where each wall of $\ddot{Y}_i$ is the preimage of a single hyperplane of X.

We define $\ddot{H}_i$ to equal $\pi_1 \ddot{Y}_i$. Note that $\pi_1 \ddot{X}$ is a normal subgroup of G, and hence each $\ddot{H}_i$ is a normal subgroup of H_i.

For any finite index normal subgroup $H_i' \subset H_i$ that is contained in $\ddot{H}_i$, we let Y_i' denote the corresponding cover of Y_i. Following Construction 9.4, we let each Y_i' have the wallspace structure induced from $\ddot{Y}_i$.

We use the notation $X^* = \langle \ddot{X} \mid gY_i' : g \in \operatorname{Aut}(\ddot{X}), 1 \leq i \leq r \rangle$. Observe that X^* corresponds to a finite covering space of $\langle X \mid gY_i' : g \in \operatorname{Aut}(X) \rangle$ (we have been cavalier about translates of cone-cells with the same base).

Theorem 3.32 applies to show that X^* is a negatively curved small-cancellation presentation and Lemma 3.70 ensures that X^* has short innerpaths. The walls of $\widetilde{X^*}$ embed in $\widetilde{X^*}$ by Theorem 5.18, and are quasi-isometrically embedded by Corollary 5.30.

By construction, the new walls embed as base-walls in X^*. Indeed, the hyperplanes of a wall of Y_i' are the entire preimage of a hyperplane of $\ddot{X}$ under the map $Y_i' \to \ddot{X}$. No self-grazing holds since we chose X to have the property that the injectivity radii of hyperplanes exceeds half the diameters of cone-pieces, and hence no base-wall of X^* can pass through the same cone-piece in two distinct hyperplanes.

$\langle X \mid \{gY_i'\} \rangle$ and hence $\langle \ddot{X} \mid \{gY_i'\} \rangle$ inherits the small-cancellation properties of $\langle X \mid \{Y_i\} \rangle$ by Lemma 9.7 and in addition, Condition 5.55.(2) is satisfied by Lemma 9.6.

The hierarchy: We now apply the cutting instructions of Chapter 10 where at each stage we alternately cut along a wall and then along various conepoints that are cutpoints. In the former case, the $B(8)$ conditions are satisfied and in the latter case, we are merely splitting along finite groups, as explained in Construction 10.11. The result after each such pair of cuts are new complexes that are smaller in the sense that they have fewer 1-cells in the cube complex. After finitely many cuts, we terminate where the underlying cube complex only contains 0-cells.

Since the $B(8)$ condition is preserved by the cuts, at each stage the walls we are cutting along are quasiconvex by Theorem 5.30 and are almost malnormal by Theorem 5.68.

We thus have an almost malnormal quasiconvex hierarchy, and so $\pi_1 X^*$ is virtually compact special and hyperbolic by Theorem 11.2. Finally, observe that $G/\langle\!\langle H_1', \ldots, H_r' \rangle\!\rangle$ contains $\pi_1 X^* = \pi_1 X / \langle\!\langle \pi_1(gY_i') : g \in \operatorname{Aut}(X), 1 \leq i \leq r \rangle\!\rangle$ as a finite index subgroup. $\qquad \Box$

12.b   Proof of the Special Quotient Theorem

Proof of Theorem 12.1. Let $\{K_1, \ldots, K_s\}$ denote the collection of infinite maximal intersections of conjugates given in the statement of Lemma 8.6. For each i, let $\mathcal{K}_i$ denote $\mathbb{C}_G(K_i)$.

By Lemma 7.14, G acts properly and cocompactly on a CAT(0) cube complex $\widetilde{X}$. Let $J \subset G$ be a finite index torsion-free normal subgroup and let $X = J \backslash \widetilde{X}$. Let $U \subset \widetilde{X}$ be a finite radius ball such that $JU = \widetilde{X}$, and such that U contains the basepoint $\tilde{x}$ of $\widetilde{X}$. The basepoint $x \in X$ is the image of $\tilde{x}$.

We apply Lemma 2.36 to obtain H_i-cocompact superconvex subcomplexes $\widetilde{Y}_i \subset \widetilde{X}$ and $\mathcal{K}_i$-cocompact superconvex subcomplexes $\widetilde{Z}_i \subset \widetilde{X}$, and such that each contains $\widetilde{U}$. We do this so that $g^{-1}\widetilde{Z}_j \subset \widetilde{Y}_i$ whenever $K_j^g \subset H_i$ (using the notation

$x^g = g^{-1}xg$). For instance, first choose $\mathcal{K}_i$-cocompact superconvex subcomplexes (hence also K_i-cocompact), and then apply Lemma 2.36 again to ensure that the H_i-cocompact superconvex subcomplexes contain the various translates required above. Let $Y_i = (J \cap H_i) \backslash \widetilde{Y}_i$ for each i.

There is an upperbound $D = D(\widetilde{X}, \{\widetilde{Z}_j\}, \{\widetilde{Y}_i\})$ on diameters of wall-pieces in $\{\widetilde{Z}_j\}$ and of cone-pieces between any G-translates $g_i \widetilde{Y}_i$ and $g_j \widetilde{Z}_j$ unless $g_j \widetilde{Z}_j \subset g_i \widetilde{Y}_i$ in which case $K_i \subset H_j^{g_i^{-1} g_j}$. The diameters of wall-pieces are bounded because of the superconvexity—and this bounds the noncontiguous cone-pieces as well. The diameters of the above contiguous cone-pieces are bounded by the proof of Lemma 3.52, since if there is an infinite contiguous cone-piece then there is an infinite intersection, but as K_j is a maximal intersection $K_j \subset H_i$. (One also sees this using fiber-products as described in Section 8.b. An infinite contiguous cone-piece corresponds to a noncontractible component in $Y_i \otimes Z_j$, which would lie in the large diagonal of a multiple fiber-product $\bigotimes Y_i$. See Corollary 8.10.)

Consider the G action on X. By postcomposing, there is a resulting action of G on the maps $Y_i \to X$, and we thus obtain a collection $\{gY_i\}$ of local-isometries to X. Note that g varies among representatives of the finitely many double cosets $J \backslash G / H$. Since $U \subset \widetilde{Y}_i$, we can choose a basepoint for each Y_i mapping to x.

Let $Y = \sqcup gY_i$. Let $\{Y_{mk}\}$ denote the collection of all possible noncontractible nondiagonal components of the m-fold fiber-product $Y \otimes_X Y \otimes_X \cdots \otimes_X Y$. Note that $\{Y_{1k}\}$ is just $\{gY_i\}$. As we made our earlier choice so that $\widetilde{U}$ lies in each $\widetilde{Z}_j$, we see that each $Y_{mk} \to X$ is surjective, and hence we can choose a basepoint of Y_{mk} that maps to the basepoint x of X. By Proposition 8.3, for each ab and cd there are finitely many double cosets $(\pi_1 Y_{ab}) g (\pi_1 Y_{cd})$ in J such that $\pi_1 Y_{ab}^g \cap \pi_1 Y_{cd}$ has infinite order. Let B be an upperbound on the infimal length of combinatorial paths representing elements in all these cosets.

By residual finiteness, for each j let $\mathcal{K}_j^\circ$ denote a finite index subgroup of $(K_j \cap J) \subset \mathcal{K}_j$ such that $\|Z_j^\circ\| > \max(24D, 8B)$ where $Z_j^\circ = \mathcal{K}_j^\circ \backslash \widetilde{Z}_j$. We apply Theorem 12.2 to $(G; \mathcal{K}_1^\circ \subset \mathcal{K}_1, \ldots, \mathcal{K}_k^\circ \subset \mathcal{K}_k)$ to obtain finite index subgroups $\mathcal{K}_1', \ldots, \mathcal{K}_k'$ contained in $\mathcal{K}_1^\circ, \ldots, \mathcal{K}_k^\circ$ (and hence contained in J) such that $\bar{G} = G / \langle\!\langle \mathcal{K}_1', \ldots, \mathcal{K}_k' \rangle\!\rangle$ is virtually special and word-hyperbolic.

Passing to the induced quotient $\bar{J}$ of $J \subset G$ we obtain our final object: a cubical presentation of the form $X^* = \langle X \mid \{g\ddot{Z}_i\} \rangle$ where $\pi_1 X = J$ as above and $\ddot{Z}_i = \mathcal{K}_i' \backslash \widetilde{Z}_i$. Note that the G-action on $\widetilde{X}$ projects to a G-action on the quotient $X = J \backslash \widetilde{X}$. The cubical presentation contains G-translates $\{g\ddot{Z}_i\}$ where g varies over representatives of the various double cosets $Jg\mathcal{K}_i'$. Note that X^* is a negatively curved small-cancellation presentation using the split-angling system by Theorem 3.32, since each piece has diameter $< \frac{1}{24}\|Z_j^\circ\| \le \frac{1}{24}\|\ddot{Z}_j\|$. Moreover, X^* has short innerpaths by Lemma 3.70.

The group $\bar{G}$ has $\bar{J} \cong \pi_1 X^*$ as a finite index subgroup, and we shall use the associated cubical presentation to verify that each $\bar{H}_i$ is quasiconvex and that $\bar{h} = \text{Height}_{\bar{G}}\{\bar{H}_1, \ldots, \bar{H}_k\} < h = \text{Height}_G\{H_1, \ldots, H_k\}$.

For each i, as $Y_i = (H_i \cap J) \backslash \widetilde{Y}_i$ the quasiconvexity of $\bar{H}_i$ will follow from the quasiconvexity of the image of the finite index subgroup $\pi_1 Y_i \to \pi_1 X$ in

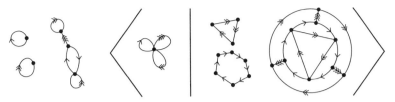

Figure 12.1. Let $\widetilde{Y}_{11}$ and $\widetilde{Y}_{12}$ and $\widetilde{Y}_{21}$ denote the universal covers of the three graphs on the left, and let $\widetilde{X}$ denote the universal cover of the bouquet of three circles. The graded cubical presentation on the right could be obtained from this initial data following Lemma 12.4.

$\bar{J} = \pi_1 X^*$. Since $\mathrm{diam}(\widetilde{Y}_i \cap g\widetilde{Z}_j) < \frac{1}{24}\|Z_j^\circ\| < \frac{1}{2}\|\ddot{Z}_j\|$ unless $g\widetilde{Z}_j \subset \widetilde{Y}_i$, we may apply Lemma 3.67 to see that the induced presentation $Y_i^* \to X^*$ has no missing shells and is thus π_1-injective by Theorem 3.68. Moreover, quasiconvexity of each $\pi_1 \widehat{Y}_i^* \to \pi_1 X^*$ holds by Corollary 3.72.

We now verify the height decrease by computing the intersection of conjugates in J and $\bar{J}$. We begin by reexpressing the intersection of conjugates in terms of associated based local-isometries.

$$J \cap \bigcap_{i=1}^{p} H_{n_i}^{g_i} = \bigcap_{i=1}^{p}(J \cap H_{n_i})^{g_i} = \bigcap_{i=1}^{p}(\pi_1 Y_{n_i})^{f_i j_i} = \bigcap_{i=1}^{p}(\pi_1(f_i^{-1}Y_{n_i}))^{j_i}$$
$$= \bigcap_{i=1}^{p}(\pi_1(Y_{1b_i}))^{j_i}.$$

Here we express the conjugator $g_i \in G$ as $g_i = f_i j_i \in f_i J$. We then use Y_{1b_i} to denote the translate of $Y_{n_{i}}$ associated to the conjugate by f_i.

Recall that we chose $\|\ddot{Z}_i\| \geq \|\ddot{Z}_i^\circ\| > 8B$, where B exceeds the minimal length representative for any double coset associated to a conjugator associated to an infinite intersection of multiple intersections. Thus the criteria of Lemma 12.10 are met, and we see by induction that $\overline{\cap_{i=1}^{p}\pi_1(Y_{1b_i})^{j_i}} = \cap_{i=1}^{p}\overline{\pi_1(Y_{1b_i})}^{j_i}$. Since intersections of conjugates in $\bar{J}$ are images of intersections of conjugates in J itself, and since the maximally infinite such intersections project to finite subgroups of $\bar{J}$ by construction, we see that $\bar{h} < h$ as claimed.

Consequently, the theorem follows by induction on the height of $\{H_1, \ldots, H_r\}$ in G. The quotient has trivial kernel in the base case where $h = 0$. $\qquad\square$

12.c   Adding Higher Grade Relators

We refer to Figure 12.1 for an example suggesting the construction in the following:

Lemma 12.4 (Adding higher grade relators and preserving quasiconvexity). *Let* $X^* = \langle X \mid \{Z_i\} \rangle$ *be a graded cubical presentation, and let* $\{Y_j \to X\}$ *and*

$\{W_k \to X\}$ be finite collections of local-isometries with compact domain. Suppose that the following holds for some $0 < \alpha < \frac{1}{14}$:

(1) X^* is compact and X^* is $C'(\alpha)$.
(2) Each $Y_j \to X$ and $W_k \to X$ is superconvex.
(3) $\pi_1 Y_j = \mathbb{C}_{\pi_1 X}(\pi_1 Y_j)$ for each j.
(4) The collection $\{\pi_1 Y_j^*\}$ is almost malnormal in $\pi_1 X^*$.
(5) For each j, k, and $g \in \pi_1 X$, either $diam(\widetilde{Y}_j^* \cap g\widetilde{W}_k^*) < \infty$ or $\widetilde{Y}_j \subset g\widetilde{W}_k$.
(6) For each i, j, and $g \in \pi_1 X$, either $|\widetilde{Z}_i \cap g\widetilde{Y}_j|_{Z_i} < \alpha \|Z_i^\circledast\|$ or $\widetilde{Z}_i \subset g\widetilde{Y}_j$ and factors through $Z_i \to Y_j$.
(7) For each i, k, and $g \in \pi_1 X$, either $|\widetilde{Z}_i \cap g\widetilde{W}_k|_{Z_i} < \alpha \|Z_i^\circledast\|$ or $\widetilde{Z}_i \subset g\widetilde{W}_k$ and factors through $Z_i \to W_k$.
(8) Each $\pi_1 Y_j^*$ is residually finite.

Then there exist finite covers $\check{Y}_j^* \to Y_j^*$ such that for any regular covers $\widehat{Y}_j^* \to Y_j^*$ that factor through $\check{Y}_j^*$ we have:

(1) $X^{\bar*} = \langle X \mid \{Z_i\}, \{\widehat{Y}_j\} \rangle$ is $C'(\alpha)$, with each $\mathrm{grade}(\widehat{Y}_j) = \max_i\{\mathrm{grade}(Z_i)\} + 1$.
(2) For each j, k, and $g \in \pi_1 X$, either $|\widetilde{Y}_j \cap g\widetilde{W}_k|_{\widehat{Y}_j} < \alpha \|\widehat{Y}_j^\circledast\|$ or $\widetilde{Y}_j \subset g\widetilde{W}_k$ and factors through $\widehat{Y}_j \to W_k$.

As usual, $\widehat{Y}_j$ denotes the cubical part of the cover $\widehat{Y}_j^* \to Y_j^*$.

Remark 12.5. Hypothesis (3) facilitates the statement of the theorem, and allows us to require symmetric cones and satisfy Convention 3.3. In the small-cancellation framework where $\alpha < \frac{1}{12}$, Hypotheses (6) and (7) imply that $Y_j^* \to X^*$ and $W_k^* \to X^*$ have no missing shells. Thus $\widetilde{Y}_j^* \to \widetilde{X}^*$ and $\widetilde{W}_k^* \to \widetilde{X}^*$ are embeddings by Theorem 3.68. Conclusion (2) is a strong form of the condition that $W_k^{\bar*} \to X^{\bar*}$ has no missing shells.

Though it can be further generalized, we have stated Lemma 12.4 in a form close to its applications in Corollary 12.7 and Theorem 12.6.

Proof. By Lemma 2.40, the superconvexity and cocompactness yields a uniform upperbound R on diameters of rectangles with base on each $\widetilde{Y}_j$. Let A be an upperbound on $\mathrm{diam}(K)$ for each component K of each $Z_i \otimes_X Y_j$. Note that $\pi_1 K \to \pi_1 Y_j^*$ has trivial image as $Y_j^* = \langle Y_j \mid \{Y_j \otimes_X Z_i\}_i \rangle$ so K is actually a cone of Y_j^*. Consequently, each $|K|_{Y_j} \le A$. By Hypothesis (4), let B be an upperbound on $\mathrm{diam}(g\widetilde{Y}_j^* \cap \widetilde{Y}_{j'}^*)$ whenever $g\widetilde{Y}_j^* \ne \widetilde{Y}_{j'}^*$. Specifically, we shall assume that for each nondiagonal component K of $Y_j \otimes_X Y_j'$, we have $|K|_{Y_j} \le B$. By compactness of Y_j and W_k, there are finitely many distinct $\pi_1 X^*$ orbits of finite diameter intersections $\widetilde{Y}_j^* \cap g\widetilde{W}_k^*$ in $\widetilde{X}^*$. Let M be an upperbound on the diameter of any such intersection.

We claim that any elevation of Y_j to W_k induces an elevation of Y_j^* to W_k^*, and we let $\bar{Y}_j^*$ be a finite cover of Y_j^* that factors through all these elevations. To verify this claim, we show that for each cone K of Y_j^*, its elevation induced by an elevation $\widehat{Y}_j$ of Y_j to W_k is an actual lift of K. Since $\alpha < \frac{1}{14}$, short innerpaths holds for X^* by Lemma 3.70, and hence Hypotheses (6) and (7) allow us to apply Lemma 3.74 to see that $\widetilde{Y}_j^*$ and $\widetilde{W}_k^*$ are convex subcomplexes of $\widetilde{X}^*$. And each $\widetilde{Z}_i^*$ is a convex subcomplex that is a copy of Z_i. An elevation of Y_j to W_k corresponds to a translate of $\widetilde{Y}_j^*$ that is contained in $\widetilde{W}_k^*$. We thus see that the cones of Y_j^* arising from $Y_j \otimes Z_i$ lift to cones of W_k^*, since they are associated to corresponding intersections of $\widetilde{W}_k^*$ with a lift of Z_i. Consequently, any cone of $\widetilde{Y}_j^*$ is a cone of $\widetilde{W}_k^*$. Hence each cone K of Y_j^* lifts to a cone in any elevation of Y_j to W_k.

By residual finiteness, choose finite covers $\check{Y}_j^* \to Y_j^*$ with $\alpha \|\check{Y}_j^*\| > \max(R, A, B, M)$, and assume, moreover, that $\check{Y}_j^*$ covers $\bar{Y}_j^*$ for each j.

Let $\widehat{Y}_j^* \to Y_j^*$ be regular covers factoring through $\check{Y}_j^* \to Y_j^*$, and let $\widehat{Y}_j$ be the cubical part of $\widehat{Y}_j^*$. Consider the cubical presentation $X^{\circledast} = \langle X \mid \{Z_i\}, \{\widehat{Y}_j\} \rangle$.

We now verify Conclusion (1). Note that $Z_i^{\circledast} = Z_i^{\circledast}$ since each grade$(Y_j) >$ grade(Z_i). Consider a path P in an intersection $\widetilde{Z}_i \cap g\widetilde{Y}_j$. By Hypothesis (6), either $\widetilde{Z}_i \subset g\widetilde{Y}_j$ and factors through $Z_i \to Y_j$ or $|\widetilde{Z}_i \cap g\widetilde{Y}_j|_{Z_i} < \alpha \|Z_i^{\circledast}\|$ in which case we have $|P|_{Z_i^{\circledast}} = |P|_{Z_i^{\circledast}} < \alpha \|Z_i^{\circledast}\| = \alpha \|Z_i^{\circledast}\|$. Furthermore, $|P|_{\widehat{Y}_j} \leq \mathrm{diam}(Z_i) < \alpha \|\check{Y}_j^*\| \leq \alpha \|\widehat{Y}_j^*\|$, where $\mathrm{diam}(Z_i) < \alpha \|\check{Y}_j^*\|$ by our choice of $\check{Y}_j^*$. A contiguous cone-piece P between $\widehat{Y}_j$ and $\widehat{Y}_{j'}$ satisfies: $|P|_{\widehat{Y}_j^*} = |P|_{\widehat{Y}_j^*} \leq B < \alpha \|\check{Y}_j^*\| \leq \alpha \|\widehat{Y}_j^*\| = \alpha \|\widehat{Y}_j^{\circledast}\|$.

We now verify Conclusion (2). If $\widetilde{Y}_j \subset g\widetilde{W}_k$ then the map $\widetilde{Y}_j \to \widetilde{W}_k$ factors through a map $\widehat{Y}_j \to W_k$ since we chose $\check{Y}_j \to W_k$ to factor through each elevation of Y_j to W_k. Otherwise, $\mathrm{diam}(\widetilde{Y}_j^* \cap g\widetilde{W}_k^*) \leq M$, and we have: $|\widetilde{Y}_j \cap g\widetilde{W}_k|_{\widehat{Y}_j^*} \leq M < \alpha \|\check{Y}_j^*\| \leq \alpha \|\widehat{Y}_j^*\| = \alpha \|\widehat{Y}_j^{\circledast}\|$. $\qquad \square$

The following gives another proof of Theorem 12.1 starting with $G = \pi_1 X$ as a consequence of Lemma 12.4 and Theorem 12.2. It can be adapted to start with a cubical presentation $\langle X \mid \{Z_i\} \rangle$ with X compact and $\pi_1 X^*$ word-hyperbolic, and each H_i represented by $Y_i \to X$ with no missing shells.

Theorem 12.6. *Let X be a virtually special compact cube complex. Suppose that $\pi_1 X$ is word-hyperbolic. Let $Y_i \to X$ be local-isometries with each Y_i compact. There exist finite covers $\widehat{Y}_i$ so that $\langle X \mid \{\widehat{Y}_i\} \rangle$ has $\pi_1 X^*$ virtually compact special and hyperbolic. Moreover, for any finite covers $\widehat{Y}_i^{\circ}$, we may assume each $\widehat{Y}_i$ factors through $\widehat{Y}_i^{\circ}$.*

More specifically, let $\{Y_{jk} \to X\}$ be a compact superconvex graded system (associated to $\{Y_i\}$). Then for each $\alpha < \frac{1}{14}$ there are finite covers $\widehat{Y}_{jk} \to Y_{jk}$, and *a graded* $C'(\alpha)$ *cubical presentation* $X^* = \langle X \mid \{\widehat{Y}_{jk}\}\rangle$, *such that* X^* *has small subcones and* $\pi_1 X^*$ *is virtually compact special and hyperbolic.*

    Moreover, given finite covers $\widehat{Y}_{jk}^{\circ} \to Y_{jk}$ *we may assume each* $\widehat{Y}_{jk}$ *factors through* $\widehat{Y}_{jk}^{\circ}$.

Note that a graded system associated to $\{Y_i\}$ is provided in Construction 8.14.

Proof. Suppose we have chosen covers $\widehat{Y}_{kj}$ for $1 \leq k < n$ so that $X_{n-1}^* = \langle X \mid \{\widehat{Y}_{kj} : 1 \leq k < n\}\rangle$ satisfies $\pi_1 X_{n-1}^*$ is virtually compact special and hyperbolic, and the hypotheses of Lemma 12.4 are satisfied for $\alpha = \frac{1}{24}$ where $\{Z_i\} = \{Y_{pq}\}_{p<n}$ and $\{Y_j\} = \{Y_{pq}\}_{p=n}$ and $\{W_k\} = \{Y_{pq}\}_{p>n}$.

    The images $\{\pi_1 Y_{nj}^*\}$ are almost malnormal in $\pi_1 X_{n-1}^*$, so we can choose regular covers $\widehat{Y}_{nj}^* \to Y_{nj}^*$ that factor through the covers supplied by Lemma 12.4 and also factor through the covers supplied by Theorem 12.2, and so that $X_n^* = \langle X \mid \{\widehat{Y}_{kj} : 1 \leq k \leq n\}\rangle$ satisfies the hypotheses of Lemma 12.4 and $\pi_1 X_n^*$ is also virtually compact special and hyperbolic. We also assume that the covers $\widehat{Y}_{kj}$ are chosen so that the hypotheses of Lemma 12.10 are ensured for all pairs Y_{pq}, Y_{rs}. Thus the intersections of conjugates of $\pi_1 Y_{pq}^*$ and $\pi_1 Y_{rs}^*$ will equal the image of the intersections of the corresponding conjugates of $\pi_1 Y_{pq}$ and $\pi_1 Y_{rs}$, and hence equal the conjugate of an image of some $\pi_1 K^*$ where K is a component of $Y_{pq} \otimes Y_{rs}$, and so $\pi_1 K$ equals a finite index subgroup of some $\pi_1 Y_{tu}$ where $t < \min(p, r)$, unless one of Y_{rs} and Y_{pq} factors through the other. This ensures Hypothesis (5) since $\pi_1 Y_{pq}^* \cap g\pi_1 Y_{rs}^* g^{-1}$ is finite precisely when $\pi_1 \widetilde{Y}_{pq}^* \cap g\widetilde{Y}_{rs}^*$ has finite diameter. This also ensures the almost malnormality of the next family $\{\pi_1 Y_{(n+1)j}^*\}$, and hence Hypothesis (4).

    Furthermore, we can choose the covers $\widehat{Y}_{nj}^* \to Y_{nj}^*$ so that each $\widehat{Y}_{nj}^*$ has small subcones (this is naturally achieved by the small-cancellation requirements).

    The covers $\widehat{Y}_{kj}$ are chosen so that they factor through all elevations of Y_{kj} to all Y_{pq} with $Y_{kj} \prec Y_{pq}$, and this ensures that Hypotheses (6) and (7) are satisfied by the quotient. Virtual specialness of the quotient ensures residual finiteness of each $\pi_1 Y_{(n+1)j}^*$ and so Hypothesis (8) persists. Hypothesis (1) persists as we used finite covers, and Hypotheses (3) and (4) are automatically inherited.

    After applying this procedure as many times as the maximal grade, we arrive at a cubical presentation $X^* = \langle X \mid \{\widehat{Y}_{ij}\}\rangle$, such that each Y_{ij} has a finite cover that is included as a relator. The result follows by letting each $\widehat{Y}_i \to Y_i$ be the cover associated to $\widehat{Y}_{ab}$, where $\pi_1 Y_{ab}$ contains a finite index subgroup that is conjugate to $\pi_1 Y_i$ in $\pi_1 X$.

    We now explain why we may assume each $\widehat{Y}_i$ factors through a chosen cover $\widehat{Y}_i^{\circ}$. Without loss of generality, we may assume $\widehat{Y}_i^{\circ} \to Y_i$ is regular. At each stage, when we choose the covers $\widehat{Y}_{pq} \to Y_{pq}$, we assume that they factor through all elevations of Y_{pq} to Y_i, or equivalently, we assume $\pi_1 \widehat{Y}_{pq} \subset \pi_1 \widehat{Y}_i^{\circ}$ whenever $[\pi_1 Y_i : \pi_1 Y_i^{\circ} \cap \pi_1 Y_{pq}] < \infty$.   □

Corollary 12.7. *Let X be a compact nonpositively curved cube complex with $\pi_1 X$ hyperbolic. Let $N_i \to X$ be the carriers of the finitely many hyperplanes of X. Assume each $\pi_1 N_i$ is virtually special. There exist finite regular covers $\widehat{N}_i$ such that (the cubical part of) $\widetilde{X}^*$ is special, where $X^* = \langle X \mid \{\widehat{N}_i\} \rangle$.*

Proof. For each i, let $\widetilde{Y}_i$ denote a superconvex $\pi_1 N_i$-cocompact subcomplex containing $\mathcal{N}_1(\widetilde{N}_i)$, and let $Y_i = \pi_1 N_i \backslash \widetilde{Y}_i$.

We will apply Theorem 12.6 with input $\{Y_i\}$. Let $Y_{p_i q_i}$ be associated to Y_i for each i. We can choose finite special covers $\{\widehat{Y}_{p_i q_i}^\circ\}$ by Corollary 6.10. The resulting cubical presentation $\langle X \mid \widehat{Y}_{ij} \rangle$ has small subcones, and hence $\widetilde{X}^*$ has well-embedded cones by Lemma 3.58.

We now verify that $\widetilde{X}^*$ is special. Consider crossing hyperplanes U_i and U_j of $\widetilde{X}^*$. Let $\widehat{N}_i = N(U_i)$ and $\widehat{N}_j = N(U_j)$, and let $\widehat{Y}_{p_i q_i}$ and $\widehat{Y}_{p_j q_j}$ denote the associated cones. Observe that $\widehat{Y}_{p_i q_i} \cap \widehat{Y}_{p_j q_j} \neq \emptyset$ since it contains $U_i \cap U_j$, and $\widehat{Y}_{p_i q_i} \cap \widehat{Y}_{p_j q_j}$ is connected as explained above. Observe that $\widehat{N}_i \cap \widehat{N}_j$ is connected by Lemma 12.8. Specialness of $\widehat{N}_i \cap \widehat{N}_j$ holds since it is a locally-convex subcomplex of the special cube complex $\widehat{Y}_{p_i q_i}$. Inter-osculation between U_i and U_j would imply inter-osculation within $\widehat{N}_i \cap \widehat{N}_j$, which is impossible. $\square$

Lemma 12.8. *Let X be a nonpositively curved cube complex. Let $C_1 \subset C_1' \subset X$ be a sequence of locally-convex subcomplexes, and suppose that C_1' deformation retracts to C_1. Define $C_2 \subset C_2' \subset X$ analogously. If $C_1 \cap C_2 \neq \emptyset$ and $C_1' \cap C_2'$ is connected, then $C_1 \cap C_2$ is connected.*

Proof. Let p and q be 0-cubes of $C_1 \cap C_2$. Choose a geodesic γ in $C_1' \cap C_2'$ that starts at p and ends at q. Then γ lies in C_i since $\widetilde{C}_i \subset \widetilde{C}_i'$ is a convex subcomplex, and $\widetilde{p}, \widetilde{q}$ lie in the convex subcomplex $\widetilde{C}_i \subset \widetilde{C}_i'$, since $\pi_1 C_i = \pi_1 C_i'$. Hence γ lies in $C_1 \cap C_2$. $\square$

12.d   Controlling Intersections in Quotient

Example 12.9. It is possible for $A \cap B = 1$ but $\bar{A} \cap \bar{B}$ to be infinite under some quotient $G \to \bar{G}$. For instance, let $G = \langle a, b \mid a = Wb \rangle$ for some small-cancellation word W. And let $\bar{G} = G / \langle\!\langle W \rangle\!\rangle$. Then usually, $\langle a \rangle \cap \langle b \rangle = 1$ but $\langle \bar{a} \rangle \cap \langle \bar{b} \rangle \cong \mathbb{Z}$. However, the intersection of the images remains trivial if we quotient by $\langle\!\langle W^n \rangle\!\rangle$ for large n. This motivates a final step in the proof of Theorem 12.1, where it is important to know that $\overline{A \cap B} = \bar{A} \cap \bar{B}$.

It is also possible for $\bar{A} \cap \bar{B}^{\bar{g}}$ to be larger than expected because of identification of conjugators under the quotient. Here is an example illustrating this:

Let $G = \langle a_1, a_2, c, d \mid c = Wd \rangle$ where W is some small-cancellation word in the generators. Let $A = \langle a_1, a_2 \rangle$ and let $B = \langle a_1^{c^{-1}}, a_2^{d^{-1}} \rangle$. Let $\bar{G} = G / \langle\!\langle W^n \rangle\!\rangle$. For most values of W, for $n = 1$ we have $\bar{A} \cap \bar{B}^{\bar{c}} = \langle \bar{a}_1, \bar{a}_2 \rangle$, but for $n \gg 1$ we have $\bar{A} \cap \bar{B}^{\bar{c}} = \langle \bar{a}_1 \rangle$.

Lemma 12.10. *Let $X^* = \langle X \mid \{Y_i\} \rangle$ be a negatively curved small-cancellation cubical presentation having tight innerpaths. Suppose each cone-piece P in Y_i satisfies $|P|_{Y_i} < \frac{1}{4} \|Y_i^{\circledast}\|$, and each wall-piece satisfies $|P|_{Y_i} < \frac{1}{8} \|Y_i^{\circledast}\|$.*

Let $A_1 \to X$ and $A_2 \to X$ be based local-isometries. Suppose X^* has small pieces relative to A_1, A_2 in the following sense: For each pair of lifts $\widetilde{A}_j, \widetilde{Y}_i$ to $\widetilde{X}$, either $|\widetilde{A}_j \cap \widetilde{Y}_i|_{Y_i} < \frac{1}{8} \|Y_i^{\circledast}\|$ or $\widetilde{Y}_i \subset \widetilde{A}_j$ and factors through a map $Y_i \to A_j$.

Let $\{\pi_1 A_1 g_i \pi_1 A_2\}$ be a collection of distinct double cosets in $\pi_1 X$. Suppose that for each chosen representative g_i and each cone Y_j we have $|g_i| < \frac{1}{8} \|Y_j^{\circledast}\|$.

Let $G \to \bar{G}$ denote the quotient $\pi_1 X \to \pi_1 X^*$. Then:

(1) *(Double coset separation)* $\overline{\pi_1 A_1 g_i \pi_1 A_2} \neq \overline{\pi_1 A_1 g_j \pi_1 A_2}$ *for $i \neq j$.*

Suppose moreover that $\{\pi_1 A_1 g_i \pi_1 A_2\}$ form a complete set of double cosets with the property that $\pi_1 A_1^{g_i} \cap \pi_1 A_2$ is nontrivial. Then:

(2) *(Square annular diagram replacement)* *If $\overline{\pi_1 A_1}^{\bar{g}} \cap \overline{\pi_1 A_2}$ is nontrivial for some $\bar{g} \in \bar{G}$, then $\overline{\pi_1 A_1 \bar{g} \pi_1 A_2} = \overline{\pi_1 A_1 g_i \pi_1 A_2}$ for some (unique) i.*

(3) *(Intersections of images)* $\overline{\pi_1 A_1^{g_i} \cap \pi_1 A_2} = \overline{\pi_1 A_1}^{\bar{g}_i} \cap \overline{\pi_1 A_2}$ *for each i.*

The initial set of hypotheses are covered under the assumption that X^* is $C'(\frac{1}{16})$, as it is a negatively curved small-cancellation presentation by Theorem 3.32, and has tight innerpaths by Lemma 3.70. Claim (1) holds under small-cancellation with short innerpaths, but the proof of Claims (2) and (3) use that X^* is negatively curved and has tight innerpaths. (Tight innerpaths can be replaced by various weaker assumptions: Whenever $QS \to Y_i$ is essential and Q lifts to A_j, and $\{\{S\}\}_{Y_i} \leq \pi$ implies that QS lifts to A_j; or $\{\{S\}\}_{Y_i} \leq \pi$ implies that $|S|_{Y_i} \leq \frac{7}{8} \|Y_i\|$.)

We will assume in the proof that each g_i is a minimal length representative for its double coset.

Proof of Claim (1). Choose a minimal complexity disk diagram $D \to X^*$ with boundary path $P = a_1 g_i a_2 g_j^{-1}$, where a_1, a_2 are closed based local geodesics in A_1, A_2 and g_i, g_j are representatives as above. If D has no cone-cells then D actually factors as a cubical diagram $D \to X \to X^*$ and hence g_i, g_j represent the same double coset in $\pi_1 X$ so $i = j$. Let us assume that this is not the case, so D contains at least one cone-cell.

Intuitively, we will now produce a maximal annular subdiagram B containing the boundary path of D such that $B \to X$ is a cubical diagram, and B is in the local convex hull of the boundary path within D. We obtain B by continually adding squares of D whose corners are already present, and adding closed edges corresponding to spurs.

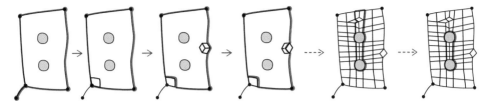

Figure 12.2. The sequence of diagrams starting with (D, P) and terminating at (E_t, B_t).

More precisely let us begin with $E_0 = D$ and $Q_0 = \partial_p D$. We will construct a sequence of paths Q_i bounding disk diagrams E_i that are contained in slight modifications D_i of D. See Figure 12.2.

For each $i \geq 0$, either:

(1) Q_i contains the outerpath of a spur of E_i.
(2) Q_i contains the outerpath of a cornsquare of E_i.
(3) Q_i contains neither, in which case the sequence terminates.

In the first case, we remove the terminal 0-cell and 1-cell of the spur from E_i to form E_{i+1} and we remove this backtrack to form Q_{i+1}. We continue to repeat the first case until there are no such backtracks remaining. We then examine if the second case holds.

In the second case, following Lemma 2.6, we first adjust the interior of E_i to obtain E_i' (and thus adjust the ambient diagram D_i to obtain D_i') so that the square actually lies along Q_i. We then remove from E_i' the open square and the corner consisting of the two open 1-cells and the open 0-cell. The path Q_{i+1} is obtained from Q_i by pushing across that square so that we replace the two edges from the outerpath corner by the opposite two edges. Note that if the outerpath of the square meeting Q_i is longer, then Q_{i+1} will obtain backtracks which we will omit when we return to the first case. After performing the second case, we return to examine if we are in the first case.

Since the number of cells in E_i is decreasing, we eventually terminate after t steps at E_t and $Q_t = \partial_p E_t$. Observe that E_t contains all cone-cells originally in D, and in particular E_t is nontrivial provided D contains at least one cone-cell.

At each stage, let B_i denote the annular subdiagram in D_i' bounded by P and Q_i. We think of the diagram B_i as the "local convex hull" of ∂P in D, but this is misleading as some hexagon moves were necessary to create it, and D is evolving.

It is important to note that at each stage, and in particular, in the final result, each dual curve in B_i emerging from an edge of Q_i terminates on P. Indeed, this holds initially and is preserved by each of the two replacement moves indicated above.

By construction E_t cannot have a cornsquare on its boundary path, nor can it have a spur. As was explained for B_i above, each dual curve emerging from a

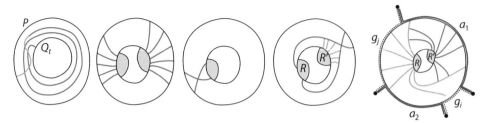

Figure 12.3. First: a dual curve from Q_t to P cannot cross itself or there will be a monogon or bigon. Second: dual curves emanating from outerpaths of shells of E are splayed—they cannot cross each other as in the third diagram. Fourth: dual curves emanating from a shell cannot cross each other around the diagram, for another shell would then have a short outerpath. Fifth: dual curves emanating from the outerpath of a shell cannot travel around the diagram and terminate on the same subpath (say a_1) of $P = a_1 g_i a_2 g_j^{-1}$.

1-cube on the boundary path of E_t must terminate on a 1-cell of the boundary of D.

By Theorem 3.46, either E_t is a single cone-cell, or E_t has at least two shells.

A dual curve emanating from an edge in Q_t cannot cross itself in B_t before terminating on P. Indeed, after crossing itself inwards it must eventually cross itself back outwards since it terminates on P. So we see that there is either a monogon or bigon dual curve subdiagram in B_t and hence a way to reduce the area, which is impossible. See the first diagram in Figure 12.3.

Consider the family of dual curves emanating from the outerpath on Q_t of a single shell R of E_t. As explained above, each such dual curve terminates on P. We claim that they cannot cross each other, and are thus splayed as in the second diagram in Figure 12.3. Indeed, if two such curves crossed each other on the same side of the annulus as in the third diagram in Figure 12.3 then there would be an outerpath of a cornsquare in B_t on R itself. This yields a complexity reduction by absorbing the square into the cone of R, and is thus impossible by the minimal area of D. This argument shows that if E_t is a single cone-cell then its emanating dual curves are splayed.

Let us now assume there are at least two shells, say R and R'. Two dual curves emanating from the outerpath of R cannot cross by going around to the other side of the annulus B_t as in the fourth diagram in Figure 12.3. For then, all the dual curves emanating from the outerpath of R' would cross one or the other of these dual curves. We thus see that the outerpath of R' is in the union of at most two contiguous wall-pieces by Lemma 3.7. Each hyperplane piece H in Y satisfies $|H|_Y < \frac{1}{4}\|Y\|$, so the outerpath would not be at least half the length of ∂R. This contradicts short innerpaths, unless $\partial_p R$ is not essential in Y, in which case D is not of minimal complexity.

In a similar manner, we see that two dual curves leaving the outerpath of the shell R cannot travel around the annulus and end on the same subpath

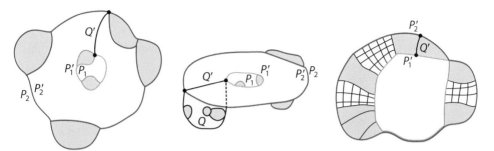

Figure 12.4. Before and after: The middle figure illustrates a typical situation where an original annulus is turned into a square annulus by pushing out cone-cells through the two bounding circles and pushing the conjugator out (so it is connected to the new conjugator by a "flap" consisting of a disk diagram in X^*). No extra flap is illustrated on the left diagram—where $Q = Q'$. The right diagram represents the final impossible annuladder we consider.

a_1, a_2, g_i, or g_j of P. See the fifth diagram in Figure 12.3 where the curves both end on a_1. For then consider another shell R', and note that the dual curves from its outerpath end entirely on two dual curves together with a_1, and hence the outerpath has length $< (\frac{1}{8} + \frac{1}{8} + \frac{1}{8})|\partial R'|$ which contradicts short innerpaths, unless $\partial_{\mathsf{p}} R$ is not essential in Y, in which case D is not of minimal complexity.

We conclude that the dual curves from the outerpath of each shell terminate in a sequence of at most four parts within a_1, a_2, g_i, and g_j. Since by hypothesis each such part is $< \frac{1}{8}$ of ∂R, we see that the outerpath of R is $< \frac{1}{2}$ of $|\partial R|$ which contradicts short innerpaths.

A similar argument works to reach a contradiction when E_t is a single cone-cell. Indeed in this case it suffices to see that each part ending on one of a_1, a_2, g_i, or g_j has length $< \frac{1}{4}\|Y^{\circledast}\|$. $\square$

Proof of Claim (2). Let $D \to X^*$ be an annular diagram for $P_1^Q = P_2$ where $P_i \to A_i$ are closed based paths that are boundary paths of D, and Q is a path in D joining the basepoints, and Q is homotopic to a based path g in X that projects to $\bar{g}$, and P_i represent nontrivial elements of $\pi_1 X^*$. We refer to Figure 12.4.

We can compress replaceable cone-cells and absorb and combine pairs while preserving the element representing the conjugating path in $\bar{G}$. Similarly, we push P_1, P_2 past any cornsquares, which could be absorbed into A_1, A_2 since $A_i \to X$ are local-isometries, and so no P_i contains the outerpath of a cornsquare in D. We may thus assume D is reduced, and more that D has no cornsquares on either boundary path. We can furthermore assume D is *minimal* in the sense that none of its cone-cells can be absorbed into A_i through P_i on either side of

∂D, and hence A_i-subpaths of such cone-cells have relatively small diameter by our hypothesis on each $A_i \to X$. Finally, we can pass to an annular subdiagram D' obtained by removing spurs, and note that spurs do not interfere with our ability to perform the above absorptions.

Let $\bar{D}'$ be the rectification of D' as in Remark 3.22. Suppose $\bar{D}'$ contains a negatively curved cone-cell (e.g., any internal cone-cell as X^* is negatively curved). Since $\chi(\bar{D}') = 0$, by Theorem 3.23, there must also be a positively curved cell, which is necessarily a shell, cornsquare, or spur. These contradict minimality as above. We therefore continue the argument under the assumption that there is no negatively curved shell and each cone-cell is external.

Consider a cone-cell R whose boundary path intersects P_i in at least two maximal subpaths. Note that R subtends a subdiagram of D' containing R that is a disk diagram containing at least two positively curved cells, and we would contradict minimality (as any such cell could be absorbed into A_i). Consider a cone-cell R whose boundary path intersects $\partial D'$ in a single subpath of P_i. As R is nonnegatively curved and cannot be absorbed into A_i, the tight innerpath hypothesis on X^* ensures that R is replaceable which contradicts minimality. We conclude that each cone-cell in D must intersect each of P_1 and P_2 in a single nontrivial subpath of its boundary.

Consequently, if D has a cone-cell then D' is a (possibly singular) annuladder by Lemma 5.60. However, this would imply that each cone-cell R is replaceable which contradicts minimality. Indeed, suppose R maps to the cone Y_i. Then each overlap S of $\partial_p R$ with P_1 or P_2 satisfies $|S|_{Y_i} < \frac{1}{4}\|Y_i^{\circledast}\|$, and the cone-piece or wall-pieces of $\partial_p R$ in the interior of D also have diameter $< \frac{1}{4}\|Y_i^{\circledast}\|$.

The only remaining possibility is that D is a square diagram. But then we have shown that Q is path-homotopic in X^* to the element Q' such that $P_1^{Q'} = P_2$ in $\pi_1 X$ where each P_i is nontrivial since its projection to $\pi_1 X^*$ is nontrivial. Consequently, Q' is in the same double coset as one of our representatives g_i.

Finally, this double coset representative is unique by Claim (1). $\square$

Proof of Claim (3). It is immediate that $\overline{\pi_1 A_1^{g_i} \cap \pi_1 A_2} \subset \left(\overline{\pi_1 A_1}^{\bar{g}_i} \cap \overline{\pi_1 A_2}\right)$ so we must only verify that $\overline{\pi_1 A_1^{g_i} \cap \pi_1 A_2} \supset \left(\overline{\pi_1 A_1}^{\bar{g}_i} \cap \overline{\pi_1 A_2}\right)$.

For each element represented by a closed path $P_2 \to A_2$ in the intersection $\overline{\pi_1 A_1}^{\bar{g}_i} \cap \overline{\pi_1 A_2}$, there is a closed based path $P_1 \to A_1$ and a closed based path $Q \to X$, such that there is an annular diagram $D \to X^*$ for $P_1^Q = P_2$. As in the proof of Claim (2), there is a sequence of removals and absorptions and replacements of cone-cells, and absorptions of cornsquares and shells to the boundary of the annulus. We can pass to a reduced and minimal (in the sense of Claim (2)) annular diagram $D' \to X$ for $P_1'^{Q'} = P_2'$ where each P_i' is a closed path in A_i that is path-homotopic to P_i in A_i^*, and likewise $Q' \to X$ is path-homotopic to Q in X^*, and such that P_1' and P_2' do not contain a subpath P of the outerpath of any cone-cell mapping to a cone Y_j with $|P|_{Y_j} \geq \frac{1}{4}\|Y_j^{\circledast}\|$.

As concluded in the proof of Claim (2), $D' \to X$ is a square annular diagram, and so we have found an element g represented by Q' such that $\bar{g} = \bar{g}_i$ and $\bar{P}_2 \in \pi_1 A_1^g \cap \pi_1 A_2$, as P_2 is path-homotopic to P_2' in A_2^*, and D' demonstrates that $P_2' \in \pi_1 A_1^g \cap \pi_1 A_2$.

If P_2 represents a nontrivial element in $\pi_1 X^*$, then it represents a nontrivial element in $\pi_1 X$, and hence so does P_2'. But then the conjugacy $P_1'^{Q'} = P_2'$ shows that g lies in one of the double cosets $\{\pi_1 A_1 g_j \pi_1 A_2\}$. The images in $\pi_1 X^*$ of these double cosets are distinct by Claim (1). However, since Q' is path-homotopic in X^* to our conjugator Q (representing $\bar{g}_i$), we see that Q' also maps to the same element $\bar{g}_i$ in $\pi_1 X^*$. Thus, for any P_2 chosen above representing a nontrivial element in $\pi_1 X^*$ in the intersection, since the associated conjugator Q' represents an element mapping to $\bar{g}_i$, Claim (1) ensures that Q' lies in $\pi_1 A_1 g_i \pi_1 A_2$.

In conclusion, each nontrivial element in the intersection of the images in $\pi_1 X^*$ (initially represented by $P_1^Q = P_2$) is the image of an element (represented by $(P_1')^{Q'} = P_2'$) in the intersection within $\pi_1 X$. And moreover, Q' represents an element of $\pi_1 A_1 g_i \pi_1 A_2$ in $\pi_1 X$. Since elements in the same double coset correspond to the same intersection of subgroups, we reach the conclusion that:

$$\text{Nontrivial elements of } \left(\overline{\pi_1 A_1}^{\bar{g}_i} \cap \overline{\pi_1 A_2} \right) \quad \subset \quad \overline{\pi_1 A_1^{g_i} \cap \pi_1 A_2}. \qquad \square$$

Remark 12.11 (Loxodromic Variant of Lemma 12.10). There is a variation of Lemma 12.10 that holds in the setting where $\pi_1 X$ is hyperbolic relative to $\{\mathrm{Stab}(\widetilde{Y}_j)\}$, and where the statement and proof of Lemma 12.10 holds with every occurrence of the word "nontrivial" replaced by the word "loxodromic." An element g of a relatively hyperbolic group is *loxodromic* if it has infinite order and does not lie in any parabolic subgroup. A subgroup is *loxodromic* if it contains a loxodromic element. For the final step of the proof of Lemma 12.10.(3), the following observation completes the proof: Any loxodromic subgroup $L \subset G$ of a relatively hyperbolic group is generated by its loxodromic elements. Indeed, if $g \in L$ is loxodromic, and $h \in L$ is parabolic or elliptic, then $g^n h \in L$ is loxodromic for $n \gg 0$, and hence $h \in \langle g, g^n h \rangle$.

To support the applicability of Parts (2) and (3) of the loxodromic version of Lemma 12.10, we note that it was proven in [HW09, Cor 8.7] that for relatively quasiconvex subgroups $A_1, A_2 \subset G$ of a relatively hyperbolic group, there are finitely many double cosets $A_1 g_i A_2$ with $A_1^{g_i} \cap A_2$ loxodromic. (This was stated there in the case that $A_1 = A_2$ but the same proof works for arbitrary A_1, A_2.)

Lemma 12.12 (Isocore variant of Lemma 12.10). *Let $\langle X \mid \{Y_j\} \rangle$ be a $C'(\frac{1}{16})$ cubical presentation. Let $A_1 \to X$ and $A_2 \to X$ be based local-isometries of non-positively curved cube complexes. Let $A_1^{\square} \subset A_1$ and $A_2^{\square} \subset A_2$ be isocores. Let $\bar{Y}_j$ be the quotient of $\widetilde{Y}_j$ by its stabilizer. (In the motivating case, $\pi_1 X$ is hyperbolic relative to $\{\pi_1 \bar{Y}_j\}$.) Let $\{\pi_1 A_1 g_i \pi_1 A_2\}$ be a collection of distinct double* *cosets. Let M_j be the maximal diameter of a piece in Y_j for each j. Suppose the following two assumptions hold for each j:*

(1) *For $\alpha, \alpha' \in \pi_1 \bar{Y}_j$ satisfying $|\alpha|, |\alpha'| \leq diam(A_1^{\square} \otimes \bar{Y}_j) + diam(A_2^{\square} \otimes \bar{Y}_j) + 2M_j + \max\{|g_i|, M_j\}$ we have: If $\pi_1 Y_j \cap (\pi_1(A_1 \otimes \bar{Y}_j)\alpha\pi_1(A_2 \otimes \bar{Y}_j)\alpha') \neq \emptyset$ then $1 \in (\pi_1(A_1 \otimes \bar{Y}_j)\alpha\pi_1(A_2 \otimes \bar{Y}_j)\alpha')$.*

(2) *If R is a shell with outerpath $Q \to A_i^{\square}$, and $QS \to Y_j$ and $|S| < 12M_j + 2\max\{|g_i|\}$, then $S \to Y_j$ is path-homotopic to $S^{\square} \to Y_j$ such that $Q \to A_i^{\square}$ extends to a closed path $QS^{\square} \to A_i^{\square}$.*

Then Conclusion 12.10.(1) holds. Moreover, if in addition $\{\pi_1 A_1 g_i \pi_1 A_2\}$ is a complete collection of intersecting conjugators, then Conclusion 12.10.(2) holds.

Proof. The proof is a variation on the proof of Lemma 12.10. The first difference in the proof is that instead of using minimal complexity diagrams, we will use a diagram D having a minimal number of cone-cells. We then assume that the paths $P_i \to A_i^{\square}$ are chosen so that their lifts $\tilde{P}_i \to \tilde{A}_i^{\square}$ are geodesics. This can always be arranged by adjusting the square complex part of D without affecting the number of cone-cells.

Consider the subdiagram $E = E_t$ surrounded by the square annulus $B = B_t$. Consider a nonnegatively curved shell R of E with $\partial_p R = QS$, if the subpath of $\partial_p R$ subtended by the dual curves of R is a geodesic, then by Lemma 14.11, $Q = S_1 Q' S_2$ where each S_i is either trivial or a wall-piece and where Q' is either trivial or a subpath of $\partial_p D$. Note that S is the concatenation of at most 8 pieces by Lemma 3.70. Hence $\partial_p R = Q'S'$ where Q' is a subpath of $\partial_p R$ and S' consists of at most 10 pieces. In particular, the above condition holds if the dual curves all end on either P_1 or P_2. Now suppose the dual curves emanating from Q either end on P_i or end on g_1, g_2, or cross a chosen dual curve ending on g_1 or g_2. Then similar reasoning shows that $Q = S_1'' Q'' S_2''$ where Q'' is a subpath of some P_i, and $|S_i''|$ is the concatenation of at most two wall-pieces and edges whose dual curves end on g_i. Hence $|S_i''| < 2M_j + |g_i|$, where M_j is an upperbound on the size of the diameter of a wall-piece for the cone Y_j associated to R. Hence $\partial_p R = Q''S''$ where $|S''| < 8M_j + 4M_j + 2\max_i\{|g_i|\}$. We refer to the left diagram of Figure 12.5.

Suppose there are dual curves in B terminating on g_1 and g_2 that emerge from edges of (one or two) outerpaths of shells in E. Then as explained above, a third nonnegatively curved shell R of E with $\partial_p R = QS$ would have $\partial_p R = Q''S''$ as above, with Q'' a subpath of P_i for some i. However, this would violate minimal complexity since, by Hypothesis (2), we could replace Q'' by a path $(S'')^{\square}$ and replace R by a square diagram between S'' and $(S'')^{\square}$. We obtain a lower complexity counterexample diagram whose boundary is still built from paths in $A_i^{\square}$.

If E is a ladder with more than one cone-cell, then the case where one shell has outgoing dual curves on both g_1, g_2 implies the same control as above on

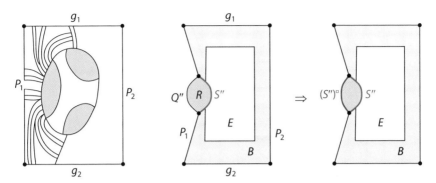

Figure 12.5. The left diagram illustrates the situation that occurs if both g_1 and g_2 have dual curves connecting to shells of D. A further shell of D would again have most of its outerpath on P_i, except for 4 wall-pieces and some edges with dual curves ending on g_1, g_2. The operation at the right illustrates a reduction, where a shell R with $\partial_{\mathsf{p}} R = Q'' S''$ is replaced by a square diagram between S'' and a path $(S'')^{\square} \to A_i^{\square}$.

the other cone-cell. We are left to consider the case where either E consists of a single cone-cell, or the two shells of E have outerpaths whose dual curves are limited to distinct g_1, g_2. We will show each cone-cell R of E can be removed leading to a diagram with fewer cone-cells and whose boundary is modified but maintains the properties enabling our argument. The argument applies to any cone-cell, but our focus will be on a cone-cell which is at the top or bottom (or both) of E.

As above, the part of $\partial_{\mathsf{p}} R$ consisting of edges whose dual curves end on P_i consists of the union of $\partial_{\mathsf{p}} R \cap P_i$ together with at most two wall pieces. Consequently, if $|\partial_{\mathsf{p}} R \cap P_1| = 0$ then $\partial_{\mathsf{p}} R = Q'' S''$ where Q'' is a subpath of P_2 and S'' is the concatenation of at most 4 wall-pieces and 2 paths that are either pieces or have length bounded by $\max_i \{|g_i|\}$. This violates minimality since, by Hypothesis (2), the path $S'' \to Y_j$ is path-homotopic to $(S'')^{\square} \to Y_j$ such that $Q'' \to A_2^{\square}$ extends to a closed path $Q''(S'')^{\square} \to A_2^{\square}$, and we may thus replace the cone-cell R by a square diagram, and replace P_2 by a path $P_2' \to A_2^{\square}$ with Q'' replacing S''. The same statement holds with the roles of P_2 and P_1 exchanged.

We may thus consider the case where $\partial_{\mathsf{p}} R \cap P_i \neq \emptyset$ for each i, and we let q_i, q_i' be endpoints of maximal subpaths ξ_i of $\partial_{\mathsf{p}} R \cap P_i$. Let μ and μ' be the subpath of $\partial_{\mathsf{p}} R$ joining $q_1 q_2$ and $q_1' q_2'$ so $\partial_{\mathsf{p}} R = \xi_1 \mu \xi_2 \mu'$. Observe that $|\mu|$ and $|\mu'|$ are each bounded by the sum of the diameters of two wall-pieces together plus either the diameter of a maximal piece or $\max_i \{|g_i|\}$. Indeed, as depicted on the left of Figure 12.6, the dual curve to the first edge of P_i above ξ_i cannot end on P_i since it is a geodesic. Consequently, all dual curves starting on μ either cross one of these dual curves or end on g_1. Similarly for μ_2, except that they could end on g_2 or the next cone-cell depending on whether the ladder is a single cone-cell. There are paths τ_i, τ_i' joining q_i, q_i' to points b_i, b_i' of $A_i^{\square} \otimes \bar{Y}$ mapping to the

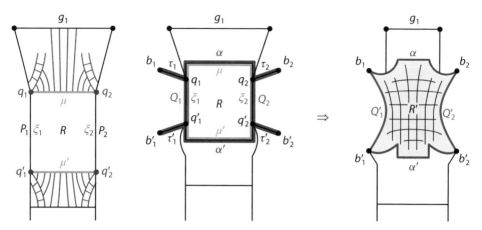

Figure 12.6. The diagram on the left indicates why μ_1 (or μ_2) have length bounded by $|g_i|$ (or a piece) plus at most two wall-pieces. The diagrams on the right illustrate the operation of replacing a cone-cell R. Add τ tails which are paths in $A_i^{\square} \otimes Y$ that travel to the basepoint of $A_i^{\square}$. The paths α, α' traveling along the tails and top and bottom of R have bounded lengths, and so by hypothesis, if $Q_1 \alpha Q_2 \alpha'$ is a closed path in $\widehat{Y}$, then there are closed paths $Q_i' \to A_i^{\square}$ such that $Q_1' \alpha Q_2' \alpha'$ is null-homotopic in X. The operation replaces R by the square diagram for $Q_1' \alpha Q_2' \alpha'$.

basepoint of $A_i^{\square}$, and with $|\tau_i|, |\tau_i'| \leq \operatorname{diam}(A_i^{\square} \otimes \bar{Y})$. Let α be the concatenation $\tau_1 \mu \tau_2$, and let α' be $\tau_1' \mu' \tau_2'$. Let ξ_i be the subpath of $P_i \cap \partial_{\mathsf{p}} R$ that joins q_i, q_i'. Let Q_i be the path from b_i to b_i' that is the concatenation $\tau_i \xi \tau_i'$.

Since $Q_1 \alpha Q_2 \alpha'$ is a closed path in Y and represents an element of $\pi_1(A_1^{\square} \otimes Y) \alpha \pi_1(A_2^{\square} \otimes Y) \alpha'$, Hypothesis (1) implies that there exist closed paths $Q_i' \to A_i^{\square} \otimes \bar{Y}$ such that $1 = Q_1' \alpha Q_2' \alpha'$. Let $R' \to Y$ be a square diagram with $\partial_{\mathsf{p}} R' = Q_1' \alpha Q_2' \alpha'$. We produce a lower complexity diagram D' by replacing R with R', and replacing each Q_i with Q_i', as on the right in Figure 12.6. We conclude that there exists a diagram with no cone-cells and so the argument is as in Lemma 12.10. $\qquad \square$

Remark 12.13 (Variant Computation of Intersection of Images). The argument to prove Lemma 12.10.(3) requires an additional assumption, and we offer two variants of this below:

(1) $((\pi_1 A_1 \cap \pi_1 \bar{Y}_j^g)/\pi_1 Y_j^g) \cap ((\pi_1 A_2 \cap \pi_1 \bar{Y}_j^g)/\pi_1 Y_j^g) = (\pi_1 A_1 \cap \pi_1 A_2 \cap \pi_1 \bar{Y}_j^g)/\pi_1 Y_j^g$.
(2) For $g, h \in \pi_1 X$, if $\pi_1 A_1^g \cap \pi_1 A_2 \cap \pi_1 \bar{Y}_j^h \neq 1$ then $\pi_1 Y_j^h \subsetneq \pi_1 A_1^g \cap \pi_1 A_2 \cap \pi_1 \bar{Y}_j^h$.

An element of $\pi_1 X$ stabilizing some $\widetilde{Y}$ is *conical* and likewise the conical elements of $\pi_1 X^*$ are those that stabilize some cone. In the motivating case, the conical elements are the parabolic elements.

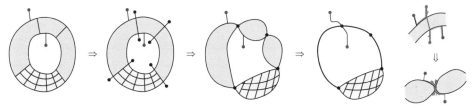

Figure 12.7. The first diagram is an annular diagram D in the simplified case where $E = D$. The second diagram has added tails so that the boundary paths along cone-cells start and end at basepoints. The double coset assumption ensures that the path joining each pair of tails is nullhomotopic in $\pi_1 X^*$. The third diagram is obtained by the operation of cutting and identifying the ends of the tails, and inserting square diagrams. The fourth is obtained by using Assumption (1) to replace each cone-cell R with $\partial_{\mathsf{p}} R = Q_1 Q_2$ by an arc $Q_3 \to A_1^{\square} \otimes A_2^{\square}$ with $Q_1 = Q_3 = Q_2$ in $\pi_1 A_i^{\square}$. The final pair of diagrams illustrate that the conjugator is unaffected by the operation, since the final and original conjugators differ by a null-homotopic path.

Assumption (1) leads to the variant conclusion: Every non-conical element in the intersection is the image of a non-conical element. Assumption (2) is easier to arrange, but leads to the weaker conclusion: An infinite/non-conical intersection of images of conjugates contains the image of an intersection that is not conical.

A minimal complexity annular diagram $D \to X^{\square}$ for $P_1^Q = P_2$ is an annuladder, where we note that the conjugator Q is allowed to vary among paths in $X^{\square}$ whose image is the same, and P_i are allowed to vary in $A_i^{\square}$ without affecting the image. Indeed, as above, we produce the annular rim B and pass to a subdiagram E. A nonnegatively curved shell leads to a lower complexity diagram, and all internal cone-cells of E are negatively curved, and so E is an annuladder by Lemma 5.60. As above, we can adjust E and hence D so that each cone-cell starts and ends at a local cutpoint.

Under Assumption (1), we can replace each cone-cell R with $\partial_{\mathsf{p}} R = Q_1 Q_2$ by an isolated path $Q_3 \to A_1^{\square} \otimes A_2^{\square}$ that has the same image in $\pi_1 A_i^{\square}$ as each Q_i. The result is a square diagram, so our conjugacy in $\pi_1 X^*$ lifts to a conjugacy in $\pi_1 X$. See Figure 12.7.

Under Assumption (2), either D is a square diagram and $\partial_{\mathsf{p}} D$ does not stabilize any $\widetilde{Y}_j$, so $\overline{\pi_1 A_1}^{\bar{g}} \cap \overline{\pi_1 A_2}$ contains an infinite image subgroup $\overline{\pi_1 A_1^g \cap \pi_1 A_2}$. Or D contains a cone-cell, in which case we see that $\overline{\pi_1 A_1^g \cap \pi_1 A_2}$ contains two distinct nontrivial conical subgroups, as one can use a pair of cones corresponding to consecutive cone-cells (if there is more than one cone-cell) in D, or a single cone together with its conjugate by a path corresponding to the remainder of D. Alternatively, we can follow the argument used under Assumption (1), except that we replace a cone-cell by a path in the intersection of the corresponding conjugates, and note that we may choose this path to represent a nontrivial element of $\pi_1 Y_j / \pi_1 \overline{Y}_j$, so the result is loxodromic.

Chapter Thirteen

Amalgams of Virtually Special Groups

13.a Virtually Special Amalgams

Theorem 13.1 (Special hyperbolic amalgam). *Let $G = A *_C B$ be an amalgamated product where G is word-hyperbolic and C is quasiconvex. If A, B are virtually compact special then so is G. A similar result holds for graphs of groups, and in particular: Let $G = A*_C$ be an HNN extension such that G is word-hyperbolic and C is quasiconvex in G. If A is virtually compact special then so is G.*

Remark 13.2. As Theorem 13.1 holds in the special case when C is malnormal by Theorem 11.2, it is attractive to attempt to deduce Theorem 13.1 from Theorem 11.2 by passing to a finite index subgroup of G whose induced graph of groups has the property that each of its edge groups is malnormal. One can do this under the additional hypothesis that C is separable.

Theorem 13.3 (Word-Hyperbolic $\mathcal{QVH}$ Special). *Every word-hyperbolic group in $\mathcal{QVH}$ is virtually compact special.*

Moreover, if it is also torsion-free, then it is the fundamental group of a compact cube complex that is virtually special.

Proof. The statement for $\mathcal{QH}$ holds by induction on the length of a hierarchy by applying Theorem 13.1. The statement likewise holds by induction for $\mathcal{QVH}$, but we also count passage to a finite index subgroup as a step in the length.

The additional conclusion in the torsion-free case holds by combining the main result with Lemma 7.14 and Corollary 6.10. $\square$

We use the notation $x^y = yxy^{-1}$ here but warn the reader that $x^y = y^{-1}xy$ is used in other places in the text.

Proof of Theorem 13.1. **Reduction to HNN case:** There is an embedding $A *_C B \to D*_C$ where $D = A * B$ and the HNN extension $D*_C$ has C embedded in the first factor of $A * B$, and has C^t embedded in the second factor. This embedding is quasiconvex, and so C is itself quasiconvex in D. A covering space argument shows that if A, B are word-hyperbolic and virtually compact special then so is $D = A * B$. Moreover, if $D*_C$ is virtually compact special, then so is the quasiconvex subgroup $A *_C B$. Thus the AFP case follows from the HNN case.

Strategy of proof: Consider $G = A*_{C^t=D}$ where A is virtually special, and G is word-hyperbolic and C is quasiconvex in G. Let $h = \text{Height}_G(C)$. We will form a quotient $G = A*_{C^t=D} \to \bar{A}*_{\bar{C}^t=\bar{D}} = \bar{G}$ where $\bar{A}$ is virtually special, $\bar{G}$ is word-hyperbolic and $\bar{C}$ is quasiconvex, and $\bar{h} = \text{Height}_{\bar{G}}(\bar{C}) < h$. The group $\bar{G}$ will therefore be virtually special by induction on height. In the base case where $h \leq 1$, the edge group is almost malnormal, and so the result holds by Theorem 11.2—or more specifically, Theorem 11.3.

The quotient $G \to \bar{G}$ will also be chosen to have the property that each intersecting conjugator k_i of C has $\bar{k}_i \notin \bar{C}$, where k is an *intersecting conjugator* if $C^k \cap C$ is infinite but $k \notin C$. Note that there are finitely many C cosets of intersecting conjugators by Proposition 8.3.

Since $\bar{G}$ is virtually compact special and word-hyperbolic, and $\bar{C}$ is quasiconvex, there is a finite quotient $\bar{G} \to Q$ such that $\bar{C}$ is separated from each $\bar{k}_i$ in Q, and hence $\phi(C)$ is separated from each $\phi(k_i)$ in the resulting composition quotient $G \xrightarrow{\phi} Q$. Let G' be the kernel of ϕ and observe that each edge group of the induced splitting of G' is almost malnormal. Consequently, G' is virtually special by Theorem 11.2.

The collection of tree stabilizers: For each i, let $\{T_{ik} : k \in J_i\}$ denote representatives of G-orbits of the finitely many i-edge subtrees with infinite pointwise stabilizer $\text{Fix}(T_{ik})$. Let $B_{ik} = \text{Stab}(T_{ik})$. Each B_{ik} acts on T_{ik} with a fixed point, and we choose one fixed vertex and declare it to be the *root* of T_{ik}. Without loss of generality we assume each T_{ik} is positioned so that its root lies at the base vertex v of T. Note that T_{ik} already has the structure as a directed graph, since it is a subgraph of the directed graph T. However, T_{ik} has an additional structure where we regard the root as the *highest* vertex and where each edge *ascends* from its lower vertex to its higher vertex (towards the root). Accordingly, for each edge e of T_{ik} ascending from the vertex a to the vertex b we have $\text{Stab}_{B_{ik}}(a) = \text{Stab}_{B_{ik}}(e)$ and $[\text{Stab}_{B_{ik}}(b) : \text{Stab}_{B_{ik}}(e)] < \infty$. For each k we let $S_{ik} \subset T_{ik}$ denote a (fundamental domain) subtree consisting of unique representatives of the B_{ik}-orbits of vertices and edges of T_{ik}. One constructs S_{ik} by beginning with the root, and at each stage, including unique representatives of the closed edges that meet vertices already included. For each $k \in J_i$ let $L_{ik} \subset G$ index the vertices of S_{ik}, so each vertex of S_{ik} equals xv for a unique $x \in L_{ik}$. For notational comfort we index the root vertex by $1 = 1_G$. Let H_{ikx} denote the subgroup of A whose conjugate H_{ikx}^x equals $A^x \cap B_{ik}$ which equals $\text{Stab}_{B_{ik}}(xv)$.

As explained above, if xv ascends to yv in S_{ik} then H_{ikx}^x is a finite index subgroup of H_{iky}^y. Hence, conjugation by xy^{-1} induces a finite index inclusion $H_{ikx}^{xy^{-1}} \hookrightarrow H_{iky}$.

The almost malnormal collection in A and the quotient $A \to \bar{A}$: It follows from $h = \text{Height}(G, C)$ that $B_{hk} = \mathbb{C}_G(\text{Fix}(T_{hk}))$. The collection $\{B_{hk} : k \in J_h\}$ is almost malnormal in G by Lemma 8.6, since it is a subcollection of a collection that is almost malnormal. Consequently, $\{H_{hkx} : k \in J_h, \ x \in L_{hk}\}$ is an almost malnormal collection of subgroups of A by Lemma 8.7 applied to $\{B_{hk}xA : x \in L_{hk}\}$ for each h.

We will apply Theorem 12.2 to A where the role of $\{H_1, \ldots, H_r\}$ is played by $\{H_{hkx}\}$. For each k, the "constraint" subgroups $\{\ddot{H}_{hkx}\}$ supplied by Theorem 12.2, embed as finite index subgroups of the root H_{hk1}. Indeed there is a sequence of monomorphisms following the ascending path in S_{hk} from xv to $1v$. In addition, we will chose "preference" subgroups $H^\circ_{hkx} \subset H_{hkx}$ below, and these preference subgroups H°_{hkx} likewise ascend to finite index subgroups of H_{hk1}. We can thus choose a finite index subgroup $H^\bullet_{hk1}$ of H_{hk1} that is contained in the intersection of all these constraint and preference subgroups. We choose H'_{hk1} to be a subgroup of $H^\bullet_{hk1}$ that is normal in H_{hk1}, and this descends to a subgroup H'_{hkx} for each x. For each directed edge (xv, yv) in S_{hk}, we thus have:

$$(H'_{hkx})^{xy^{-1}} = H'_{hky}.$$

The torsion-free finite index subgroup A^φ: The group A acts properly and cocompactly on a CAT(0) cube complex $\tilde{X}$. Let A^φ denote a torsion-free finite index normal subgroup of A, and let $X = A^\varphi \backslash \tilde{X}$. The subgroup A^φ will facilitate our cubical small-cancellation theory computations below. Let $U = (C \cap A^\varphi) \backslash \tilde{U}$ and $V = (D \cap A^\varphi) \backslash \tilde{V}$. Note that it is possible that $(\pi_1 U)^t \neq \pi_1 V$.

The preference subgroups: We will choose the subgroups $H^\circ_{hkx} \subset H_{hkx}$ to be small enough that:

(1) Each H°_{hkx} lies in A^φ.
(2) For each intersecting conjugator g of C in G, the quotient $G \to \bar{G}$ maps g to an intersecting conjugator $\bar{g}$ of $\bar{C}$ in $\bar{G}$. We ensure this below by requiring that for certain $a_s \in A$, we have $\bar{a}_s \notin \bar{C}$ and $\bar{a}_s \notin \bar{D}$ in $A \to \bar{A}$.
(3) Double cosets $H_{ikx} a H_{\ell j y}$ corresponding to intersecting conjugators in A of the various H_{ikx} are separated by $A \to \bar{A}$ where $1 \le i \le h, \ k \in J_i, \ x \in L_{hk}$ and $1 \le \ell \le h, \ j \in J_\ell, y \in L_{\ell j}$.
(4) Both $\bar{C}$ and $\bar{D}$ are quasiconvex in $\bar{A}$.

The conjugation isomorphism $C^t = D$ projects to $\bar{C}^{\bar{t}} = \bar{D}$: We now show how to verify that the kernel of $C \to \bar{C}$ maps isomorphically to the kernel of $D \to \bar{D}$ under the map $C \to D$, and hence the conjugation isomorphism $C^t = D$ projects to $\bar{C}^{\bar{t}} = \bar{D}$. We begin with some notation: Let (xv, yv) denote an edge of S_{hk} that is directed in T from xv to yv. The element $x^{-1}y$ is of the form $\alpha t \beta$ for some $\alpha, \beta \in A$ where $\alpha = \alpha(x, y)$ and $\beta = \beta(x, y)$. There are two points here:

Firstly, as we later confirm, we have the following presentations for $\bar{C}$ and $\bar{D}$:

$$\bar{C} = C / \langle\!\langle\; (H'_{hkx})^{\alpha} \;:\; k \in J_h,\; (xv, yv) \in \text{Di-Edges}(S_{hk}),\; \alpha = \alpha(x, y) \;\rangle\!\rangle$$

$$\bar{D} = D / \langle\!\langle\; (H'_{hky})^{\beta^{-1}} \;:\; k \in J_h,\; (xv, yv) \in \text{Di-Edges}(S_{hk}),\; \beta = \beta(x, y) \;\rangle\!\rangle$$

Secondly, the map $C \to D$ induced by conjugation by t, sends $(H'_{hkx})^{\alpha}$ to $(H'_{hky})^{\beta^{-1}}$. Indeed, $(H_{hkx})^{xy^{-1}} \subset H_{hky}$ holds by construction, and our "descending choice" of the H'_{hkx} ensures that $(H'_{hkx})^{xy^{-1}} = H'_{hky}$. Since $xy^{-1} = \alpha t \beta$ we have: $((H'_{hkx})^{\alpha})^t = ((H'_{hkx})^{xy^{-1}})^{\beta^{-1}} = (H'_{hky})^{\beta^{-1}}$. Thus the conjugation isomorphism $C^t = D$ maps the generators of kernel$(C \to \bar{C})$ isomorphically to the generators of kernel$(D \to \bar{D})$. Thus $C^t = D$ projects to an isomorphism $\bar{C}^{\bar{t}} = \bar{D}$ as claimed.

The associated superconvex subcomplexes: Let $\{H_{ikx} : x \in L_{ik}\}$ denote representatives of the conjugacy classes in A of intersections with conjugates of B_{ik}. Specifically, each $H_{ikx} = B_{ik}^{x^{-1}} \cap A$ for some $x \in G$. By Lemma 2.36, each H_{ikx} acts properly and cocompactly on a superconvex basepoint containing subcomplex $\widetilde{Y}_{ikx}$. Choosing $H_{111} = C$ and $H_{11t} = D$ we let $\widetilde{U} = \widetilde{Y}_{111}$ and $\widetilde{V} = \widetilde{Y}_{11t}$. We require that for $i > i'$ if H_{ikx} is conjugate into $H_{i'k'x'}$ in A, then the corresponding translate of $\widetilde{Y}_{ikx}$ lies in $\widetilde{Y}_{i'k'x'}$. This property can be arranged by making the choices with $i = h$ first and then proceeding towards $i = 1$ to make the choices inclusively at each stage. By cocompactness and the pigeon-hole principle (or an equivariant version of Lemma 8.9), there is a constant M such if $\varnothing(a\widetilde{Y}_{ijk} \cap a'\widetilde{Y}_{i'j'k'}) > M$ then this intersection of A-translates equals some $a''\widetilde{Y}_{i''j''k''}$ with $i'' \geq i, i'$ and moreover if $i'' = i$ then $a''\widetilde{Y}_{i''j''k''} = a\widetilde{Y}_{ijk}$ and similarly for $i'' = i'$. By Lemma 2.40, there is a constant R bounding the diameters of base rectangles on these superconvex subcomplexes of $\widetilde{X}$.

Computing presentations for $\bar{C}$ and $\bar{D}$: Assume each $H^{\circ}_{hjk} \subset H^{\circ}_{hjk}$ and assume each H°_{hjk} has $\|H^{\circ}_{hjk} \backslash \widehat{Y}_{hjk}\| > 24 \max(R, M)$. Under these assumptions, we verify that our claimed presentations for $\bar{C}$ and $\bar{D}$ are correct. The normal subgroup $N = \ker(A \to \bar{A})$ we obtain is $N = \langle\!\langle \{\pi_1 \widehat{Y}_{hjk}\} \rangle\!\rangle_A$ and is contained in the normal subgroup A° of A since each $\pi_1 \widehat{Y}_{hjk} = H'_{hjk} \subset H^{\circ}_{hjk} \subset A^{\circ}$. The presentation $X^* = \langle X \mid a\widehat{Y}_{hjk} : a \in A \rangle$ accurately determines the presentation for $\bar{A}^{\circ}$. Indeed, $\langle\!\langle \{\pi_1 \widehat{Y}_{hjk}\} \rangle\!\rangle_A$ is a subgroup of A° since each $\pi_1 \widehat{Y}_{hjk}$ is, and moreover it is the smallest normal subgroup of A° containing each $\pi_1 \widehat{Y}_{hjk}$ that is also invariant under the A-action by conjugation. Accordingly, $\langle\!\langle \{\pi_1 \widehat{Y}_{hjk}\} \rangle\!\rangle_A = \langle\!\langle (\pi_1 \widehat{Y}_{hjk})^a : a \in A \rangle\!\rangle_{A^{\circ}}$ which equals $\ker(\pi_1 X \to \pi_1 X^*)$.

The key to computing $C \cap N$ is to first observe that $C \cap N = C \cap (A^{\circ} \cap N) = (C \cap A^{\circ}) \cap N = \pi_1 U \cap N$. Applying Lemma 3.67 and Theorem 3.68, we use no missing shells to compute an induced presentation U^* from $U \to X$ and X^*. It is of the form $U^* = \langle U \mid c\widehat{Y}_{hjk} : c \in C \rangle$. And consequently $C \cap N = \langle\!\langle \pi_1 \widehat{Y}_{hjk} \rangle\!\rangle_C$.

Likewise $D \cap N = (D \cap A^{\varphi}) \cap N = \pi_1 V \cap N$, and there is an induced presentation $V^* = \langle V \mid d\widehat{Y}_{hjk} : d \in D \rangle$, and we see that $D \cap N = \langle\!\langle \pi_1 \widehat{Y}_{hjk} \rangle\!\rangle_D$.

Separating intersecting conjugators: By Proposition 8.3, there are finitely many double cosets CgC in G such that $C^g \cap C$ is infinite. Each representative g has a normal form $a_0 t^{\pm 1} a_1 t^{\pm 1} \cdots t^{\pm 1} a_m$ for some $m \geq 0$ depending on g where $\{a_s : 0 \leq s \leq m\}$ are elements of A. Note the possiblity that $g = a_0$ in which case $a_0 \notin C$. To be in normal form means that the above concatenation does not contain $t a_s t^{-1}$ with $a_s \in C$ or contain $t^{-1} a_s t$ with $a_s \in D$. Since C and D are separable subgroups of A, there is a finite index normal subgroup $A^\diamond \subset A$ such that varying over the finitely many intersecting conjugator representatives $\{g\}$ and the finitely many relevant elements $\{a_s\}$ in the normal form of each g, we have $a_s C \not\subset A^\diamond C$ and $a_s D \not\subset A^\diamond D$ as appropriate. We add the intersections $H^\diamond_{hjx} = H_{hjx} \cap A^\diamond$ to our list of preferences. Thus $H^\diamond_{hkx} \subset H^\diamond_{hjx}$ ensures that in the homomorphism $A \to \bar{A}$, we have $\bar{a}_s \notin \bar{C}$ whenever $a_s \notin C$ and $\bar{a}_s \notin \bar{D}$ whenever $a_s \notin \bar{D}$. Consequently, in the quotient $A *_{C^t = D} \to \bar{A} *_{\bar{C}^t = \bar{D}}$, the normal form of g maps to a normal form of $\bar{g}$ and hence our intersecting conjugators of C in G map to intersecting conjugators of $\bar{C}$ in $\bar{G}$.

Reduction in height: There is an equivariant map $T \to \bar{T}$ from the G-action on the Bass-Serre tree T of $A *_C$ to the $\bar{G}$ action on the Bass-Serre tree $\bar{T}$ of $\bar{A} *_{\bar{C}}$. Consider the finitely many subgroups $\{H_{ikx} : 1 \leq i \leq h, k \in J_i, x \in L_{ik}\}$.

By Proposition 8.3, there are finitely many double cosets $H_{ikx} a H_{\ell jy}$ in A of intersecting conjugators with the property that $H_{ikx} \cap H^a_{\ell jy}$ is infinite. We will show below that for suitable $A \to \bar{A}$, if $\bar{H}_{ikx} \cap \bar{H}^{\bar{a}}_{\ell jy}$ is infinite, then $\bar{H}_{ikx} \bar{a} \bar{H}_{\ell jy}$ is the image of one of the finitely many $H_{ikx} a H_{\ell jy}$ above. Consequently any finite subtree $\bar{F} \subset \bar{T}$ with infinite pointwise stabilizer can be lifted to a finite subtree $F \subset T$ with infinite pointwise stabilizer and $\mathrm{Fix}_{\bar{G}}(\bar{F}) = \overline{\mathrm{Fix}_G(F)}$. Consequently any such subtree $\bar{F}$ has at most h edges, since F does as $h = \mathrm{Height}_G(C)$. Moreover, if F has h edges and infinite pointwise stabilizer then $\mathrm{Stab}_G(F)$ is conjugate to B_{hk} for some k, and so $\mathrm{Stab}_{\bar{G}}(\bar{F})$ is finite.

Towards height control: We will choose $A \to \bar{A}$ so that $\overline{H_{ikx}} \cap \overline{H_{\ell jy}}^{\bar{a}}$ is commensurable with $\overline{H_{ikx} \cap H^a_{\ell jy}}$ for each double coset $H_{ikx} a H_{\ell jy}$. Indeed, the commensurabilities below are immediate. The equality holds by Lemma 12.10.(3) applied to $A^{\varphi} \to \bar{A}^{\varphi} = \pi_1 \bar{X}^*$ with respect to $\{H^{\varphi}_{ikx} : i < h\}$. The extra preference is that each $\|\widehat{Y}_{hkx}\| > 8\mu$ where μ is an upperbound on the lengths of based combinatorial paths in X representing elements that are representatives of the finitely many intersecting conjugators of these pairs of subgroups in A.

$$\overline{H_{ikx}} \cap \overline{H_{\ell jy}}^{\bar{a}} \sim \overline{H^{\varphi}_{ikx}} \cap \overline{H^{\varphi}_{\ell jy}}^{\bar{a}} = \overline{H^{\varphi}_{ikx} \cap (H^{\varphi}_{\ell jy})^a} \sim \overline{H_{ikx} \cap H^a_{\ell jy}}.$$

Moreover each intersecting conjugator in $\bar{A}$ is the image of an intersecting conjugator in A in the sense that if $\overline{H_{ikx}} \cap \overline{H_{\ell jy}}^{\bar{b}}$ is infinite then $\bar{b} = \bar{a}'$ for some a' in a double coset $H_{ikx} a H_{\ell jy}$ of intersecting conjugators. For this we consider the larger finite set of subgroups $\{(H^\wp_{ikx})^g\}$ where g varies over representatives of the cosets $\{gA^\wp\}$ in A. We apply Lemma 12.10.(2) to a complete set of double cosets $\{(H^\wp_{ikx})^g q (H^\wp_{\ell jy})^f\}$ representing intersecting conjugators. Observe that $b = g\beta$ for one of our coset representatives g and some $\beta \in A^\wp$. Thus:

$$\overline{(H^\wp_{ikx})} \cap \overline{(H^\wp_{\ell jy})}^{\bar{b}} = \overline{(H^\wp_{ikx})} \cap \overline{(H^\wp_{\ell jy})}^{\overline{g\beta}} = \overline{(H^\wp_{ikx})} \cap \overline{(H^\wp_{\ell jy})}^{\bar{g}^{\bar{\beta}}}.$$

Thus by Lemma 12.10.(2), we have $\bar{\beta} = \bar{c}'$ for some $c' \in H^\wp_{ikx} c (H^\wp_{\ell jy})^g$ where c and hence c' is an intersecting conjugator so $H^\wp_{ikx} \cap \left((H^\wp_{\ell jy})^g\right)^{c'}$ is infinite. Consequently $H_{ikx} \cap \left((H_{\ell jy})^{gc'}\right)$ is infinite so $gc' \in H_{ikx} a (H_{\ell jy})$ for one of the known intersecting conjugators a. Letting $a' = gc'$ we see that $\bar{b} = \bar{a}'$ with the claimed property.

Hyperbolicity of $\bar{G}$ and quasiconvexity of $\bar{C}$: Observe that $\bar{A}$ is hyperbolic by Theorem 12.2, and $\bar{C}, \bar{D}$ are quasiconvex in $\bar{A}$ since $\bar{C}^\wp$ and $\bar{D}^\wp$ are quasiconvex in $\bar{A}^\wp$ by Corollary 3.72. Consequently the finite height of $\bar{C}$ in $\bar{G}$ implies that $\bar{G}$ is hyperbolic by the "no long annulus" criterion of [BF92], and then $\bar{C}$ is quasiconvex in $\bar{G}$ by [Mit04]. $\qquad \square$

Remark 13.4. An alternative way to structure the proof is to note that the procedure can also be iterated until, at the last step, C and D are quotiented to a finite group, and we have separated C from the intersecting conjugators in a group that is an HNN extension of a virtually compact special hyperbolic group along a finite subgroup. This accomplishes the objective which enables the strategy—but without induction on height to know that $\bar{G}$ is virtually special. Instead the onus is on a stronger form of the special quotient theorem.

Problem 13.5. Let G act properly on a CAT(0) cube complex. Is $G \cong \pi_1 X^*$ where X^* is a cubical small-cancellation presentation? If G is word-hyperbolic and acts properly and cocompactly, can X^* be chosen so that $\tilde{X}$ is δ-hyperbolic? Perhaps the canonical associated cube of spaces or some variation of it works. The grading should be in reverse order of the dimension.

Chapter Fourteen

Large Fillings Are Hyperbolic and Preservation of Quasiconvexity

We now give a self-contained explanation of the word-hyperbolicity of hyperbolic fillings of cubulated groups that are relatively hyperbolic. This is a limited special case of results obtained by Osin and Groves-Manning [Osi07, GM08] but there is a fairly natural proof within our framework. We also prove the persistence of quasiconvexity of certain subgroups under suitably large fillings.

14.a Hyperbolic Fillings

We will need the following result suggested by Figure 14.1. It can be deduced from [DS05, Thm 1.12].

Proposition 14.1. *Let G be hyperbolic relative to subgroups $P_1, \ldots, P_r$. Suppose G acts properly and cocompactly on the geodesic metric space $\widetilde{X}$, and each P_i cocompactly stabilizes a nonempty connected subspace $\widetilde{Z}_i$. There exists a constant κ with the following property:*

Let γ_1, γ_2 be a pair of geodesics in $\widetilde{X}$ with the same endpoints. Then there is a sequence of cosets $\{g_i P_{n_i}\}$ such that γ_1, γ_2 asynchronously κ-fellow travel relative to $\{g_i \widetilde{Z}_{n_i}\}$ in the sense that for $0 \leq t_1 \leq |\gamma_1|$ and reparametrization θ, either $\mathsf{d}\big(\gamma_1(t), \gamma_2(\theta(t))\big) \leq \kappa$, or $\gamma_1(t), \gamma_2(\theta(t))$ both lie in $\mathcal{N}_\kappa(g_i \widetilde{Z}_{n_i})$, where $g_i \widetilde{Z}_{n_i}$ varies with t.

With a bit of care, the proofs have a sparse generalization and we have separately sketched the details in that setting. In addition to the propositions stated here, we will also use the "isometric core property" and Lemma 7.37 to support the proof in the cosparse case.

Theorem 14.2. *Suppose that X is a nonpositively curved cube complex such that:*

(1) *$G = \pi_1 X$ is hyperbolic relative to subgroups $P_1, \ldots, P_r$.*
(2) *G acts cosparsely on $\widetilde{X}$ relative to the quasiflats $F_1, \ldots, F_k$.*

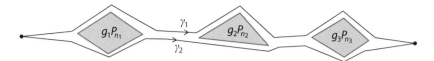

Figure 14.1. γ_1 and γ_2 fellow travel relative to parabolic subgroups.

For each i, there is a finite subset $S_i \subset P_i - \{1_G\}$ such that if P_i' is a hyperbolic-index normal subgroup of P_i that is disjoint from S_i, then the quotient $G/\langle\!\langle P_1', \ldots, P_r' \rangle\!\rangle$ is word-hyperbolic.

Remark 14.3. Theorem 14.2 applies in the relatively hyperbolic case provided an isometric cocompact subcomplex $\widetilde{X}_o \subset \widetilde{X}$ exists. It is only to guarantee its existence through Lemma 7.37 that the sparse hypothesis is required here.

Proof in compact case. By Lemma 2.36, for each i, let $\widetilde{Y}_i$ be a superconvex subcomplex that P_i acts on cosparsely. The superconvexity and malnormality of the collection $\{P_i\}$ ensures that the wall-pieces and cone-pieces between the $\widetilde{Y}_i$ are of uniformly bounded diameter $\leq B$. Let S_i consist of the nontrivial elements of P_i with translation distance $< \frac{1}{24}B$ in $\widetilde{Y}_i$.

Let P_i' be a hyperbolic-index normal subgroup disjoint from S_i for each i. Let $\widehat{Y}_i = P_i' \backslash \widetilde{Y}_i$, so the cubical presentation $\langle X \mid \widehat{Y}_1, \ldots, \widehat{Y}_r \rangle$ satisfies the $C'(\frac{1}{24})$ small-cancellation condition.

Consider geodesics γ_1, γ_2 in $\widetilde{X}^*$. Let $E \to X^*$ be a minimal complexity disk diagram between γ_1 and γ_2, with the following properties: There are geodesics $\lambda_i \to E$ with the same endpoints as γ_i such that each pair λ_i, γ_i bounds a square subdiagram D_i of E, and $E = D_1 \cup_{\lambda_1} D \cup_{\lambda_2} D_2$, and where D has minimal complexity among all possible such choices.

Note that γ_i, λ_i are geodesics in $\widetilde{X}$, for otherwise γ_i would not be a geodesic in $\widetilde{X}^*$.

We will show that D is α-thin and show that each D_i is β-thin, and consequently E is $(\alpha + 2\beta)$-thin. Since α, β will depend only on X and the choices of the P_i', we see that $\widetilde{X}^*$ has $(\alpha + 2\beta)$-thin bigons. We thus conclude that the 1-skeleton of $\widetilde{X}^*$ has $(\alpha + 2\beta)$-thin geodesic bigons, and is thus δ-hyperbolic by Proposition 4.6.

Observe that there is no shell in D whose outerpath is a subpath of either λ_i for this would contradict that λ_i is a geodesic, indeed X^* is $C'(\frac{1}{24})$ so has short innerpaths by Lemma 3.70. Observe that there is no cornsquare in D whose outerpath is on λ_i, for we could then push the square out to D_i, and reduce the complexity of D.

We conclude from Theorem 3.43 that D is either an arc, or is a ladder or single cone-cell with a (possibly trivial) arc attached at each end. Moreover, we

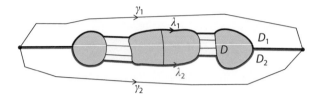

Figure 14.2. The diagram E between γ_1, γ_2 must decompose as the union of a "stemmed ladder" D surrounded by two square diagrams D_1, D_2.

note that the square parts of D are actually grids, since a square within a shard of D could be passed into D_1 or D_2, thus violating our minimal choice of D. We refer the reader to Figure 14.2 for a picture of a typical D within E.

Let α be a constant so that for any pair of geodesics in one of the finitely many $\widehat{Y}_j$, if the endpoints of these geodesics are within B of each other then the geodesics are within α of each other, and note that $\alpha \geq B$.

As in Definition 3.42.(2), each λ_i is a concatenation $\lambda_{i1}\lambda_{i2}\ldots\lambda_{i\ell}$, where λ_{1j} and λ_{2j} are on opposite sides of successive cone-cells and grids within the ladder. Observe that the distance between endpoints in λ_{1j} and λ_{2j} is bounded by the maximal diameter B of a piece. Consequently, the geodesics $\lambda_{1j}, \lambda_{2j}$ are accordingly either B-fellow travelers within a grid, or α-fellow travelers within the cone that is a translate of $\widehat{Y}_{n_j}$. Thus λ_1, λ_2 are in α-neighborhoods of each other.

By Proposition 14.1, there exists μ such that γ_i, λ_i must μ-fellow travel relative to translates $g_j \widetilde{Y}_{n_j}$. Consider subpaths γ_i', λ_i' that bound a geodesic rectangle in $\mathcal{N}_\mu(\widetilde{Y})$ whose left and right sides have length $\leq \mu$. There exists δ such that the finitely many spaces $P_i' \backslash \mathcal{N}_\mu(\widetilde{Y}_i)$ are δ-hyperbolic (they are just thickenings of the $\widehat{Y}_i$). Let β be such that for any pair of geodesics in one of these $\widehat{Y}_i$ thickenings, if the endpoints are μ-close then the geodesics β-fellow travel, and note that $\beta \geq \mu$. Then γ_i, λ_i must β-fellow travel since they piecewise β-fellow travel. $\square$

Proof in cosparse case. By Lemma 7.37, there is a connected π_1-surjective subcomplex $X_o \subset X$ such that $\widetilde{X}_o \subset \widetilde{X}$ is an isometric embedding, and moreover, the complementary components $\widetilde{X} - \widetilde{X}_o$ are contained in unique quasiflats $g_j \widetilde{F}_j$ of $\widetilde{X}$. Let $\widetilde{Y}_i^o$ denote the intersection $\widetilde{X}_o \cap \widetilde{Y}_i$, and we will later let $\widehat{Y}_i^o$ denote the quotient of $\widetilde{Y}_i^o$ corresponding to $\widehat{Y}_i$.

As in the compact case, we consider a pair of geodesics γ_1, γ_2 with the same endpoints, but now assume that they lie in $\widetilde{X}_o^*$, which denotes the preimage of X_o in $\widetilde{X}^*$. We then consider a minimal complexity $E = D_1 \cup_{\lambda_1} D \cup_{\lambda_2} D_2 \to \widetilde{X}^*$. As before D is a ladder. By the minimality of E, the boundary path of each cone-cell C of D lies in $\widetilde{X}_o^*$, and so can be regarded as a path in $\widetilde{X}_o$. Indeed, for an edge e in $\partial_{\mathsf{p}}C$, either e is on ∂E which implies that e is in $\widetilde{X}_o$ since each

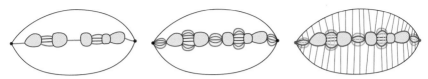

Figure 14.3. $\widetilde{X}_o^*$ has thin bigons for large girth hyperbolic fillings.

γ_i is in $\widetilde{X}_o$, or there is a subsequent cell s or C' meeting C along e. If it is a neighboring cone-cell C' then e lies in $g\widetilde{Y} \cap g'\widetilde{Y}'$ which lies in $\widetilde{X}_o^*$. If it is a square s, then either $s \subset \widetilde{X}_o^*$ or s lies in $\widetilde{Y} - \widetilde{X}_o$ and so could be absorbed into C.

We now consider the (possibly degenerate) rectangular parts of D. Their top and bottom boundaries are geodesics in $\widetilde{X}$ that are path-homotopic to paths in $\widetilde{X}_o$. We replace E by a diagram E' that contains these $\widetilde{X}_o$ paths as illustrated in Figure 14.3. This gives us geodesics λ_1', λ_2' with the same endpoints as λ_1, λ_2 that bound diagrams D_1' and D_2' within E', and also a diagram D' that is a thickening of the ladder D.

D_1', D_2' are thin by Proposition 14.1 applied to $\widetilde{X}_o$ as before. And D' is thin as before, by breaking it up into cone-cells and grids, each of which has a geodesic top and bottom in $\widehat{Y}_{n_j}^o$ or in $\widetilde{X}_o$ respectively. As in the compact case, we regard each cone-cell as mapping to a word-hyperbolic $\widehat{Y}_{n_j}^o$. However, the grids are treated again as we treated D_1, D_2. $\square$

Remark 14.4. It is likely that there is a more general version of Theorem 14.2 that takes as input a small-cancellation cubical presentation X^* with $\pi_1 X^*$ relatively hyperbolic.

The following provides a simple criterion to ensure that the image of a quasi-convex subgroup is quasiconvex. In Theorem 14.10, we describe a more general criterion that applies in the cosparse case and does not require the cones to correspond to finite index subgroups of the parabolics.

Theorem 14.5. *Continuing with the notation of Theorem 14.2. Let X be compact, and let H be a relatively quasiconvex subgroup of $\pi_1 X$ that is full. There exist (slightly larger) finite subsets $S_i^+ \subset P_i - \{1\}$ such that H projects to a quasiconvex subgroup of $\pi_1 X^*$ provided that the following hold for all i and $g \in \pi_1 X$:*

(1) Each $P_i' \subset P_i$ is a finite index subgroup disjoint from S_i^+.
(2) $P_i' \subset (H^g \cap P_i)$ whenever $H^g \cap P_i$ is nontrivial.

Proof. By Lemma 2.37, let $\widetilde{A}$ be an H-cocompact superconvex core, and let $A = H\backslash\widetilde{A}$. There exists a uniform L such that for each g, i either $\mathrm{diam}(\widetilde{A} \cap g\widetilde{Y}_i) \le L$ or $g\widetilde{Y}_i \subset \widetilde{A}$ and hence factors through a map $Y_i \to A$ by Condition (14.5). We can thus choose S_i^+ to ensure that $\|Y_i\| > 4|K|_{Y_i}$ whenever K is a contractible component of $Y_i \otimes A$. Hence the induced presentation $A^* \to X^*$ has no missing shells by Lemma 3.67 and is thus π_1-injective by Theorem 3.68. Suppose that cone-pieces in X^* are small in the sense that $|M|_{Y_i} < \frac{1}{8}\|Y_i\|$ whenever M is a cone-piece between distinct $\widetilde{Y}_i, g\widetilde{Y}_j$. Then $\widetilde{A}^* \to \widetilde{X}^*$ is a convex subcomplex by Lemma 3.74.   □

14.b Quasiconvex Image

We now describe a variant of the no missing shells notion from Definition 3.61 that functions in conjunction with isocores to give a variant of Corollary 3.72.

Definition 14.6 (No Missing Teleshells). Let X^* be a small-cancellation cubical presentation. Let $A^\square \to X$ be a map of cube complexes whose elevation to $\widetilde{X}$ is an isometric embedding.

The map $A^\square \to X$ has *no missing [nonnegatively curved] teleshells* with respect to X^* if the following holds for each [nonnegatively curved] shell R in a reduced diagram $D \to X^*$. Suppose $\partial_{\mathsf{p}} R = QS$ with outerpath Q and innerpath S, and where $Q = S_1 Q^\square S_2$ and $Q^\square \to X$ factors as $Q^\square \to A^\square \to X$ and S_1, S_2 are possibly trivial wall-pieces. Then $S_2 S S_1$ is path-homotopic in the associated cone-cell Y to a path $S^\square$ such that $Q^\square \to A^\square$ extends to a closed path $Q^\square S^\square \to A^\square$.

Lemma 14.7. *Suppose X^* is $C'(\frac{1}{18})$. Then $A^\square \to X^*$ has no missing teleshells provided the injectivity radius of each component $A^\square \otimes Y_i \subset Y_i$ is greater than $10M_i$ where M_i is the maximal diameter of a piece in Y_i. Similarly, at $C'(\frac{1}{20})$ it has no missing nonnegatively curved teleshells.*

Proof. By Lemma 3.70, the path S is the concatenation of at most 7 pieces, and hence $S_2 S S_1$ is the concatenation of at most 9 pieces. By hypothesis, there is a geodesic $S^\square$ in the appropriate component of $A^\square \otimes Y_i$ with the same endpoints as $S_2 S S_1$. Note that $(S^\square)^{-1} S_2 S S_1$ is a closed path in Y_i that is the concatenation of a path of length $< 9M_i$ together with a path that is the concatenation of at most 9 pieces. Hence $S_2 S S_1$ is path-homotopic in Y_i to $S^\square$. The analogous reasoning holds for nonnegatively curved shells with $18, 7, 9$ replaced by $20, 8, 10$.   □

Lemma 14.8. *Let $\langle X \mid \{Y_i\}\rangle$ be $C'(\frac{1}{20})$. Let $A^\square \to X$ be such that its elevation to $\widetilde{X}$ is isometrically embedded. Suppose each Y_i is locally finite. Suppose $A^\square \otimes Y_i$ is compact for each i, and that $\pi_1 J \subset \pi_1 Y_i$ is separable for each component J of $A^\square \otimes Y_i$. Then there exist finite covers $\widehat{Y}_i'$ such that for any further covers $\widehat{Y}_i$ with $\pi_1\widehat{Y}_i \lhd \mathrm{Stab}(\widetilde{Y}_i)$ we have: $X^* = \langle X \mid \{\widehat{Y}_i\}\rangle$ is a $C'(\frac{1}{20})$ cubical presentation, and $A^\square$ has no missing nonnegatively curved teleshells with respect to X^*.*

Proof. Consider a path $\beta \to Y_i$ that is the concatenation of at most 10 pieces. (In particular, this covers the case where $\beta = S_2 S S_1$ where S_1, S_2 are possibly trivial wall-pieces and S is (an innerpath consisting of) the concatenation of at most 8 pieces.) Moreover, assume the endpoints of β lie on vertices in the image of J. By local finiteness of Y_i and compactness of J, there are finitely many such paths β. Each such β is either path-homotopic in Y_i to a path in J, or not. In the latter case, by separability of $\pi_1 J$, there is a finite cover of Y_i such that $J \subset Y_i$ lifts to an embedded subcomplex, but β does not lift to a path whose endpoints lie on the lift of J. Finally, we choose $\widehat{Y}_i' \to Y_i$ to be a finite cover that factors through all the finite covers above as β varies. Consider covers $\widehat{Y}_i \to \widehat{Y}_i'$, such that $\pi_1 \widehat{Y}_i$ is normal in $\operatorname{Stab} \widetilde{Y}_i$. The cubical presentation X^* has the desired property since if $S_1 S S_1 \to \widehat{Y}_i$ has endpoints on a component $\widehat{J}$ of $A^{\square} \otimes \widehat{Y}_i$, then $S_1 S S_2$ is homotopic into $\widehat{J}$, for otherwise our separability choice on the projection $\beta \to Y$ of $S_1 S S_2$ would show that $S_1 S S_2$ cannot have endpoints on $\widehat{J}$. $\square$

Remark 14.9. Isocores are used in the proof of Theorem 15.10, where A is a hyperplane carrier and $A^{\square}$ is induced from a compact isocore $X^{\square} \subset X$ of Lemma 7.37. As each $X^{\square} \otimes Y_i$ is compact, we see that each component of $A^{\square} \otimes Y_i$ is compact.

Theorem 14.10 (QI Embedding). *Let $X^* = \langle X \mid \{Y_i\} \rangle$ be a small-cancellation presentation with tiny innerpaths. Let $A^{\square} \to X$ be a map of cube complexes such that the universal cover of $A^{\square}$ lifts to an isometric embedding in $\widetilde{X}$. Suppose $A^{\square} \to X^*$ has no missing teleshells. Suppose each component $A^{\square} \otimes Y_i$ embeds in Y_i by a uniform quasi-isometry. Then the elevation $\widetilde{A}^{\square} \to \widetilde{X}^*$ is a quasi-isometric embedding.*

Proof. Let p, q be points of $\widetilde{A}^{\square}$. Let $D \to \widetilde{X}^*$ be a disk diagram such that $\partial_p D = \gamma^{-1}\alpha$ where α is a geodesic from p to q in $\widetilde{X}$, and where γ is a geodesic from p to q in $\widetilde{A}^{\square}$. Moreover, choose D to be of minimal complexity among all such possible choices with p, q fixed. Let $\beta \to D$ be a path in the "convex hull" of γ within D, so β and γ are separated by a square diagram $D_\gamma \to D$, and every dual curve in D_γ has one end on β and another on γ. Moreover, suppose D_γ has maximal area with this property within D (after possibly shuffling). Let D_α be the subdiagram between α and β.

We will apply Theorem 3.43 to see that D_α is a single cone-cell or a ladder whose two features of positive curvature have outerpaths containing p, q. To do this, we will verify that D_α has no cornsquare or shell whose outerpath lies on α or β. No cornsquare or shell has outerpath in α, since in the cornsquare case we could push a square past α to obtain a lower complexity diagram with a different choice of α, and the shell case would violate that α is a geodesic because of short innerpaths. No cornsquare in D_α has outerpath in β by maximality of D_γ. Finally, suppose there is a shell R with $\partial_p R = QS$ and outerpath Q a subpath of β. By Lemma 14.11, we see that $\partial_p R = Q'S'$ where $S' = S_1 S S_2$ and S_1, S_2 are (possibly trivial) wall-pieces and where Q' is a subpath of γ (as Q' cannot be

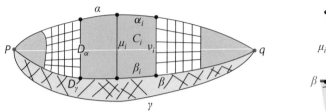
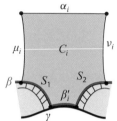

Figure 14.4. The geodesics $\alpha \to \widetilde{X}^*$ and $\gamma \to \widetilde{A}^{\square}$ are compared by considering a minimal complexity diagram D between them, and dividing D into subdiagrams D_α, D_γ where D_γ is a largest square diagram among all possible such minimal D. We deduce that D_α is a ladder. We compare α and β by comparing them within each cone-cell C_i. This uses that $\beta_i' \subset \partial_{\mathsf{p}} C_i$ is a subpath of γ where β_i is the concatenation of β_i' and two wall-pieces.

trivial). By no missing teleshells, S' is path-homotopic to $S^{\square}$ in the cone-cell Y carrying R and the path $Q'S^{\square}$ is a closed path in $A^{\square}$ and hence in $\widetilde{A}^{\square}$ since $Q'S^{\square}$ is a closed path in Y. We know that $|Q'| > |S'|$ by tiny innerpaths. Since S' and $S^{\square}$ are path-homotopic in Y, they have the same endpoints in $\widetilde{X}$, and so a geodesic in $\widetilde{A}$ and hence $\widetilde{A}^{\square}$ with the same endpoints as $S^{\square}$ has length $\leq |S'|$. This contradicts that γ is a geodesic in $A^{\square}$.

Since pieces between cone-cells are uniformly finite, we show, in a fashion similar to the proof of Theorem 14.2, that $|\gamma| = |\beta|$ is bounded by a linear function of $|\alpha|$. Indeed, each pseudo-grid of D_α is actually a grid, and each cone-cell C_i of D_α has $\partial_{\mathsf{p}} C_i = \mu_i \alpha_i \nu_i^{-1} \beta_i^{-1}$ where α_i, β_i are subpaths of α, β and the "rungs" μ_i, ν_i are the sides of (possibly degenerate) grids. We refer to Figure 14.4. Thus, to see that α, β uniformly fellow travel, it suffices to show that each α_i, β_i uniformly fellow travel in the cone Y_i supporting the cone-cell C_i, and similarly, to see that there is a uniform relationship between $|\alpha|, |\beta|$, we prove the same for each $|\alpha_i|, |\beta_i|$ and verify that each $|\beta_i| \geq 1$. By Lemma 14.11, the path β_i is of the form $S_1 \beta_i' S_2$ where S_1, S_2 are trivial or wall-pieces, and where β_i' is a subpath of γ and hence a path in $\widetilde{A}^{\square}$. If β_i' is trivial, then α_i has the same endpoints as the concatenation of four pieces, so either $\alpha_i \to \widetilde{X}^*$ is not a geodesic or D is not minimal as C_i is replaceable since $|\partial_{\mathsf{p}} C_i| \ll \|Y_i\|$. Thus $\alpha_i \to Y_i$ and $\beta_i' \to Y_i^{\square} = A^{\square} \otimes Y_i$ are geodesics whose endpoints lie at most two pieces away from each other. Finally, our hypothesis that each embedding $Y_i^{\square} \subset Y_i$ is a uniform quasi-isometry completes the proof. $\qquad \square$

Lemma 14.11. *Let $D \to X^*$ be a reduced diagram. Let R be a cone-cell of D with $\partial_{\mathsf{p}} R = QS$. Let P be the smallest subpath of $\partial_{\mathsf{p}} D$ that contains the endpoints of the dual curves emanating from Q. Suppose that the subdiagram of D*

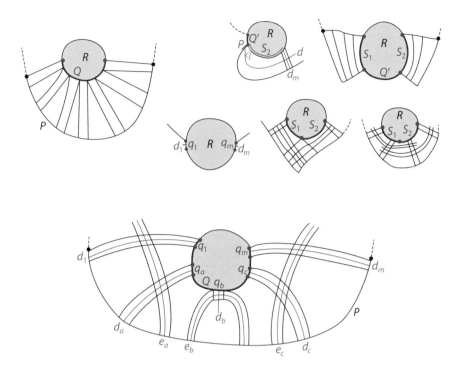

Figure 14.5. The diagram on the left illustrates the hypothetical scenario where all dual curves starting at edges of $Q \subset \partial_{\mathsf{p}} R$ end on a subpath $P \subset \partial_{\mathsf{p}} D$ such that no dual curve starts and ends on P. The five diagrams in the middle illustrate conclusion scenarios, the most typical of which is at the upper right. The diagram on the right illustrates the main idea of the argument—if the rightmost left-exiter is left of the leftmost right-exiter, then all edges between them are on P since otherwise a dual curve starts and ends on P.

bounded by Q, P, and rectangles carrying the first and last dual curves between them is a square diagram. Suppose no dual curve in D has both endpoints on P and bounds a square subdiagram (in particular, this holds if $P \to X$ lifts to a geodesic $\widetilde{P} \to \widetilde{X}$). Then $Q = S_1 Q' S_2$ where each S_i is either trivial or a wall-piece and where Q' is either trivial or is a subpath of P and hence a subpath of $\partial_{\mathsf{p}} D$.

Proof. We refer to Figure 14.5. Express $Q = q_1 q_2 \cdots q_m$ as a concatenation of edges. Let d_i denote the dual curve emanating from q_i. Call q_i a *left-exiter* if d_i crosses a dual curve e_i that intersects d_1. Similarly, call q_i a *right-exiter* if d_i crosses a dual curve e_i that intersects d_m. We claim that if d_i is nontrivial then q_i is either a left-exiter or a right-exiter. Indeed, for each dual curve e_i crossed by d_i, if e_i had an endpoint on ∂R then there would be a cornsquare on R

which contradicts that D is reduced, and e_i cannot have both endpoints on P by hypothesis. Note that q_1 is a left-exiter if d_1 is nontrivial, and q_m is a right-exiter if d_m is nontrivial. If d_1 is trivial then there is no left-exiter, and if d_m is trivial then there is no right-exiter. If d_1, d_m are both trivial then $Q = P$. If exactly one of d_1, d_m is trivial then either $Q = P$ or Q consists of a single wall-piece. To see this, suppose d_1 is trivial but $Q \neq P$, and consider the first edges p, q where P, Q differ. Let d be the dual curve emanating from p and note that d cannot terminate on P by hypothesis, and cannot terminate on Q or there would be a bigon in D or a cornsquare on R (consider the second dual curve crossing the square where d terminates on R). Hence d crosses d_m and we see that Q is a single wall-piece. Let q_a, q_c be the rightmost left-exiter and the leftmost right-exiter, so a is maximal among left-exiters, and c is minimal among right-exiters. If $c \leq a$ then Q is the concatenation of at most two wall-pieces. If $a < c$, then let $Q' = q_{a+1} \cdots q_{c-1}$. We conclude that Q' is a subpath of P since d_b is trivial for $a < b < c$. $\qquad\square$

Chapter Fifteen

Relatively Hyperbolic Case

15.a   Introduction

The main result of this section is Theorem 15.1 which generalizes Theorem 13.3 to a setting that includes many groups that are hyperbolic relative to abelian subgroups.

For a finite volume hyperbolic 3-manifold M with at least one cusp, Corollary 17.13 shows that $\pi_1 M$ has a hierarchy of the type specified in Theorem 15.1. We now state Theorem 15.1, but prove it near the end of Section 15.f.

Theorem 15.1. *A group G is virtually compact [sparse] special provided:*

(1) *G is hyperbolic relative to virtually abelian groups.*
(2) *G splits as a graph Λ of groups with relatively quasiconvex edge groups.*
(3) *Each vertex group G_v is virtually compact [sparse] special respectively.*
(4) *G_e has trivial intersection with each $\mathbb{Z}^2 \subset \pi_1(\Lambda - e)$, for each edge e of Λ.*

Furthermore, G is π_1 of a compact cube complex that is virtually special provided:

(3') *Each vertex group G_v is π_1 of a compact cube complex that is virtually special.*

Moreover, if rank$(P) = 2$ *for each parabolic subgroup, and each vertex group is virtually special strongly sparse, then G acts strongly cosparsely on a CAT(0) cube complex and has a finite index torsion-free subgroup such that the quotient is special.*

We clarify that Condition (4) is with respect to both the initial and terminal inclusions $G_e \subset \pi_1(\Lambda - e)$. Furthermore, when $\Lambda - e$ has two components, the two inclusions $G_e \to \pi_1(\Lambda - e)$ have different targets since $\pi_1(\Lambda - e)$ depends on the basepoint. As G_e is relatively quasiconvex, Condition (4) is equivalent to requiring that G_e is word-hyperbolic and has no accidental parabolics.

Definition 15.2 (Accidental Parabolic). Let G be hyperbolic relative to peripheral subgroups $\{P_i\}$. Suppose G splits as a graph of groups. Let e be an edge in the Bass-Serre tree T of the splitting. An element $g \in G_e = \text{Stab}(e)$ is an *accidental parabolic* if g is an infinite order element in some conjugate P of a parabolic subgroup, but e lies outside the *associated P-invariant subtree* of the Bass-Serre T of G, which either consists of a P-invariant line upon which P acts nontrivially, or consists of the largest subtree that P fixes.

From the viewpoint of the graph of groups, an accidental parabolic corresponds to an infinite order parabolic element of an edge group such that the parabolic subgroup isn't "essentially split" by that edge. This notion agrees with the notion in Definition 17.11 of an accidental parabolic in an incompressible surface S by considering the splitting of $\pi_1 M$ along $\pi_1 S$.

A natural simplification of Theorem 15.1 is the following:

Corollary 15.3. *Let G be torsion-free and hyperbolic relative to virtually abelian subgroups. Suppose G splits as a graph of groups where:*

(1) *each vertex group is word-hyperbolic and virtually compact special,*
(2) *each edge group is quasiconvex,*
(3) *no edge group has an accidental parabolic.*

Then G is the fundamental group of a virtually special compact cube complex.

Remark 15.4. I believe Condition (15.3) can be dropped, and that there is a more direct proof of Corollary 15.3 than deducing it from Theorem 15.1. However, although the general form of Theorem 7.59 in [HW10b] can be applied directly without Condition (15.3), one obtains a proper action on a CAT(0) cube complex that is not necessarily locally finite. Note that the definition of "sparse" used in [HW10b] omits the local finiteness properties required in Definition 7.21.

Lemma 17.20 shows that a compact hyperbolic 3-manifold with $\partial M \neq \emptyset$ has a finite cover $\widehat{M}$ such that $\pi_1 \widehat{M}$ has a splitting satisfying the hypothesis of Corollary 15.3.

We expect a more complete generalization of Theorem 13.3 would be as follows:

Conjecture 15.5. *Let G be (torsion-free and) hyperbolic relative to virtually abelian subgroups. Suppose G has a quasiconvex hierarchy terminating at finite groups. Then G is virtually cosparse/compact special.*

As we are not dealing with word-hyperbolic groups, we emphasize that having a "quasiconvex hierarchy" means that each edge group arising in the hierarchy is quasi-isometrically embedded in G at each stage.

A potential direct approach towards Conjecture 15.5 would require reworking the results in [HW12] and [HW15] so that they apply in a relatively hyperbolic context. The main obstacle towards a relatively hyperbolic generalization of [HW12] is to provide a relatively hyperbolic version of the "trivial wall projection lemma." What is missing from [HW15] is a cubulation criterion that allows more flexibility in the interaction between the edge groups and the parabolic subgroups. In particular, we would need to drop the aparabolicity requirement on the edge groups. We explain this route towards proving Conjecture 15.5 in Remark 15.23.

We will follow a less direct approach that circumvents these obstacles at the expense of developing some additional tools to prove (a limited form of) the relatively hyperbolic case as a consequence of the hyperbolic case. We outline the main points in (1)–(4) below. We emphasize that for both (2) and (3) below we prove separability in a relatively hyperbolic group by quotienting to a virtually compact special hyperbolic group. This separability is used in (4) to pass to a finite cover with an easier hierarchy.

(1): Theorem 15.6 provides a virtually special parabolic filling result analogous to Theorem 12.2 except that only parabolic subgroups are filled.

(2): Theorem 15.8 gives a separability result for graphs of relatively hyperbolic groups whose vertex groups are virtually special. This is proven in [HW18] following a generalization of the proof scheme of Theorem 13.1.

(3): We use the separability to pass to a finite index subgroup that has a revised splitting along a "cage" having controlled interactions with the parabolic subgroups. This better splitting is aparabolic on at least one side, and so we are able to apply Theorem 7.59′ to cubulate.

(4): In Theorem 15.10, we show that in a relatively hyperbolic case, a sparse/compact cube complex with a hierarchy is virtually special by combining the parabolic special quotient result in Theorem 15.6 with the double coset criterion for separability of Proposition 6.9.

15.b Parabolic Fillings That Are Virtually Special

We now describe a variant of Theorem 12.2 for relatively hyperbolic groups that is restricted to quotienting only by finite index subgroups of the parabolic groups. Our focus, which is Theorem 15.6, provides a special case of Conjecture 20.1. (The full case of Conjecture 20.1 can probably be proven by combining Theorem 15.6 with Theorem 12.2 as in the proof of Theorem 12.1.) The reader not yet familiar with sparseness can assume the cube complexes are compact.

A normal subgroup $A \subset B$ has *virtually cyclic index* if B/A is virtually cyclic.

Theorem 15.6. *Let X be a virtually special cube complex that is sparse relative to $F_1, \ldots, F_r$, and let $P_i = \pi_1 F_i$ for each i, so $G = \pi_1 X$ is hyperbolic relative to $\{P_i\}$. There exist finite index subgroups $P_1^o, \ldots, P_r^o$ such that for any normal finite index or virtually cyclic index subgroups $P_i^c \subset P_i^o$ the quotient group $G/\langle\!\langle P_1^c, \ldots, P_r^c \rangle\!\rangle$ is a word-hyperbolic group virtually having a quasiconvex hierarchy terminating in finite groups. Hence the quotient group is virtually compact special.*

Specifically, let $\widetilde{E}_i$ denote the superconvex hull of the universal cover $\widetilde{F}_i$ of F_i for each i, and let $E_i^c = P_i^c \backslash \widetilde{E}_i$ for each i. Then the cubical presentation $X^* = \langle X \mid E_1^c, \ldots, E_r^c \rangle$ is $B(6)$ and X^* has a finite cover with a hierarchy.

As a consequence of Theorem 4.1, we see that each P_i/P_i^c embeds in $\pi_1 X^*$.

Proof. The P_i-action on $\widetilde{F}_i$ extends to an action on $\widetilde{E}_i$. The definition of sparse implies that distinct translates of the $\widetilde{F}_i$ are isolated in the sense that their intersection has diameter bounded by a uniform constant, and after passing to $\widetilde{E}_i$ these intersections likewise have bounded diameter and each $\widetilde{E}_i$ is moreover isolated from external hyperplanes, since any large overlap would be absorbed in $\widetilde{E}_i$. There is thus an upperbound D on diameters of wall-pieces in the $\widetilde{E}_i$ and on diameters of contiguous cone-pieces between distinct translates of the $\widetilde{E}_i, \widetilde{E}_j$.

Let $\widehat{X} \to X$ be a finite degree special cover, so that each induced cover $\widehat{E}_i$ has systole exceeding $12D$. Let $\ddot{X} \to \widehat{X}$ be the finite cover corresponding to the wall homomorphism $\#_{\mathrm{W}} : \pi_1 \widehat{X} \to \mathbb{Z}_2^{\mathrm{W}}$ that counts (modulo 2) the number of times a path passes through a hyperplane. Let $\ddot{E}_i$ be the induced covers, and as in Section 9.a, there is a wallspace structure on each $\ddot{E}_i$ induced by the hyperplanes of $\widehat{X}$. Accordingly, $\langle \ddot{X} \mid \{g \ddot{E}_i : g \in \mathrm{Aut}(\ddot{X})\} \rangle$ is both $B(6)$ and has a hierarchy. For any further cover $E_i^c \to \dot{E}_i$, we give E_i^c the induced wallspace structure of Construction 9.4. Letting $X^* = \langle X \mid \{E_i^c\} \rangle$, hyperbolicity holds for $\pi_1 X^*$ by Theorem 14.2.

The hierarchical $B(6)$ condition holds for the finite cover $\langle \ddot{X} \mid \{g \ddot{E}_i : g \in \mathrm{Aut}(\ddot{X})\} \rangle$ of X^* and hence for $\langle \ddot{X} \mid \{g E_i^c : g \in \mathrm{Aut}(\ddot{X})\} \rangle$. Letting $P_i^o = \pi_1 \ddot{E}_i$ provides the claimed finite index subgroups for each i. The cosparseness of the action of $\pi_1 X$ on $\widetilde{X}$ provides two new technicalities when we transition from the hierarchy for X^* to the hierarchy for $\pi_1 X^*$. Firstly, we must verify that the walls we are cutting along are quasi-isometrically embedded. Secondly, the splitting at conepoints are not along finite groups.

For each wall W in $\widetilde{X}^*$, we have $N(W) \to \widetilde{X}^*$ is a quasi-isometric embedding by Corollary 5.30, and indeed $T(W) \to \widetilde{X}^*$ is an isometric embedding by Theorem 5.29. As in the proof of Theorem 14.2 in the sparse case, we employ Lemma 7.37 to see that $\widetilde{X}$ contains a cocompact isocore $\widetilde{X}^{\square}$ containing GK. Let $\widetilde{X}^{\square *}$ be the image of $\widetilde{X}^{\square}$ in $\widetilde{X}^*$. And let $N^{\square} = N(W) \cap \widetilde{X}^{\square *}$, so the part of $N^{\square}$ outside GK is contained in the union of the W-intersecting translates of $P_i^c \backslash \widetilde{E}_i^{\square}$ where each $\widetilde{E}_i^{\square} = \widetilde{E}_i \cap \widetilde{X}^{\square}$ is the induced P_i-cocompact isometrically embedded subcomplex of $\widetilde{E}_i$. Note that $N^{\square}$ is $\mathrm{Stab}(W)$-cocompact. To see that $N^{\square} \subset N(W)$ isometrically embeds we observe that each $P_i^c \backslash \widetilde{E}_i^{\square} \subset P_i^c \backslash \widetilde{E}_i$ is isometrically embedded. Indeed, any geodesic $\sigma \to P_i^c \backslash \widetilde{E}_i$ has the same endpoints as a geodesic $\sigma^{\square} \to P_i^c \backslash \widetilde{E}_i^{\square}$, since $\widetilde{E}_i$ is P_i-invariant and isometrically embedded, and so the lift $\widetilde{\sigma} \to \widetilde{E}_i$ has the same endpoints as a geodesic $\widetilde{\sigma}^{\square} \to \widetilde{E}_i^{\square}$ which projects to a $\sigma^{\square} \to P_i^c \backslash \widetilde{E}_i^{\square}$.

Virtual compact specialness holds for $\pi_1 X^*$ since the hierarchy for $\langle \ddot{X} \mid \{g E_i^c : g \in \mathrm{Aut}(\ddot{X})\} \rangle$ arising from the sequence of base-walls in the hierarchy for $\widehat{X}$ provides a hierarchy for X^*. A slight difference in the virtually cyclic index case arises in Construction 10.11, since the splittings at the conepoints can be over virtually cyclic groups instead of finite groups. We can then conclude virtual specialness by applying Theorem 13.3. Alternatively, by first ensuring the extra ingredients in Theorem 12.2 providing a malnormal quasiconvex hierarchy, we could instead apply the weaker Theorem 11.2. $\qquad\square$

Implicit in the statement of Theorem 15.6 is the following variation on omnipotence (see Section 16.b) that is related to the work of Behrstock-Neumann on quasi-isometric classification of 3-manifolds.

Corollary 15.7. *Let M be a compact hyperbolic 3-manifold and let ∂M contain one or more tori denoted by $\partial_1, \ldots, \partial_r$. There exist finite covers $\partial_i^{\circ} \to \partial_i$ such that for any further finite covers $\partial_i^c \to \partial_i^{\circ}$, there is a finite cover $\widehat{M} \to M$ such that for each i, the cover of ∂_i induced by $\widehat{M}$ is isomorphic to ∂_i^c.*

As is clear from the proof, the same statement holds for an arbitrary dimension hyperbolic manifold M, provided that $\pi_1 M = \pi_1 X$ where X is sparse and virtually special.

Proof. We use that $\pi_1 M = \pi_1 X$ where X is a virtually special compact cube complex, as proven in Theorem 17.14.

We can assume without loss of generality that M is orientable and hence each ∂_i is a torus. Indeed, if $\widehat{M} \to M$ is a finite cover, and ∂_{ij} are the various preimages in $\widehat{M}$ of some ∂_i, then we can choose ∂_i° to be a finite cover factoring through all covers ∂_{ij}° as j varies. Thus the statement for $\widehat{M}$ implies the statement for M itself.

For each i, let $P_i = \pi_1 \partial_i$. By Theorem 15.6, there are finite index subgroups P_i° such that $\pi_1 M / \langle\!\langle P_1^c, \ldots, P_r^c \rangle\!\rangle$ is virtually special for any finite index subgroups $P_i^c \subset P_i^{\circ}$. Moreover, we can assume that $\langle X \mid E_1^{\circ}, \ldots, E_r^{\circ} \rangle$ satisfies the $C'(\frac{1}{12})$ small-cancellation condition (and in fact, such an assumption enables the proof of Theorem 15.6). It follows that $\langle X \mid E_1^c, \ldots, E_r^c \rangle$ is also $C'(\frac{1}{12})$ for any finite index normal subgroups $P_i^c \subset P_i$ with $P_i^c \subset P_i^{\circ}$. It is here that we use that each P_i is free-abelian, and hence $E_i^c \to E_i$ is regular, so pieces between E_i^c and itself already arise between E_i° and itself.

Let $X^* = \langle X \mid E_1^c, \ldots, E_r^c \rangle$. Observe that E_i^c lifts to an embedding in $\widetilde{X}^*$ by Theorem 4.1. It follows that the kernel of $P_i \to \pi_1 X^*$ equals P_i^c. Alternatively, this follows from Theorem 3.68 since the map $\langle E_i \mid E_i^c \rangle \to \langle X \mid E_1^c, \ldots, E_r^c \rangle$ has no missing shells.

Since $\pi_1 X^*$ is virtually special, it is virtually torsion-free. Let $\pi_1 X^* \to Q$ be a finite quotient whose kernel is torsion-free. The kernel of the composition $\pi_1 M \to \pi_1 X^* \to Q$ corresponds to a cover $\widehat{M} \to M$ whose boundary tori are ∂_i^c as desired. $\qquad\square$

15.c Separability for Relatively Hyperbolic Hierarchies

The following is proven in [HW18, Thm 5.18].

Theorem 15.8. *Let G be hyperbolic relative to virtually abelian subgroups and suppose G splits as a finite graph of groups whose edge groups are relatively quasiconvex and whose vertex groups are fundamental groups of virtually special sparse cube complexes. Then each relatively quasiconvex subgroup of a vertex group of G is separable.*

Relative quasiconvexity of all vertex groups is equivalent to relative quasiconvexity of all edge groups in the statement of Theorem 15.8. See [BW13, Lem 4.9].

The proof begins with the construction of Remark 15.9 which describes how to quotient $G \to \bar{G}$ to a graph of hyperbolic special groups. One could complete the proof along these lines by verifying that $\bar{G}$ is word-hyperbolic, that each $\bar{G}_e$ is quasiconvex, that $\bar{H} \subset \bar{G}$ is quasiconvex, and $\bar{g} \notin \bar{H}$ for a chosen $g \notin H$. However, the proof in [HW18] instead follows a generalization of the proof of Theorem 13.1 using an argument hinging on "finite generalized height."

Remark 15.9 (Quotienting consistently)**.** Let G be as in the statement of Theorem 15.8. One can choose quotients $G_v \to \bar{G}_v$ that are virtually compact special hyperbolic, and that consistently induce quotients $G_e \to \bar{G}_e$ of the edge groups, so there is an induced quotient $G \to \bar{G}$ splitting as a graph of virtually compact special hyperbolic groups. Each parabolic subgroup P of G intersects the vertex groups G_v in finitely many conjugacy classes of infinite subgroups denoted by $\{P_{vi}\}$. By Theorem 15.6, there are finite index subgroups P_{vi}^o of these P_{vi}, so that any further normal finite index subgroups P_{vi}^c yield virtually compact special hyperbolic quotients $\bar{G}_v = G_v / \langle\!\langle \{P_{vi}^c\} \rangle\!\rangle$. Note that P can have multiple intersections (i.e., with more than one vertex group and/or conjugacy class within a vertex group).

When P is elliptic relative to Γ, we let P^o be a finite index subgroup such that $(P^o \cap P_{vi}) \subset P_{vi}^o$ for each P_{vi}. In this case, P acts cocompactly on a finite diameter subtree $T_P \subset T$ of the Bass-Serre tree associated to the graph of groups Γ. The quotient $\bar{T}_P = P \backslash T_P$ has vertex groups corresponding to the various P_{vi}, and descendant vertex groups in the rooted tree $\bar{T}_P$ are subgroups of their ancestors, and this enables us to arrange that $(P^o \cap P_{vi}) \subset P_{vi}^o$ for each P_{vi}.

When P is not elliptic, P acts nontrivially by translations on a line $L_P \subset T$, and all vertices of T with nontrivial P-stabilizer lie in a tree $T_P \subset T$ containing L_P such that T_P lies in a finite neighborhood of L_P and such that P acts cocompactly on T_P. The vertex groups of $P \backslash T_P$ are the groups $\{P_{vi}\}$. The kernel of the action of P on L_P is a codimension-1 infinite cyclic index normal subgroup P^N that equals each edge group of $\bar{L}_P = P \backslash L_P$. When $\bar{L}_P$ is a circle its vertex groups are isomorphic to its edge groups, but when $\bar{L}_P$ is an arc its boundary vertex groups contain P^N as an index 2 subgroup. In any case, regarding $\bar{L}_P$ as the "root" of the finite graph $\bar{T}_P$, the descendant vertex groups are again subgroups of their ancestors. Hence each P_{vi} contains an index ≤ 2 subgroup that lies in P^N. We may again choose P^o to be a characteristic finite index subgroup of P^N with $(P^o \cap P_{vi}) \subset P^o_{vi}$ for each P_{vi}.

15.d   Residually Verifying the Double Coset Criterion

Theorem 15.10. *Let X be a compact nonpositively curved cube complex. Then X is virtually special provided the following hold:*

(1) *$\pi_1 X$ is hyperbolic relative to the collection $\{P_i\}$ of virtually abelian subgroups stabilizing the quasiflats in the action of $\pi_1 X$ on $\widetilde{X}$.*
(2) *X has a hierarchy that terminates in cube complexes with virtually abelian fundamental groups.*

When $\pi_1 X$ is hyperbolic, Theorem 15.10 is a geometric special case of Theorem 13.3. The proof appears to work with compactness relaxed to strong cosparseness, and is written in that generality, except that in the hierarchy, it is possible that an infinite graph of groups arises, but finite generation implies that a finite subgraph of groups contains the entire group.

Proof. Let us first describe our strategy for showing that $G = \pi_1 X$ is virtually special. Let U_1, U_2 be a pair of immersed hyperplanes whose basepoints map to the same point of X. We will produce a quotient $G \to \bar{G}$ where $\bar{G}$ is hyperbolic and virtually compact special, each $\overline{\pi_1 U_i}$ is quasiconvex, and $\{\overline{\pi_1 U_1} \bar{g}_i \overline{\pi_1 U_2}\}$ are also distinct. Since quasiconvex double cosets are separable for word-hyperbolic groups with separable quasiconvex subgroups [Git99a, Min04], we see that there is a finite quotient $\bar{G} \to Q$ where these double cosets are separated, and this separates our original double cosets of G. Consequently, X is virtually special by Proposition 6.9. (In fact, it appears the images of our subgroups correspond directly to a double hyperplane coset in a word-hyperbolic virtually compact special cube complex, in which case the separability would follow from the "only if" part of Proposition 6.9.)

With this strategy in mind, let us examine how to choose $\bar{G}$. For each P_i, let $\widetilde{E}_i$ denote the superconvex hull of $\widetilde{F}_i$ in $\widetilde{X}$ and note that $\widetilde{E}_i \subset (\widetilde{F}_i \cup GK)$ for each i. Let $E_i = P_i \backslash \widetilde{E}_i$. We will let $\widehat{E}_i$ denote a high-systole regular cover of E_i, or equivalently, a quotient of $\widetilde{E}_i$ by a normal subgroup P^c_i of P_i that is either of finite index or virtually cyclic index. Our quotient $\bar{G}$ will be of the form $\pi_1 X^*$ where $X^* = \langle X \mid \{\widehat{E}_i\} \rangle$, and where we will describe how to choose appropriate $\widehat{E}_i$ below.

Let $A_i = N(U_i)$ be the immersed carrier for $i \in \{1, 2\}$, so we have local-isometries $A_i \to X$ with $\pi_1 A_i$ mapping isomorphically to $\pi_1 U_i$. Let $X^\square \subset X$ be a compact isocore provided by Lemma 7.37. Following Remark 7.40, let $A_i^\square \subset A_i$ and $E_j^\square$ be the induced isocores for each i and j.

By Lemma 14.8, there are finite covers of each E_i, so that for any further regular covers $\widehat{E}_i$, the maps $A_i^\square \to X^*$ have no missing teleshells. Hence the image of each $\pi_1 A_i^\square$ will be quasiconvex in $\pi_1 X^*$ by Theorem 14.10.

Condition 12.12.(2) can also be ensured for regular covers $\widehat{E}_i$ factoring through some finite cover by the same argument used to prove Lemma 14.8.

For each j, let L_j be an upperbound on the diameters of components of $A_i^\square \otimes E_j$. For instance, we could let L_j be the product of the numbers of 0-cells in $A_i^\square$ and $E_j^\square$. For each j, let M_j be the maximal diameter of a piece in E_j in $\langle X \,|\, \{E_j\} \rangle$. Let $\pi_1 A_1 g_1 \pi_1 A_2$ and $\pi_1 A_1 g_2 \pi_1 A_2$ be a pair of distinct double cosets. By separability of double cosets in $P_j = \pi_1 E_j$, we can choose the regular cover $\widehat{E}_j$ so that for the finitely many closed based paths α, α' in $E_j^\square$ satisfying $|\alpha|, |\alpha'| \leq 2L_j + 2M_j + \max\{|g_i|, M_j\}$ we have: if $1 \notin \pi_1 B_1 \alpha \pi_1 B_2 \alpha'$ then $\pi_1 \widehat{E}_j \cap (\pi_1 B_1 \alpha \pi_1 B_2 \alpha') = \emptyset$.

The above choice of $\{\widehat{E}_i\}$ ensures that Condition 12.12.(1) holds. Thus Lemma 12.12 applies to see that $\overline{\pi_1 A_1 \bar{g}_1 \pi_1 A_2}$ and $\overline{\pi_1 A_1 \bar{g}_2 \pi_1 A_2}$ have distinct images in $\pi_1 X^*$. Moreover, the same statement holds for further covers of the E_i.

By induction on the length of the hierarchy, the vertex group(s) corresponding to the cube complex obtained by cutting along the first hyperplane U are virtually special. The base case of the induction is when we arrive at a cube complex that is virtually abelian, and therefore virtually special by Proposition 6.9.

Following the construction of Remark 15.9, we apply Theorem 15.6 to choose normal subgroups P_i^o that are of finite or virtually cyclic index in P_i depending on whether or not P_i is elliptic with respect to the splitting of the first hyperplane. These choices ensure that for any finite index characteristic subgroups $P_i^c \subset P_i^o$, the result will yield cubical small-cancellation word-hyperbolic quotients of the vertex groups that are consistent on the edge groups so that we obtain a quotient to a graph of word-hyperbolic groups. We emphasize that we are not quotienting by a finite index subgroup of entire parabolic subgroups that are not elliptic. This avoids "damaging" the stable letters of our graph of groups—as we aim to provide a graph of quotiented groups.

By Theorem 14.2, the group $\bar{G}$ is word-hyperbolic since we have performed a sufficiently large hyperbolic-index filling of each peripheral subgroup.

Our final requirement on the choices of $\widehat{E}_i$ is to choose them so that $\overline{\pi_1 U} \subset \bar{G}$ is quasiconvex. Letting $A = N(U)$, and $A^\square$ be induced from $X^\square$, this uses the same argument applying Lemma 14.10 that was used earlier for U_1, U_2.

Finally, the group $\bar{G}$ is a hyperbolic group that splits as an HNN extension or amalgamated free product over a quasiconvex subgroup with virtually special hyperbolic vertex groups. Hence $\bar{G}$ is virtually special by Theorem 13.1.

Consequently, our double quasiconvex cosets can be separated in a finite quotient of $\bar{G}$. □

Remark 15.11. The argument used to prove Theorem 15.10, also proves that $H_1 H_2$ is separable whenever each $H_i \subset \pi_1 X$ is a full f.g. quasi-isometrically embedded subgroup. Indeed, a compact isocore $A_i^\square$ for each H_i is produced by applying Lemma 7.37 to the sparse core A_i provided by Proposition 2.31. This yields isocores in X by composing $A_i^\square \to A_i \to X$. The cocompactness of each $A_i^\square \otimes E_j$ is satisfied as before so Lemma 14.8 can be invoked. We revisit this in Theorem 15.13.

Theorem 15.10 is a bit neater to prove under the additional direct assumption that hyperplanes are separable. In that case one can use a cofinite filling of the peripheral subgroups instead of a cocyclic filling. The vertex groups are virtually special by induction on the dimension, and so Theorem 15.8 implies that the edge groups are separable, and we can use this to pass to a large cover relative to the initial splitting. This would enable a total filling of the vertex peripheral subgroups.

Corollary 15.12. *Suppose X is sparse and has closeable quasiflats (e.g., X is strongly sparse), and suppose X has embedded 2-sided hyperplanes. Then X is virtually special.*

Proof. By Theorem 7.54, there is a local-isometry $X \to X'$ where X' is compact, $\pi_1 X'$ is hyperbolic relative to virtually abelian subgroups, and X' has a hierarchy terminating in cube complexes with virtually abelian π_1. Indeed, each hyperplane of X extends to an embedded 2-sided hyperplane in X', and so the components remaining after cutting along these extended hyperplanes are subcomplexes of the closed quasiflats, and hence have virtually abelian π_1. Thus X' is virtually special by Theorem 15.10. Hence X is virtually special. □

Theorem 15.13 (Sparse Separability). *Let X be sparse and special, and let $H \subset \pi_1 X$ be a quasi-isometrically embedded f.g. subgroup. Then H is a virtual retract of $\pi_1 X$ and hence H is separable in $\pi_1 X$.*

Let H_1, H_2 be quasi-isometrically embedded f.g. full subgroups of $\pi_1 X$. Then each double coset $H_1 H_2$ is separable.

By Lemma 7.56, there is a local-isometry $X \to X'$ such that X' has the same properties as X but is also compact. A quasi-isometrically embedded f.g. subgroup of $\pi_1 X$ maps to a quasi-isometrically embedded subgroup of $\pi_1 X'$. As in Remark 15.11, the proof of Theorem 15.10 explains that (single and) double full quasiconvex cosets are separable in $\pi_1 X'$. Hence they are separable in $\pi_1 X$. This explanation is heavier than necessary, and we describe more basic proofs for the separability of H as a virtual retract, and also for the separability of $H_1 H_2$ when each H_i is full.

Proof. We first describe the virtual retraction proof. Extend $H \subset \pi_1 X'$ to a full quasiconvex subgroup by amalgamating with finite index subgroups of the parabolic subgroups where necessary, and note that there is a retraction $H^+ \to H$ induced by retractions of (the abelian) parabolic subgroups. The subgroup H^+ is represented by a based local-isometry $Y \to X'$ with Y compact by Proposition 2.31. We apply Proposition 6.3 to obtain a finite cover $\mathsf{C}(Y \to X')$ that retracts to Y. Thus $\pi_1 Y$ is a virtual retract, and since $\pi_1 Y$ retracts to H we see that H is a virtual retract of $\pi_1 X'$ and hence of $\pi_1 X$. This mode of proof was first introduced in a primitive form in [Wis00].

We now prove double separability under the additional assumption that each H_i is full, and X is compact so we can assume $X' = X$. Let $Y_1^+ \to X'$ be obtained as above for H_1, and let $X'_\ominus$ be the mapping cylinder of $Y_1^+ \to X'$. Note that $X'_\ominus$ can be proved to be virtually special using the canonical completion $\mathsf{C}(Y_1^+ \to X')$ to find a finite cover of $X'_\ominus$ without inter-osculations. Similarly, let $Y_2^+ \to X'_\ominus$ be associated to H_2, and by choosing a sufficient cubical thickening, we can assume that some edge of Y_2^+ traverses the hyperplane of $X'_\ominus$ associated to Y_1^+. Let $X'_\oplus$ be the mapping torus of $Y_2^+ \to X_\ominus$. Applying canonical completion to an elevation of Y_2^+ to a special cover of $X'_\ominus$, we can likewise show that $X'_\oplus$ is virtually special. By our thickened choice of Y_2^+, the hyperplanes associated to (finite index subgroups of) H_1 and H_2 cross each other. This double coset is separable by Proposition 6.9.

Note that to extend this argument to obtain separability of $H_1 H_2$ without the additional assumptions that H_1, H_2 are full and X is compact, we would need to also check that for each $g \notin \pi_1 H_1 \pi_1 H_2$, we can choose each $Y_i^+ \to X'$ sufficiently so that $g \notin H_1^+ H_2^+$ where $H_i^+ = \pi_1 Y_i^+$. $\square$

The following generalizes the compact hyperbolic case in Corollary 6.10.

Theorem 15.14. *Let G be hyperbolic relative to virtually abelian subgroups. Suppose G acts properly and cosparsely on a CAT(0) cube complex $\widetilde{Y}$, and suppose that G has a torsion-free finite index subgroup G' such that $G' \backslash \widetilde{Y}$ is special. Then for any proper cosparse action of G on a CAT(0) cube complex $\widetilde{X}$ with closeable quasiflats, there exists a torsion-free finite index subgroup G'' with $G'' \backslash \widetilde{X}$ special.*

Proof. Let $(\widetilde{F}_i, P_i)$ be representatives of the finitely many (non-cocompact) quasiflats of $\widetilde{X}$, and let $(\widetilde{F}_i', P_i')$ be the cocompact quasiflat that extends $(\widetilde{F}_i, P_i)$. For each i, let $A_i' \subset P_i'$ be a finite index free-abelian subgroup. By Corollary 6.8, $A' \subset G'$ is separable, so we may choose $K \subset G$ to be a finite index normal subgroup with $K \subset G'$, and with $B_i = (K \cap P_i') \subset A_i'$ for each i. By separability in abelian groups, we choose finite index subgroups $B_i' \subset A_i'$ such that B_i is a direct factor of B_i', and note that B_i is also a direct factor of any subgroup of B_i' containing it.

Let $X = K \backslash \widetilde{X}$. Apply Theorem 7.54, to $(\widetilde{X}, K)$ together with the various cocompact extensions $(\widetilde{F}_i', B_i')$ of $(\widetilde{F}_i, B_i)$, as well as the additional quasiflats obtained by passing to the finite index subgroup $K \subset G$, and note that these also have the same direct factor property. We obtain a local-isometry $X \to \dot{X}$ to a compact nonpositively curved cube complex $\dot{X}$, such that $\dot{G} = \pi_1 \dot{X}$ is hyperbolic relative to abelian subgroups. Moreover, $\pi_1 \dot{X}$ splits as a tree of groups with central vertex group isomorphic to $\pi_1 X$, and where the leaf vertex groups are free-abelian, and the edge groups are direct factors of them. Applying Lemma 15.21 several times, we see that $\dot{G}$ is the fundamental group of a virtually special sparse cube complex. Consequently, each hyperplane of $\dot{X}$ has separable π_1 by Theorem 15.13. Thus as $\dot{X}$ is compact, $\dot{X}$ has a finite cover $\widehat{\dot{X}}$ with embedded 2-sided hyperplanes. Hence $\widehat{\dot{X}}$ is virtually special by Theorem 15.10. Thus X has a finite cover admitting a local-isometry to a special cube complex, and is hence virtually special. $\square$

15.e   Relative Malnormality and Separability

We now revisit the notions of height from Section 8.a in the relatively hyperbolic case. This will be used in the proof of Theorem 15.1.

Definition 15.15 (Relative Height and Malnormality). Let G be relatively hyperbolic. The *relative height* of $H \subset G$ is the smallest number h such that for any $h + 1$ distinct cosets $\{Hg_1, \ldots, Hg_{h+1}\}$ the intersection $\cap_i H^{g_i}$ is either parabolic or elliptic. The subgroup H of G is *relatively malnormal* if $H \cap gHg^{-1}$ is either elliptic or parabolic for each $g \in G - H$. Thus H is relatively malnormal exactly when its relative height is 0 or 1. In particular, any malnormal subgroup and any parabolic or elliptic subgroup is relatively malnormal.

The following was proven in [HW09, Cor 8.4]:

Proposition 15.16. *Let $H \subset G$ be a relatively quasiconvex subgroup of a relatively hyperbolic group. Then H has finite relative height.*

The following was proven in [HW09, Thm 9.3]:

Proposition 15.17. *Let H be a separable, relatively quasiconvex subgroup of the relatively hyperbolic group G. Then there is a finite index subgroup K of G containing H such that H is relatively malnormal in K.*

Definition 15.18 (Full). Let G be hyperbolic relative to $\{P_i\}$. A subgroup $H \subset G$ is *full* if for each parabolic subgroup P_i^g, either $H \cap P_i^g$ is finite or $H \cap P_i^g$ is of finite index in P_i^g. Similarly, H is *completely full* if either $H \cap P_i^g$ is finite or $H \cap P_i^g = P_i^g$.

15.f The Hierarchy in the Relatively Hyperbolic Setting

The goal of this section is to prove Theorem 15.1.

Cages and expanded edge groups: To avoid cubulation technicalities having to do with parabolics, the proof of Theorem 15.1 utilizes splittings that are slightly different from the sequence of splittings along edge groups. Let G be a relatively hyperbolic group that splits as a graph Γ of groups. Let e be an edge of Γ and let $E = G_e$ be the associated edge group. The *expanded edge group* $E^+ = \langle g : \exists e \neq 1 \text{ with } e^g \in E \rangle$ is the subgroup generated by E together with all elements that conjugate some nontrivial element of E to another. Note that $E \subset E^+$ and E contains each parabolic subgroup having infinite intersection with E. We refer to Figure 15.1 for a picture suggesting what E^+ looks like in a setting where vertex groups do not contain an entire aparabolic subgroup. By first passing to an appropriate finite index subgroup, E^+ is actually generated by E together with all parabolic subgroups P with $E \cap P$ infinite.

In our case of interest, as all the parabolic subgroups are of the form $\mathbb{Z}^2$ mapping to an immersed circle in Γ, the induced structure of E^+ as a graph of groups has "initial" and "terminal" vertex groups isomorphic to an edge group E, and a collection of arcs starting and ending on these initial and terminal vertex groups. The internal vertex and edge groups of each such arc are all isomorphic to the same subgroup of a parabolic group which is associated to as many arcs as the number of its conjugates intersecting E. Finally, there is a homomorphism $E^+ \to G$ which is induced by a map of underlying graphs of groups. See Figure 15.1. For simplicity below, we will refer to each such arc of the underlying graph of E^+ as an edge.

The *cage* K associated to the edge e is the graph of groups obtained from the graph of groups for E^+ by removing the open edge e associated to E itself. The splitting of K has two distinguished vertex groups—the initial and terminal image groups of E, and has arcs with parabolic stabilizers that start and end on these respectively. We will not consider the cage K in the disconnected case.

The proof of Theorem 15.1 begins by passing to a sufficient finite index subgroup whose cages are controlled as follows:

Lemma 15.19. *Let G be hyperbolic relative to virtually abelian subgroups. Suppose G splits as a finite graph Λ of groups whose edge groups are relatively quasiconvex and whose vertex groups are virtually fundamental groups of special sparse cube complexes.*

There exists a finite index subgroup G' of G whose induced splitting Λ' has vertex groups that are fundamental groups of sparse special cube complexes. Moreover, Λ' has the following property: For each connected subgraph $\Gamma \subset \Lambda'$, and for each edge e' of Γ, the expanded edge group $(E')^+$ associated to e has the property that $(E')^+ \subset \pi_1 \Gamma$ is malnormal and relatively quasiconvex.

Consequently, the cage K associated to e has the property that $K \subset \pi_1(\Gamma - e)$ is malnormal and relatively quasiconvex.

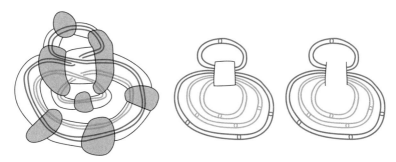

Figure 15.1. A graph of spaces on the left (peripherals are highlighted on inside), an expanded edge space in the middle, and its associated cage on the right. In this example, the two parabolic subgroups are represented by immersed tori. The bold vertical lines in the cage correspond to the initial and terminal images of the edge space, the small circles in the cage correspond to places where the parabolic torus passes through vertex spaces, and the cylinders in the cage correspond to places where the parabolic torus passes through edge spaces.

Furthermore, each expanded edge group E^+ is a subgraph of groups of a finite index subgroup of a multiple HNN extension $\langle G_{\bar{e}}, \{t_1, \ldots, t_m\} \mid Z_i^{t_i} = Z_i : 1 \leq i \leq m \rangle$ where $G_{\bar{e}}$ is the edge group of G corresponding to the image $\bar{e}$ of e in Λ, and each stable letter t_i centralizes a codimension-1 subgroup Z_i of $G_{\bar{e}}$.

Moreover, if each edge group has no accidental parabolics, and all maximal parabolic subgroups are virtually $\mathbb{Z}^2$, then the cage K associated to e is aparabolic in the relatively hyperbolic structure on $\pi_1(\Gamma - e)$ that omits cyclics.

In the statement of Lemma 15.19, if E is aparabolic then K consists of two copies of E (one initial and one terminal), and the conclusion applies to each copy. However we will only employ K when E has infinite intersection with some maximal parabolic subgroup, and hence assuming there are no accidental parabolics, we find that e does not separate Γ.

Proof. By Theorem 15.8, G is residually finite, and hence by possibly passing to a finite index subgroup, we may assume that each maximal parabolic subgroup is abelian. Similarly, since each vertex group has a torsion-free finite index subgroup, by passing to a finite index subgroup, we may assume that G is torsion-free. And we may assume that each vertex group is the fundamental group of a sparse special cube complex. Each edge group of G is separable by Theorem 15.8. By Lemma 15.17, each edge group E is relatively malnormal within a finite index subgroup of G. By replacing G with the intersection of finitely many finite index subgroups of G, we may thus assume that each edge group is relatively malnormal.

By separability, we may pass to a finite index subgroup such that for each parabolic subgroup P the axis $L \subset \tilde{T}$ that it stabilizes in the Bass-Serre tree projects to an immersed circle $P\backslash L$ in the underlying graph. Here we are using separability of edge groups in (the whole group and hence in the) vertex groups, and that distinct translates of an edge at a vertex of T correspond to distinct cosets in the vertex group. There are finitely many elements to separate, and this immersed property is stable under further covers. Then the parabolic subgroups (that aren't elliptic with respect to T) project to immersed circles in the underlying graph of the splitting of a finite index subgroup. By separability of cyclic subgroups of the underlying free group, we may pass to a further finite index subgroup so that they are all embedded. Each E^+ will then have the simpler structure that it is a multiple HNN extension of E. Moreover, it arises as a subgraph of groups of a multiple HNN extension of a finite index subgroup of an original edge group. This enables the multiple HNN extension claim. (For simplicity, we assume that no E has an accidental parabolic in a maximal parabolic that lies entirely in a vertex group. That assumption can be circumvented by applying the argument to a new splitting where a pair of adjacent vertex groups are both extended by adding the parabolics so the edge group is full with respect to each vertex.)

Let $\{P_i\}$ be representatives of the maximal parabolic subgroups of G having infinite intersection with E, and note that $\{P_i\}$ is finite since E is relatively quasiconvex. By [MP09, Prop 5.11], there exist finite index subgroups $P_i^m \subset P_i$ that contain $P_i \cap E$ such that the subgroup $M = \langle E, \cup P_i^m \rangle$ is full and relatively quasiconvex. We may thus pass to a further finite index subgroup G', induced by a finite cover Λ' of the underlying graph as above, so that each $G' \cap P_i \subset P_i^m$. Note that $M \subset G'$ is completely full since we include the entire parabolic subgroup intersecting E.

Let Λ' be the graph of the induced splitting of G', let e' be an edge of Λ', let E' be its edge group, and let $(E')^+$ be the associated expanded edge group. Let K' be the cage of e', and let $V' = \pi_1(\Lambda' - e')$. The relative quasiconvexity of K' holds as follows: Relative quasiconvexity of $(E')^+ \subset G'$ holds because $(E')^+$ is the fundamental group of a subgraph of groups of a finite index subgroup of a conjugate of some M above and is hence also relatively quasiconvex. Relative quasiconvexity of $E' \subset G'$ implies the relative quasiconvexity of $V' \subset G'$ by [BW13, Lem 4.9]. Hence $K' = ((E')^+ \cap V') \subset G'$ is relatively quasiconvex, since relative quasiconvexity is preserved by intersection (see, e.g., [Hru10]). Malnormality of $(E')^+ \subset G'$ is proven below. Malnormality of $K' \subset V'$ follows from malnormality of $(E')^+ \subset G'$, by intersecting with $V' \subset G'$. Finally, these two properties continue to hold after deleting further edges of Λ'.

We now verify malnormality of $(E')^+ \subset G'$ using the notation $A \subset B$. Suppose $b^{-1}ab = a^b \in A$ for some $b \in B$. Firstly, suppose a is elliptic. As elliptic elements of A are conjugate in A to the vertex group $E' \subset A$, there exists $c \in A$ with $a^{bc} \in E'$. Hence $bc \in E'$ as E' is malnormal, so $bc \in A$ so $b \in A$. Secondly, suppose a is hyperbolic. If $a \in P'$ is parabolic, then a^b is parabolic, and hence $a^b \in (P'')^c$ for some $c \in A$ and maximal parabolic subgroup of A, since A is completely full. But nonconjugate parabolic subgroups of A are nonconjugate in B, since they correspond to distinct lines in the Bass-Serre tree T. Hence $a^{bc^{-1}} \in P^d$ for some

$d \in A$, so $bc^{-1}d^{-1} \in P \subset A$, so $b \in A$. Thirdly, suppose a is hyperbolic but not parabolic. Since we may assume the underlying circles of the P_i^m in Λ' are much larger than their pairwise overlaps (which are uniformly bounded), there is only one way to decompose the axis of a hyperbolic element of A as the product of parabolic subpaths (and elliptic transitions). Let L be the axis for a, so $b^{-1}L$ is the axis for a^b. Since A has a unique vertex group, there exists $c, d \in A$ such that dbc is elliptic, and $a^{dbc} \in A$. Indeed, choose $c \in A$ so that $c^{-1}b^{-1}L$ passes through the base vertex of A, and then choose $d \in \langle a \rangle$ so that $c^{-1}b^{-1}d^{-1}$ actually fixes the base vertex. Let $p \in A$ be the parabolic element corresponding to the first parabolic subpath of the axis L of a. Then dbc conjugates p to another parabolic element of A. Hence $dbc \in A$ by the second case above. Hence $b \in A$. $\quad\square$

Lemma 15.20. *If an amalgamated free product $G = E *_B A$ satisfies the following conditions then G is virtually compact special.*

(1) *E has a finite index subgroup that is π_1 of a compact special cube complex.*
(2) *A is a f.g. free-abelian group and B embeds as a direct factor so $A = B \times \mathbb{Z}^m$.*
(3) *There does not exist a free-abelian subgroup $B' \subset E$ with $[B : B \cap B'] < \infty$ and $\operatorname{rank}(B') > \operatorname{rank}(B)$.*

Moreover, if E is $\pi_1 X$ where X is a virtually special compact cube complex, then $G = \pi_1 Z$ where Z is a virtually special cube complex. The inclusion $E \subset G$ is represented by an injective local-isometry $X \to Z$. The inclusion $A \subset G$ is represented by a local-isometry $Y \times (S^1)^m \to Z$ where $Y \to X$ is a local-isometry from a compact nonpositively curved cube complex and $\pi_1 Y$ maps to B.

Note that Condition (15.20) always holds when $B \subset E$ is a maximal abelian subgroup of a raag. See [WW17, Cor 5.4].

Proof. The first statement holds by m applications of [WW17, Thm 5.5] which proves the above statement for $m = 1$. The second statement holds as follows. Let $E = \pi_1 X$ with X compact and virtually special, and by Lemma 2.38, let $Y \to X$ be a local-isometry with Y compact and $\pi_1 Y$ mapping to B. We then use m applications of [WW17, Lem 5.1] which has the following special case: Let $\phi : Y \to X$ be a local-isometry of nonpositively curved cube complexes where Y is compact and X is special. Then the cube complex Z below is virtually special, and $B = \pi_1(Y \times \{\frac{1}{2}\})$:

$$Z = \big(X \cup (Y \times [0,1])\big) / \big\{ (y,0) \sim \phi(y) \sim (y,1) : \forall y \in Y \big\} \qquad\square$$

The following variant of Lemma 15.20 can also be proven by capitalizing on Lemma 7.51.

Lemma 15.21. *Let $G = E *_B A$ where E has a finite index subgroup that is the fundamental group of a sparse special cube complex, and where B is a maximal* *parabolic subgroup of E, and B embeds as a direct factor of $A = B \times \mathbb{Z}^m$. Then G has a finite index subgroup that is the fundamental group of a sparse special cube complex. Moreover, if E is the fundamental group of a sparse cube complex that is virtually special, then so is G.*

Proof. We shall assume that A is infinite, for otherwise the result can be obtained using free products. Let E' be a finite index normal subgroup of E such that $E' = \pi_1 X'$ where X' is sparse and special. Let $G' \subset G$ be the finite index subgroup induced by E' and the retraction $G \to E$ that collapses $1_B \times \mathbb{Z}^m$. Note that G' can be built from E' by a sequence of finitely many amalgams with free-abelian subgroups $A_i = B_i \times \mathbb{Z}^m$ of E' along maximal abelian subgroups $B_i \subset E'$. Let $X' \to R$ be the local-isometry of Proposition 6.2, and let $B'_i \subset \pi_1 R$ be a maximal abelian subgroup containing B_i. By Lemma 2.38, let $F'_i \to R$ be a local-isometry with $\pi_1 F'_i = B'_i$ and with F'_i compact. For each i, we choose a B_i-invariant translate of $\widetilde{F}_i$, and let $\widetilde{F}_i = \widetilde{F}'_i \cap \widetilde{X}'$, and note that we may assume that the sparse structure of $\widetilde{X}'$ is obtained from translates of the various $\widetilde{F}_i$ together with quasiflats associated to the other maximal parabolic subgroups. Let R^+ be the cube complex obtained from R by attaching a copy of $F'_i \times T^m$ for each i, where T^m is the standard cubical m-torus with 0-cube p, and where we are identifying $F'_i \times \{p\}$ with $F'_i \to R$. Let $F_i = B_i\backslash\widetilde{F}_i$ for each i, and consider the local-isometry $F_i \to X'$. Let X^+ be obtained from X' by attaching a copy of $F_i \times T^n$ using the map $F_i \times \{p_i\} \to F_i \to X'$ for each i, and note that $G' \cong \pi_1 X^+$. There is a natural local-isometry $X^+ \to R^+$. Hence X^+ is virtually special since R^+ is by Lemma 15.20.

If $E = \pi_1 X$ where X is a sparse nonpositively curved cube complex, then the map $X' \to R$ can be chosen to be Q-invariant where $Q = \text{Aut}(X' \to X)$, and we can assume that Q permutes $\{F'_i\}$ and $\{F_i\}$. Hence $Q\backslash X^+$ is a nonpositively curved virtually sparse cube complex containing X and justifying the final claim.   □

Proof of Theorem 15.1. **Outline of induction:** In Part 1, we apply Lemma 15.19 to pass to a finite index subgroup G' whose induced graph of groups Λ' has the property that its vertex groups are special, its expanded edge groups are relatively quasiconvex and malnormal, and each cage is relatively quasiconvex and malnormal and aparabolic in its associated subgraph of groups. Moreover, the analogous property holds for any subgraph of groups of Λ'. We shall prove the theorem by induction on the number of edges in a subgraph $\Gamma \subset \Lambda'$ and note that the base case holds by Condition (3) when Γ is a single vertex. Let $\Gamma \subset \Lambda'$ be a connected subgraph, and let e be an edge of Γ. By induction, the theorem holds for $\pi_1(\Gamma - e)$. In general $\Gamma - e$ might have two components in which case the theorem holds inductively for each, but for simplicity below we assume there is one component.

We shall use the following notation for each Γ and e: Let $J = \pi_1(\Gamma)$, let E be the edge group at e in J, let E^+ be the expanded edge group at e in J, let K be the cage associated to E, and let $V = \pi_1(\Gamma - e)$.

At each stage: Either e is separating and $E = E^+$ and we use the ordinary splitting of J over E corresponding to a graph of groups splitting along a specific edge group. Or e is non-separating and we use a splitting $\pi_1\Gamma = E^+ *_K V$ along a cage, as described in Part 2.

The virtual specialness of V holds by induction, and the expanded edge group E^+ is virtually compact special by Lemma 15.20. In Part 3, we apply Theorem 7.59 to this splitting to obtain a nonpositively curved cube complex. And in Part 4, we verify its cocompactness. In Part 5, under the assumption that the vertex groups are virtually compact, we reach the conclusion that this cube complex is compact. In Part 6, we address virtual specialness as follows: The subcomplex associated to V has a finite cover with embedded hyperplanes by Theorem 15.13, and so we can apply Theorem 15.8 to pass to a finite cover of the cubulation of J with a hierarchy. In Part 7, we explain how to reach the conclusion that G is π_1 of a compact nonpositively curved cube complex, by assuming its vertex groups have this property, and making earlier choices to ensure the compatibility of cubulations of conjugate tori in G'. Note that Part 7 can be skipped on the first reading, and the subsequent sections can be read independently. Finally, we apply Theorem 15.10 to obtain virtual specialness. The sparse cases are proven in Parts 7 and 8 as consequences of the compact case, but most of the proof could be directly adapted to those cases.

1. Controlled cages: By Lemma 15.19, there is a finite index subgroup G' of G whose induced splitting along Λ' has the following property: For each connected subgraph $\Gamma \subset \Lambda'$ and each edge e of Γ, its associated expanded edge group E^+ is a malnormal, relatively quasiconvex subgroup of $J = \pi_1\Gamma$. Furthermore its associated cage K embeds as a malnormal, relatively quasiconvex subgroup $K \subset V = \pi_1(\Gamma - e)$, and moreover $K \subset V$ is aparabolic.

When $E = E^+$, the subgroup $E \subset J$ is malnormal, aparabolic, and relatively quasiconvex, and we will use the usual splitting of J along E. In the typical case, when $E^+ \neq E$, we will use an alternate splitting that we describe below. Note that when $E^+ \neq E$ there is a maximal parabolic subgroup P having infinite intersection with E, and so Condition (4) implies that P splits nontrivially along $P \cap E$, and more specifically, there is an edge $\tilde{e}$ in the Bass-Serre tree that projects to e, and P acts nontrivially on a line containing $\tilde{e}$. In particular, it follows that $\Gamma - e$ is connected, and the two vertex groups of K are connected by at least one edge group.

2. Splitting along cages: The cage K splits as a graph of groups whose two vertex groups are the isomorphic copies of E corresponding to the two inclusions of E in its initial and terminal vertex groups, and the edge groups of K are the intersections between K and peripheral subgroups.

We now describe the splitting $J = E^+ *_K V$ as an amalgamated free product along the cage K associated to e. One vertex group of this splitting is the group $V = \pi_1(\Gamma - e)$. The other vertex group is the expanded edge group E^+. The edge group is the cage K.

We claim that $J \cong E^+ *_K V$ where the inclusions of K are the natural ones. We verify this by rethinking it geometrically as follows. Let $\phi: Y^+ \to Z$ be an inclusion of graphs of spaces, where Y^+ is a graph of spaces corresponding to E^+ with Y corresponding to the edge e, and Z is a graph of spaces corresponding to the splitting of J over Γ. The mapping cylinder M_ϕ deformation retracts to X by pushing $Y^+ \times [0, 1]$ forwards to $Y^+ \times \{1\} \subset Z$, so $\pi_1 M_\phi \cong \pi_1 Z$. There is another deformation retraction which pushes $Y \times [0, 1]$ upwards to $Y \times \{0\}$. The result is a graph of spaces associated to the amalgamated product $E^+ *_K V$.

3. The cubulation: By induction on the number of edges in Γ', the group V acts properly and cocompactly [cosparsely] on a CAT(0) cube complex $\widetilde{X}_V$.

As $E \subset V$ is relatively quasiconvex and aparabolic, Proposition 2.31 implies that E acts freely and cocompactly on a convex subcomplex $\widetilde{X}_e \subset \widetilde{X}_V$. Consequently, by Lemma 15.20, E^+ is the fundamental group of a virtually special compact nonpositively curved cube complex.

Let V' be a finite index torsion-free subgroup of V such that $V' \backslash \widetilde{X}_V$ is special. Let $K' = K \cap V'$.

Let $(E^+)' \subset E^+$ be a finite index normal subgroup such that $((E^+)' \cap K) \subset K'$. Indeed, $K' \subset E^+$ is separable by Theorem 15.13, and so we initially find a finite index subgroup of E^+ containing K' but not containing any of the finitely many nontrivial cosets of K' in K. We then let $(E^+)'$ be a finite index normal subgroup of E^+ contained in the initial finite index subgroup.

We apply Lemma 7.30 to obtain a free cocompact action of E^+ on a CAT(0) cube complex $\widetilde{X}'_e$ such that the stabilizer of each hyperplane is a subgroup of $(E^+)'$. The advantage of $\widetilde{X}'_e$ is that for each hyperplane U' of $\widetilde{X}'_e$, we have $(K \cap \mathrm{Stab}(U')) \subset V'$, and so Hypothesis 7.59.(9') holds. Indeed, for each hyperplane of $\widetilde{X}'_e$ equipped with the E^+ action, the induced wall for K has stabilizer lying in V' which is sparse and special. As $K \subset V$ is quasiconvex and aparabolic and $K \cap \mathrm{Stab}(U')$ is quasiconvex, Theorem 6.25 applies to extend the $(K \cap \mathrm{Stab}(U'))$-wall to a relatively quasiconvex wall for V.

Since $K \subset V$ is malnormal and K is relatively quasiconvex in $J = E^+ *_K V$, we can apply the variant of Theorem 7.59 indicated by Restatements (5'), (6'), (8'), and (9'). Note that the hyperplane stabilizers provided by Theorem 7.59 are relatively quasiconvex.

4. Cocompactness: We now verify J-cocompactness of the dual cube complex obtained in Part 3 where we assume cocompactness of the cube complexes of the factors of the amalgam $J = E^+ *_K V$. As explained in Remark 6.27, the new walls crossing V do not provide new commensurability classes of codimension-1 subgroups in the parabolic subgroups of V, and so we will not "overcubulate" the parabolic subgroups. Hence the relative cocompactness provided by Theorem 7.9 is actual cocompactness, since the cubulations $C_\star(P_i)$ of P_i that arise in the vertex groups remain cocompact after adding the new walls.

We shall examine this by understanding the new wallspace and applying Remark 7.10. We regard J as acting on a tree of wallspaces $\widetilde{X}$, whose vertex spaces are CAT(0) cube complexes that are copies of $\widetilde{X}_V$ and $\widetilde{X}'_e$, and as explained in Remark 7.60, the walls are of two types: Walls that are dual to the edges of the Bass-Serre tree of the splitting, and walls whose intersection with a vertex space is either $\emptyset$ or consists of a single hyperplane or a hypothesized extended wall.

There are two types of maximal parabolic subgroups: Those conjugate into V and those conjugate into E^+. In each case we will show that $C_\star(P)$ is P-cocompact. Hence the entire CAT(0) cube complex is cocompact by Theorem 7.9.(1).

Since $K \subset V$ is aparabolic, each parabolic subgroup P of V has finite coarse overlap with K. Consequently, $C_\star(P)$ is dual to the wallspace whose walls consist precisely of the extended walls and original hyperplanes of $\widetilde{X}_V$ that have infinite diameter intersection with $\widetilde{F}^{+r}$ for some uniform r, where $\widetilde{F} \subset \widetilde{X}_V$ is a P-invariant convex subcomplex. We now verify that $C_\star(P)$ is P-cocompact. There are finitely many P-orbits of new walls and hyperplane carriers intersecting $\widetilde{F}^{+r}$. Each hyperplane carrier is a convex cocompact subcomplex, and by Remark 6.27 each new wall is represented by a *trace* consisting of a convex cocompact subcomplex of $\widetilde{F}^{+r}$. A cube of $C_\star(P)$ corresponds to a collection of pairwise crossing walls in this wallspace, and this yields a collection of pairwise intersecting carriers and/or traces. Note that two walls cross if and only if they cross within the vertex space, or cross beyond an adjacent edge space that they both travel through. Hence, since there is a uniform upperbound on the coarse overlap between $\widetilde{F} \subset \widetilde{X}_V$ and any edge space adjacent to $\widetilde{X}_V$, we see that there is a uniform upperbound on the distance within $\widetilde{F}^{+r}$ between any two such walls that both cross $\widetilde{F}$ and enter the same edge space. Consequently, after taking a uniform cubical thickening of each trace, we see by Lemma 2.10 that for any finite cardinality collection of pairwise intersecting walls, there is a point in their mutual intersection. Hence by the local finiteness in $\widetilde{F}^{+r}$ of the full collection of such thickened carriers and traces, we see that there is an upperbound on the cardinality of any pairwise intersecting collection, and thus a bound on the number of orbits of such collections. Hence each such $C_\star(P)$ has finitely many orbits of cubes, and is thus P-cocompact by Remark 7.11. (If we had chosen $\widetilde{F}$ to be superconvex, then we would know that any wall having infinite coarse overlap with $\widetilde{F}$ must already have infinite intersection with $\widetilde{F}$ and not just $\widetilde{F}^{+r}$.)

Let P be a peripheral subgroup within an E^+ vertex group. Let $\widetilde{F} \subset \widetilde{X}'_e$ be a superconvex P-cocompact convex subcomplex. By construction, any wall of the tree of spaces $\widetilde{X}$ is either disjoint from $\widetilde{X}'_e$ or intersects $\widetilde{X}'_e$ in a hyperplane. Suppose some wall W in the tree of spaces satisfies $\mathrm{diam}(\mathcal{N}_j(W) \cap \widetilde{F}) = \infty$ for some $j > 0$ (technically we need each halfspace to have infinite diameter coarse intersection, but these are equivalent when $\mathrm{rank}(P) > 1$). Then for some uniform r we have $\mathrm{diam}(\mathcal{N}_r(W) \cap \widetilde{F}) = \infty$. Moreover, either $W \cap \widetilde{F}$ is a hyperplane of $\widetilde{F}$, or W is a wall that does not intersect $\widetilde{X}'_e$ but is stabilized by an infinite subgroup $Z \subset P \cap K$. Since $\mathrm{rank}(P) = 2$ we see that Z stabilizes a hyperplane U of $\widetilde{F}$, by Lemma 15.20. Consequently, for each r there exists s such that the coarse intersection $\mathcal{N}_r(W) \cap \widetilde{F}$ lies in $U^{+s} \subset \widetilde{F}$ for some hyperplane U of $\widetilde{F}$.

Consider two walls W_1, W_2 that cut $\widetilde{F}$ and also cross each other in the entire tree of spaces, then $\mathcal{N}_m(W_1) \cap \widetilde{F} \cap W_2 \neq \emptyset$ for some $m = m(\widetilde{X})$ by [HW14, Lem 8.11]. Letting $W_i' = \mathcal{N}_r(W_i)$ be thickened walls, and applying the same result, we find that if W_1', W_2' cut $\widetilde{F}$ and also cross each other in the entire tree of spaces, then $\mathcal{N}_s(W_i') \cap \widetilde{F} \cap W_2' \neq \emptyset$ for some $s = s(\widetilde{X}, m)$. However, each $\mathcal{N}_s(W_i') \cap \widetilde{F} = \mathcal{N}_{r+s}(W_i) \cap \widetilde{F}$ lies in $U_i^{+t} \subset \widetilde{F}$ for some hyperplane U_i of $\widetilde{F}$, and some t depending on $r + s$. Thus, for any collection of pairwise crossing walls $\{W_i\}$ that have infinite coarse intersection with $\widetilde{F}$, their neighborhoods $\mathcal{N}_{r+s}(W_i)$ pairwise intersect within $\widetilde{F}$. Consequently $\{U_i^{+t}\}$ pairwise intersect. Lemma 2.10 implies that any finite subcollection of $\{U_i^{+t}\}$ must mutually intersect. Local finiteness gives an upperbound on the cardinality of such a collection. We conclude that there are finitely many P-orbits of cubes in $C_\star(P)$ when P is a peripheral subgroup that lies in E^+.

6. Virtual specialness: Let $\widetilde{C}$ be the CAT(0) cube complex dual to our wallspace structure on $\widetilde{X}$, let $C = J \backslash \widetilde{C}$, and let U be a hyperplane dual to the wall associated to the edge group K. Consider the cube complex $C - N^o(U)$ obtained by removing the open carrier of U. Let C_V and C_{E^+} be the components with $\pi_1 C_V = V$ and $\pi_1 C_{E^+} = E^+$. Since V and E^+ are virtually [cosparse] special groups, by Theorem 15.14, both C_V and C_{E^+} have finite special covers, and hence finite covers whose hyperplanes are embedded. We may thus apply Theorem 15.8 to choose a (regular) cover $\widehat{C}$ of C such that the covers of C_V and C_E factor through these finite covers. Theorem 15.10 applies to show that $\widehat{C}$ is virtually special.

5. Compatible parabolic cubulations and G-cocompactness: We now describe how to ensure that the cocompact cubulation of G' will induce a cocompact cubulation of G. This will require making choices in the cubulations of the peripheral subgroups $\{P^g \cap G'\}_{g \in G}$ so that they are compatibly cubulated. By replacing it with the intersection of its conjugates, we may assume G' is normal, which simplifies the discussion of compatibility since the various conjugates $P^g \cap G'$ are all conjugate to the same finite index subgroup $P \cap G'$ of P. For each representative of a conjugacy class of a maximal parabolic subgroup P of G that is not elliptic in the splitting, let R be the cyclic subgroup of P that is elliptic in the action on the Bass-Serre tree. If P is a Klein bottle group then let $\dot{P}$ be a $\mathbb{Z}^2$ index 2 subgroup of P containing R, and if P is abelian then let $\dot{P} = P$. Let α, β be a basis for $\dot{P}$ with $R = \langle \alpha \rangle$. We will use this basis to organize our choices below.

For each edge $\bar{e}$ of Λ, let $E_{\bar{e}}$ be its edge group, let $\{P_a\}_{a \in A_{\bar{e}}}$ be representatives of the finitely many distinct $E_{\bar{e}}$-conjugates of maximal parabolic subgroups of G having infinite intersection with $E_{\bar{e}}$. For each $a \in A_{\bar{e}}$, let $\dot{P}_a \subset P_a$ be an index 2 subgroup with $\dot{P}_a \cong \mathbb{Z}^2$ and $\dot{P}_a \cap E_{\bar{e}} = P_a \cap E_{\bar{e}}$ (note that P_a could be isomorphic to a Klein bottle group). Let $D_{\bar{e}}$ be the group obtained by amalgamating $E_{\bar{e}}$ with each $\dot{P}_a$ along $\dot{P}_a \cap E_{\bar{e}}$. By Lemma 15.20, $D_{\bar{e}}$ is π_1 of a compact nonpositively curved cube complex $X_{\bar{e}}^+$ that is virtually special, and moreover each $\dot{P}_a \cap E_{\bar{e}}$ is the stabilizer of a hyperplane. The cubulation of a torus $\dot{P}_a$ arising from $X_{\bar{e}}^+$

is the product of a cubulation of a cyclic subgroup $\dot{P}_a \cap E_{\bar{e}}$ of $\pi_1 X_{\bar{e}}$ through Proposition 2.31, and a (standard) cubulation of a circle. Our choice of basis α, β for each $\dot{P}$, gives a basis α_a, β_a for each P_a which we now employ: Choose the isomorphism from $D_{\bar{e}}$ to $\pi_1 X_{\bar{e}}^+$ so that it is the identity on each $\pi_1 X_{\bar{e}}$, and so that it sends the horizontal generator β_a of $\dot{P}_a$ to an element stabilizing a hyperplane. It follows that $X_{\bar{e}}^+$ has compatibly cubulated tori, in the sense that if P_a and P_b are conjugate to the same parabolic subgroup $P \subset G$, then the cubulations of $\dot{P}_a$ and $\dot{P}_b$ have the same stabilizers of P-essential hyperplanes (after conjugating to P). By Proposition 2.31, cocompactness holds for any quasiconvex subgroup E_e^+ of D_e with the property that the intersection of E_e^+ with each $\dot{P}$ is either trivial, or finite index, or is commensurable with $\dot{P} \cap G_e$. The cubulation of P obtained by combining the cubulations of $\dot{P}_a$ and $\dot{P}_b$ (or finite index subgroups of these), will still be P-cocompact by Lemma 7.12, and likewise for any finite set of such subgroups and cubulations.

As we proceed through the sequence of splittings along edge groups of Λ' to build G' in Part 3, an expanded edge group E_e^+ associated to an edge e of a subgraph $\Gamma \subset \Lambda'$ will map to a subgroup $E_{\bar{e}}^+ \subset D_{\bar{e}}^+$ associated to the image $\bar{e}$ of e in Λ. As above, $E_e^+ = \pi_1 X_e^+$ where X_e^+ is compact and $X_e^+ \to X_{\bar{e}}^+$ is a local-isometry.

Observe that compatibility of the cubulations holds for non-elliptic $\mathbb{Z}^2$ subgroups of G' that are conjugate to finite index subgroups of the same peripheral subgroup of G. Indeed, although the cubulation of each non-elliptic parabolic arises from a choice at $X_{\bar{e}}^+$, all choices are compatible because of the basis chosen for maximal abelian subgroups of representatives of non-elliptic maximal parabolics of G, and the isomorphism $D_{\bar{e}} \to X_{\bar{e}}^+$ was chosen to send basis elements to hyperplane stabilizers. Furthermore, as explained in the G'-cocompactness part of the proof, the new walls in the cubulation of the amalgam, do not create additional commensurability classes of hyperplanes for the cubulations of the elliptic parabolic subgroups.

For each vertex group of G, we choose a fixed cubulation, and use that to cubulate the finite index subgroups of it in the vertex groups of G'. This will ensure that elliptic parabolic subgroups of G' that are conjugate to finite index subgroups of the same parabolic subgroup of G are cubulated compatibly.

It was shown in Part 4 that any nontrivial P^g-stabilizer of a hyperplane in the cubulation of a peripheral subgroup P^g is commensurable with a cyclic subgroup generated by one of its chosen basis elements when P is non-elliptic, and when P is elliptic, the P^g-stabilizer is commensurable with the intersection of the P^g stabilizers of finitely many P^g-essential hyperplanes in the original cocompact cubulation. Consequently, compatibility holds for the cubulations of peripheral subgroups of G' and so G is the fundamental group of a compact nonpositively curved cube complex by Lemma 7.17.

7. Deducing the cosparse case from the cocompact case: The finite index subgroup $G' \subset G$ obtained at the beginning of the proof, has the property that each vertex group G_v of $G' = \pi_1 X_v$ where X_v is sparse and special. Lemma 7.56 provides a local-isometry $X_v \to \bar{X}_v$ where $\bar{X}_v$ is compact, and $\pi_1 \bar{X}_v$ is special and hyperbolic relative to abelian subgroups. Moreover $\pi_1 \bar{X}_v$ retracts to $\pi_1 X_v$. Let $\bar{G}'$ be the graph of groups obtained by extending each G_v to $\bar{G}_v$, and note that the retractions $\bar{G}_v \to G_v$ induce a retraction $\bar{G}' \to G'$. Consequently G' is a quasi-isometrically embedded subgroup of $\bar{G}'$, and hence relatively quasiconvex. Moreover, Condition (2) persists since relatively quasiconvexity is transitive. We explain how to ensure Condition (4) below. Virtual compact specialness holds for $\bar{G}'$ by the cocompact case proven above. Hence $\bar{G}'$ has a finite index subgroup that is sparse and special by Lemma 2.31.

To ensure that Condition (4) persists we will ensure that for each G_v, no (non-cocompact) quasiflat of $\widetilde{X}_v$ has rank 1 stabilizer. That way, we can omit the rank 1 parabolic subgroups, and so if an edge group G_e contains an infinite parabolic in G_v' then it already had an infinite parabolic in G_v. Lemma 7.43 provides the desired cosparse action whose (rank ≥ 2) quasiflats are the same as the originals, and are hence closeable. Virtual specialness persists by Theorem 15.14.

8. Deducing the strongly sparse case from the sparse case: Since G has a finite index subgroup that is sparse, Lemma 7.34 implies that G acts cosparsely on a CAT(0) cube complex $\widetilde{A}$. Since all parabolic subgroups have rank at most 2, Lemma 7.44 implies that G acts strongly cosparsely on a CAT(0) cube complex $\widetilde{B}$. We now show how to produce a finite index torsion-free subgroup J' such that $\widehat{B} = J' \backslash \widetilde{B}$ is special. Let $J \subset G$ be a finite index subgroup so that $A = J \backslash \widetilde{A}$ is special. Then $B = J \backslash \widetilde{B}$ has embedded 2-sided hyperplanes, by the construction used to prove Lemma 7.44. (Lemma 7.44 could have been applied directly to the finite special cover A to produce B having embedded 2-sided hyperplanes.) The quasiflats of J are closeable by Lemma 7.51, and consequently Corollary 15.12 provides a finite special cover $\widehat{B} \to B$.   □

Remark 15.22 (Verifying Finiteness Properties in the Sparse Case). The proof of Theorem 15.1 deduces the cosparse cases as consequences of the cocompact case. We could have instead verified the Ball-WallNbd separation and WallNbd-WallNbd separation properties to see that the dual $\widetilde{C}$ to the wallspace of $\widetilde{X}$ is [strongly] cosparse via Theorem 7.28. Note that we must first truncate $\widetilde{X}_v$ using a convex core supplied by Lemma 7.42 so that $\widetilde{X}$ is cocompact and hence a geometric wallspace. The resulting new wallspace has the same dual as the old wallspace in the strongly cosparse case, but can have a different dual in the cosparse case as the core of $\widetilde{X}_V$ doesn't preserve intersection of walls.

Cosparse case: Consider a wall W in the tree of spaces. Let P be a parabolic subgroup in E^+. If W intersects $\widetilde{X}_e'$ then $\mathrm{Stab}_P(W)$ is the intersection of P-essential walls by Lemma 7.23. If W does not intersect $\widetilde{X}_e'$ then $\mathrm{Stab}_P(W)$ lies in the edge group associated to the edge between W and the E^+ vertex. Since $K \cap P$ is trivial or cyclic, $\mathrm{Stab}_P(W)$ is either trivial or cyclic, and in the cyclic case, $\mathrm{Stab}_P(W)$ equals the stabilizer of a P-essential hyperplane by the construction of Lemma 15.20. Let P be a parabolic subgroup of V. If $W \cap \widetilde{X}_V = \emptyset$, then $\mathrm{Stab}_P(W)$ is trivial since $K \subset V$ is aparabolic. Otherwise, $W \cap \widetilde{X}_V$ is a hyperplane or a trace. If $W \cap \widetilde{X}_V$ is a hyperplane then $\mathrm{Stab}_P(W)$ equals the intersection of stabilizers of P-essential hyperplanes by Lemma 7.23. If $W \cap \widetilde{X}_V$ is a trace, then as explained in Remark 6.27, the construction of Theorem 6.25 ensures that $\mathrm{Stab}_P(W \cap \widetilde{X}_V)$ is commensurable with the intersection of the P-stabilizers of finitely many P-essential hyperplanes. Thus Ball-WallNbd separation will hold for each peripheral subgroup by Lemma 7.29.

Strongly cosparse case: By Corollary 7.24, the $\mathrm{rank}(P) = 2$ hypothesis implies that for a parabolic subgroup P and a new wall W, we have $\mathrm{Stab}_P(W)$ is either trivial, or commensurable with P, or commensurable with $\mathrm{Stab}_P(W_1)$ where W_1 is a P-essential wall already appearing in $\widetilde{X}_V$. Consequently, WallNbd-WallNbd separation will hold by Lemma 7.31 as in Remark 7.35.

Remark 15.23 (Relaxing Aparabolicity)**.** Consider the situation that arises if we drop Condition (4) which ensures that each inclusion of $K \to V$ is aparabolic with respect to the relatively hyperbolic structure on V that ignores elementary parabolic subgroups. The proof could proceed without this provided we can generalize Theorem 7.59 by dropping Conditions 7.59.(6)/(6′) so that aparabolicity of the edge group is relaxed.

If edge groups were permitted to intersect parabolics in $\mathbb{Z}^n$ subgroups with $n > 1$, then we would need to be able to choose the cubulations of the nonelementary parabolic subgroups of E^+ consistently with their cubulations in $\widetilde{X}_V$. Furthermore, the extension property provided by Theorem 6.25 requires that the walls be represented using local-isometries with compact domain. However, most codimension-1 subgroups of $\mathbb{Z}^n$ do not have such representations once the cubical structure is determined.

Problem 15.24. Let G be hyperbolic relative to virtually abelian subgroups, and suppose G is virtually special. Does G (virtually) have a hierarchy terminating at the parabolic subgroups?

Chapter Sixteen

Largeness and Omnipotence

In this section we examine some properties of virtually special groups that generalize well-known properties of free groups.

16.a   Virtual Separation and Largeness

Lemma 16.1. *Let $D \subset X$ be a separating hyperplane in a connected special cube complex, and suppose that neither $D \to X_+$ nor $D \to X_-$ is π_1-surjective, and so $\pi_1 X$ splits nontrivially along $\pi_1 D$. There is a finite cover $\widehat{X} \to X$ such that no component of the preimage $\widehat{D}$ separates $\widehat{X}$.*

Proof. Let $\alpha_+ \to X_+$ and $\alpha_- \to X_-$ be closed based paths representing elements outside of $\pi_1 D$. Use the separability of $\pi_1 D$ in $\pi_1 X$ to pass to a finite quotient of $\pi_1 X$ such that $\bar{\alpha}_+ \notin \overline{\pi_1 D}$ and likewise such that $\bar{\alpha}_- \notin \overline{\pi_1 D}$. Let $\widehat{X}$ denote the corresponding finite cover. Let $D_1, \ldots, D_r$ denote the components of the preimage of D, and consider the *dual graph* Γ whose vertices correspond to components of $\widehat{X} - \cup_i D_i$ and whose edges correspond to components D_i. Note that Γ is connected since $\widehat{X}$ is connected. By construction, each vertex has valence ≥ 2. And by regularity, $\pi_1 X$ acts transitively on the edges of Γ. If some edge separated, then each edge would separate, and so the finite regular tree would be an n-star for some n. Consequently, no D_i separates. $\square$

Definition 16.2. Let X be a connected nonpositively curved cube complex, and let D be a hyperplane of X that is embedded and 2-sided, so $N^o(D) \cong D \times (-\frac{1}{2}, +\frac{1}{2})$. Let $Y = X - N^o(D)$ and let $D^- \to Y$ and $D^+ \to Y$ be the local-isometries $D \times \{\pm\frac{1}{2}\} \to Y$ that are induced by $N(D) \to Y$. If D is separating then we let $Y = Y^- \sqcup Y^+$ where $D^{\pm}$ map to $Y^{\pm}$.

We say D is *co-large* if either:

(1) D is separating and $[\pi_1 Y^+ : \pi_1 D^+] \geq 3$ and $[\pi_1 Y^- : \pi_1 D^-] \geq 2$ or vice versa.
(2) D is not separating and at least one of $D^+ \to Y$ or $D^- \to Y$ is not π_1-surjective.

Lemma 16.3. *Let D be a co-large hyperplane of a special cube complex X. There is a finite cover $\widehat{X} \to X$ such that the dual graph to the set of hyperplanes in the preimage of D has negative euler characteristic. Consequently, $\pi_1 \widehat{X}$ surjects onto a rank 2 free group.*

Proof. This is similar to the proof of Lemma 16.1. When D is non-separating and say $\alpha \in \pi_1 Y^+ - \pi_1 D^+$, then we choose a finite regular cover $\widehat{X}$ such that $\bar{\alpha} \notin \overline{\pi_1 D}$. Let $D_1, \ldots, D_r$ denote the hyperplanes of $\widehat{X}$ mapping to D. Then the associated dual graph has each vertex of valence ≥ 3, since each component of $\widehat{X} - \cup_i D_i$ has at least one edge for a D^- type side, and at least two edges corresponding to D^+ type sides that are connected by a lift of α.

In the case where D is separating, we let $\alpha_0, \alpha_1, \alpha_2$ be representatives of distinct nontrivial cosets of $\pi_1 D^+$ in $\pi_1 Y^+$, and we let β_0, β_1 be representatives of distinct nontrivial cosets of $\pi_1 D^-$ in $\pi_1 Y^-$. We choose a finite quotient of $\pi_1 X$ maintaining the distinctness of the three cosets and the two cosets, and we let $\widehat{X}$ be the corresponding finite cover. In the dual graph, each vertex corresponding to a component of the preimage of Y^+ has valence at least three, and each vertex corresponding to a component of the preimage of Y^- has valence at least two.

Thus the dual graph has negative euler characteristic in both the separating and non-separating cases, and the quotient of $\widehat{X}$ to the dual graph yields a non-cyclic free quotient. $\qquad\square$

The punchline of this section is the following strengthening of the Tits alternative for groups acting properly on finite dimensional CAT(0) cube complexes [SW05]. Our argument follows a plan similar to that used in [SW05] but avoids the algebraic torus theorem.

A group is *large* if it has a finite index subgroup with a rank 2 free quotient.

Theorem 16.4. *Let X be a finite dimensional virtually special cube complex. Either $\pi_1 X$ is virtually abelian or X has a finite cover with a co-large hyperplane. Consequently, either $\pi_1 X$ is virtually abelian or $\pi_1 X$ is large.*

Proof. By possibly replacing X with a finite special cover, we assume without loss of generality that X is special.

Observe that the second claim follows from the first claim, since a co-large hyperplane D implies the largeness of $\pi_1 X$ by Lemma 16.3 and the separability of $\pi_1 D$ which holds by Corollary 6.7.

We prove the first claim by induction on $\dim(X)$. Suppose that no hyperplane of X is co-large and that $\pi_1 X$ is nontrivial. Let g be a nontrivial element of $\pi_1 X$, and let $\tilde{\gamma}$ be a geodesic in $\widetilde{X}$ stabilized by g. Let D be a hyperplane of X whose universal cover lifts to a hyperplane $\widetilde{D}$ intersecting $\tilde{\gamma}$ at a point. It follows that $\pi_1 X$ has a nontrivial splitting along $\pi_1 D$.

By induction either $\pi_1 D$ is virtually abelian or D has a co-large hyperplane. However any co-large hyperplane E of D would extend to a co-large hyperplane Y of X, since if E is non-separating then Y is non-separating, and the index of $\pi_1 Y$ on a side is bounded below by the index of $\pi_1 E$ on that side. Hence, since we assumed that no co-large hyperplane Y exists, we may proceed under the assumption that $\pi_1 D$ is virtually abelian.

Either $\pi_1 D$ has index 2 on both sides in the separating case, or $\pi_1 D$ has index 1 on each side in the non-separating case. These two situations lead to short exact sequences of the form: $1 \to \pi_1 D \to \pi_1 X \to \mathbb{Z}$ or $1 \to \pi_1 D \to \pi_1 X \to \mathbb{Z}_2 * \mathbb{Z}_2 \to 1$. In each case we see that $\pi_1 X$ is virtually solvable. Finally, a virtually solvable group acting properly-discontinuously and semi-simply on a CAT(0) space must be virtually abelian [BH99]. $\qquad\qquad\qquad\square$

Perhaps Theorem 16.4 can be strengthened to: If X is special and finite dimensional then either $\pi_1 X$ is abelian or $\pi_1 X$ has a noncyclic free quotient. In particular, I suspect that $G \cong \mathbb{Z}^n$ whenever G is both special and virtually abelian. Note that the Klein bottle group is virtually special and virtually abelian, but not special as it is not residually torsion-free nilpotent.

Consideration of the proof of Theorem 16.4 suggests the following might hold.

Conjecture 16.5. *Let X be a finite dimensional nonpositively curved cube complex. Suppose that $\pi_1 D$ is virtually abelian for each hyperplane D of X. Then $\pi_1 X$ is either A or A-by-S where A is virtually abelian and S is virtually a free or surface group.*

16.b   Omnipotence

The goal of this section is to discuss some material related to omnipotence:

An ordered set of elements $\{g_1, \ldots, g_r\}$ is *independent* in G if each g_i has infinite order, and the subgroups $\langle g_i \rangle$, $\langle g_j \rangle$ do not have conjugates with nontrivial intersection for $i \neq j$. A group G is *omnipotent* if for each independent set of elements $\{g_1, \ldots, g_r\}$ there is a number $K \geq 1$ such that for each set of positive natural numbers $\{n_1, \ldots, n_r\}$ there is a quotient $G \to \bar{G}$ where $\bar{g}_i$ has order $n_i K$ for each i.

In [Wis00] we proved that free groups are omnipotent and we now generalize this to:

Theorem 16.6. *Let G be a virtually compact special word-hyperbolic group. Then G is omnipotent.*

Proof. For each i, let H_i be a maximal virtually cyclic subgroup containing g_i. Thus $\{H_1, \ldots, H_r\}$ is an almost malnormal collection of subgroups of G. By Theorem 12.2, there exist finite index normal subgroups $\ddot{H}_i \subset H_i$, such that for any finite index subgroups $H_i' \subset \ddot{H}_i$, the quotient $\bar{G} = G/\langle\!\langle\{H_i'\}\rangle\!\rangle$ is virtually compact special. There exists K such that for each i, we have $g_i^K \in \ddot{H}_i$. For instance we could let $K = \mathrm{lcm}\{[H_i : \ddot{H}_i]\}$. Let $H_i' = \langle g_i^{n_i K} \rangle$.

Let $\bar{H}_i$ denote the image of H_i in $\bar{G}$, and note that $\bar{H}_i = H_i/H_i'$—which holds by an application of Theorem 3.68 as arose in the proof of Theorem 12.2. Let $\bar{g}_i$

be the image of g_i in $\bar{G}$, and note that $\bar{g}_i$ has order $n_i K$ by choice of H_i'. As $\bar{G}$ is residually finite, we let $\bar{G} \to Q$ be a finite quotient such that $\langle \bar{g}_i \rangle$ embeds in Q for each i. Then the quotient $G \to \bar{G} \to Q$ provides the finite quotient where the image of each g_i has the desired order. $\qquad\qquad\qquad\qquad\qquad\qquad$ $\square$

Remark 16.7. When G is hyperbolic relative to abelian subgroups, the analogous result holds with maximal abelian subgroups replacing cyclic subgroups. Indeed, this is an application of Theorem 15.6.

The proof of Theorem 16.6 uses hyperbolicity of G to ensure that cyclic subgroups have associated compact cores. We can avoid applying Theorem 12.2 in the proof of Theorem 16.6 if we can obtain trivial wall projections of elevations of these cores, and then apply Lemma 16.9. This gives an elementary proof that is a more direct generalization of the proof in [Wis00].

Conjecture 16.8. *Let G be a virtually compact special word-hyperbolic group. Let $g_1, \ldots, g_r$ be an independent set of elements. There exists $N_1, \ldots, N_r$ such that for any $n_i \geq N_i$ the group $\bar{G} = G / \langle\!\langle g_1^{n_1}, \ldots, g_r^{n_r} \rangle\!\rangle$ satisfies:*

(1) *$\bar{G}$ is virtually torsion-free.*
(2) *$\bar{G}$ is residually finite.*
(3) *$\bar{G}$ virtually splits as a graph of groups.*
(4) *$\bar{G}$ is virtually special.*

We note that $\bar{G}$ acts properly on a CAT(0) cube complex for large N_i (see Section 5.d). This already follows from [Wis04] when G is free, which is the most important test case for Conjecture 16.8. The euler characteristic calculation for a finite index torsion-free subgroup shows that for large n_i, the virtual torsion-freeness yields virtually large first betti number and hence some splittings. One hopes that this might be enough to get a hierarchy, which would be a quasiconvex hierarchy by [MW08].

Lemma 16.9 (Surviving together in free-abelianization). *Let G be virtually compact special and word-hyperbolic. Let $\sigma_1, \ldots, \sigma_k$ be elements of infinite order in G. There exists a finite index characteristic subgroup G' such that letting $n = [G : G']$, the elements $\sigma_1^n, \ldots, \sigma_k^n$ (and hence their G-conjugates) have nontrivial image in $G' \to \mathsf{H}^1(G')$.*

Proof. Let G act properly and cocompactly on the CAT(0) cube $\widetilde{X}$, and let J be a finite index characteristic subgroup such that $X = J\backslash\widetilde{X}$ is special and compact. See Lemma 7.14.

By Proposition 2.31, for each i, let $\widetilde{Y}_i$ be a $\langle \sigma_i \rangle$-cocompact convex subcomplex of $\widetilde{X}$, and let $Y_i = (J \cap \langle \sigma_i \rangle)\backslash\widetilde{Y}_i$ so there is a local-isometry $Y_i \to X$. For each i, let $\widehat{X}_i = \mathsf{C}(Y_i \to X)$, so there is a retraction $\widehat{X}_i \to Y_i$ by Proposition 6.3. Let $\widehat{X}$ denote a finite cover factoring through each $\widehat{X}_i$ and with $G' = \pi_1 \widehat{X}$ characteristic in G, and let $n = [G : G']$. Then for each i, the composition $\widehat{X} \to \widehat{X}_i \to Y_i$ shows that σ_i^n is nontrivial in $\mathsf{H}^1(\widehat{X})$.   □

Lemma 16.9 generalizes as follows:

Proposition 16.10. *Let G be f.g. and virtually special. Let $A_1, \ldots, A_k$ be virtually abelian subgroups of G. There exists a finite index (characteristic) subgroup $G' \subset G$, such that for each $1 \le i \le k$ and $g \in G$, the subgroup $G' \cap A_i^g$ maps isomorphically to its image under the homomorphism $G' \to \mathsf{H}^1(G')$.*

Proof. Let X be a special cube complex with $\pi_1 X$ isomorphic to a finite index subgroup of G. Let $X \to C(\Gamma)$ be the local-isometry of Lemma 6.2. Let $R = C(\Gamma)$, and note that we can choose X so that R is compact since $\pi_1 X$ is f.g. By Lemma 2.38, for every maximal abelian subgroup M of $\pi_1 R$, there exists a local-isometry $Y \to R$ with Y compact and $\pi_1 Y$ mapping to M. For each i, let B_i be a free-abelian finite index subgroup of $A_i \cap \pi_1 X$. Let M_i be a maximal abelian subgroup of $\pi_1 R$ containing the image of B_i. Let $Y_i \to R$ be a local-isometry with Y_i compact and $\pi_1 Y_i$ mapping to M_i. Applying Proposition 6.3 for each i, let $R_i = \mathsf{C}(Y_i \to R)$ and note that the retraction $\mathsf{C}(Y_i \to R) \to Y_i$ ensures that M_i injects in $\mathsf{H}^1(\mathsf{C}(Y_i \to R))$. Let X_i be the finite cover of X induced by $\mathsf{C}(Y_i \to R) \to R$, and note that it follows that B_i injects in $\mathsf{H}^1(X_i)$. Let $G_i \subset G$ be the finite index subgroup corresponding to $\pi_1 X_i$. Let G' be a finite index characteristic subgroup of G containing each G_i. Note that $B_i \cap G_i$ injects in $\mathsf{H}^1(G_i) = \mathsf{H}^1(X_i)$. Hence $A_i \cap G'$ injects in $\mathsf{H}^1(G')$ for each i, so $A_i \cap [G', G'] = \{1_G\}$. The same holds for each $A_i^g \cap G'$ as G' is characteristic. Indeed, $(A_i^g \cap G') \cap [G', G'] = A_i^g \cap [G', G'] = (A_i \cap [G', G'])^g = (\{1_G\})^g = \{1_G\}$.   □

Lemma 16.11. *Let G be hyperbolic and virtually compact special. Let $g_1, \ldots, g_k$ be a finite set of infinite order elements of G. Suppose G has a quasiconvex codimension-1 subgroup H such that no g_i^n is conjugate into H for any $n > 0$. Then G has a finite index characteristic subgroup G' such that there is a rank 2 free quotient $G' \to F$, such that letting $n = [G : G']$ each G-conjugate of each g_i^n has nontrivial image in F.*

Proof. By Lemma 7.14, we can assume that G acts properly and cocompactly on a CAT(0) cube complex $\widetilde{X}$ and that the stabilizer of one of the hyperplanes $\widetilde{U}$ of $\widetilde{X}$ is H, and that $\widetilde{X} - \widetilde{U}$ consists of two components that do not lie in any finite neighborhood of $\widetilde{U}$.

Let J be a finite index characteristic subgroup of G that acts freely and cocompactly on $\widetilde{X}$. By separability of H, we may moreover assume that $X = J \backslash \widetilde{X}$ has the additional property that $U = (H \cap J) \backslash \widetilde{U}$ is embedded and 2-sided, and that $(H \cap J)$ is malnormal in J. The last claim holds by Proposition 15.17, which is essentially a consequence of separability and Proposition 8.3 in this case. Let $\{A_i \to X : 1 \le i \le r\}$ be local-isometries representing the distinct cyclic subgroups arising as intersections $\langle gg_ig^{-1} \rangle \cap \pi_1 X$ as $g \in G$ and i vary.

Apply Theorem 12.2 to $(\pi_1 X, \pi_1 U)$ to obtain a quotient with cubical presentation $\langle X \mid \widehat{U} \rangle$. Moreover, we choose the finite regular cover $\widehat{U}$ large enough so that each $A_i^* \to X^*$ has no missing shells and is hence π_1-injective by Theorem 3.68.

Note that $\pi_1 X^*$ splits as an amalgamated free product or HNN extension along $\pi_1 U / \langle\!\langle \pi_1 \widehat{U} \rangle\!\rangle$ with virtually compact special hyperbolic vertex groups. By residual finiteness, we pass to a torsion-free finite index subgroup of $\pi_1 X^*$ whose induced splitting is along a graph of groups with trivial edge groups and with vertex groups that are virtually compact special and hyperbolic. Let $\sigma_1, \ldots, \sigma_m$ be closed paths that represent generators of the various intersections of conjugates of $\pi_1 A_i^*$ with this finite index subgroup. By possibly passing to a further cover using residual finiteness, we can assume without loss of generality that each σ_i is either elliptic or projects to an immersed circle in the underlying graph of the splitting.

By Lemma 16.9 applied to each vertex group, we can pass to a further finite cover such that all vertex groups have the property that elliptic σ_i survive together in a cyclic quotient. In conclusion, we have passed to a finite index subgroup, such that there is a homomorphism from our graph of virtually compact special groups to a corresponding graph with $\mathbb{Z}$ vertex groups and trivial edge groups, and such that each hyperbolic σ_i projects to a circle and hence embeds in the free group, and each elliptic σ_i projects to a nontrivial element of a vertex group and hence embeds. □

Chapter Seventeen

Hyperbolic 3-Manifolds with a Geometrically Finite Incompressible Surface

Let M be a finite volume hyperbolic 3-manifold with an incompressible geometrically finite 2-sided surface S. There exists a (topological) hierarchy for M that begins by cutting the 3-manifold along S and proceeds by cutting along further incompressible surfaces until only balls remain. It is known that geometrical finiteness is equivalent to quasiconvexity (even when there are cusps) [Hru10]. All further cuts in the hierarchy correspond algebraically to splittings along quasiconvex subgroups, as it is a theorem of Thurston's that f.g. subgroups of a geometrically finite infinite volume group are themselves geometrically finite [Can94]. In the hyperbolic case, the 3-manifold hierarchy induces a quasiconvex hierarchy of $\pi_1 M$, and we thus have:

Theorem 17.1. *Let M be a closed hyperbolic 3-manifold with an incompressible surface S such that $\pi_1 S \subset \pi_1 M$ is geometrically finite. Then $\pi_1 M$ is the fundamental group of a compact special cube complex.*

We note that the earliest versions of this text contained proofs that $\pi_1 M$ is (strongly) cosparse and virtually compact special. As I revised and finalized the manuscript and carefully traced the cubulations of parabolic subgroups in the proof of Theorem 15.1, I found that with a bit of care, the argument actually shows that $\pi_1 M$ equals the fundamental group of a compact cube complex that is virtually special. For practical purposes and applications, there is very little difference between compact virtual specialness, and virtual compact specialness or even cosparseness.

A group G is *residually finite rational solvable* (RFRS) if there is a decreasing sequence of finite index normal subgroups $G = G_0 > G_1 > G_2 > \cdots$ such that $\cap G_i = \{1_G\}$ and such that for each i we have $G_{i+1} > K_i$ where G_i/K_i is torsion-free abelian. To absorb the definition, imagine repeatedly passing to a finite index subgroup by pulling back a large index subgroup of the free-abelianization at each stage.

The following two results are proven in [Ago08]:

Lemma 17.2. *Every right-angled Artin group is (virtually) RFRS.*

Theorem 17.3. *Let M be a compact irreducible 3-manifold with $\chi(M) = 0$. If $\pi_1 M$ is virtually RFRS then M has a finite cover that is a surface bundle over a circle.*

Corollary 17.4. *Every closed hyperbolic 3-manifold M with a geometrically finite incompressible surface satisfies:*

(1) $\pi_1 M$ *is virtually a subgroup of a right-angled Artin group (right-angled Coxeter group).*
(2) $\pi_1 M$ *is virtually fibered.*
(3) $\pi_1 M$ *is subgroup separable.*

Proof. Since $\pi_1 M$ has a quasiconvex hierarchy, we see that $\pi_1 M$ is virtually special by Theorem 13.3.

Since $\pi_1 M$ has a finite index subgroup that lies in a raag it is RFRS by Lemma 17.2. Agol's virtual fibering criterion [Ago08] applies to the finite cover $\widehat{M}$ with $\pi_1 \widehat{M}$ special.

A consequence of the tameness theorem [Ago04, CG06] is that subgroups of $\pi_1 M$ are either geometrically finite or virtual fibers. The latter are easily seen to be separable using an index 2 normal subgroup with $\mathbb{Z}$ quotient. We proved in [HW08] that quasiconvex subgroups of a compact word-hyperbolic special cube complex are separable, as they are virtual retracts. But geometrically finite subgroups are precisely the same as quasiconvex subgroups [Hru10]. $\square$

We will handle the case of a cusped hyperbolic 3-manifold in Section 17.c. Repeating the theme in Problem 15.4 in a special case:

Problem 17.5. Let M be a hyperbolic 3-manifold with an incompressible geometrically finite subgroup. Does M (virtually) have a hierarchy where the tori don't get cut until the last steps?

Problem 17.6. Show that every fibered hyperbolic 3-manifold has a finite cover with an incompressible geometrically finite surface.

17.a   Some Background on 3-Manifolds

For simplicity, we assume all manifolds are orientable. An *incompressible surface* $S \subset M$ is a properly embedded 2-sided surface whose components are π_1-injective, and such that $\pi_1 M$ has a nontrivial splitting along $\pi_1 S$. A 3-manifold M is *Haken* if it is aspherical and contains an incompressible surface. A *hierarchy* for M is a sequence $M_0, M_1, \ldots, M_t$ where $M_0 = M$ and $M_t = \sqcup B_i$ is a union of 3-balls, and for each i, the 3-manifold M_{i+1} is obtained from M_i by cutting along an incompressible surface S_i in the sense that each component of

M_{i+1} is the closure of a component of $M_i - S_i$. We refer to [Hem04, Thm 13.3] for a proof of Haken's result that:

Proposition 17.7. *Every compact Haken 3-manifold M has a hierarchy.*

Of course, a hierarchy for M induces a hierarchy for $\pi_1 M$. When M is hyperbolic, the hierarchy for M induces a quasiconvex hierarchy of $\pi_1 M$ provided that the first cut is along an incompressible geometrically finite surface S. Indeed, all further splittings in the hierarchy of M starting with S must be along geometrically finite surface subgroups because of the following result of Thurston's [Can94]:

Proposition 17.8. *Let M be a geometrically finite infinite volume hyperbolic 3-manifold. Let H be a f.g. subgroup of $\pi_1 M$. Then H is geometrically finite.*

Our source of geometrical finiteness for the first cut is the following result of [CS84, Thm 1+2]:

Theorem 17.9 (Culler-Shalen). *Let M be a compact, connected, orientable 3-manifold. Suppose a component of ∂M is not simply-connected. Suppose $M \not\cong D^2 \times S^1$ and $M \not\cong S^1 \times S^1 \times I$. Suppose $\mathsf{H}_1(\partial M, \mathbb{Q}) \to \mathsf{H}_1(M, \mathbb{Q})$ is surjective. Then M contains a separating incompressible surface S that is not boundary parallel and that has nonempty boundary. Suppose moreover that $\partial M = T_1 \sqcup \cdots \sqcup T_k$ is the union of one or more tori. Then we can assume that ∂S has a circle on T_1.*

Corollary 17.10. *Let M be a hyperbolic 3-manifold with $\partial M \neq \emptyset$. Then there is an incompressible surface $S \subset M$ with $\pi_1 S \subset \pi_1 M$ geometrically finite.*

Proof. First suppose $\mathsf{H}_1(\partial M, \mathbb{Q}) \to \mathsf{H}_1(M, \mathbb{Q})$ is surjective. Let S be the surface provided by Theorem 17.9. Note that S cannot be a virtual fiber, since it is separating. Hence each component of S is geometrically finite, as first proved in [Bon86].

Now suppose $\mathsf{H}_1(\partial M, \mathbb{Q}) \to \mathsf{H}_1(M, \mathbb{Q})$ is not surjective. Choose a map $\phi: M \to S^1$ such that each component of ∂M is null-homotopic. Homotope ϕ to be transversal to a point p of S^1, and consider the surface $P = \phi^{-1}(p)$. After performing compressions along disks, we obtain a non-separating incompressible surface S. A component of S is geometrically finite, since $\partial S = \emptyset$. $\square$

Let M be a 3-manifold such that ∂M contains several tori $\{T_i\}$. A *Dehn filling* of M is a new 3-manifold

$$\bar{M} = \left(M \sqcup \bigsqcup_i K_i \right) / (\partial K_i = T_i)$$

that is obtained by gluing a solid torus $K_i \cong B^2 \times S^1$ to M by identifying ∂K_i with T_i for each i. The filling is *large* if the curve ∂B^2 corresponds to an element of $\pi_1 T_i$ that is long with respect to a prescribed metric, as measured by the distance between the endpoints of a lift to $\widetilde{T}_i$. This serves to exclude a certain finite list of problematic fillings of each torus, and ensures that hyperbolicity of the resulting manifold is preserved by Thurston's hyperbolic Dehn surgery theorem.

17.b   Aparabolic Hierarchy

Definition 17.11. Let S be an incompressible surface in a hyperbolic 3-manifold M. An *accidental parabolic* σ in S is a closed essential path in S that is homotopic to a path in a torus of ∂M but not homotopic to a path in ∂S. Cf. Definition 15.2.

Lemma 17.12 (No accidental parabolics). *Let $S \subset M$ be an incompressible surface. Then there is an incompressible surface S' with $\partial S \subset \partial S'$ such that S' has no accidental parabolics. Moreover, if M is hyperbolic and S is geometrically finite then S' is geometrically finite. Moreover, each boundary component that is separated by S is separated by S'.*

Proof. Suppose S contains a circle C that is not null-homotopic and not homotopic to ∂S, and suppose C is homotopic to a circle C' in a torus $T \subset \partial M$. By the Annulus Theorem [JS79, Sec 3], there is a properly embedded thickened annulus $A \times [0,1] \subset M$ such that $A \times \{0\}$ is a regular neighborhood of C and $A \times \{1\}$ is a regular neighborhood of C'. Let $S' = (S - A \times \{0\}) \cup (\partial A \times [0,1])$. Note that $\chi(S') = \chi(S)$ but S' is smaller, in the sense that the sum of the genus of its components has decreased. Consequently, this procedure can only be implemented finitely many times. Finally, note that S' is homotopic to the π_1-injective subsurface that is the closure of $S - A \times \{0\}$. Furthermore the splitting of $\pi_1 M$ along the new surface refines the splitting along the old surface in the sense that there is a $\pi_1 M$-equivariant map from the new Bass-Serre tree T' to the old tree T (corresponding to folding together pairs of edges associated to the cut surface). Thus, if $\pi_1 M$ is hyperbolic, and each component of S is geometrically finite, then the same holds for S'. And likewise, the fundamental group of a boundary component that is split by the action on T is also split by the action on T'. We $\pi_1 M$-invariantly direct the edges of T, and pull this back to T'. Separation is equivalent to the property that each element has zero total translation length where directed edges are counted with multiplicity. This property transfers from T to T'. $\qquad\square$

Corollary 17.13. *Let M be a compact hyperbolic 3-manifold with nonempty boundary. Then M has a hierarchy $M_0, M_1, \ldots, M_t$ such that for each i, the cutting* *surface* S_{i+1} *of* M_i *is geometrically finite and contains no accidental parabolic in* M_i.

Moreover, if ∂M *contains a surface that is not a torus, then we may assume that the first incompressible surface* S_1 *cutting* M *has a circle on each torus of* ∂M, *so each component of* $M - S_1$ *is atoroidal.*

Proof. The case where some boundary component is not a torus follows from Lemma 17.19. Indeed, the surface S it provides has a boundary circle on each torus. Moreover, S and all subsequent incompressible surfaces of the hierarchy of Proposition 17.7 starting with S are geometrically finite by Proposition 17.8. More specifically, each component of $M - S$ is atoroidal, since the tori of M are cut by ∂S.

We now focus on the case where ∂M is the union of one or more tori. By Corollary 17.10, M has a geometrically finite incompressible surface S. By Lemma 17.12, there is a geometrically finite incompressible surface S' with no accidental parabolics such that $\partial S \subset \partial S'$. The remainder of the hierarchy after cutting along S' is handled by the initial part of the proof since after cutting along S', some component of the boundary is not a torus. □

17.c   Virtual Specialness of Hyperbolic 3-Manifolds with Boundary

Theorem 17.14. *Let* M *be a finite volume cusped hyperbolic 3-manifold. Then* $\pi_1 M$ *is the fundamental group of a virtually special compact cube complex.*

More generally, let M *be an arbitrary compact atoroidal 3-manifold with* $\partial M \neq \emptyset$. *Then* $\pi_1 M$ *is the fundamental group of a virtually special compact cube complex.*

Proof. We focus first on the finite volume case. By Corollary 17.13, $\pi_1 M$ has a hierarchy where at each stage, the edge groups are relatively quasiconvex and have no accidental parabolics. Thus $\pi_1 M$ is the fundamental group of a virtually special compact cube complex by induction on the length of the hierarchy, using Theorem 15.1 at each stage.

We now prove the second statement in the infinite volume case. Corollary 17.13 gives a splitting of $\pi_1 M$ as a graph of word-hyperbolic 3-manifold groups with geometrically finite edge groups and no accidental parabolics. Each of the vertex groups has a quasiconvex hierarchy by Propositions 17.7 and 17.8, and so Corollary 15.3 applies. □

The infinite volume case of the second statement of Theorem 17.14 can be deduced from the first statement of Theorem 17.14. In each case, it is useful to note that the fundamental group $\pi_1 M_c$ of a compact atoroidal 3-manifold M_c embeds as a geometrically finite subgroup of the fundamental group $\pi_1 M_f$ of a finite volume hyperbolic 3-manifold. The first option is to note that the conclusion of Lemma 17.20 holds for M_c as a consequence of it holding for M_f. We then proceed with the original proof. The second option is to first choose an alternate representation so that $\pi_1 M_c$ has no accidental parabolics. We then use the above embedding $\pi_1 M_c \subset \pi_1 M_f$ as a geometrically finite subgroup. Proposition 2.31 provides a nonempty $\pi_1 M_c$-cocompact convex subcomplex $\widetilde{Y} \subset \widetilde{X}$ of the universal cover of X. This is where we use that $\pi_1 M_c$ has no accidental parabolics, which ensures that $\pi_1 M_c \subset \pi_1 M_f$ is full. Finally, the quotient $Y = \pi_1 M_c \backslash \widetilde{Y}$ is special since the map $Y \to X$ is a local-isometry and X is special by Proposition 6.2.

Problem 17.15. Let M be a hyperbolic 3-manifold. Does $\pi_1 M$ have a finite index subgroup that is the fundamental group of a 3-dimensional nonpositively curved cube complex?

Problem 17.16. Let M be a hyperbolic 3-manifold with nonempty boundary. Does $\pi_1 M$ have a finite index subgroup that is π_1 of a 2-dimensional nonpositively curved cube complex?

In general, the 3-manifold itself may not be homeomorphic to a nonpositively curved cube complex. Indeed, Li showed that if M is orientable and irreducible with ∂M a torus, and M contains no closed nonperipheral embedded incompressible surfaces, then only finitely many Dehn fillings of M yield 3-manifolds homeomorphic to nonpositively curved cube complexes [Li02]. However, for some valid 3-dimensional cases see the work of [AR90]. A motivating 2-dimensional case is Weinbaum's observation that the Dehn complex of a prime alternating link is nonpositively curved (see [Wei71] and [Wis06]).

Although we have not insisted on using only surfaces to cubulate, this can be concluded as well using the following:

Lemma 17.17. *Let H be a f.g. (codimension-1) subgroup of a 3-manifold group $G = \pi_1 M$. There exists a collection of immersed incompressible surfaces $S_1, \ldots, S_r$ in M with each $\pi_1 S_i \subset H$, such that any group element cut by an H-wall is also cut by a translate of the $\pi_1 S_i$-wall associated to $\widetilde{S}_i$ for some i.*

We count essential embedded spheres as incompressible surfaces here. Note that when M is hyperbolic and H is geometrically finite, then each $\pi_1 S_i$ is geometrically finite as well.

Proof. Let N be a compact core of the cover $\widehat{M}$ corresponding to H, and let $S'_1, \ldots, S'_{r'}$ be the distinct surfaces in the closure of $\partial N - \partial \widehat{M}$. Any H-wall is coarsely the same as an H-wall obtained by partitioning the components of $\widehat{M} - N$. An element g with geodesic axis $\widetilde{A}$ in $\widehat{M}$ that is cut by the H-wall has the property that the immersed line $\widetilde{A} \to \widehat{M}$ is cut by one of the surfaces S'_i in the sense that one end of $\widetilde{A}$ lives on one end of $\widehat{M} - S'_i$ and the other in one of the remaining ends. Consequently, g is also cut by the $\pi_1 S_i'$-wall associated to $\widetilde{S}_i'$. Let $S_1, \ldots, S_r$ denote the incompressible surfaces obtained from $S_1', \ldots, S_{r'}'$ by performing a sequence of disc compressions. Then $S_1, \ldots, S_r$ have the desired property. Indeed, if one or two surfaces are obtained from a third surface using a disc compression then every geodesic cut by the parent is cut by a child. $\square$

Combining Theorems 17.1 and 17.14 with Lemma 17.17, we obtain:

Corollary 17.18. *Let M be a compact hyperbolic 3-manifold containing a geometrically finite incompressible surface, e.g., when $\partial M \neq \emptyset$. There exists a collection of geometrically finite immersed incompressible surfaces $S_1, \ldots, S_r$ in M such that $\pi_1 M$ acts freely and cocompactly on the associated dual CAT(0) cube complex.*

Following the reasoning that supports Corollary 17.18, I expect that if we had started with a [strongly] cosparse action of $\pi_1 M$ on a CAT(0) cube complex, then we also obtain a [strongly] cosparse action of $\pi_1 M$ dual to the surfaces alongside the corresponding codimension-1 subgroups. The key additional consideration is to verify the behavior of the cubulated parabolics, but the essential and inessential hyperplanes of the parabolics arise from surfaces alongside codimension-1 subgroups that intersect the parabolics.

17.d   Cutting All Tori with First Surface

The goal of this section, achieved in Lemma 17.20, is to produce a finite cover $\widehat{M}$ and a splitting of $\pi_1 \widehat{M}$ as a graph of groups where the edge groups are quasiconvex and the vertex groups are word-hyperbolic. This is compatible with Corollary 15.3. In my initial planning, this provided a shortcut towards cubulation that sidesteps malnormality of edge groups by substituting aparabolicity on one side. In retrospect, the quasiconvex hierarchy of Corollary 17.13 jives with Theorem 15.1, and that is used in the proof of Lemma 17.20. However, with a bit more care, one can arrange the proof of Lemma 17.20 to depend only on the virtual specialness in the special case where some component of ∂M is not a torus.

Lemma 17.19. *Let M be a compact 3-manifold with ∂M containing tori $T_1, \ldots, T_k$ and $k \geq 1$. There exists an incompressible surface $S \subset M$ such that each T_i contains a circle τ_i of ∂S.*

Proof. The map $\mathsf{H}_1(T_i) \to \mathsf{H}_1(M)$ has infinite image for each i. Indeed, this follows from an easy homology computation (e.g., fill all other boundary components and then apply [Hat07, Lem 3.5]). Note that for any finite collection of nontrivial elements in a free-abelian group, there is an infinite cyclic quotient in which they all survive. Consequently, there is a map $\phi \colon M \to S^1$ such that

$T_i \to S^1$ is essential for each i. It is then a standard argument (see, e.g., [Hem04]) that we can homotope ϕ so that it is transversal to a point of S^1, and let S' be the preimage surface, and then compress to obtain an incompressible surface S such that $\partial S = \partial S'$ and hence each $\partial S \cap T_i$ consists of one or more essential circles as claimed. $\qquad \square$

It will be useful to also know that (each component of) the incompressible surface S produced is geometrically finite. However, the surface S of Lemma 17.19 is sufficient if the goal is merely to show that M virtually fibers. Indeed, we stop here if S is a virtual fiber, and otherwise S is geometrically finite in which case by Lemma 17.11 we may assume S has no accidental parabolics. Then we may apply Corollary 15.3 to the quasiconvex hierarchy obtained after cutting along S from Proposition 17.7 and 17.8. Hence $\pi_1 M$ is virtually special and so M virtually fibers by Theorem 17.3.

We shall now provide a geometrically finite strengthening of Lemma 17.19. I expect this could be done more directly with constructions involving surfaces in 3-manifolds, and perhaps without taking a finite cover, but it is interesting that we can proceed with Proposition 17.7, Proposition 17.8 and Theorem 17.9 together with some special cube complex gymnastics: Note that the proof uses only that there is a word-hyperbolic quasiconvex codimension-1 subgroup in a virtually special group that is hyperbolic relative to virtually abelian subgroups. Only at the last step is a hyperbolic 3-manifold property recalled, to ensure a geometrically finite surface dual to a rank 2 free quotient.

Lemma 17.20. *Let M be a compact hyperbolic 3-manifold with nonempty boundary. There exists a finite cover $\widehat{M} \to M$ and an incompressible surface $S \subset \widehat{M}$ whose components are geometrically finite, such that each torus $\widehat{T}_i$ of $\widehat{M}$ contains a circle $\tau_i \subset \partial S$.*

Proof. Suppose first that $\mathrm{vol}(M) = \infty$. Each boundary torus has nontrivial image in $\mathsf{H}^1(M)$, and hence we can choose a homomorphism to $\mathbb{Z}$ where they all map nontrivially. Choose a map $M \to S^1$ inducing this homomorphism, and let N' be the preimage of a regular point. Repeatedly compressing it, leads to an incompressible surface with the same boundary (counted with multiplicity) which is geometrically finite by Proposition 17.8 since ∂M contains a surface of genus ≥ 2.

We now assume that $\mathrm{vol}(M) < \infty$, and ∂M is the union of tori $T_1, \ldots, T_r$ with $r \geq 1$. We will establish the following below:

Claim:

(1) There is a quotient $\pi_1 M \to G$ with G hyperbolic and virtually compact special.
(2) G has a quasiconvex codimension-1 subgroup H such that:
(3) for each component T_i of ∂M, the image $\pi_1 T_i \to G$ is virtually an infinite cyclic subgroup $\langle g_i \rangle$ that is not virtually conjugate into H.

We may thus apply Lemma 16.11 to (G, H) and the various $\{g_i\}$ to obtain a finite cover $\widehat{M} \to M$ and a surjection $\pi_1 \widehat{M} \to F_2$ to a free group, such that each torus of $\widehat{M}$ maps to an infinite cyclic subgroup. This homomorphism is induced by a map from $\widehat{M}$ to a bouquet of circles, and we can assume the map is transversal to the barycenters of edges. The preimage of the barycenters yields a 2-sided surface R' that intersects each boundary torus of $\widehat{M}$ in a nonzero number of circles, up to multiplicity. Moreover, keeping track of the normal vectors, we see that after a sequence of compressions, we obtain an incompressible geometrically finite surface R that likewise intersects each boundary torus of $\widehat{M}$, in the same number of circles (up to multiplicity). Finally, each component of R is geometrically finite since R is dual to a noncyclic free quotient, and hence Proposition 17.8 applies.

By Theorem 17.14, there is a finite cover of M whose fundamental group is isomorphic to that of a compact special cube complex. As we will take a further finite cover anyhow after verifying the claim, to contain notation, we assume without loss of generality that $\pi_1 M \cong \pi_1 X$ where X is a compact special cube complex. By the construction in the proof of Theorem 15.1, we may assume that a hyperplane U of X is associated to a geometrically finite surface S with no accidental parabolics (e.g., as provided by Theorem 17.9 after modification via Lemma 17.12).

For each i, let $P_i = \pi_1 T_i$. The quotient $\pi_1 M \to G$ will be of the form $\pi_1 M / \langle\!\langle \{P_i'\} \rangle\!\rangle$ where each $P_i' \subset P_i$ is a subgroup with P_i/P_i' virtually infinite cyclic. Let $g_i \in G$ generate the maximal cyclic subgroup in image$(P_i \to G)$ for each i. The subgroup H is the image of $\pi_1 S = \pi_1 U$. We now describe how to choose $\{P_i'\}$ to ensure the desired properties.

Let $T_1, \ldots, T_\ell$ denote the tori such that $P_1, \ldots, P_\ell$ have conjugates with nontrivial intersection with $J = \pi_1 U$. Note that it is possible that there are no such tori. More specifically, as J is relatively quasiconvex, it is hyperbolic relative to finitely many conjugates in J of infinite intersections with the parabolic subgroups above. A simple way to see this uses Bowditch's fine graphs (see, e.g., [MPW11]). Let K be the full quasiconvex subgroup obtained from J by amalgamating a sufficient finite index subgroup $\dot{P}_i^{g_i}$ of $P_i^{g_i}$ with J along $J \cap P_i^{g_i}$. Here one uses separability of $J \cap P_i^{g_i}$ in $P_i^{g_i}$ to facilitate the choice of $\dot{P}_i^{g_i}$ (see, e.g., [MP09, Thm 1.1]). Moreover, for sufficient choices of $\dot{P}_i^{g_i}$, we can assume that the only conjugates of the various P_i subgroups having nontrivial intersection with K are those already represented by the $P_i^{g_i}$ above. (It is possible for J to nontrivially intersect multiple conjugates of P_i, but we will suppress this to conserve notation. In the 3-manifold setting, these intersections correspond to the same subgroup of P_i.)

There are finitely many double cosets $\{P_i g_{ij} K\}$ corresponding to infinite intersections $P_i^{g_{ij}} \cap K$. These intersecting conjugates are precisely those that occur above in the production of K from J, as a conjugate of P_i intersects K infinitely if and only if it intersects J infinitely.

By Proposition 2.31, let $Y \to X$ be a local-isometry with Y compact and $\pi_1 Y$ mapping to K. For each i, let $\widetilde{E}_i \to X$ be a P_i-cocompact superconvex complex provided by Lemma 2.37, and let $E_i = \pi_1 T_i \backslash \widetilde{E}_i$. We will choose virtually cyclic index subgroups $\widehat{P}_i$ of P_i for $1 \leq i \leq r$ such that $\widehat{P}_i \subset \dot{P}_i$ for $1 \leq i \leq \ell$, and we let $\widehat{E}_i = \widehat{P}_i \backslash E_i$, and we let $X^* = \langle X \mid \widehat{E}_1, \ldots, \widehat{E}_r \rangle$ be the associated cubical presentation.

Theorem 15.6 provides finite covers P_i° so that when $\widehat{P}_i \subset P_i^\circ$ are virtually cyclic index subgroups contained in them, then X^* above is a $B(8)$ cubical presentation with $\pi_1 X^*$ word-hyperbolic and virtually the fundamental group of a compact special cube complex. As explained in the proof of Theorem 15.6, there is a uniform bound on the wall-pieces and cone-pieces of the $\widetilde{E}_i$, and this enables the other choices below.

Theorem 14.5 explains how to make the choices so that the image $\pi_1 Y^*$ of K in $\pi_1 X^*$ is quasiconvex. Note that the fullness of K and compactness of Y enable us to bound the diameters of intersections of translates $\widetilde{Y} \cap h\widetilde{E}_i$ unless $h\widetilde{E}_i \subset \widetilde{Y}$ in which case the inclusion factors through a map $\dot{E}_i \to Y$. Moreover, the map between induced presentations $Y^* \to X^*$ is π_1-injective provided that we have short innerpaths which is enabled when X^* has a sufficiently strong small-cancellation—e.g., $C'(\frac{1}{14})$ by Lemma 3.70. Let H be the image of J in $\pi_1 X^*$. Then $H \subset \pi_1 Y^* \subset \pi_1 X^*$ is quasiconvex since each of the above inclusions is quasiconvex, as $\pi_1 Y^*$ splits as a multiple HNN extension over H where all the edge groups are finite.

Lemma 12.10.(2) and (3), ensure that the conjugates of $\pi_1 E_i^*$ having infinite intersection with $\pi_1 Y^*$ arise precisely from conjugates of $\pi_1 E$ having infinite intersection with $\pi_1 Y$, in the sense that the conjugator arises from one in the list above, and the intersection of the images is the image of the intersection. To satisfy the hypotheses of Lemma 12.10, the systoles $\|\widehat{E}_i\|$ must be sufficiently large compared to the pieces, the diameters of overlaps $\widetilde{Y} \cap h\widetilde{E}_i$, and the lengths of the representatives of double cosets of intersecting conjugators listed above. Thus there are no intersections in $\pi_1 X$ between $\pi_1 Y^*$ and conjugates of $\pi_1 E_{\ell+1}^*, \ldots, \pi_1 E_r^*$. For $i \leq \ell$, infinite intersection of conjugates of images of $\pi_1 E_i^*$ with $\pi_1 Y^*$ arises from a conjugate that is virtually contained inside $\pi_1 Y^*$. But these cannot have infinite intersection with H, as they are virtually generated by elements that are hyperbolic in the splitting of $\pi_1 Y^*$ as a multiple HNN extension of the image of H with finite edge groups.

To see that H has codimension-1 in G, observe that H is the stabilizer of a wall in $\widetilde{X}^*$ associated to the hyperplane U. That wall is deep since the tori crossed by K are hyperbolic in the splitting of $\pi_1 Y^*$ above. The case where $\partial S = \emptyset$ arises from a situation where S and hence U is non-separating, and so U will be non-separating in the cubical part of X^*. $\qquad\square$

Remark 17.21 (Alternate 3-Manifold Route). Although the quotient $\pi_1 M \to G$ and its properties were primarily understood using cubical small-cancellation theory, it appears that it can be proven entirely through a 3-manifold argument. For sufficiently large Dehn fillings of $T_{r+1}, \ldots, T_k$, one obtains a hyperbolic manifold $\bar{M}$ with one cusp T_1. This is a consequence of Thurston's Dehn surgery theorem, which uses the Gromov-Thurston 2π-Theorem—see [BH96]. From the details of the proof, the core circle of each filling is homotopic to a closed geodesic, which is disjoint from a locally-convex subspace deformation retracting to N, and hence the core circle is not homotopic into N in the resulting negatively curved Riemannian manifold. At the next stage of the construction, one fills the tori $\{T_1, \ldots, T_r\}$ to obtain $\bar{\bar{N}}$, using disks that are degree n covers of the boundary circles of the $\{N \cap T_i\}$, for some sufficiently large n. The resulting hyperbolic orbifold $\bar{\bar{N}}$ has $\mathbb{Z}_n$ orbifold loci at the core circles of the filled tori. The incompressible surface N and the subsequent $\bar{N}$ is not a virtual fiber, since it is embedded and separating.

A key point in the proof of Lemma 17.20 was the production of a codimension-1 quasiconvex subgroup H such that certain subgroups do not have conjugates that nontrivially intersect H. We raise the possibility of generalizing this as follows:

Problem 17.22 (Missing wall). Let $\sigma_1, \ldots, \sigma_k$ be infinite order elements of a virtually compact special word-hyperbolic group G. Does there exist a quasiconvex codimension-1 subgroup $H \subset G$ such that no σ_i has a power conjugate into H?

Let $F \subset G$ be an infinite index quasiconvex subgroup (e.g., $\langle \sigma_1^n, \ldots, \sigma_k^n \rangle$ for some large n). Does there exist a quasiconvex codimension-1 subgroup H such that no nontrivial element of F is conjugate to an element of H?

The following variant of Problem 17.22 has an affirmative answer when there is only one cusp [MZ08]. Presumably the multiple cusp case can be deduced from the single cusp case, or can be reproven using the same methodology.

Problem 17.23. Let M be a cusped hyperbolic 3-manifold. Does M contain the fundamental group of a closed surface containing no nontrivial parabolic elements?

Chapter Eighteen

Limit Groups and Abelian Hierarchies

The goal of this section is to examine some simple quasiconvex hierarchies of specific interest. We apply the results of Chapter 15 to prove that limit groups are virtually compact special in Section 18.a. We prove that relatively hyperbolic groups with abelian hierarchies are virtually sparse special in Section 18.b.

18.a   Limit Groups

A group J is *fully residually free* if for each finite set of nontrivial elements $\{j_1, \ldots, j_k\}$ there is a free quotient $J \to \bar{J}$ such that $\bar{j}_i$ is nontrivial for each $i \in \{1, \ldots, k\}$. A *limit group* is a f.g. fully residually free group. It was shown in [KM98] that every limit group is a subgroup of a group G_r constructed as follows:

(1) G_0 is a trivial group.
(2) For each $i \geq 0$, we have $G_{i+1} \cong G_i *_{C_i} A_i$ where C_i is a malnormal abelian subgroup of G_i and $A_i = C_i \times B_i$ is a finite rank free-abelian group.

Remark 18.1. Considering the corresponding topological space (a bouquet of circles with a sequence of tori attached to it along maximal pre-existing tori), we see that each maximal torus in the resulting space could instead already be attached at the first stage that it has a rank ≥ 2 subtorus attached along a cyclic root. We may thus assume that each subgroup C_i above is either cyclic or trivial.

Lemma 18.2. *Each group G_r above is the fundamental group of a compact non-positively curved cube complex that is virtually special.*

Proof. This holds by Lemma 15.20 via induction on the hierarchy length. $\square$

As f.g. subgroups of G_r quasi-isometrically embed [Kap02] (see also Lemma 18.12), we can apply Proposition 2.31 to obtain the following consequence:

Corollary 18.3. *Every limit group is the fundamental group of a virtually special sparse cube complex.*

Sela gave the following "ω-residually free tower" description of groups with the same elementarily theory as free groups in [Sel03]:

Theorem 18.4 (Sela). *The class $\mathcal{E}$ of elementarily free groups is the class of groups that contains free products of f.g. free groups and closed surface groups with $\chi \leq -2$, and is closed under the following two constructions:*

(1) *If $G \in \mathcal{E}$ then $G *_Z \mathbb{Z}^n \in \mathcal{E}$ provided Z is a cyclic subgroup with $Z \subset G$ maximal abelian and $Z \subset \mathbb{Z}^n$ a direct factor.*

(2) *If $G \in \mathcal{E}$ then $G' \in \mathcal{E}$ provided $G' = \pi_1 Y'$ where $Y' = Y \cup S$ and $G = \pi_1 Y$ and:*

    (a) *S is a compact surface with S either a punctured torus or $\chi(S) \leq -2$,*

    (b) *$Y \cap S = \partial S = \sqcup_{i=1}^r C_i$,*

    (c) *each $C_i \rightarrow Y$ is π_1-injective,*

    (d) *there is a retraction $Y' \rightarrow Y$ such that the image of $\pi_1 S$ is not abelian.*

Moreover, limit groups are precisely the f.g. subgroups of groups in $\mathcal{E}$.

This leads to the following stronger result:

Theorem 18.5. *Every elementarily free group is the fundamental group of a compact cube complex that is virtually special.*

Proof. We will verify that the property of being the fundamental group of a compact cube complex that is virtually special is preserved by the two constructions indicated in Theorem 18.4.

The first case is covered by Lemma 15.20.

In the second case, G' splits as an HNN extension $K *_{Z^t=Z'}$ where Z is an (aparabolic) malnormal cyclic subgroup of $\pi_1 G'$ that lies in $\pi_1 S$, and where $K = G * F$ is the free product of G and a f.g. free group F. We refer to Figure 18.1 to see how to cut S. Note that K is π_1 of a compact cube complex that is virtually special, since we can add a bouquet of circles to the virtually special cube complex for G. Consequently, G' is virtually compact special by Theorem 15.1. We will use this below to prove the virtual specialness of a compact cube complex X with $\pi_1 X = G$.

We describe two ways to see that $G' = \pi_1 X$ where X is a compact nonpositively curved cube complex. The first approach uses the above HNN extension, and the second approach uses the retraction.

The HNN case of Theorem 7.59 applies to $K *_{Z^t=Z'}$ since Z is malnormal. Each maximal parabolic subgroup P is conjugate to a subgroup of $K = G * F$, and hence into G. By Remark 6.27, for any wall W, the subgroup $\mathrm{Stab}_P(W)$ is commensurable with $\cap \mathrm{Stab}_P(W_i)$ where each W_i is a P-essential wall that extends a P-essential hyperplane of the cubulation of $G * F$. Hence the resulting

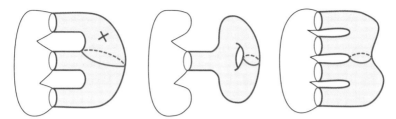

Figure 18.1. G' splitting along a malnormal cyclic subgroup when S is a projective plane with three boundary circles, when S is a punctured torus, and when ∂S has at least four circles.

dual for P is cocompact by Lemma 7.12, and so by Remark 7.10, the entire dual is cocompact by Theorem 7.9. Virtual specialness of X holds by Theorem 15.14 since we proved above that G' is virtually compact special.

We now describe an alternative approach: The retraction $G' \to G$ induces a retraction $(Y \cup S) \to Y$, which is essential on each circle. The hyperplanes of Y pull back to provide a system of immersed walls in $Y \cup S$. We augment these with a collection of finitely many embedded circles including circles parallel to ∂S. As f.g. subgroups of limit groups are quasi-isometrically embedded [Kap02], all walls have relatively quasiconvex stabilizers. The result provides a compact cube complex X, since we have added no new commensurability classes of P-essential walls for each maximal parabolic subgroup P, so Theorem 7.9, Remark 7.10, and Lemma 7.12 apply to give cocompactness as above.

X is virtually special by Theorem 15.14 as before. However, the alternative cubulation admits a different approach towards virtual specialness: We show directly that X has a finite cover with embedded 2-sided hyperplanes, and so X is virtually special by applying Corollary 15.12 to this finite cover. To see this, observe that the hyperplanes associated to the circles in S are already 2-sided and embedded, and if $\widehat{Y} \to Y$ is a finite special cover, then we can let $\widehat{X} \to X$ be the cover induced by $\widehat{Y}$ together with the retraction map $(Y \cup S) \to Y$. And then let $\bar{Y} \to Y$ be a finite regular cover factoring through $\widehat{Y}$. Then all hyperplanes of $\bar{Y}$ are embedded, since they are associated to embedded 2-sided immersed walls in the corresponding cover $(\overline{Y \cup S})$. $\qquad\square$

Theorem 18.6 (Limit Groups)**.** *Every limit group H is the fundamental group of a compact nonpositively curved cube complex X that is virtually special.*

It is unclear if passing to a finite index subgroup is necessary:

Problem 18.7. Is every limit group isomorphic to π_1 of a compact special cube complex?

Proof of Theorem 18.6. By Theorem 18.4, $H \subset G$ where G has an ω-residually free tower. The proof is by induction on the length of the tower. As in the proof of Theorem 18.5, either $G = K *_Z \mathbb{Z}^n$ or $G = K *_{Z^t = Z'}$ where in the former case Z is maximal cyclic and malnormal in K which has a lower length tower, and in the latter case Z is malnormal and $K = G' * F$ where G' has a lower length tower. Observe that H has a splitting as a finite graph of groups whose edge groups are cyclic or trivial. Using this splitting, there are two (similar) arguments depending on the above case:

When $L = K *_{Z^t = Z'}$, each H_v is a f.g. subgroup of K. By induction, each H_v is the fundamental group of a compact cube complex that is virtually special. The graph of groups for H shows that H can be assembled by using a sequence of HNN extensions or amalgamated products along malnormal aparabolic cyclic subgroups. As explained in the proof of Theorem 18.5 the HNN case preserves compactness and virtual specialness, and the amalgamated free product case is proven likewise.

When $L = K *_Z \mathbb{Z}^n$, each vertex group H_v is a f.g. subgroup of K or $\mathbb{Z}^n$ and we may assume all edge groups at an $H_v \subset \mathbb{Z}^n$ subgroup are conjugate to each other in H_v. (If an edge group is nontrivial, we can pull a maximal cyclic subgroup from H_v into each of its edge groups and neighboring vertex groups to assume all its edge groups are the same.) Let $L' \subset L$ be the subgroup associated to the same graph of groups except that we minimize each $\mathbb{Z}^n$ vertex subgroup H_v by replacing H_v with the cyclic (or trivial) group H'_v that is isomorphic to the edge groups at H_v. Then L' can be assembled by a using a sequence of HNN extensions and/or amalgamated products along malnormal aparabolic cyclic subgroups, so L' is the fundamental group of a compact cube complex that is virtually special as above. Then L can be built from L' by repeatedly amalgamating with H_v over H'_v for each minimized vertex group. Thus L is the fundamental group of a compact cube complex that is virtually special by Lemma 15.20. $\square$

In [KM98] it was shown that every limit group can be produced from free-abelian groups and free groups using a *separated cyclic hierarchy* whose split-tings are either free products, or amalgamated free products over an infinite cyclic subgroup that is malnormal on at least one side, or an HNN extension that conjugates two cyclic groups that do not have intersecting conjugates, and such that one of them is malnormal. This is apparent from the proof of Theorem 18.6 which uses a hierarchy with the stronger property that edge groups are malnormal on each side except when being amalgamated with a $\mathbb{Z}^n$ subgroup.

The following is a special case of Theorem 18.15 which is the main goal of Section 18.b:

Corollary 18.8. *Every group with a separated cyclic hierarchy is π_1 of a strongly sparse cube complex that is virtually special.*

Problem 18.9. Is every group with a separated cyclic hierarchy virtually a limit group?

18.b   Abelian Hierarchies

Definition 18.10. G has an *abelian hierarchy* if G has a hierarchy where at each step, the edge group C in $A *_C B$ or $A*_{C^t=C'}$ is a f.g. free-abelian group.

Remark 18.11. Let G be hyperbolic relative to free-abelian subgroups. Each infinite abelian subgroup of G is contained in a unique maximal abelian subgroup of G. Consequently, if G also has an abelian hierarchy then G has a hierarchy where at each stage we have either:

(1) An amalgamated free product (AFP) $A *_1 B$ or HNN extension $A*_{1^t=1}$ where 1 is the trivial subgroup.
(2) An AFP $A *_Z P$ where P is a maximal abelian subgroup of G and Z is a maximal cyclic subgroup of both A and P.
(3) An AFP $A *_P B$ where P is a maximal abelian subgroup of G, and hence of A and B.
(4) An HNN extension $A*_{P^t=P'}$ where P and P' are maximal abelian subgroups of G, and $\{P, P'\}$ form a malnormal pair in A.

The following is proven in [Dah03, BW13]:

Lemma 18.12. *Let G be hyperbolic relative to free-abelian subgroups. If G has an abelian hierarchy then every f.g. subgroup of G is quasi-isometrically embedded.*

Lemma 18.13 (Selfish parabolics). *Let X be a [strongly sparse] compact non-positively curved cube complex. Suppose $\pi_1 X$ is hyperbolic relative to virtually abelian subgroups $P_1, \ldots, P_r$. There exists a cube complex X' with the following properties:*

(1) *X' is [strongly sparse] compact.*
(2) *X' has the same collection of commensurability classes of stabilizers of essential and inessential hyperplanes for each P_i.*
(3) *No hyperplane $\widetilde{U}$ is essential for both $P_i^{g_i}$ and $P_j^{g_j}$ unless $P_i = P_j$ and $g_i \widetilde{U} = g_j \widetilde{U}$.*
(4) *X' is virtually special if X is virtually special.*

Note that we permit some of the $\{P_i\}$ to be cyclic.

Proof. We first describe the proof in the case where each P_i is free-abelian.

We begin by describing a certain subdivision of the universal cover $\widetilde{X}$, and a useful $\pi_1 X$-equivariant notation for some of its hyperplanes. For each i, let m_i be the number of distinct P_i-orbits of hyperplanes that essentially cut P_i. We subdivide $\widetilde{X}$ precisely $2 \sum_{i=1}^{m_i}$ times, and use the notation $\widetilde{U}_{\pm pq}$ for the hyperplanes in the subdivision obtained from the original hyperplane $\widetilde{U}$, where $2 = |\{-, +\}|$ and $1 \le p \le r$, and $1 \le q \le m_i$.

By Lemma 2.37, let $\widetilde{F}_i$ be the superconvex subcomplex associated to P_i for each i. Consider the cubical presentation $\langle X \mid \widetilde{F}_1, \ldots, \widetilde{F}_r \rangle$. Let B_i be an upper-bound on the diameter of all pieces of $\widetilde{F}_i$ in X^*. Note that Lemma 7.37 provides a $\pi_1 X$-cocompact isometrically embedded subcomplex $\widetilde{X}^\square \subset \widetilde{X}$ that contains all intersections of cones and all wall-pieces, and hence the bound follows from the P_i-cocompactness of the $\widetilde{F}_i \cap \widetilde{X}^\square$ and their superconvexity.

For each i, let $z_i \in P_i$ be an element such that no z_i^n stabilizes a hyperplane $\widetilde{U}$, unless a finite index subgroup of P_i lies in $\mathrm{Stab}(\widetilde{U})$. We now define an exotic wallspace structure on each $\widetilde{F}_i$ as follows: Declare $\widetilde{U}_{\pm pq}$ to be a solo wall if either: $\widetilde{U}$ is P-inessential, or $p = i$ and q corresponds to $\widetilde{U}$. Otherwise, declare that $\widetilde{U}_{-pq}$ belongs in a wall together with $z_i^{m_i} \widetilde{U}_{+pq}$, where $m_i > 0$ is later chosen to be sufficiently large.

Apply Corollary 5.45 to obtain a free action of $\pi_1 X^* = \pi_1 X$ on the resulting CAT(0) cube complex. Condition 5.44.(2) always holds when cones are contractible. Condition 5.44.(3) is immediate when $m_i > 4B_i$. Condition 5.44.(1) is satisfied as follows: Firstly, X^* has short innerpaths by Lemma 3.70 since X^* is $C'(\frac{1}{14})$ as it is $C'(\alpha)$ for each $\alpha > 0$ (or alternatively, since X^* has contractible cones, and hence no essential cone-cells). Secondly, X^* is $B(6)$ since Conditions 5.1.(3) and (4) hold when $m_i > 7B_i$ by Remark 5.4, and the other $B(6)$ conditions are immediate in our setting.

By Theorem 5.18, for each wall W, the map $N(W) \to \widetilde{X}^*$ is an embedding. We emphasize that $\pi_1 X = \pi_1 X^*$ and that the cubical part of $\widetilde{X}^*$ equals $\widetilde{X}$. By Corollary 5.30, the map $N(W) \to \widetilde{X}^*$ is a quasi-isometric embedding. However, since $N(W)$ can contain cones that are entire quasiflats $\widetilde{Y}_i$ that don't lie in a finite neighborhood of the union of hyperplanes of W, the action of $\mathrm{Stab}(W)$ on $N(W)$ might not be cocompact, so we must justify that $\mathrm{Stab}(W) \subset \pi_1 X$ is a quasi-isometric embedding. The justification is supplied by Lemma 18.14 for $m_i \gg 0$. We note as well that when all $m_i \gg 0$, the various subgroups that occur as P-stabilizers of walls of $\widetilde{Z}^*$ are the same as the various subgroups occurring as P stabilizers of hyperplanes of $\widetilde{X}$.

We now show that the finiteness properties of the dual $C(\widetilde{X}^*)$ replicate the finiteness properties of $\widetilde{X}$. Note that $\widetilde{X}^*$ is not a geometric wallspace, since its walls are formed from equivalence classes of hyperplanes of $\widetilde{X}$. However, we can extend it into a geometric wallspace $\widetilde{Z}^*$ as follows: For each P_i, let $\widetilde{E}_i \subset \widetilde{F}_i$ be a P_i-invariant convex cocompact subspace, and attach a copy of $\widetilde{E}_i \times I$ by identifying $\widetilde{E}_i \times \{0\}$ with $\widetilde{E}_i \subset \widetilde{X}$. A hyperplane $\widetilde{U}$ of $\widetilde{X}$ is either disjoint from $\widetilde{E}_i$, or contains $\widetilde{E}_i$, or separates $\widetilde{E}_i$ and intersects it in a codimension-1 subspace. Hence when a wall intersects $\widetilde{F}_i$ in a single hyperplane, we either position all of $\widetilde{E}_i \times I$ on one side of the wall, or position it entirely within the wall, or cut it by the wall, according to the cases above. When a wall intersects $\widetilde{F}_i$ in two hyperplanes, then we are necessarily in the cutting case above, and we join these two hyperplanes by a "bracket" in $\widetilde{E}_i \times I$ (this is called a "turn" in [HW15]). Finally, observe that two walls cross in $\widetilde{X}^*$ if and only if they cross in $\widetilde{Z}^*$, and hence they have the same dual CAT(0) cube complex. However $\widetilde{Z}^*$ has the advantage that its walls are "geometric," as we can use the subspace metric of the associated CAT(0) metric space. Moreover, if we follow the above procedure starting instead with the $\pi_1 X$-cocompact convex subspace $\widetilde{X}^\square$, then we may also assume that there is a $\pi_1 X$-cocompact action.

Lemma 7.23 asserts that the P-stabilizer of each hyperplane of $\widetilde{X}$ is commensurable with the intersection of P-stabilizers of P-essential hyperplanes. Hence the P-stabilizer of each wall of $\widetilde{Z}^*$ is commensurable with the intersection of P-stabilizers of P-essential walls. Hence $\widetilde{Z}^*$ has Ball-WallNbd separation with respect to $\widetilde{F}$ by Lemma 7.29. [Let P' be a finite index abelian subgroup of a maximal parabolic, and let U_1, U_2 be hyperplanes of $\widetilde{X}$. Lemma 7.33 asserts that $\mathrm{Stab}_{P'}(U_1) \, \mathrm{Stab}_{P'}(U_2)$ is commensurable with the intersection of P-stabilizers of P-essential hyperplanes. The analogous statement thus holds for $\widetilde{Z}^*$. Hence Lemma 7.31 asserts that WallNbd-WallNbd separation holds for $\widetilde{Z}^*$ with respect to $\widetilde{F}$.]

Theorem 7.28 now applies to $\widetilde{Z}^*$ to give [strong] cosparseness of the dual.

Cocompactness holds by showing that the cubulation of $\widetilde{F}_i$ with its new wallspace structure is cocompact if $\widetilde{F}_i$ is cocompact. To verify that, we observe that a pairwise intersecting collection of walls in $\widetilde{X}$ (technically, of pairs of halfspaces, and singles of halfspaces) will correspond to a pairwise intersecting collection of hyperplanes and "sectors" associated to pairs of hyperplanes. Moreover, we can assume they intersect in some thickening $\widetilde{F}_i^{+s}$. But if $\widetilde{F}_i$ is cocompact, then so is $\widetilde{F}_i^{+s}$, and there are finitely many such intersections since there are finitely many in each parallelism class that can intersect.

Virtual specialness is preserved by Theorem 15.14.

The general case is treated similarly: subdivide further (add an additional factor for the index of a maximal free-abelian subgroup in P_i) and use a collection of conjugates of z_i by representatives of these cosets. $\qquad \square$

Lemma 18.14. *Within the framework of Lemma 18.13, if each $m_i \gg 0$ then each* $\mathrm{Stab}(W) \subset \pi_1 X$ *quasi-isometrically embeds.*

Proof. We give two proofs, the first of which depends on an external basic result about piecewise geodesics in CAT(0) spaces with isolated flats. The second is closer to the framework of this text. Both proofs employ the cocompact convex subspace $\widetilde{S} \subset \widetilde{X}$ of Lemma 7.37.

For the first proof, for each hyperplane V of W, let $\widetilde{S}_V = \widetilde{S} \cap \widetilde{\mathcal{N}}_r(V)$, and for each cone $\widetilde{Y} \subset N(W)$, let $\widetilde{S}_{\widetilde{Y}} = \widetilde{S} \cap \mathcal{N}_r(V) \cap \mathcal{N}_r(V') \cap \widetilde{Y}$ or $\widetilde{S}_{\widetilde{Y}} = \widetilde{S} \cap \mathcal{N}_r(V)$ depending on whether $\widetilde{Y}$ contains two or one hyperplanes of W, and where $\widetilde{S}$ and r are large enough to ensure the intersections above are nonempty. We will apply [HW15, Thm 2.3], which holds for $m_i \gg 0$. Piecewise geodesics in $\bigcup_{V \in W} \widetilde{S}_V \cup \bigcup_{\widetilde{Y} \in N(W)} \widetilde{S}_{\widetilde{Y}}$ that project to geodesics in the tree Γ_W of Definition 5.16 and that have the property that the $\widetilde{S}_V$ segments travel as quickly as possible towards their $\widetilde{S}_Y$ neighbors will obey the alternating and diverging conditions of [HW15,

Thm 2.3] and are hence uniform quasi-geodesics in $\widetilde{S}$, so consideration of the piecewise geodesics between points in an orbit shows that $\mathrm{Stab}(W)$ is quasi-isometrically embedded in $\pi_1 X$.

For the second proof, we will show below that a finite neighborhood of the union of carriers of hyperplanes of W contains a convex subcomplex $\widetilde{Y}_W$. Indeed, $\widetilde{Y}_W$ may be defined to be the intersection of all minor halfspaces containing all hyperplanes of W. Thus $\widetilde{S}_W = \widetilde{S} \cap \widetilde{Y}_W$ is convex and $\mathrm{Stab}(W)$-cocompact, and hence $\mathrm{Stab}(W)$ is quasi-isometrically embedded.

A $\pi_1 Y_i$-minimal subcomplex of the strong quasiflat $\widetilde{Y}_i$ is a subcomplex $\widetilde{Y}_i^\diamond \subset \widetilde{Y}_i$ that is a product $\prod \widetilde{Y}_{ik}$ where each factor Y_{ik} is either compact or is a quasiline, and hence has compact hyperplanes. We choose $z_i^{m_i}$ to preserve the factors and act properly (and hence cocompactly) on each of the finitely many factors.

Local finiteness of the collection of translates of $\widetilde{Y}_i$ in $\widetilde{X}$ holds by Definition 7.21.(2). Let d be an upperbound on the number of distinct translates of the various $\widetilde{Y}_i$ containing a point q. Let $K \geq 0$ be such that for any i, and any geodesic $\omega \to \widetilde{Y}_i$, at most K of the edges of ω are dual to nonessential hyperplanes of quasiline factors of $\widetilde{Y}_i$. For instance, we can let $K = K'' + 2K'$ where K'' is an upperbound on the sum of the number of hyperplanes in each compact factor of $\widetilde{Y}_i^\diamond$ and $\widetilde{Y}_i \subset \mathcal{N}_{2K'}(\widetilde{Y}_i^m)$ for each i. Let $L \geq 1$ have the following property for any cone $\widetilde{Y}_i$ and any wall W intersecting $\widetilde{Y}_i$ in two distinct hyperplanes $V, V' = z^m V$: There does not exist a collection of disjoint essential hyperplanes $U_1, \ldots, U_L$ in $\widetilde{Y}_i$ such that each U_ℓ is disjoint from V, and each U_ℓ either separates V, V' or crosses V'. We can choose L to exceed the following for each such $V, V', \widetilde{Y}_i$: the number of hyperplanes separating V, V' in their associated quasiline plus the number of hyperplanes in that quasiline intersecting V'.

Consider the Ramsey number $R = R(L + dK, \dim(X) + 1)$, so any geodesic of length R in $\widetilde{X}$ has at least $L + dK$ edges dual to pairwise disjoint hyperplanes. Let $M = \cup_{w \in W} N(w)$. We claim that $\mathrm{hull}(M)$ lies in $\mathcal{N}_R(M)$. Choose $p \in \widetilde{X}^0$ with $\mathsf{d}(p, M) \geq R$. We will show that $p \notin \mathrm{hull}(M)$ by showing that some hyperplane of $\widetilde{X}$ separates p from M. Let V be a hyperplane such that $\mathsf{d}(p, N(V)) = \mathsf{d}(p, M)$. Consider a geodesic σ between p and $N(V)$, and let $q \in N(V)$ be the other endpoint of σ. By Lemma 2.18, each hyperplane dual to an edge of σ separates $p, N(V)$. By our choice of R, there are $L + dK$ disjoint hyperplanes dual to edges of σ. And by discarding at most K hyperplanes for each of the at most d quasiflats containing q, we can choose $U_1, \ldots, U_L$ to be a sequence of disjoint hyperplanes dual to edges of σ, and ordered in proximity to V, and having the property that each U_ℓ is essential for any quasiflat containing q that U_ℓ intersects. We will show that M lies in a single halfspace of U_L, and hence $p \notin \mathrm{hull}(M)$.

Suppose either U_L separates V from another hyperplane V'' of W, or U_L crossed some hyperplane V'' of W. Consider the alternating sequence of hyperplanes and cones in $N(W)$ starting with V and ending with V'', and assume

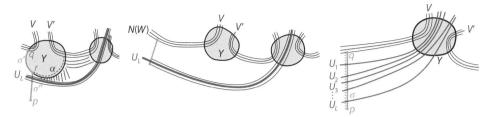

Figure 18.2. Left: If U_L intersects W beyond $V' \cap \widetilde{Y}$ then $|\alpha|$ and hence $|\sigma'|$ is $\leq 2\,\mathrm{diam}(\mathrm{piece})$ which is impossible. Center: If U_L intersects W beyond $\widetilde{Y}$, then there would be an impossible disk diagram with a highly negatively curved external cone-cell, but few positively curved features. Right: If U_L intersects $\widetilde{Y}$ and either separates V, V' or intersects V' then the property of the sequence $U_1, \ldots, U_{L+1}$ is violated.

this sequence and V'' are chosen to minimal length with this property. Note that this sequence corresponds to the vertices of a geodesic in Γ_W. We now refer to Figure 18.2 for the various cases to consider. Let $\widetilde{Y}$ be the first cone in the sequence and let V' be the hyperplane in the sequence after V.

The first case is where $q \in \widetilde{Y}$. In this case we rechoose σ to be of the form $\sigma = \sigma'\sigma''$ where σ' is a path in $\widetilde{Y}$ with endpoints q, q', and each hyperplane dual to an edge of σ'' separates $\widetilde{Y}, p$. However, we retain the sequence $U_1, \ldots, U_L$. If U_L crosses both V' and $\widetilde{Y}$, then our choice of L is violated by V, V' and the sequence $\{U_1, \ldots, U_L\}$. If U_L crosses V' but does not cross $\widetilde{Y}$, or U_L crosses beyond V' in the sequence, then considering the geodesic $\alpha \to \widetilde{Y}$ from p' to $N(V' \cap \widetilde{Y})$, we see that α consists of at most one or two pieces. Since σ' minimizes distance to M we see that $|\sigma'| \leq |\alpha|$ and so there is a path $\sigma'\alpha$ in $\widetilde{Y}$ between $N(V \cap \widetilde{Y})$ and $N(V' \cap \widetilde{Y})$ whose length is bounded by at most four pieces. This contradicts the construction of the pair V, V'.

The second case is where $q \notin \widetilde{Y}$. If U_L crosses both V' and $\widetilde{Y}$, then our choice of L is violated as in the previous case. Otherwise, U_L crosses V' but does not cross $\widetilde{Y}$, or U_L crosses further in the sequence. (Note that U_L cannot cross $\widetilde{Y}$ and cross a cone or hyperplane in the sequence beyond V' unless U_L also crosses V' within $\widetilde{Y}$, for otherwise U_L would be nonessential in $\widetilde{Y}$, and was discarded earlier.) In these cases, there is an impossible disc diagram as suggested in the center of Figure 18.2. Indeed, this diagram could have at most three features of positive curvature: One where U_L crosses a cone or hyperplane of $N(W)$ and one at each endpoint of σ. Note that σ cannot contain an outerpath of a shell because each cone-cell is replaceable in our setting, and moreover, we can choose σ so that all cornsquares are passed through it, and likewise pass cornsquares through the carriers of hyperplanes. Such a diagram cannot exist since there is a cone-cell arising from $\widetilde{Y}$ that is highly negatively curved, since its curvature is in proportion with $m_i \gg 0$. $\qquad\square$

Theorem 18.15. *Suppose G is hyperbolic relative to free-abelian subgroups $\{P_1, \ldots, P_r\}$, and G has an abelian hierarchy terminating in groups that are fundamental groups of virtually special strongly sparse cube complexes. Then G is the fundamental group of a strongly sparse cube complex X that is virtually special.*

Corollary 18.16. *Suppose G has an abelian hierarchy, and is hyperbolic relative to free-abelian subgroups $\{P_1, \ldots, P_r\}$. Then G is π_1 of a strongly sparse cube complex that is virtually special.*

Proof of Theorem 18.15. By Remark 18.11, there is a hierarchy, where each P_i is added by amalgamating along a maximal cyclic subgroup before any further amalgams or HNN extensions occur. For each P_i, let $\{Z_{ij}\}$ be the finitely many cyclic subgroups of P_i along which P_i is amalgamated. Note that P_i can appear in many places within the hierarchy, and these are eventually amalgamated with each other. Let $\widetilde{F}_i$ be a strong quasiflat upon which P_i acts. Specifically, we cubulate P_i with a system of finitely many orbits of walls associated to codimension-1 subgroups, such that each Z_{ij} is commensurable with the intersection of the stabilizers of finitely many of these, and moreover, each $\langle Z_{ij}, Z_{ik} \rangle$ is likewise commensurable with the intersection of the stabilizers of finitely many of these chosen essential walls. Let $F_i = P_i \backslash \widetilde{F}_i$.

We now follow the hierarchy of Remark 18.11, and describe how to cubulate the resulting AFP/HNN at each stage. Case (1) where the edge group is trivial is handled by gluing two cube complexes together along an edge or attaching an edge to a single cube complex. Case (2) is handled as follows: Let X be the cube complex constructed for A, and let F be the cube complex for P. Choose a closed geodesic in F and in P which represents the conjugacy class of a generator of Z. Note that by subdividing, we may assume this circle has the same length ℓ on each side. Let $C = S_\ell \times [0, 1]$ be a cylinder whose circle S_ℓ has ℓ-edges. Form the quotient $Y = F \cup C \cup P$ by attaching the two boundary circles of C to the chosen closed geodesics using maps $S_\ell \times \{0\} \to F$, and $S_\ell \times \{1\} \to X$. The immersed hyperplanes of F and P extend to immersed hyperplanes in Y and there is an additional "vertical" hyperplane in Y at $S_\ell \times \{\frac{1}{2}\}$. These immersed hyperplanes lift to walls in $\widetilde{Y}$ (which is probably not nonpositively curved). However, the group $A *_Z P$ acts freely by Lemma 5.42 since hyperbolic elements are cut by "vertical" walls and elliptic elements are cut by walls in the vertex spaces that are the extensions of original hyperplanes cutting them. Note that a wall intersects a vertex space in at most one hyperplane, since the edge spaces have connected intersection with hyperplanes (cf. [HW10b, Lem 5.10]).

We now turn to Case (4) and note that Case (3) holds using a similar (and slightly simpler) proof. By applying Lemma 18.13 we may assume that $A = \pi_1 X$ where the nonpositively curved cube complex X has the property that no immersed hyperplane cuts both P and P'. By the flat torus theorem [BH99], let

$\widetilde{T} \subset \widetilde{X}$ be a P-cocompact convex flat, and likewise define $\widetilde{T}' \subset \widetilde{X}$, and let $T \to X$ and $T' \to X$ be the associated quotients.

By possibly subdividing, we can assume that $\widetilde{T}, \widetilde{T}'$ do not lie in hyperplanes, and moreover, by convexity, for each hyperplane $\widetilde{U}$ of $\widetilde{X}$, we have $\widetilde{U} \cap \widetilde{T}$ is either empty or is a codimension-1 subflat, and similarly for $\widetilde{T}'$. By our inductive construction of $\widetilde{X}$, the families of such codimension-1 subflats are in correspondence with each other, in the sense that the same parallelism classes arise. By our earlier application of Lemma 18.13 we see from Property (3) that no immersed hyperplane of X crosses both $\widetilde{T}$ and $\widetilde{T}'$, nor does such a hyperplane cross in more than one subtorus.

We may assume that X has the additional property that P-stabilizers and P'-stabilizers of essential hyperplanes that are commensurable are actually equal. (This will come at the expense of allowing multiple immersed hyperplanes intersecting T along immersed subtori that are homotopic to each other. But we will later subdivide anyway.) We achieve this additional property as follows: Let $\{K_j\}$ be the various maximal codimension-1 subgroups that are commensurable with essential hyperplane stabilizers of P, P' (identified via conjugation by t). For each j, choose a finite index subgroup $\dot{K}_j \subset K_j$ that lies in each P and P' stabilizer that is commensurable with K_j. By separability, we choose P_1, P_1' so that they lie in $P^\circ, (P')^\circ$ and so that $P_1^t = P_1'$ and so that their intersections with K_j lie in $\dot{K}_j$, for each j. By Theorem 15.6, there exists subgroups $P^\circ, (P')^\circ$ such that for any further finite index subgroups of $P^c, (P')^c$ of P, P', there is a normal finite index subgroup G' that intersects P, P' in $P^c, (P')^c$. Finally, by Lemma 7.30, there is a strongly cosparse action of G whose hyperplane stabilizers are the intersections of previous hyperplane stabilizers with G'. Hence we may assume that commensurable hyperplane subtori are equal.

We now subdivide X so that there is also a one-to-one correspondence between the numbers of each type of subtorus in T and in T'. Let $C = T'' \times [0, 1]$ be a cylinder, where T'' is a torus with $\pi_1 T'' = P$, and let $Y = X \cup C$ where we attach the boundary of C to Y by identifying $T'' \times \{0\}$ with T and identifying $T'' \times \{1\}$ with T'. We choose a bijection between hyperplanes having the same type of immersed subtorus intersection, and extend across $T'' \times [0, 1]$ to obtain immersed walls. Additionally, we have the immersed wall $T'' \times \{\frac{1}{2}\}$. The group $A*_{P^t = P'}$ acts freely as above. All the walls have stabilizers that are f.g. by construction, and are relatively quasiconvex since they are quasi-isometrically embedded by [BW13, Thm 2.6]. The Ball-WallNbd and WallNbd-WallNbd separation properties hold by Lemma 7.31 and our choice of the essential walls of the various P_i. Thus the cubulation is cosparse by Theorem 7.28 (note that there is a cocompact geometric wallspace if we substitute an isocore $X^\square$ for X using Lemma 7.37 or as specifically supplied by Lemma 7.42). Virtual specialness holds by Theorem 15.14. $\square$

Remark 18.17. If G is hyperbolic relative to virtually abelian groups and has a virtually abelian hierarchy terminating in virtually strongly sparse special groups, then G has a finite index subgroup that is π_1 of a strongly sparse special cube complex. This holds by induction using Theorem 18.15 since at each stage we can quotient by free-abelian subgroups of the amalgamated parabolics (and the other parabolics) using Theorem 15.6, to obtain a graph of virtually special groups along finite subgroups. An induced finite index subgroup is then an amalgam along free-abelian subgroups.

Problem 18.18. Suppose G is hyperbolic relative to virtually abelian subgroups, and suppose G has a (virtually) abelian hierarchy. Does G virtually have a compact cubulation?

Chapter Nineteen

Application Towards One-Relator Groups

19.a   Overview

We now provide an application resolving the conjecture of Gilbert Baumslag on the residual finiteness of one-relator groups with torsion.

In analogy with Haken 3-manifolds, every one-relator group has a hierarchy terminating in a group isomorphic to $F_r * \mathbb{Z}_n$ for some r, n.

We describe the hierarchy in Section 19.b, and prove in Lemma 19.8 that it is a quasiconvex hierarchy when $n \geq 2$. Without the torsion assumption, there are simple examples where the hierarchy is not quasiconvex, as described in Example 19.3.

Theorem 19.1. *The Magnus-Moldavanskii hierarchy is quasiconvex for any one-relator group with torsion.*

Combining Theorem 19.1 with Theorem 13.3, we obtain the following result strongly affirming Baumslag's conjecture:

Corollary 19.2. *Every one-relator group with torsion has a finite index subgroup that is the fundamental group of a compact special cube complex.*

We now describe a word-hyperbolic one-relator group whose Magnus hierarchy is not a quasiconvex hierarchy.

Example 19.3. Consider the presentation $\langle a, b, c, t \mid abc^{-1}, a^t = b, b^t = c \rangle$. Its group is an HNN extension of the free group $\langle a, b, c \mid abc^{-1} \rangle \cong \langle a, b \mid - \rangle \cong F_2$, and in fact, a semi-direct product $F_2 \rtimes_\phi \mathbb{Z}$ using the automorphism induced by: $\phi(a) = a^t = b$ and $\phi(b) = b^t = c = ab$.

By choosing words with no repeated letters (or even primitive words), that are complicated relative to the subscript shift isomorphism between Magnus subgroups, we can obtain a small-cancellation group, with similar behavior.

For instance, letting x^y denote yxy^{-1} the HNN extension: $\langle a_1, \ldots, a_{r+1}, t \mid a_1 \cdots a_{r+1}, a_i^t = a_{i+1} : 1 \leq i \leq r \rangle$ is isomorphic to the one-relator group: $\langle a_1, t \mid a_1 a_1^t a_1^{t^2} \cdots a_1^{t^r} \rangle$ which equals $\langle a_1, t \mid a_1 t a_1 t^2 a_1 t^3 \cdots a_1 t^r \rangle$ which is a $C'(\frac{1}{6})$ small-cancellation group for $r > 21$. Indeed, pieces have length at most $2r$ but the word has length $r + \frac{r(r+1)}{2} = \frac{r^2+3r}{2} > 12r$. (In fact the group is $C'(\frac{1}{4}) - T(4)$ when $r > 13$.)

Since the HNN extension is a semi-direct product, the vertex group and hence the edge groups are normal of infinite index and thus cannot be quasiconvex. It may be that some alternate one-relator presentation provides a quasiconvex mal-normal hierarchy, but I reckon there are examples that have no quasiconvex hierarchy. Nevertheless, I expect the following holds:

Conjecture 19.4. *Every word-hyperbolic one-relator group has a finite index subgroup with a quasiconvex hierarchy.*

19.b The Magnus-Moldavanskii Hierarchy

Construction 19.5 (Magnus-Moldavanskii construction)**.** Following [LS77], we shall now describe a variation on Moldavanskii's variant of Magnus's inductive construction of one-relator groups.

We start with a one-relator group $\langle S \mid W^n \rangle$ where the generating set is S, and W is cyclically reduced word in the generators that is not equal to a proper power. We define the *repetition complexity* of W to be $|W|$ minus the number of distinct letters that occur in W. Equivalently, this is the sum of the number of times that letters occurring in W are repeated, so the complexity of $aba^{-1}bbc$ would be $1 + 2 + 0 = 3$. Having a given presentation in mind, we refer to the *complexity* of a one-relator group as the repetition complexity of the relator in this presentation.

Note that when the complexity is zero, and more generally, when some generator appears exactly once in W, then the group presented by $\langle S \mid W^n \rangle$ is isomorphic to a virtually free group $F * \mathbb{Z}_n$ where the rank of F equals $|S| - 1$.

Assuming that no generator appears exactly once, we show that $G' = G * \mathbb{Z}$ splits as an HNN extension $K *_M$ of a one-relator group K over a Magnus subgroup. The complexity of K is less than the complexity of G, and so this process stops after finitely many steps. It is clear that this induces an actual hierarchy (that terminates at finite cyclic groups), since a splitting of $G * \mathbb{Z}$ induces a splitting of G, and if it is a quasiconvex splitting, then it induces a quasiconvex splitting. The free factor with $\mathbb{Z}$ is an artifice that facilitates the combinatorial group theory description. The stable letter t of the HNN extension conjugates one Magnus subgroup to another by a "subscript shift." A *Magnus subgroup* of a one-relator group $\langle S \mid W^n \rangle$ is a subgroup generated by a subset of the generators omitting some generator appearing in the relator W^n. Magnus's "Freiheitsatz" theorem states that these specific generators freely-generate the Magnus subgroup of a one-relator group. We state a general formulation of the Freiheitsatz in Proposition 19.10.

Let us now describe this process a bit more carefully. We add a new further generator t to the presentation, so that the resulting finitely presented group G'

is isomorphic to $G * \langle t \rangle$. We let $\bar{S}$ denote a new set of generators in one-to-one correspondence with the generators in S, so $s \leftrightarrow \bar{s}$. We will choose an integer p_s for each $s \in S$, and perform a substitution $s \mapsto \bar{s}t^{p_s}$, that rewrites the relator W as a word $\bar{W}'$ in terms of $\bar{S} \cup \{t\}$, so we have a new presentation $\langle t, \bar{s} \in \bar{S} \mid \bar{W}' \rangle$ for G'. There are Tietze transformations justifying this: we first add generators $\bar{s}$ with relators $s^{-1}\bar{s}t^{p_s}$, and then rewrite the relator by substituting $\bar{s}t^{p_s}$ for each s in W, and finally, we remove the old generators s and relators $s^{-1}\bar{s}t^{p_s}$.

For each generator x of a free group F, there is a homomorphism $\#_x : F \to \mathbb{Z}$ induced by sending x to 1 and all other generators to 0. For a word V, the value $\#_x(V)$ is the exponent sum of the letter x in the word V.

We shall now assume that the integers p_s are chosen so that the resulting word satisfies $\#_t(\bar{W}) = 0$. Indeed, assuming that W contains at least two letters, there is always a way of doing this: If $\#_a(W) = 0$ for some generator $a \in S$, then we let $p_a = 1$ and let $p_s = 0$ for all $s \neq a$. Otherwise, we can choose $a, b \in S$ with $\#_a(W) \neq 0$ and $\#_b(W) \neq 0$, and we then define $p_a = \#_b(W)$ and define $p_b = -\#_a(W)$, and define $p_s = 0$ for all $s \neq a, b$.

In this way, our word W in S becomes a word $\bar{W}'$ in $\bar{S} \cup \{t\}$.

For each $\bar{s} \in \bar{S}$ let $L_{\bar{s}} \leq R_{\bar{s}}$ be the smallest and greatest values of $\#_t(U)$ where U is a prefix of $\bar{W}'$ preceding an occurrence of $\bar{s}^{\pm 1}$ so $U\bar{s}^{\pm 1}V = \bar{W}'$. In the degenerate case where $\bar{W}'$ contains no occurrence of $\bar{s}^{\pm 1}$ then we assign $L_{\bar{s}}, R_{\bar{s}} = 0$.

We introduce *new generators* $\bar{s}_i : L_{\bar{s}} \leq i \leq R_{\bar{s}}$, and *new relators* $\bar{s}_i = t^i \bar{s} t^{-i}$: $L_{\bar{s}} \leq i < R_{\bar{s}}$. Adding these generators and relators, we thus obtain a new presentation for G'. Finally, since $\#_t(\bar{W}') = 0$, we see that $\bar{W}'$ is freely equivalent to a word $\bar{W}''$ in $(t^i \bar{s} t^{-i}) : \bar{s} \in \bar{S}$, and we use the introduced relators to rewrite $\bar{W}''$ as a word $\bar{W}$ in our new generators $\bar{s}_i$.

Exchanging the relations $\{\bar{s}_i = t^i \bar{s} t^{-i} : L_{\bar{s}} \leq i < R_{\bar{s}}\}$ for relations of the form $\bar{s}_i^t = \bar{s}_{i+1}$ (here we use the notation $x^y = y^{-1}xy$), the resulting presentation for G' is the following:

$$\langle t, \bar{s}_i \ : \ L_{\bar{s}} \leq i \leq R_{\bar{s}}, \ \bar{s} \in \bar{S} \mid \bar{W}^n, \bar{s}_{i+1}^t = \bar{s}_i \ : \ L_{\bar{s}} \leq i \leq R_{\bar{s}}, \ \bar{s} \in \bar{S} \rangle$$

Thus G' is an HNN extension of the one-relator group K presented by $\langle \bar{S} \mid \bar{W}^n \rangle$ with stable letter t, and Magnus subgroups $M_+^t = M_-$ where: $M_- = \langle \bar{s}_i : L_{\bar{s}} \leq i < R_{\bar{s}} \rangle$, and $M_+ = \langle \bar{s}_i : L_{\bar{s}} < i \leq R_{\bar{s}} \rangle$. We note that when G and hence G' is f.g. then so is K.

Observe that $|W| = |\bar{W}|$ and the number of generators occurring in $\bar{W}$ exceeds the number occurring in W by $\sum_{\bar{s} \in \bar{S}} R_{\bar{s}} - L_{\bar{s}}$. Recall that the complexity is the difference between the relator length and the number of occurring generators, and hence the complexity decreases by $\sum_{\bar{s} \in \bar{S}} R_{\bar{s}} - L_{\bar{s}}$. If t appears at least once in $\bar{W}'$ we can hope that $R_{\bar{s}} > L_{\bar{s}}$ for some $\bar{s}$, and hence the complexity will decrease. To this end we must choose our p_s integers a bit more carefully, and will explain how to do so below.

Recapitulation and geometric interpretation: We began with a one-relator group G with presentation $\langle S \mid W \rangle$. We define $G' = G * \langle t \rangle$, which obviously has the presentation $\langle S, t \mid W \rangle$. Using different generators $\bar{s} = st^{-p_s}$, and rewriting

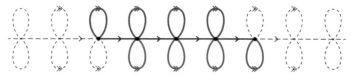

Figure 19.1. The subcomplex $Y^1 \subset \widehat{X}$ for the word $\bar{W}' = at^3bt^{-2}at^3bt^{-4}$ is illustrated in bold. The reader should check that the lift of $\bar{W}'$ at the basepoint only passes through two b edges and two a edges, but we include their intermediate $\mathbb{Z}$-translates.

W in terms of these new generators, we obtain a new presentation for G' of the form $\langle \bar{S}, t \mid \bar{W}' \rangle$. The word $\bar{W}'$ has occurrences of $\bar{s}$ in one-to-one correspondence with occurrences of s in W, but there are additional occurrences of t. We then express G' as an HNN extension whose base group K is a new one-relator group. Let X denote the standard 2-complex of the presentation $\langle \bar{S}, t \mid \bar{W}' \rangle$. Let $\widehat{X} \to X$ be the $\mathbb{Z}$-covering space induced by $t \mapsto 1, \bar{s} \mapsto 0$, and note that this extends the homomorphism $\phi \colon G \to \mathbb{Z}$ induced by $s \mapsto p_s$.

Consider the based lift of the closed path $\bar{W}'$ to $\widehat{X}$. Let Y denote the subcomplex of $\widehat{X}$ that contains the closure of the based lift of the 2-cell of X, and also contains all $\mathbb{Z}$-translates of a 1-cell $\bar{s}$ that lie between the first and last such $\bar{s}$ 1-cell that $\bar{W}'$ passes through. We refer the reader to Figure 19.1.

The group $K = \pi_1 Y \subset \pi_1 \widehat{X}$. To obtain a presentation for K we contract the t-tree within Y, and our $\bar{s}_i$ generators correspond to the loops attached to this t-tree. The new relator $\bar{W}$ is closely related to our original word W, but with some of its generators subscripted—the temporary t-letters have now disappeared. More precisely, regarding a t-path as a maximal tree in Y^1, there is a natural generator system for $\pi_1 Y^1$ consisting of conjugates of the original generators of $\pi_1 X^1$. Namely: $\{\bar{s}_i = t^{-i} \bar{s} t^i : L_{\bar{s}} \leq i \leq R_{\bar{s}}\}$. We also include a single loop $\bar{s}_0$ for each generator of $\bar{S}$ that doesn't appear in $\bar{W}'$.

Finally, the complexity of the presentation for K decreases precisely if $\bar{W}$ contains more distinct generators than occurred in W. This happens precisely if the closed path $\bar{W}'$ passes through two non-t 1-cells of $\widehat{X}$ that map to the same 1-cell of X. This occurs precisely if $\phi \colon G \to \mathbb{Z}$ "separates two occurrences of a generator" which we will define and examine in the explanation below of how to make the choices so that the complexity decreases.

Complexity reduction: We will now show that for suitable choices of the integers p_s above, the new one-relator group K has lower complexity than the original one-relator group G.

A homomorphism $\phi \colon F \to \mathbb{Z}$ with $\phi(W) = 0$ corresponds to a choice of integers $p_s = \phi(s)$. Consider two occurrences of a generator s in W corresponding to a subword of the form $s^{\pm 1} U s^{\pm 1}$ in W. We say that ϕ *separates* these two occurrences provided that either: $\phi(sU) \neq 0$ in case of sUs; or $\phi(s^{-1}U) \neq 0$ in case of $s^{-1}Us^{-1}$; or $\phi(U) \neq 0$ in case of sUs^{-1} or $s^{-1}Us$. We refer the reader to the later geometric interpretation below. When ϕ separates occurrences of s, we have $R_{\bar{s}} > L_{\bar{s}}$ and hence this choice yields a complexity reduction.

Suppose there is a generator a with $\#_a(W) = 0$. If some pair of successive appearances of $a^{\pm 1}$ in W have the same sign, then they are separated by $\#_a$. So let us consider the alternating case where $W = aA_1a^{-1}A_2aA_3\cdots$ and each A_i is nonempty and contains no $a^{\pm 1}$. If some generator b occurs in both A_i, A_j with $i \not\equiv_2 j$ then the $\#_a$ homomorphism separates these two occurrences of b. So, let us assume that no generator appears in both an odd and an even syllable.

We call b a *zero letter* of W if $\#_b(W) = 0$ and call b a *nonzero letter* of W if $\#_b(W) \neq 0$. Considering $aA_1a^{-1}A_2aA_3\cdots$, we see that either each A_i contains at least one nonzero letter b, or some A_i consists entirely of zero letters.

Suppose some A_i consists entirely of zero letters, and consider an innermost pair $b^{\pm 1}Cb^{\pm 1} \subset A_i$ with the property that $b \notin C$. We can assume that C is nonempty for in the case bb or $b^{-1}b^{-1}$ the $\#_b$ homomorphism separates the occurrences of b. Now for any $c \in C$, we have $\#_c(C) \neq 0$ so $\#_c$ separates the surrounding occurrences of $b^{\pm 1}$.

Let us therefore consider the case where each A_i contains a nonzero letter. There is thus an "even" nonzero letter b in A_2 and an "odd" nonzero letter c in A_1, and we use the $\phi = w_c\#_b - w_b\#_c$ homomorphism, where we use the notation $w_x = \#_x(W)$. Let A_i be a syllable with $\#_b(A_i) \neq 0$, and note that $c \notin A_i$ since i is even, so ϕ separates the two occurrences of a around A_i.

We now assume that $\#_a(W) \neq 0$ for each letter a occurring in W. Suppose some letter occurs with both signs. Then we can choose $a^\epsilon Ba^{-\epsilon}$ such that $a^{\pm 1}$ does not occur in B, and such that $|B|$ is minimal among all such choices with a, B allowed to vary. Observe that $\#_b(B) \neq 0$ for any letter b occurring in B, for then b^{+1}, b^{-1} both occur in B and this would violate the minimality of $|B|$.

We then use the homomorphism $w_a\#_b - w_b\#_a$ which separates the two occurrences of a around B.

Without loss of generality, we may assume that each occurring letter occurs only positively. Express W as $aA_1aA_2aA_n$ where each A_i is nonempty and has no occurrence of a. If $\#_b(A_i) = 0$ but b occurs in W, then $w_a\#_b - w_b\#_a$ separates the a's around aA_ia. If $\#_b(A_i) \geq 2$, but b occurs in W, then $w_a\#_b - w_b\#_a$ separates the occurrences of b in A_i. We thus have $\#_b(A_i) = 1$ for each b occurring in W.

We may thus assume that each A_i contains each letter occurring in W, and moreover it occurs exactly once. It follows that $|A_i|$ is constant, so each occurrence of a within W occurs in the same position modulo the total number of letters occurring in W. The same statement applies for each letter occurring in W. Consequently, $W = U^n$ where each letter occurs exactly once in U.

19.c   Quasiconvexity Using the Strengthened Spelling Theorem

Let X be a staggered 2-complex. Such 2-complexes are discussed in Section 19.d, but in particular, we have in mind the motivating case where X is the standard 2-complex of $\langle a, b, \dots \mid W^n \rangle$. In this case, a 2-cell in a disk diagram $D \to X$ is *extreme* if more than $\frac{n-1}{n}$ of its boundary path is a subpath of the boundary path $\partial_p D$ of D. Such a subpath is the *outerpath* of the extreme 2-cell. More generally, for a staggered 2-complex X we say r is an *extreme 2-cell* if this outerpath has length exceeding $\frac{n-1}{n}|\partial_p r|$ where n is the *exponent* of r.

The Newman spelling theorem (see [LS77] and the references therein and [HW01] for the staggered case) states that a reduced diagram $D \to X$ that is nontrivial, spurless, and not a single 2-cell, must contain at least two extreme 2-cells. The following amplification of this is proven in [LW13]:

Proposition 19.6. *Let X be a staggered 2-complex. Let $D \to X$ be a reduced spurless disk diagram with an exponent n internal 2-cell r (meaning that $\partial r \cap \partial D \subset D^0$). Then D contains at least $2n$ extreme 2-cells.*

Proposition 19.6 is most interesting when each 2-cell of X has exponent ≥ 2, as arises in the motivating case of a one-relator group with torsion.

Lemma 19.7 (Fellow traveling). *Let X be a staggered 2-complex with torsion, whose 2-cell attaching maps have the form $W_i^{n_i}$ with $n_i \geq 2$. Let $M = \max\{n_i|W_i|\}$ and let $\kappa = \frac{3}{2}M$.*

Let $D \to X$ be a reduced diagram between λ and γ. Suppose that γ lifts to a geodesic in the 1-skeleton of $\widetilde{X}$. Suppose λ and γ are immersed combinatorial paths, and that neither contains a subpath that is the outerpath of an extreme 2-cell.

Then $\gamma \subset \mathcal{N}_\kappa(\lambda)$.

Proof. Without loss of generality, we can assume that D is spurless. Indeed, the only possible spurs occur at an initial or terminal agreement between λ and γ, and removal of these does not affect the claim. By Proposition 19.6, D cannot contain an exponent n internal 2-cell, because then D would have $2n$ extreme 2-cells, but D can accommodate at most two such 2-cells—one at each end, by hypothesis on λ, γ.

Claim: For each open 2-cell r in D, the intersection $\partial r \cap \gamma$ is connected.

Indeed, otherwise r would subtend a subdiagram D' with $D' \neq \bar{r}$, so by the spelling theorem, D' has at least two extreme 2-cells which is impossible by hypothesis on γ.

A *bridge* in D is a pair of 2-cells $a \neq b$ such that $\bar{a} \cap \bar{b} \neq \emptyset$ and $\partial a, \partial b$ both intersect γ. See Figure 19.2. The *base vertex* p of the bridge is the endpoint of $\bar{a} \cap \bar{b}$ on the λ-side of the bridge. It is conceivable that $\bar{a} \cap \bar{b}$ is disconnected, in which case we choose the lowest such point.

Observe that the subpath $\gamma' \subset \gamma$ subtended by the bridge satisfies $|\gamma'| \leq M$ and so $\gamma' \subset \mathcal{N}_M(p)$. Indeed, γ is a geodesic, and thus so is γ'. The endpoints of γ' are joined by a pair of subpaths of $\partial a, \partial b$, which are of length $\leq \frac{1}{2}M$. Thus $|\gamma'|$ is bounded by the sum of these lengths.

A bridge is *outermost* if its subtended path γ' is not properly contained in the path subtended by another bridge.

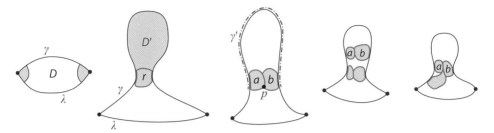

Figure 19.2. Pac-man: The first figure shows the only two places where an extreme cell could lie, the second figure shows the subdiagram D' subtended by r, the third figure illustrates a bridge, and the fourth and fifth figures illustrate bridges that are not outermost.

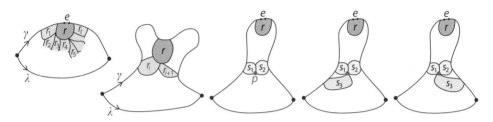

Figure 19.3.

Let e be an edge on γ and suppose that e is not on λ. Then e lies on some 2-cell r of D. Either:

(1) $\partial r \cap \lambda$ contains a 1-cell,
(2) some 2-cell r' is incident with both r and λ [this becomes case (1) when $r' = r$], or
(3) no 2-cell is incident with both r and λ.

We consider Case (3). Since there are no internal 2-cells in D, each 2-cell around r must have a 1-cell on γ. Traveling around r, we denote the 2-cells $r_1, r_2, \ldots, r_t$, where $\partial r_1 \cap \gamma$ precedes $\partial r_2 \cap \gamma$ in the orientation on γ as in the first diagram in Figure 19.3.

Since r_1 intersects γ before r, and r_t intersects γ after r, there must be some i such that r_i intersects γ before r and r_{i+1} intersects γ after r. We thus find that r and hence e lies in a subdiagram subtended by a bridge as in the second diagram in Figure 19.3.

As in the third diagram in Figure 19.3, let s_1, s_2 be 2-cells of an outermost bridge subtending a subdiagram containing r, and let p be the basepoint of this bridge. Either:

(1) $p \in \lambda$,
(2) p lies on a 2-cell containing a 1-cell of λ, or
(3) p lies on a 2-cell s_3 distinct from s_1, s_2, such that s_3 contains a 1-cell of γ.

Note that Case (3) is impossible because then, as in the fourth and fifth diagrams of Figure 19.3, either s_3, s_2 or s_1, s_3 would form an even larger bridge contradicting that s_1, s_2 is maximally outermost.

We are thus left with Cases (1) and (2).

We conclude that e lies in $\mathcal{N}_M(p)$ and p lies in $\mathcal{N}_{\frac{1}{2}M}(\lambda)$, so e lies in $\mathcal{N}_\kappa(\lambda)$. Thus γ lies in $\mathcal{N}_\kappa(\lambda)$ as claimed.   □

Lemma 19.8. *K is quasiconvex in G [or $G * \mathbb{Z}$].*

Proof. Suppose that $G = \langle t, a_1, a_2, \cdots \mid W^n \rangle$ such that $\#_t(W) = 0$ where $\#_x(Y)$ denotes the exponent sum of the letter x in the word W. Let X be the standard 2-complex of the presentation for G, and let $\widehat{X} \to X$ be the $\mathbb{Z}$-cover corresponding to the quotient induced by $a_i \mapsto 0$ and $t \mapsto 1$. Let Y denote the subcomplex of $\widehat{X}$ that equals the closure of the based lift of the 2-cell of X. By the Moldavanskii-Magnus construction, $\pi_1 Y \cong K$.

Note that $Y \subset \widehat{X}$ is π_1-injective. When $n \geq 2$ this follows from the Newman spelling theorem [LS77, HW01]. Indeed, any other lift of W is shifted over from the based lift. Consequently $Y \to X$ has no missing extreme cell bounded by a path W^n, since W^n would travel $2n$-times through a lift of a translate of a t-edge that is not in Y.

Let $\widetilde{Y} \subset \widetilde{X}$ be a component of the preimage of Y in $\widetilde{X}$. Note that $\widetilde{Y}$ also has no missing W^{n-1} paths. Let γ be a geodesic whose endpoints lie on $\widetilde{Y}$. Let D be a minimal area diagram between γ and a path $\lambda \to \widetilde{Y}$ with the same endpoints, such that $\mathsf{Area}(D)$ is minimal among all possible choices of λ. Observe that D has no extreme cells whose outerpaths lie in λ or γ, for then we could find a smaller area diagram in the first case, and contradict that γ is a geodesic in the second. Consequently, Lemma 19.7 shows that $\gamma \subset \mathcal{N}_\kappa(\lambda) \subset \mathcal{N}_\kappa(\widetilde{Y})$.

The Magnus subgroup corresponding to the edge group of the HNN extension is the intersection of two conjugates of K and is hence itself quasiconvex. (But the proof we gave can be applied to it directly.)   □

19.d   Staggered 2-Complex with Torsion

A 2-complex X is *staggered* if there is a linear order on a subset of the 1-cells, and a linear ordering on the 2-cells, such that each 2-cell has an ordered 1-cell in its attaching map and $\max(a) < \max(b)$ and $\min(a) < \min(b)$ whenever $a < b$ are 2-cells. Here we let $\max(a)$ denote the greatest 1-cell in ∂a and let $\min(a)$ denote the least. The spelling theorem [HP84, HW01] shows that a finite staggered 2-complex has word-hyperbolic fundamental group provided that the attaching map of each 2-cell equals w^n for some nontrivial closed immersed path w, and some $n \geq 2$. We refer to such a 2-complex as a *staggered 2-complex with torsion*.

Finally, for each finite staggered 2-complex X, either X has a single 2-cell, in which case $\pi_1 X$ is virtually special by Corollary 19.2, or else $\pi_1 X$ splits as an amalgamated product of staggered 2-complexes with fewer 2-cells, where, as we shall explain, the amalgamated subgroup is quasiconvex and malnormal. It follows that $\pi_1 X$ is virtually compact special by Theorem 11.2. We thus obtain the following result:

Theorem 19.9. *Every staggered 2-complex with torsion has virtually special π_1.*

By initially adding some additional unordered 1-cells, we can assume that all 0-cells are connected by a path of unordered 1-cells. This has the effect of adding a free factor to the fundamental group and doesn't affect the existence of a quasiconvex hierarchy, but facilitates its description by maintaining connectedness. The splitting now arises geometrically as follows: One factor is π_1 of the 2-complex consisting of the closure of the top 2-cell together with the unordered 1-skeleton, and the second factor is π_1 of the 2-complex consisting of the closure of all other 2-cells together with the unordered 1-skeleton.

Let X be a staggered 2-complex. A *Magnus subcomplex $Z \subset X$* is a connected subcomplex with the following properties:

(1) If C is a 2-cell of X, and all the ordered boundary 1-cells of C lie in Z, then $C \subset Z$.
(2) The ordered 1-cells of X contained in Z form an interval.

Note that the intersection of Magnus subcomplexes is a Magnus subcomplex. The following generalization of the Freiheitsatz was shown in [HW01, Thm 6.1] (though the result is incorrectly stated there) for Magnus subcomplexes as defined above:

Proposition 19.10. *If Z is a Magnus subcomplex of a staggered 2-complex, then $Z \to X$ is π_1-injective.*

We augment Proposition 19.10 with the following two statements, the second of which generalizes Newman's result that Magnus subgroups of one-relator groups with torsion are malnormal.

Lemma 19.11. *Let $Z \subset X$ be a Magnus subcomplex of a staggered 2-complex with torsion. Then*

(1) $\widetilde{Z} \to \widetilde{X}$ *is convex.*
(2) $\pi_1 Z$ *is a malnormal subgroup of $\pi_1 X$.*

The following proof presumes familiarity with the mode of proof in [HW01] rather than its main theorem.

Proof of convexity. Let γ be a geodesic in $\widetilde{X}$ such that $\gamma \cap \widetilde{Z}$ consists precisely of the endpoints of γ. Let $D \to \widetilde{X}$ be a disk diagram between γ and a path $\lambda \to \widetilde{Z}$, and assume that D is of minimal area among all such (D, λ) with γ fixed, so in particular D is reduced. We can assume that D does not map entirely into $\widetilde{Z}$, so can assume that some 2-cell of D maps to a 2-cell above or below the Magnus complex Z—and shall assume that it maps above, without loss of generality.

We choose a maximal cyclic tower lift $\phi: D \to T$, and then consider the greatest 2-cell r in T and note that by Howie's Lemma [How87], $\partial_{\mathsf{p}} r$ is of the form $(Qe)^n$ where Q does not traverse e, and n is the exponent of r, and no other 2-cell is incident with e. The 1-cells in $\phi^{-1}(e)$ are all in ∂D, for otherwise there would be a cancellable pair. Moreover, since r is above Z, we see that $\phi^{-1}(e)$ lies entirely in γ. However, if $n \geq 2$, we see that consideration of a single 2-cell in $\phi^{-1}(r)$ shows that γ is not a geodesic, which is a contradiction. $\square$

Proof of malnormality. Consider a reduced annular diagram $A \to X$ where the two boundary paths of A map to Z. We will show that any such A maps to Z, and since Z is a connected subcomplex of X, this shows that $\pi_1 Z$ is malnormal.

Choose a shortest path $\gamma \to A$ whose endpoints lie on the disjoint circles in ∂A. Choose $d \in \mathbb{Z} = \mathrm{Aut}(\widetilde{A} \to A)$ sufficiently large that the translates $0\widetilde{\gamma}$ and $d\widetilde{\gamma}$ in $\widetilde{A}$ do not both intersect a common 2-cell.

Cutting $\widetilde{A}$ along $0\widetilde{\gamma}$ and $d\widetilde{\gamma}$, we obtain a disk diagram $D \to X$ with $\partial_{\mathsf{p}} D = \lambda_1 \gamma \lambda_2^{-1} \gamma^{-1}$ where λ_1, λ_2 each travel d times around the distinct boundary cycles of A.

As in the proof of convexity, A and hence D must contain a 2-cell that does not map to Z. Consideration of a maximal tower lift shows that the preimage of a highest 2-cell in a maximal tower lift has all of its highest 1-cells on ∂D (since D is reduced), and hence on γ, γ^{-1} since λ_1, λ_2 map to Z. Moreover, our choice of d ensures that the 2-cell has all its highest 1-cells on either γ or on γ^{-1}. When the exponent $n \geq 2$, we can thus travel along the innerpath of this 2-cell to find a shorter path in $\widetilde{A}$, and hence in A between the two boundary paths.

The base case, where A is singular and $|\gamma| = 0$, yields a contradiction, since there is no room for a highest 2-cell in D, so there was a cancellable pair of 2-cells. $\square$

Chapter Twenty

Problems

Generalize this to relatively hyperbolic CAT(0) cube complexes. Are they always virtually special? Do they have pseudograph CAT(0) quotients? Perhaps there is an argument by induction on dimension and/or depth. An important test case are the negatively curved square complexes.

Generalize the virtual special quotient theorem so that there exists finite index subgroups H_i' such that $G/\langle\!\langle H_i''\rangle\!\rangle$ is virtually special for any finite index subgroups $H_i'' \subset H_i'$. This appears to work in the cyclic case.

Let H be a codimension-1 subgroup of G. Does there exist a finite index subgroup $H' \subset H$ such that $H/\langle\!\langle H'\rangle\!\rangle$ is also codimension-1 in $G/\langle\!\langle H'\rangle\!\rangle$?

The following should be proven following the proof of Theorem 12.1, by repeatedly using a relatively malnormal special quotient theorem in the sparse case. Alternatively, the graded small-cancellation approach should work (assuming G initially has this structure).

Conjecture 20.1 (Relatively hyperbolic virtually special quotient theorem). *Let G be hyperbolic relative to virtually abelian subgroups. And suppose G is virtually sparse special. Let $H_1, \ldots, H_k$ be quasiconvex subgroups. And let $H_i^o \subset H_i$ be finite index subgroups. There exist finite index subgroups $H_i' \subset H_i$ such that $G/\langle\!\langle H_1', \ldots, H_k'\rangle\!\rangle$ is virtually sparse special.*

Problem 20.2. Suppose G is hyperbolic relative to virtually abelian subgroups. Suppose G is virtually sparse special. Is G virtually compact special?

Finding general conditions under which the following problem has an affirmative answer will yield many applications.

Problem 20.3 (Virtual Haken problem for cube complexes). Let X be a compact nonpositively curved cube complex. Does there exist a finite cover $\widehat{X}$ with the property that the immersed hyperplanes of $\widehat{X}$ are actually embedded?

The following example shows that there exists a nonpositively curved cube complex with no finite cover whose hyperplanes embed. Hence there are certainly limits to the positive prospects for Problem 20.3.

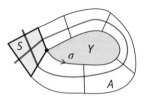

Figure 20.1. The extra hyperplane crosses itself in any finite cover.

Example 20.4. Let Y denote a compact nonpositively curved cube complex with no finite cover but with $\pi_1 Y$ nontrivial. We can choose Y so that $\text{link}(y)$ is a complete bipartite graph for each $y \in Y^0$ (see [BM97, Wis07]). We can choose a closed path $\sigma \to Y$ such that the lift $\widetilde{\sigma} \to \widetilde{Y}$ is an isometry (the path σ travels through ends of 1-cells in the same class of the bipartite structure in each link). Let $A = [-\frac{1}{2}, \frac{1}{2}] \times [0, n]$ be a flat strip of length $n = |\sigma|$, and glue A to Y along σ by identifying σ with $\{-\frac{1}{2}\} \times [0, n]$. Finally, add an extra square S at the basepoint along the two 1-cells $[-\frac{1}{2}, \frac{1}{2}] \times \{0\}$ and $[-\frac{1}{2}, \frac{1}{2}] \times \{n\}$ corresponding to the first and last 1-cells of A. Let X be the resulting complex which is heuristically illustrated in Figure 20.1. Since X deformation retracts to Y, it has no finite covers. Since there are no corners of squares along the interior of σ we see that X is a nonpositively curved square complex. However, the hyperplane of X containing $\{0\} \times [0, n]$ self-crosses within the square S, and hence within X.

Conjecture 20.5. *Let G be a word-hyperbolic group acting properly and cocompactly on a CAT(0) cube complex C. Then G has a finite index subgroup F acting specially on C.*

Problem 20.6 (Codimension-1 subgroups in $C(6)$ groups). Is there an example of a $C(6)$ presentation with infinite fundamental group but with no codimension-1 subgroup? Is there such an example with Property-(T)? Can such examples be identified using a generalized or appropriately aimed version of the spectral gap criterion [Żuk96].

References

[Ago04] Ian Agol. Tameness of hyperbolic 3-manifolds. Preprint, 2004.

[Ago08] Ian Agol. Criteria for virtual fibering. *J. Topol.*, 1(2):269–284, 2008.

[AR90] I. R. Aitchison and J. H. Rubinstein. An introduction to polyhedral metrics of nonpositive curvature on 3-manifolds. In *Geometry of low-dimensional manifolds, 2 (Durham, 1989)*, pages 127–161. Cambridge Univ. Press, Cambridge, 1990.

[Arz01] G. N. Arzhantseva. On quasiconvex subgroups of word hyperbolic groups. *Geom. Dedicata*, 87(1–3):191–208, 2001.

[AS83] K. I. Appel and P. E. Schupp. Artin groups and infinite Coxeter groups. *Invent. Math.*, 72(2):201–220, 1983.

[BF92] M. Bestvina and M. Feighn. A combination theorem for negatively curved groups. *J. Differential Geom.*, 35(1):85–101, 1992.

[BH96] Steven A. Bleiler and Craig D. Hodgson. Spherical space forms and Dehn filling. *Topology*, 35(3):809–833, 1996.

[BH99] Martin R. Bridson and André Haefliger. *Metric spaces of non-positive curvature*. Springer-Verlag, Berlin, 1999.

[BL77] D. C. Brewster and J. C. Lennox. Maximal torsion free subgroups of polycyclic by finite groups. *Arch. Math. (Basel)*, 29(1):39–40, 1977.

[BM97] Marc Burger and Shahar Mozes. Finitely presented simple groups and products of trees. *C. R. Acad. Sci. Paris Sér. I Math.*, 324(7):747–752, 1997.

[Bon86] Francis Bonahon. Bouts des variétés hyperboliques de dimension 3. *Ann. of Math. (2)*, 124(1):71–158, 1986.

[BW13] Hadi Bigdely and Daniel T. Wise. Quasiconvexity and relatively hyperbolic groups that split. *Michigan Math. J.*, 62(2):387–406, 2013.

[Can94] Richard D. Canary. Covering theorems for hyperbolic 3-manifolds. In *Low-dimensional topology (Knoxville, TN, 1992)*, Conf. Proc.

Lecture Notes Geom. Topology, III, pages 21–30. Internat. Press, Cambridge, MA, 1994.

[CD95] Ruth Charney and Michael W. Davis. Finite $K(\pi,1)$s for Artin groups. In *Prospects in topology (Princeton, NJ, 1994)*, volume 138 of *Ann. of Math. Stud.*, pages 110–124. Princeton Univ. Press, Princeton, NJ, 1995.

[CG06] Danny Calegari and David Gabai. Shrinkwrapping and the taming of hyperbolic 3-manifolds. *J. Amer. Math. Soc.*, 19(2):385–446 (electronic), 2006.

[CN05] Indira Chatterji and Graham Niblo. From wall spaces to CAT(0) cube complexes. *Internat. J. Algebra Comput.*, 15(5–6):875–885, 2005.

[CS84] M. Culler and P. B. Shalen. Bounded, separating, incompressible surfaces in knot manifolds. *Invent. Math.*, 75(3):537–545, 1984.

[Dah03] François Dahmani. Combination of convergence groups. *Geom. Topol.*, 7:933–963 (electronic), 2003.

[DLS91] Carl Droms, Jacques Lewin, and Herman Servatius. Tree groups and the 4-string pure braid group. *J. Pure Appl. Algebra*, 70(3):251–261, 1991.

[DS05] Cornelia Druţu and Mark Sapir. Tree-graded spaces and asymptotic cones of groups. *Topology*, 44(5):959–1058, 2005. With an appendix by Denis Osin and Mark Sapir.

[ECH$^+$92] David B. A. Epstein, James W. Cannon, Derek F. Holt, Silvio V. F. Levy, Michael S. Paterson, and William P. Thurston. *Word processing in groups.* Jones and Bartlett Publishers, Boston, MA, 1992.

[Git99a] Rita Gitik. On the profinite topology on negatively curved groups. *J. Algebra*, 219(1):80–86, 1999.

[Git99b] Rita Gitik. Ping-pong on negatively curved groups. *J. Algebra*, 217 (1):65–72, 1999.

[GM08] Daniel Groves and Jason Fox Manning. Dehn filling in relatively hyperbolic groups. *Israel J. Math.*, 168:317–429, 2008.

[GMRS98] Rita Gitik, Mahan Mitra, Eliyahu Rips, and Michah Sageev. Widths of subgroups. *Trans. Amer. Math. Soc.*, 350(1):321–329, 1998.

[Gro87] M. Gromov. Hyperbolic groups. In *Essays in group theory*, volume 8 of *Math. Sci. Res. Inst. Publ.*, pages 75–263. Springer, New York, 1987.

[Gro03]   M. Gromov. Random walk in random groups. *Geom. Funct. Anal.*, 13(1):73–146, 2003.

[Hag07]   Frédéric Haglund. Isometries of CAT(0) cube complexes are semi-simple, pages 1–15, 2007. arXiv:0705.3386.

[Hag08]   Frédéric Haglund. Finite index subgroups of graph products. *Geom. Dedicata*, 135:167–209, 2008.

[Hag14]   Mark F. Hagen. Cocompactly cubulated crystallographic groups. *J. Lond. Math. Soc. (2)*, 90(1):140–166, 2014.

[Hat07]   Allen Hatcher. Notes on basic 3-manifold topology. https://www.math.cornell.edu/ hatcher/3M/3M.pdf, 2007.

[Hem04]   John Hempel. *3-manifolds*. AMS Chelsea Publishing, Providence, RI, 2004. Reprint of the 1976 original.

[HK05]   G. Christopher Hruska and Bruce Kleiner. Hadamard spaces with isolated flats. *Geom. Topol.*, 9:1501–1538 (electronic), 2005. With an appendix by the authors and Mohamad Hindawi.

[How87]   James Howie. How to generalize one-relator group theory. In S. M. Gersten and John R. Stallings, editors, *Combinatorial group theory and topology*, pages 53–78, Princeton Univ. Press, Princeton, NJ, 1987.

[HP84]   J. Howie and S. J. Pride. A spelling theorem for staggered generalized 2-complexes, with applications. *Invent. Math.*, 76(1):55–74, 1984.

[HP98]   Frédéric Haglund and Frédéric Paulin. Simplicité de groupes d'automorphismes d'espaces à courbure négative. In *The Epstein birthday schrift*, pages 181–248 (electronic). Geom. Topol., Coventry, 1998.

[Hru10]   G. Christopher Hruska. Relative hyperbolicity and relative quasiconvexity for countable groups. *Algebr. Geom. Topol.*, 10(3):1807–1856, 2010.

[Hum94]   S. P. Humphries. On representations of Artin groups and the Tits conjecture. *J. Algebra*, 169:847–862, 1994.

[HW99]   Tim Hsu and Daniel T. Wise. On linear and residual properties of graph products. *Michigan Math. J.*, 46(2):251–259, 1999.

[HW01]   G. Christopher Hruska and Daniel T. Wise. Towers, ladders and the B. B. Newman spelling theorem. *J. Aust. Math. Soc.*, 71(1):53–69, 2001.

[HW08]   Frédéric Haglund and Daniel T. Wise. Special cube complexes. *Geom. Funct. Anal.*, 17(5):1551–1620, 2008.

[HW09]   G. Christopher Hruska and Daniel T. Wise. Packing subgroups in relatively hyperbolic groups. *Geom. Topol.*, 13(4):1945–1988, 2009.

[HW10a] Frédéric Haglund and Daniel T. Wise. Coxeter groups are virtually special. *Adv. Math.*, 224(5):1890–1903, 2010.

[HW10b] Tim Hsu and Daniel T. Wise. Cubulating graphs of free groups with cyclic edge groups. *Amer. J. Math.*, 132(5):1153–1188, 2010.

[HW12]   Frédéric Haglund and Daniel T. Wise. A combination theorem for special cube complexes. *Ann. of Math. (2)*, 176(3):1427–1482, 2012.

[HW14]   G. C. Hruska and Daniel T. Wise. Finiteness properties of cubulated groups. *Compos. Math.*, 150(3):453–506, 2014.

[HW15]   Tim Hsu and Daniel T. Wise. Cubulating malnormal amalgams. *Invent. Math.*, 199:293–331, 2015.

[HW18]   Jingyin Huang and Daniel T. Wise. Stature and separability in graphs of groups. Preprint, pages 1–31, 2018.

[JS79]   William H. Jaco and Peter B. Shalen. Seifert fibered spaces in 3-manifolds. *Mem. Amer. Math. Soc.*, 21(220):viii+192, 1979.

[Kap02] Ilya Kapovich. Subgroup properties of fully residually free groups. *Trans. Amer. Math. Soc.*, 354(1):335–362 (electronic), 2002.

[KM98]   O. Kharlampovich and A. Myasnikov. Irreducible affine varieties over a free group. II. Systems in triangular quasi-quadratic form and description of residually free groups. *J. Algebra*, 200(2):517–570, 1998.

[KS96]   Ilya Kapovich and Hamish Short. Greenberg's theorem for quasiconvex subgroups of word hyperbolic groups. *Canad. J. Math.*, 48(6):1224–1244, 1996.

[Lea13] Ian J. Leary. A metric Kan-Thurston theorem. *J. Topol.*, 6(1):251–284, 2013.

[Li02]   Tao Li. Boundary curves of surfaces with the 4-plane property. *Geom. Topol.*, 6:609–647 (electronic), 2002.

[LS77]   Roger C. Lyndon and Paul E. Schupp. *Combinatorial group theory*. Springer-Verlag, Berlin, 1977. Ergebnisse der Mathematik und ihrer Grenzgebiete, Band 89.

[LW13]   Joseph Lauer and Daniel T. Wise. Cubulating one-relator groups with torsion. *Math. Proc. Cambridge Philos. Soc.*, 155(3):411–429, 2013.

[Mas08]   Joseph D. Masters. Kleinian groups with ubiquitous surface subgroups. *Groups Geom. Dyn.*, 2(2):263–269, 2008.

[Min04]   Ashot Minasyan. Separable subsets of gferf negatively curved groups. Preprint, 2004.

[Mit04]   Mahan Mitra. Height in splittings of hyperbolic groups. *Proc. Indian Acad. Sci. Math. Sci.*, 114(1):39–54, 2004.

[Mou88]   Gábor Moussong. *Hyperbolic Coxeter Groups*. PhD thesis, Ohio State University, 1988.

[MP09]   Eduardo Martínez-Pedroza. Combination of quasiconvex subgroups of relatively hyperbolic groups. *Groups Geom. Dyn.*, 3(2):317–342, 2009.

[MPW11]   Eduardo Martínez-Pedroza and Daniel T. Wise. Relative quasiconvexity using fine hyperbolic graphs. *Algebr. Geom. Topol.*, 11(1): 477–501, 2011.

[MV95]   John Meier and Leonard VanWyk. The Bieri-Neumann-Strebel invariants for graph groups. *Proc. London Math. Soc. (3)*, 71(2):263–280, 1995.

[MW02]   Jonathan P. McCammond and Daniel T. Wise. Fans and ladders in small cancellation theory. *Proc. London Math. Soc. (3)*, 84(3):599–644, 2002.

[MW08]   Jonathan P. McCammond and Daniel T. Wise. Locally quasiconvex small-cancellation groups. *Trans. Amer. Math. Soc.*, 360(1):237–271 (electronic), 2008.

[MZ08]   Joseph D. Masters and Xingru Zhang. Closed quasi-Fuchsian surfaces in hyperbolic knot complements. *Geom. Topol.*, 12(4):2095–2171, 2008.

[Nic04]   Bogdan Nica. Cubulating spaces with walls. *Algebr. Geom. Topol.*, 4:297–309 (electronic), 2004.

[NR03]   G. A. Niblo and L. D. Reeves. Coxeter groups act on CAT(0) cube complexes. *J. Group Theory*, 6(3):399–413, 2003.

[Oll06]   Yann Ollivier. On a small cancellation theorem of Gromov. *Bull. Belg. Math. Soc. Simon Stevin*, 13(1):75–89, 2006.

[Osi06]   Denis V. Osin. Relatively hyperbolic groups: intrinsic geometry, algebraic properties, and algorithmic problems. *Mem. Amer. Math. Soc.*, 179(843):vi+100, 2006.

[Osi07]   Denis V. Osin. Peripheral fillings of relatively hyperbolic groups. *Invent. Math.*, 167(2):295–326, 2007.

[OW11]   Yann Ollivier and Daniel T. Wise. Cubulating random groups at density less than 1/6. *Trans. Amer. Math. Soc.*, 363(9):4701–4733, 2011.

[Pap95]   P. Papasoglu. Strongly geodesically automatic groups are hyperbolic. *Invent. Math.*, 121(2):323–334, 1995.

[Rat94]   John G. Ratcliffe. *Foundations of hyperbolic manifolds*, volume 149 of *Graduate Texts in Mathematics*. Springer-Verlag, New York, 1994.

[RS87]   Eliyahu Rips and Yoav Segev. Torsion-free group without unique product property. *J. Algebra*, 108(1):116–126, 1987.

[Sag95]   Michah Sageev. Ends of group pairs and non-positively curved cube complexes. *Proc. London Math. Soc. (3)*, 71(3):585–617, 1995.

[Sag97]   Michah Sageev. Codimension-1 subgroups and splittings of groups. *J. Algebra*, 189(2):377–389, 1997.

[Sel03]   Z. Sela. Diophantine geometry over groups. II. Completions, closures and formal solutions. *Israel J. Math.*, 134:173–254, 2003.

[SW05]   Michah Sageev and Daniel T. Wise. The Tits alternative for CAT(0) cubical complexes. *Bull. London Math. Soc.*, 37(5):706–710, 2005.

[SW15]   Michah Sageev and Daniel T. Wise. Cores for quasiconvex actions. *Proc. Amer. Math. Soc.*, 143(7):2731–2741, 2015.

[Wei71]   C. M. Weinbaum. The word and conjugacy problems for the knot group of any tame, prime, alternating knot. *Proc. Amer. Math. Soc.*, 30:22–26, 1971.

[Wis00]   Daniel T. Wise. Subgroup separability of graphs of free groups with cyclic edge groups. *Q. J. Math.*, 51(1):107–129, 2000.

[Wis02]   Daniel T. Wise. The residual finiteness of negatively curved polygons of finite groups. *Invent. Math.*, 149(3):579–617, 2002.

[Wis03]   Daniel T. Wise. Sixtolic complexes and their fundamental groups. Preprint, 2003.

[Wis04]   Daniel T. Wise. Cubulating small cancellation groups. *GAFA, Geom. Funct. Anal.*, 14(1):150–214, 2004.

[Wis06] Daniel T. Wise. Subgroup separability of the figure 8 knot group. *Topology*, 45(3):421–463, 2006.

[Wis07] Daniel T. Wise. Complete square complexes. *Comment. Math. Helv.*, 82(4):683–724, 2007.

[Wis12] Daniel T. Wise. *From riches to raags: 3-manifolds, right-angled Artin groups, and cubical geometry*, volume 117 of *CBMS Regional Conference Series in Mathematics*. Published for the Conference Board of the Mathematical Sciences, Washington, DC, 2012.

[Wis14] Daniel T. Wise. Cubular tubular groups. *Trans. Amer. Math. Soc.*, 366(10):5503–5521, 2014.

[WW17] Daniel T. Wise and Daniel J. Woodhouse. A cubical flat torus theorem and the bounded packing property. *Israel J. Math.*, 217(1):263–281, 2017.

[Żuk96] Andrzej Żuk. La propriété (T) de Kazhdan pour les groupes agissant sur les polyèdres. *C. R. Acad. Sci. Paris Sér. I Math.*, 323(5):453–458, 1996.

Index